Lecture Notes in Networks and Systems

Volume 606

The series "Lecture Notes in Networks and Systems" publishes the latest developments in Networks and Systems—quickly, informally and with high quality. Original research reported in proceedings and post-proceedings represents the core of LNNS.

Volumes published in LNNS embrace all aspects and subfields of, as well as new challenges in, Networks and Systems.

The series contains proceedings and edited volumes in systems and networks, spanning the areas of Cyber-Physical Systems, Autonomous Systems, Sensor Networks, Control Systems, Energy Systems, Automotive Systems, Biological Systems, Vehicular Networking and Connected Vehicles, Aerospace Systems, Automation, Manufacturing, Smart Grids, Nonlinear Systems, Power Systems, Robotics, Social Systems, Economic Systems and other. Of particular value to both the contributors and the readership are the short publication timeframe and the world-wide distribution and exposure which enable both a wide and rapid dissemination of research output.

The series covers the theory, applications, and perspectives on the state of the art and future developments relevant to systems and networks, decision making, control, complex processes and related areas, as embedded in the fields of interdisciplinary and applied sciences, engineering, computer science, physics, economics, social, and life sciences, as well as the paradigms and methodologies behind them.

Indexed by SCOPUS, INSPEC, WTI Frankfurt eG, zbMATH, SCImago.

All books published in the series are submitted for consideration in Web of Science.

For proposals from Asia please contact Aninda Bose (aninda.bose@springer.com).

K. Ashoka Reddy · B. Rama Devi · Boby George ·
K. Srujan Raju · Mathini Sellathurai
Editors

Proceedings of Fourth International Conference on Computer and Communication Technologies

IC3T 2022

Springer

Editors
K. Ashoka Reddy
Kakatiya Institute of Technology
and Science
Warangal, India

Boby George
Department of Electrical and Electronics
Engineering
Indian Institute of Technology
Chennai, Tamil Nadu, India

Mathini Sellathurai
Department of Signal Processing
Heriot-Watt University
Edinburgh, UK

B. Rama Devi
Department of Electronics
and Communication Engineering
Kakatiya Institute of Technology
and Science
Warangal, India

K. Srujan Raju
Department of Computer Science
and Engineering
CMR Technical Campus
Hyderabad, Telangana, India

ISSN 2367-3370 ISSN 2367-3389 (electronic)
Lecture Notes in Networks and Systems
ISBN 978-981-19-8562-1 ISBN 978-981-19-8563-8 (eBook)
https://doi.org/10.1007/978-981-19-8563-8

This Springer imprint is published by the registered company Springer Nature Singapore Pte Ltd.
The registered company address is: 152 Beach Road, #21-01/04 Gateway East, Singapore 189721, Singapore

Preface

The Springer Scopus indexed 4th International Conference on Computer and Communication Technologies (IC3T) 2022 organized by Department of ECE, KITSW, was held in Warangal, Telangana, during July 29–30, 2022. Warangal, the home of a major state university amid pleasant surroundings, was a delightful place for the conference. The 230 scientific participants, 102 of whom were students, had many fruitful discussions and exchanges that contributed to the success of the conference. Participants from eight countries and 14 states of India made the conference truly international in scope. The 62 abstracts that were presented on the two days formed the heart of the conference and provided ample opportunity for discussion. This change, allowing the conference to end with invited talks from the industry experts, was a departure from the format used at previous IC3T conferences. The abstracts were split almost equally between the five main conference areas, i.e., image processing and communications system, VLSI, wireless networks, Internet of Things (IoT) and machine learning (ML), and the posters were distributed across the days of the conference, so that approximately equal numbers of abstracts in the different areas were scheduled for each day. Of the total number of presented abstracts, 50 of these are included in this proceedings volume, the first time that abstracts have been published by IC3T. There were four plenary lectures covering the different areas of the conference: Dr. Suresh Chandra Satapathy (Professor, KIIT, Bhubaneswar) talked on Social Group Optimization and its applications to Image Processing, Dr. Mathini Sellathurai, (Professor, Heriot-Watt University, UK) on Sustainable 6G: Energy Efficient Auto encoder-Based Coded Modulation Designs for Wireless Networks, Dr. K. Srujan Raju (Professor, CMR Technical Campus, Hyderabad), on optimizing solutions in a variety of disciplines of Computer Science and Engineering, and Dr. K. Ashoka Reddy (Professor, KITS Warangal) on Applications of Biomedical Signal processing in Current Engineering Technologies. Two eminent speakers from industry gave very illuminating public lectures that drew many people from the local area, as well as conference participants: Vijay Kumar Gupta Kopuri (CEO, Kwality Photonics Pvt. Ltd.) on "Manufacturing of LEDs and Compound Semiconductors" and K. Jagadeshwar Reddy (Managing Director, Elegant Embedded Solutions Pvt. Ltd.) on "optimized solutions adopted in Embedded System Design in Current Scenario." These public talks were very accessible to a general audience. In addition, notably, this was the third conference at KITSW, and a formal session was held the first day to honor the event as well as those who were instrumental in initiating the conference.

Generous support for the conference was provided by Captain V. Lakshmikantha Rao, Honorable Ex. MP (Rajya Sabha), Former Minister, and Chairman, KITS, Warangal. The funds were sizeable, timely, greatly appreciated, and permitted us to support a significant number of young scientists (postdocs and students) and persons from developing/disadvantaged countries. Nevertheless, the number of requests was far greater than the total support available (by about a factor of five!), and we had to turn down many financial requests. We encourage the organizers of the next IC3T to seek a higher level of funding for supporting young scientists and scientists from developing/disadvantaged countries. All in all, the Springer Scopus Indexed 4th IC3T 2022 in Warangal was very successful. The plenary lectures and the progress and special reports bridged the gap between the different fields of Computers and Communication Technology, making it possible for non-experts in a given area to gain insight into new areas. Also, included among the speakers were several young scientists, namely postdocs and students, who brought new perspectives to their fields. The next IC3T will take place in Warangal in 2023 and trend to be continued every year. Given the rapidity with which science is advancing in all of the areas covered by IC3T 2022, we expect that these future conferences will be as stimulating as this most recent one was, as indicated by the contributions presented in this proceedings volume.

We would also like to thank the authors and participants of this conference, who have considered the conference above all hardships. Finally, we would like to thank all the reviewers, session chairs and volunteers who spent tireless efforts in meeting the deadlines and arranging every detail to make sure that the conference runs smoothly.

Warangal, India	Dr. K. Srujan Raju
Warangal, India	Dr. Mathini Sellathurai
Chennai, India	Dr. K. Ashoka Reddy
Hyderabad, India	Dr. Boby George
Edinburgh, UK	Dr. B. Rama Devi
September 2022	

Contents

Editors and Contributors

About the Editors

Dr. K. Ashoka Reddy is acting as Principal, Kakatiya Institute of Technology and Science, Warangal (KITSW). He studied Bachelor of Engineering in Electronics and Instrumentation Engineering at KITSW and received B.Tech. degree in 1992 from Kakatiya University, Warangal, Telangana. He received M.Tech. degree in 1994 from Jawaharlal Nehru Technological University, Kakinada (JNTUK), Andhra Pradesh. He did research on Pulse Oximeters and received Ph.D. in Electrical Engineering in 2008 from Indian Institute of Technology Madras (IITM), Chennai, India. He received Innovative Project Award in 2008 from Indian National Academy of Engineering (INAE) for his Ph.D. work. His teaching and research interests include signal processing and instrumentation. He has authored over 15 research papers in refereed journals and 50 papers in conferences proceedings. He received Rs.10 lakh funding under RPS from AICTE, New Delhi, during 2013–2016; received 7.5 lakhs from AICTE in 2017. He is a reviewer for *IEEE Transactions on Measurements and Instrumentation*, *IEEE transactions on Biomedical Engineering*, and *IEEE Sensors Journal*. He is also a member of IEEE, a life member of ISTE, a member of IETE and a member of CSI.

Dr. B. Rama Devi received the Ph.D. from JNTUH College of Engineering, Hyderabad, Telangana, India, in April 2016. She completed M.Tech.—Digital Communication Engineering from Kakatiya University, Warangal in 2007. She joined the faculty of Electronics and Communication Engineering, KITSW in 2007. Currently, she is working as Professor and Head, Department ECE. She published more than 40 papers in various journals and conferences, and filed four patents. She published three books and acted as Session Chair for various international conferences. Her areas of interest include wireless communication, wireless networks, Signal processing for communications, medical body area networks, and Smart grid. She is an active reviewer for *IEEE Transactions on Vehicular Technology (TVT)*, Elsevier, *Wireless Personal Communications* and Springer journals.

Dr. Boby George, received the M.Tech. and Ph.D. degrees in Electrical Engineering from the Indian Institute of Technology (IIT) Madras, Chennai, India, in 2003 and 2007, respectively. He was a Postdoctoral Fellow with the Institute of Electrical Measurement and Measurement Signal Processing, Technical University of Graz, Graz Austria from 2007 to 2010. He joined the faculty of the Department of Electrical Engineering, IIT Madras in 2010. Currently, he is working as a Professor there. His areas of interest include magnetic and electric field-based sensing approaches, sensor interface circuits/signal conditioning circuits, sensors and instrumentation for automotive and industrial applications. He has co-authored more than 75 IEEE transactions/journals. He is an Associate Editor for *IEEE Sensors Journal*, *IEEE Transactions on Industrial Electronics*, and *IEEE Transactions on Instrumentation and Measurement.*

Dr. K. Srujan Raju is currently working as Dean Student Welfare and Heading Department of Computer Science and Engineering and Information Technology at CMR Technical Campus. He obtained his Doctorate in Computer Science in the area of Network Security. He has more than 20 years of experience in academics and research. His research interest areas include computer networks, information security, data mining, cognitive radio networks and image processing and other programming languages. Dr. Raju is presently working on two projects funded by Government of India under CSRI and NSTMIS, has also filed seven patents and one copyright at Indian Patent Office, edited more than 14 books published by Springer Book *Proceedings of AISC series, LAIS series* and other which are indexed by Scopus, also authored books in *C Programming and Data Structure, Exploring to Internet, Hacking Secrets*, contributed chapters in various books and published more than 30 papers in reputed peer-reviewed journals and international conferences. Dr. Raju was invited as Session Chair, Keynote Speaker, a Technical Program Committee member, Track Manager and a reviewer for many national and international conferences also appointed as subject Expert by CEPTAM DRDO—Delhi and CDAC. He has undergone specific training conducted by Wipro Mission 10X and NITTTR, Chennai, which helped his involvement with students and is very conducive for solving their day-to-day problems. He has guided various student clubs for activities ranging from photography to Hackathon. He mentored more than 100 students for incubating cutting-edge solutions. He has organized many conferences, FDPs, workshops and symposiums. He has established the Centre of Excellence in IoT, Data Analytics. Dr. Raju is a member of various professional bodies, received Significant Contributor Award and Active Young Member Award from Computer Society of India and also served as a Management Committee member, State Student Coordinator and Secretary of CSI—Hyderabad Chapter.

Dr. Mathini Sellathurai is currently the Dean of Science and Engineering and the head of the Signal Processing for Intelligent Systems and Communications Research Group, Heriot-Watt University, Edinburgh, UK and leading research in signal processing for Radar and wireless communications networks. Professor Sellathurai has 5 years of industrial research experience. She held positions with

Bell-Laboratories, New Jersey, USA and with the Canadian (Government) Communications Research Centre, Ottawa Canada. She was an Associate Editorship for the *IEEE Transactions on Signal Processing* (2005–2018) and IEEE Signal Processing for Communications Technical Committee member (2013–2018). She was an organizer for the IEEE International Workshop on Cognitive Wireless Systems, IIT Delhi, India, 2009, 2010 and 2013; and the General Chair of the 2016 IEEE Workshop on Signal Processing Advances in Wireless Communications (SPAWC), Edinburgh, UK. She is also a peer review college member and Strategic Advisory Committee member of Information and Communications Technology of Engineering and Physical Sciences Research Council, UK. Professor Sellathurai has published over 200 peer reviewed papers in leading international journals and IEEE conferences as well as a research monograph. She was the recipient of the IEEE Communication Society Fred W. Ellersick Best Paper Award in 2005, Industry Canada Public Service Awards for her contributions in science and technology in 2005 and awards for contributions to technology Transfer to industries in 2004. She was also the recipient of the Natural Sciences and Engineering Research Council of Canada's doctoral award for her Ph.D. dissertation Her research has been funded by UK Engineering Physical Sciences Research Council titled "A Unified Multiple Access Framework for Next Generation Mobile Networks By Removing Orthogonality"; "Large Scale Antenna Systems Made Practical: Advanced Signal Processing for Compact Deployments"; "A Systematic Study of Physical Layer Network Coding: From Information-theoretic Understanding to Practical DSP Algorithm Design"; "Advanced Signal Processing Techniques for Multi-user Multiple-input Multiple-output Broadband Wireless Communications"; "Bridging the gap between design and implementation of soft-detectors for Turbo-MIMO wireless systems"; "Signal Processing Techniques to Reduce the Clutter Competition in Forward Looking Radar".

Contributors

Aarti Lovely Professional University, Phagwara, India

Syed Jaffar Abbas Department of CSE, Jharkhand Rai University, Ranchi, Jharkhand, India

Edison Abrigo Escuela Superior Polítecnica de Chimborazo, Telecommunications, Riobamba, Ecuador

Saket Acharya Department of Computer Science and Engineering, Manipal University Jaipur, Dahmi Kalan, India

Vivekanand Aelgani CMR College of Engineering & Technology, Kandlakoya, Telangana, India

Najeema Afrin Department of CSE, CMRTechnical Campus, Hyderabad, Telangana, India

Kacham Akanksha BVRIT HYDERABAD College of Engineering for Women, Hyderabad, Telangana, India

D. G. Anand Rajiv Gandhi Institute of Technology, Bengaluru, India

V. Anantha Natarajan CSE, Sree Vidyanikethan Engineering College, Tirupati, AP, India

Naga Yamini Anche Qualcomm India Pvt. Ltd, Hyderabad, India

Sharvani Sharma Annavajjula Department of ECE, Kakatiya Instistute of Technology and Science, Warangal, Telangana, India

Radhika Arumalla Department of Information Technology, B.V. Raju Institute of Technology, Narsapur, Hyderabad, Telangana, India

S. Arun Kumar Department of Computer Science and Engineering, Bethesda Institute of Technology and Science, Gwalior, India

J. Avanija Department of CSE, Sree Vidyanikethan Engineering College, Tirupati, Andhra Pradesh, India

Usha Rani Badavath Department of ECE, Kakatiya Instistute of Technology and Science, Warangal, Telangana, India

Tushar Baliyan Department of Computer Science and Engineering, CMR Institute of Technology, Bengaluru, India

Kiran Kumar Bejjanki Department of IT, Kakatiya Institute of Technology and Science, Warangal, India

Yash Bhardwaj Jodhpur Institute of Engineering and Technology, Mogra, India

Roheet Bhatnagar Department of Computer Science and Engineering, Manipal University Jaipur, Dahmi Kalan, India

Y. Bhavani Department of IT, Kakatiya Institute of Technology and Science, Warangal, India

K. Bhavya Electrical and Electronics Engineering Department, BVRIT HYDERABAD College of Engineering for Women, Hyderabad, Telangana, India

Kanneganti Bhavya Sri School of Computer Science and Engineering, VIT-AP University, Amaravati, India

Akash Kumar Bhoi Sikkim Manipal University, Gangtok, Sikkim, India;
KIET Group of Institutions, Delhi-NCR, Ghaziabad, India;
AB-Tech eResearch (ABTeR), Sambalpur, Burla, India;
Victoria University, Melbourne, Australia;
Research Associate at IIST, National Research Council (ISTI-CNR), Pisa, Italy

Sreeya Bhupathiraju Department of ECE, BVRIT HYDERABAD College of Engineering for Women, Bachupally, Hyderabad, India

Rama Devi Boddu Department of ECE, Kakatiya Insistute of Technology and Science, Warangal, Telangana, India

Vamsi Krishna Bunga Andhra University, Visakhapatnam, India

Raman Chahar Delhi Technological University, New Delhi, India

R. Chandana Computer Science and Engineering, MSRIT, Bengaluru, India

Piyush Chauhan Informatics Cluster, University of Petroleum and Energy Studies (UPES), Dehradun, Uttarakhand, India

Kevin Chiguano Escuela Superior Polítecnica de Chimborazo, Telecommunications, Riobamba, Ecuador

Amitava Choudhury Pandit Deendayal Energy University, Gandhinagar, Gujarat, India

Tanupriya Choudhury School of Computer Science, University of Petroleum and Energy Studies (UPES), Bidholi, Dehradun, Uttarakhand, India;
Informatics Cluster, University of Petroleum and Energy Studies (UPES), Dehradun, Uttarakhand, India

Andrea Liseth Coro Escuela Superior Polítecnica de Chimborazo, Telecommunications, Riobamba, Ecuador

Soham Das Department of ECE, Dr. B.R. Ambedkar National Institute of Technology, Jalandhar, India

K. Deepika Department of Information Technology, Kakatiya Institute of Technology and Science, Warangal, India

Vinay Kumar Deolia Department of ECE, GLA University, Mathura, UP, India

Ramesh Deshpande Associate Professor, Department of ECE, B V Raju Institute of Technology, Hyderabad, India

U. M. Fernandes Dimlo Department of CSE, Sreyas Institute of Engineering and Technology, Hyderabad, Telangana, India

S. Divya Meena School of Computer Science and Engineering, VIT-AP University, Amaravati, India

Shital Dongre Vishwakarma Institute of Technology, Pune, India

Raja Ram Dutta BIT Mesera, Ranchi, Jharkhand, India

Amitesh Kumar Dwivedi Department of Computer Science and Engineering, Manipal University Jaipur, Rajasthan, India

M. Dyva Sugnana Rao BVRIT HYDERABAD College of Engineering for Women, Bachupally, Hyderabad, India

Tabeen Fatima Department of Computer Science and Engineering, CMR Technical Campus, Hyderabad, Telangana, India

Binish Fatimah CMR Institute of Technology, Bengaluru, Karnataka, India

Venkat Sai Kedari Nath Gandham Department of ECE, Kakatiya Instistute of Technology and Science, Warangal, Telangana, India

A. Geetha CSE, CMR Technical Campus, Hyderabad, India

G. Geetha School of Engineering and Technology, Sri Padmavati Mahila Visvavidyalayam, Tirupati, India

K. Ghamya CSE, Sree Vidyanikethan Engineering College, Tirupati, India

M. N. Giri Prasad Academics and Audit, JNTUA, Anantapuramu, Andhra Pradesh, India;
Department of ECE, JNTUA, Anantapuramu, Andhra Pradesh, India

Sruthi Priya Godishala Department of ECE, Kakatiya Instistute of Technology and Science, Warangal, Telangana, India

Maina Goni Department of ECE, Kakatiya Instistute of Technology and Science, Warangal, Telangana, India

Sam Goundar School of Information Technology, RMIT University, Melbourne, Australia

Babita Gupta BVRIT HYDERABAD College of Engineering, Hyderabad, India

Pooja Gupta Department of Computer Science, Banasthali Vidyapith, Banasthali, India

Suneet K. Gupta Bennett University, Greater Noida, Uttar Pradesh, India

Juturu Harika Department of Information Technology, VR Siddhartha Engineering College, Vijayawada, India

Nandula Haripriya BVRIT HYDERABAD College of Engineering for Women, Bachupally, Hyderabad, India

Ahatsham Hayat University of Madeira, Funchal, Portugal

Sumanth Indrala Department of ECE, Kakatiya Instistute of Technology and Science, Warangal, Telangana, India

D. B. V. Jagannadham Department of ECE, Gayatri Vidya Parishad College of Engineering, Madhurawada, Visakhapatnam, India

Sanyam Jain Maharaja Agrasen Institute of Technology, Delhi, India

Nidhi Jani Pandit Deendayal Energy University, Gandhinagar, Gujarat, India

Jyothi Jarugula CSE, VNITSW, Guntur, India

M. Jaya Pooja Sri Electrical & Electronics Engineering Department, BVRIT HYDERABAD College of Engineering for Women, Hyderabad, Telangana, India

Dabbara Jayanayudu Department of EECE, GST, GITAM (Deemed to be University), Visakhapatnam, India

G. S. Jayant CMR Institute of Technology, Bengaluru, Karnataka, India

Smita Jolania IET-DAVV, Indore, India

Ramya Jothikumar Department of Electronics and Communication Engineering, Sri Manakula Vinayagar Engineering College, Puducherry, India

A. Jyothi Babu Department of MCA, Sree Vidyanikethan Engineering College, Tirupati, India

Rathlavath Kalavathi Research Scholar, Department of Computer Science and Engineering, Osmania University, Hyderabad, India

Shivaprasad Kaleru Juniper Networks Inc, Sunnyvale, CA, USA

P. Kamakshi Department of Information Technology, Kakatiya Institute of Technology and Science, Warangal, India

Apparao Kamarsu IT, MA&UD, Mangalagiri, Andhra Pradesh, India

Lalit Kane School of Computer Science, University of Petroleum and Energy Studies, Dehradun, Uttarakhand, India

G. C. R. Kartheek Department of Computer Science and Engineering, CMR Institute of Technology, Bengaluru, India

G. Karthika Department of Computer Science and Engineering, GITAM School of Technology, GITAM (Deemed to be University), Visakhapatnam, Andhra Pradesh, India

G. Karuna CSE Department, GRIET, Hyderabad, India

M. Karuppasamy Department of Computer Applications, Kalasalingam Academy of Research and Education, Srivilliputhur, TamilNadu, India

Nitesh Kashyap Department of ECE, Dr. B.R. Ambedkar National Institute of Technology, Jalandhar, India

Mahesh Babu Katta BVRIT HYDERABAD College of Engineering for Women, Hyderabad, Telangana, India

G. Kavya Electrical & Electronics Engineering Department, BVRIT HYDERABAD College of Engineering for Women, Hyderabad, Telangana, India

A. Keerthi Reddy Department of ECE, BVRIT HYDERABAD College of Engineering for Women, Bachupally, Hyderabad, India

Sudarshana Kerenalli Department of CSE, SoT, GITAM University, Bengaluru, Karnataka, India

Soumya Suvra Khan Department of CSE, Meghnad Saha Institute of Technology (MSIT), Kolkata, West Bengal, India

Mamta Khosla Dr. B.R. Ambedkar National Institute of Technology, Jalandhar, Punjab, India

Madhu Khurana University of Gloucestershire, Cheltenham, UK

Paleti Krishnasai Department of Computer Science and Engineering, Design and Manufacturing, Indian Institute of Information Technology, Kancheepuram, Chennai, Tamil Nadu, India

Ch Anil Kumar BVRIT HYDERABAD College of Engineering for Women, Bachupally, Hyderabad, India

Jitendra Kumar Department of ECE, GLA University, Mathura, UP, India

Katakam Ananth Yasodharan Kumar School of Computer Science and Engineering, VIT-AP University, Amaravati, India

Voruganti Naresh Kumar Department of CSE, CMR Technical Campus, Hyderabad, Telangana, India

Mohit Lalwani Vishwakarma Institute of Technology, Pune, India

Urvi Latnekar Bennett University, Noida, India

Bagam Laxmaiah Department of Computer Science and Engineering, CMR Technical Campus, Hyderabad, Telangana, India

Sivaprasad Lebaka Department of Electronics and Communication Engineering, Visvesvaraya Technological University, Belagavi, Karnataka, India

Pappala Lokesh Department of Information Technology, VR Siddhartha Engineering College, Vijayawada, India

Vijay Souri Maddila Department of Computer Science and Engineering, GITAM School of Technology, GITAM (Deemed to be University), Visakhapatnam, Andhra Pradesh, India

K. Reddy Madhavi CSE, Sree Vidyanikethan Engineering College, Tirupati, AP, India

Hussain Falih Mahdi Department of Computer and Software Engineering, University of Diyala, Baquba, Iraq

K. Maheswari Department of CSE, CMR Technical Campus, Hyderabad, Telangana, India

Arfa Mahwish Department of IT, CMR Technical Campus, Hyderabad, Telangana, India

Devendra Mandava School of Computer Science and Engineering, VIT-AP University, Amaravati, India

Suneetha Manne Department of Information Technology, VR Siddhartha Engineering College, Vijayawada, India

M. Meenakshi Department of Electronics and Instrumentation Engineering, Dr. Ambedkar Institute of Technology, Bangalore, India

Hussain Falih Mehdi Department of Computer and Software Engineering, University of Diyala, Baquba, Iraq

Vidyanand Mishra School of Computer Science, University of Petroleum and Energy Studies, Dehradun, Uttarakhand, India

Vaka Murali Mohan Malla Reddy College of Engineering for Women, Medchal, Hyderabad, Telangana, India

Sachi Nandan Mohanty Department of Computer Science, Singidunum University, Belgrade, Serbia

Suraya Mubeen Department of ECE, CMR Technical Campus, Hyderabad, Telangana, India

Nagasai Mudgala Department of Information Technology, Sultan Qaboos University, Al Khoudh, Muscat, Oman

Shaik Munawar Annamalai University, Chidambaram, India

C. Mylarareddy Department of CSE, SoT, GITAM University, Bengaluru, Karnataka, India

J. Naga Vishnu Vardhan Department of ECE, BVRIT HYDERABAD College of Engineering for Women, Bachupally, Hyderabad, India

T. Nagasai Anjani kumar Department of IT, Kakatiya Institute of Technology and Science, Warangal, India

Jonnadula Narasimharao Department of Computer Science and Engineering, CMR Technical Campus, Hyderabad, Telangana, India

V. A. Narayana CMR College of Engineering & Technology, Kandlakoya, Telangana, India

B. Narendra Kumar Rao Mohan Babu University, Tirupati, Andhra Pradesh, India

Kanna Naveen Department of Information Technology, Sultan Qaboos University, Al Khoudh, Muscat, Oman

Sanjana S. Nazare Department of Computer Science and Engineering, CMR Technical Campus, Hyderabad, Telangana, India

Satyam Nigam Electronics and Communication Department, Netaji Subhas University of Technology, Dwarka, Delhi, India

Swati Nigam Department of Computer Science, Banasthali Vidyapith, Banasthali, India

Dushmanta Kumar Padhi Department of Computer Science and Engineering, School of Engineering and Technology, GIET University, Gunupur, India

Neelamadhab Padhy Department of Computer Science and Engineering, School of Engineering and Technology, GIET University, Gunupur, Odisha, India

K. Padma Mayukha Department of ECE, BVRIT HYDERABAD College of Engineering for Women, Bachupally, Hyderabad, India

Lopamudra Panda School of Computer Science and Engineering, VIT-AP University, Amaravati, India

Harshbardhan Pandey CMR Institute of Technology, Bengaluru, Karnataka, India

Vedant Pandey Department of Computer Science and Engineering, CMR Institute of Technology, Bengaluru, India

Laveesh Pant CMR Institute of Technology, Bengaluru, Karnataka, India

Ganesh Panuganti Department of ECE, Kakatiya Instistute of Technology and Science, Warangal, Telangana, India

Ashok Patel Department of Computer Science, Florida Polytechnic University, Lakeland, FL, USA

Sibo Prasad Patro Department of CSE, Giet University, Gunupur, Odisha, India

C. S. Pavan Kumar Department of Information Technology, Sultan Qaboos University, Al Khoudh, Muscat, Oman

Sathwik Preetham Pendhota Department of ECE, Kakatiya Instistute of Technology and Science, Warangal, Telangana, India

A. Prabhu Department of CSE, CMR Technical Campus, Hyderabad, Telangana, India

Nitesh Pradhan Department of Computer Science and Engineering, Manipal University Jaipur, Rajasthan, India

Devarakonda Venkata Sai Pranav CSE, Sree Vidyanikethan Engineering College, Tirupati, India

K. Prasad Babu Department of ECE, JNTUA, Anantapuramu, Andhra Pradesh, India

Kamuju Sri Satya Priya BVRIT HYDERABAD College of Engineering for Women, Bachupally, Hyderabad, India

R. Priyakanth BVRIT HYDERABAD College of Engineering for Women, Hyderabad, Telangana, India

B. Pujitha Electrical & Electronics Engineering Department, BVRIT HYDERABAD College of Engineering for Women, Hyderabad, Telangana, India

Prashanth Ragam School of Computer Science and Engineering, VIT-AP University, Vijayawada, Andhra Pradesh, India

K. Srujan Raju CSE, CMR Technical Campus, Hyderabad, Telangana, India

Boddu Rama Devi Department of ECE, Kakatiya Instistute of Technology and Science, Warangal, Telangana, India

Ashish Raman Department of ECE, Dr. B.R. Ambedkar National Institute of Technology, Jalandhar, Punjab, India

Luis Ramirez Escuela Superior Polítecnica de Chimborazo, Telecommunications, Riobamba, Ecuador

Banothu Ramji Department of CSE (DS), CMR Technical Campus, Hyderabad, Telangana, India

Nakkeeran Rangasamy Department of Electronics Engineering, School of Engineering and Technology, Pondicherry University, Puducherry, India

D. Sandhya Rani Department of CSE, CMR Technical Campus, Hyderabad, Telangana, India

Kshitiz Rathore Dr. B.R. Ambedkar National Institute of Technology, Jalandhar, Punjab, India

Umashankar Rawat Department of Computer Science and Engineering, Manipal University Jaipur, Dahmi Kalan, India

Nagaraju Rayapati CSE, Sree Vidyanikethan Engineering College, Tirupati, AP, India

K. Reddy Madhavi CSE, Sree Vidyanikethan Engineering College, Tirupati, AP, India;
Mohan Babu University, Tirupati, Andhra Pradesh, India

L. Shashank Reddy Sreyas Institute of Engineering and Technology, Hyderabad, India

M. Sujith Reddy Sreyas Institute of Engineering and Technology, Hyderabad, India

Kanaka Durga Returi Malla Reddy College of Engineering for Women, Medchal, Hyderabad, Telangana, India

Guruswamy Revana BVRIT HYDERABAD College of Engineering, Hyderabad, India

M. Rishitha Sreyas Institute of Engineering and Technology, Hyderabad, India

Rahul Roy Department of Information Technology, Sultan Qaboos University, Al Khoudh, Muscat, Oman

V. Ruchitha Electrical & Electronics Engineering Department, BVRIT HYDERABAD College of Engineering for Women, Hyderabad, Telangana, India

Rahul Deo Sah Department of CA & IT, Dr. SPM University, Ranchi, Jharkhand, India

N. M. Sai Krishna BVRIT HYDERABAD College of Engineering for Women, Hyderabad, Telangana, India

M. Sai Priyanka Department of ECE, BVRIT HYDERABAD College of Engineering for Women, Bachupally, Hyderabad, India

A. Sai Suneel School of Engineering and Technology, Sri Padmavati Mahila Visvavidyalayam, Tirupati, India

Nagesh Salimath Department of CSE, Poojya Dodappa Appa College of Engineering, Kalaburagi, India

Sandhyarani Department of CSE (Data Science), CMR Technical Campus, Hyderabad, Telangana, India

J. Sangeetha Computer Science and Engineering, MSRIT, Bengaluru, India

Thumuluri Sai Sarika Department of Computer Science and Engineering, BVRIT HYDERABAD College of Engineering for Women, Hyderabad, India

Mayukh Sarkar Department of Computer Science and Engineering, Motilal Nehru National Institute of Technology Allahabad, Prayagraj, India

Madipally Sai Krishna Sashank Department of Computer Science and Engineering, GITAM School of Technology, GITAM (Deemed to be University), Visakhapatnam, Andhra Pradesh, India

Ankit Saxena Medicaps University, Indore, India

Priyanshi Shah Pandit Deendayal Energy University, Gandhinagar, Gujarat, India

M. Shanmuga Sundari Department of Computer Science and Engineering, BVRIT HYDERABAD College of Engineering for Women, Bachupally, Hyderabad, India

Anubhav Sharma CMR Institute of Technology, Bengaluru, Karnataka, India

Durgansh Sharma School of Business and Management, Christ University, Ghaziabad, India

Hitesh Kumar Sharma School of Computer Science, University of Petroleum and Energy Studies (UPES), Bidholi, Dehradun, Uttarakhand, India

J. Sheela School of Computer Science and Engineering, VIT-AP University, Amaravati, India

K. Shilpa Department of CSE, CMR Technical Campus, Hyderabad, Telangana, India

Mohamed Ghouse Shukur Department of Computer Science, King Khalid University, Abha, Saudi Arabia

Roohi Sille Systemics Cluster, University of Petroleum and Energy Studies (UPES), Dehradun, Uttarakhand, India

Ravi Sindal IET-DAVV, Indore, India

Bhupesh Kumar Singh B. S. Anangpuria Educational Institutes, Alampur, Faridabad, India

Rajiv Singh Department of Computer Science, Banasthali Vidyapith, Banasthali, India

Tarun Singh Department of Computer Science and Engineering, CMR Institute of Technology, Bengaluru, India

Y. Sirisha Sreyas Institute of Engineering and Technology, Hyderabad, India

Nitesh Sonawane Vishwakarma Institute of Technology, Pune, India

Vedaraju Nithya Sree BVRIT HYDERABAD College of Engineering for Women, Bachupally, Hyderabad, India

K. E. Sreenivasa Murthy Department of ECE, RECW, Kurnool, Andhra Pradesh, India

N. Sridevi Department of Electronics and Instrumentation Engineering, Dr. Ambedkar Institute of Technology, Bangalore, India

Cheekati Srilakshmi Department of Information Technology, VR Siddhartha Engineering College, Vijayawada, India

E. Srinath Department of CSE, Keshav Memorial Institute of Technology, UGC Autonomous, Hyderabad, Telangana, India

K. Srinivas CSE, CMR Technical Campus, Hyderabad, India

Vootla Srisuma Department of IT, CMR Technical Campus, Hyderabad, Telangana, India

Gaurav Srivastava Department of Computer Science and Engineering, Manipal University Jaipur, Rajasthan, India

G. Sruthi Department of Information Technology, Kakatiya Institute of Technology and Science, Warangal, India

T. Subburaj Department of Computer Applications, Rajarajeswari College of Engineering, Bangalore, India

P. Subhashitha Electrical and Electronics Engineering Department, BVRIT HYDERABAD College of Engineering for Women, Hyderabad, Telangana, India

M. Sudheer Kumar KPIT Technologies, Bangalore, Karnataka, India

A. Ch. Sudhir Department of EECE, GST, GITAM (Deemed to be University), Visakhapatnam, India

B. Sujatha Electrical and Electronics Engineering Department, BVRIT HYDERABAD College of Engineering for Women, Hyderabad, Telangana, India

Saba Sultana Department of CSE, CMR Technical Campus, Hyderabad, Telangana, India

Gurram Sunitha CSE, Sree Vidyanikethan Engineering College, Tirupati, AP, India

K. Suthendran Department of Information Technology, Kalasalingam Academy of Research and Education, Srivilliputhur, Tamilnadu, India

Debabrata Swain Pandit Deendayal Energy University, Gandhinagar, Gujarat, India

M. Swamy Das Department of Computer Science and Engineering, Chaitanya Bharati Institute of Technology, Hyderabad, India

A. Swathi Sreyas Institute of Engineering and Technology, Hyderabad, India

V. Swathi Sreyas Institute of Engineering and Technology, Hyderabad, India

Naresh Tangudu IT Department, Aditya Institute of Technology and Management, Tekkali, Srikakulam, AP, India

S. Tejaswi Sreyas Institute of Engineering and Technology, Hyderabad, India

Richa Tengshe CMR Institute of Technology, Bengaluru, Karnataka, India

Bryan Tite Escuela Superior Polítecnica de Chimborazo, Telecommunications, Riobamba, Ecuador

Pragati Tripathi Department of ECE, GLA University, Mathura, UP, India

Chandrasekhar Uddagiri Department of Computer Science and Engineering, BVRIT HYDERABAD College of Engineering for Women, Hyderabad, India

Raheem Unnisa Department of Computer Science and Engineering, CMR Technical Campus, Hyderabad, Telangana, India

Kunamsetti Vaishnavi Department of Computer Science and Engineering, BVRIT HYDERABAD College of Engineering for Women, Hyderabad, India

Dhaval Vasava Pandit Deendayal Energy University, Gandhinagar, Gujarat, India

B. Vikas Department of Computer Science and Engineering, GITAM School of Technology, GITAM (Deemed to be University), Visakhapatnam, Andhra Pradesh, India

S. Vikram Sindhu Department of Computer Science and Engineering, School of Engineering and Technology, GIET University, Gunupur, Odisha, India

Rajeswari Viswanathan BVRIT HYDERABAD College of Engineering, Hyderabad, India

R. Hitesh Sai Vittal Hyundai Mobis, Hyderabad, Telangana, India

Aditya Wanjari Vishwakarma Institute of Technology, Pune, India

Anushka Wankhade Vishwakarma Institute of Technology, Pune, India

Chintalacheri Charan Yadav CSE, Sree Vidyanikethan Engineering College, Tirupati, India

Shresthi Yadav Vishwakarma Institute of Technology, Pune, India

Mohamed Yasin Department of Math and Information Technology, Sultan Qaboos University, Alkhoudh, Muscat, Oman

Vamsidhar Yendapalli Department of CSE, SoT, GITAM University, Bengaluru, Karnataka, India

Plant Diseases Detection Using Transfer Learning

S. Divya Meena, Katakam Ananth Yasodharan Kumar, Devendra Mandava, Kanneganti Bhavya Sri, Lopamudra Panda, and J. Sheela

Abstract For several decades, plants have demonstrated to be an effective assessment of human life in a variety of fields. Plant pathogens are currently wreaking havoc on our farming sector. As an outcome, farmers are expected to make a loss. The precise and rapid diagnosis of plant diseases can aid in the development of an early treatment method, reducing enormous economic pain. Manually detecting plant diseases necessitates specialist knowledge of plant diseases, which is difficult, time consuming, and exhausting. In this research, a profound technique for detecting and classifying plant diseases from leaf images captured at multiple resolutions is provided. The inclusion of deep learning networks on top of basic models for effective feature learning is the main goal of this study. Various leaves image datasets from different countries are used to train a dense deep neural networks architecture. Because there were not enough photos, the gathered image data was augmented. In this experiment, 70,874 data shots were utilized to fit the classifier and 17,856 data images have been used to analyze it. The suggested CNN architecture can accurately categorize miscellaneous varieties of plant leaves, according to experimental evidence, and provide the possible remedies. In terms of accuracy, precision, recall, and F-score, a comparison of MobileNetV1 and ResNet34 structure was conducted, and the results revealed that the ResNet34 model is an efficient approach for disease categorization. The results of the experiments show that on

---

S. Divya Meena · K. A. Y. Kumar · D. Mandava · K. Bhavya Sri · L. Panda · J. Sheela (✉)
School of Computer Science and Engineering, VIT-AP University, Amaravati, India
e-mail: sheela.j@vitap.ac.in

S. Divya Meena
e-mail: divyameena.s@vitap.ac.in

K. A. Y. Kumar
e-mail: yasodharan.19bce7187@vitap.ac.in

D. Mandava
e-mail: devendra.19bce7500@vitap.ac.in

K. Bhavya Sri
e-mail: bhavya.19bce7315@vitap.ac.in

L. Panda
e-mail: lopamudra.19BCE7179@vitap.ac.in

K. A. Reddy et al. (eds.), *Proceedings of Fourth International Conference on Computer and Communication Technologies*, Lecture Notes in Networks and Systems 606,
https://doi.org/10.1007/978-981-19-8563-8_1

photographs with complicated backgrounds, an average test accuracy of 98.91% may be achieved. The processing time is 0.067 s with considerable precision for an individual plant leaf image, indicating that it is both real time and practicable.

Keywords Plant diseases classification · Deep learning · Leaf classification · ResNet34 · MobileNetV1

1 Introduction

Agriculture is vital in every economy since it plays a major role to supply food and income to a large portion of the population. Plant diseases have become a major source of concern in this agricultural sector because they generally cause crop damage, leading to a reduction in the availability and value of food produced [TMLAI]. Crop product quantity and quality have a direct impact on people's daily living conditions [1]. The worsening diversity has changed the environmental structure to a greater extent in the past years, which has paved the way for a widespread outbreak of agricultural diseases and pests. Even pathologists and agriculturists may not be able to spot illnesses that have afflicted plants just by looking at disease-affected leaves due to the development of a varied diversity of crop products. Visual inspection is still the major method of disease detection in rural parts of poor nations [2], but it involves various processes which are unsuitable, and also making it unfeasible for various medium-sized farms all throughout the world [3]. According to the Food and Agriculture Organization of the United Nations (FAO), around 40% of pants are lost each year due to sickness and pests [4]. The prevention of crop losses is becoming a center of research today, as studied by the study in [5] since it is a highly connected issue with climate change food standards and protection of the environment. As a result, it is so important and impactful in productivity that it has become a key instrument in precision agriculture (PA) [6]. It is easy to determine the crop's insufficiency just with the help of the image of the leaf. As a result, finding a quick, easy-to-use, and low-cost method to automatically detect plant illnesses is critical and realistic.

To control the problem, an automatic, less expensive, and accurate detection system for predicting the illness from images of any parts of plants and suggesting an apt pesticide as a possible solution is essential [7]. Even while image processing techniques are effective in detecting illnesses of the plant, they are subject to differences in leaf pictures due to form, texture, image noise, texture, and other factors. Machine learning techniques may also be used to classify plant diseases utilizing a variety of feature sets. Several steps have to be followed prior to extracting the features, efficiently, for example, enhancement of pictures, segmentation, and color modification. Before efficiently extracting features, preprocessing is required, such as picture enhancement, color modification, and segmentation [8]. Many classifiers, such as random forest, support vector machine, artificial neural network, deep neural network model, and others, may be employed after feature extraction. Traditional machine learning algorithms for disease diagnosis are very challenging to implement. As a result, deep learning methods can assist in overcoming these challenges in developing a better and expert system for agricultural growth. Many deep learning concepts have been applied to the agricultural area in recent years to solve problems, including insect identification, fruit detection, plant leaf categorization, and fruit disease detection, among others. To design a plant disease detection system, pictures of other parts of the plants can also be taken. But the most common and easiest portion of a plant to detect sickness of a particular plant is its leaves. As a result, we have used the leaves as a sample in this study to identify diseased crops [7].

This paper proposes a system based on deep learning for detecting and classifying plant diseases. The MobileNet and ResNet model are used to analyze performance on a minimal memory-efficient interface. The MobileNetV1 infrastructure can attain better accuracy rates compared to ResNet while decreasing the number of parameters and computations. Images of plant leaves from 14 crops were considered in 38 different classes depending on their state of fitness and illness categories. To ensure broad implementation of the suggested paradigm, publicly accessible resource includes pictures from multiple countries' archives. To establish a strong foundation, the photographs include both laboratory and field images. The following point discusses the contribution of this work:

- A Transfer learning concept is employed in the proposed methodology by fitting the data into four CNN models, and the proposed system classifies the leaves depending on the disease.
- With the emergence of smart applications, a very simple web application has been designed to give enhanced farming platform and assistance for recognizing plant pathogens.
- For numerous iterations, a vast dataset of gathered photos in diverse characteristics is being used to analyze deep learning designs.
- Different neural network models are compared in these studies.

The following is the paper's structure: Section 2 dealt with the literature review, Sect. 3 with the proposed work methodology, Sect. 4 with the framework analysis and datasets, Sect. 5 with the results, and Sect. 6 with the conclusion and future research recommendations.

2 Literature Review

Regardless of the fact that many researchers have been focusing on machine learning methods to detect a vast range of materials, very few studies are done on utilizing a MobileNet network either with transfer learning or without to predict illnesses of plants. Pest analysis demands the statistical study of vast amounts of data to determine the association of multiple components in order to acquire the guideline for protection. Hand identification techniques have a plethora of issues such as being only applicable to limited size plantations. On the one hand, the experience of employees varies a lot, resulting in inappropriate data on plant diseases and pests, negating the project, and resulting in agronomic losses. Most of the other research in this area, that has arisen on prevailing deep networks, has been constrained to the actions of computer systems with tremendous amassment responsiveness and also asset simulation in the majority of cases. SVM model which requires constructed features to differentiate the classes is used to recognize many plant diseases for a long time, including grape leaf diseases, palm oil leaf diseases, potato blight illnesses, and so on. Singh et al. [9] demonstrated a CNN model, i.e., multilayer CNN, to classify the leaves of mango tree produced by bacterial blight with the help of dataset containing real-time images with both afflicted and uninfected leaves. Mohanty et al. [10] employed deep learning architecture for diagnosing 26 infections in 14 different crops with 99.35% estimated accuracy. Barbedo [11] explored the challenges of detecting plant disease using visible range imagery. The author has examined numerous issues connected to plant disease recognition in this work. La et al. [12] described a new method for detecting rice illnesses. To diagnose ten rice ailments, 500 pictures of good and sick rice leaves were used. When utilizing tenfold cross-validation, the suggested CNN achieves good accuracy.

Ma et al. [13] used a deep learning model for common side effect categorization to classify four diseases of cucumber as follows: downy mildew, anthracnose, target leaf spots, and powdery mildew. Dan et al. [14] used an updated version of the MobileNet V2 algorithm for photo recognition to assess 11 different *Lycium barbarum* illnesses and pests. For this experiment, a total of 1955 photographs were taken, which were subsequently spatially metamorphosed into 18,720. Their recommended solution, SEMobileNet V2, has a 98.23% accuracy rate, which is greater than previous testing in this sector. The deep neural network is also trained by Rangarajan et al. [15]. The GoogleNet, AlexNet, and VGGNet, three excellent deep learning architecture systems are used to detects various kind of plant disease and their combined method was able to reach an overall accuracy of 80%.

A smaller amount of time and effort has gone into determining the extent of stress, which Kranz [16] and Bock et al. [17] believe is critical for controlling pests infestations, estimating harvest, and suggesting control remedies, as well as understanding fundamental biological processes like coevolution and plant disease causation [17]. This input is severely limited owing to a scarcity of reliable information that includes these crucial data. All of the prior research and published findings are encouraging, but more inventive and improved solutions in the field of plant identification are still needed. Disease detection and categorization with high accuracy are utilizing sophisticated neural network designs. Such automated analysis methods should indeed be analyzed with a large number of crops in diverse classes and scanning circumstances to increase their durability and efficacy. As a result, our algorithms suggested in this research increase both the efficiency and classification accuracy for plant disease photos by referencing the research and pertinent data. To solve the stated difficulties, this research provides a pest image processing approach based on an enhanced transfer learning model. Our technology takes advantage of the MobileNet and ResNet networks to deliver exact and effective pest-type identification. The proposed deep learning-based method beats the existing methods, with good accuracy.

3 Proposed Model

Models based on CNN are the deep learning methods that are used for identifying and classifying the diseases of the plant based on the images. CNN models are used for most image processing works, because of their accuracy in predicting the classes of the image. CNN (Fig. 1) models consist of several layers, and numerous designs have been implemented to obtain accurate results. In this project, various CNN models are used. The workflow of the project is shown in Fig. 1.

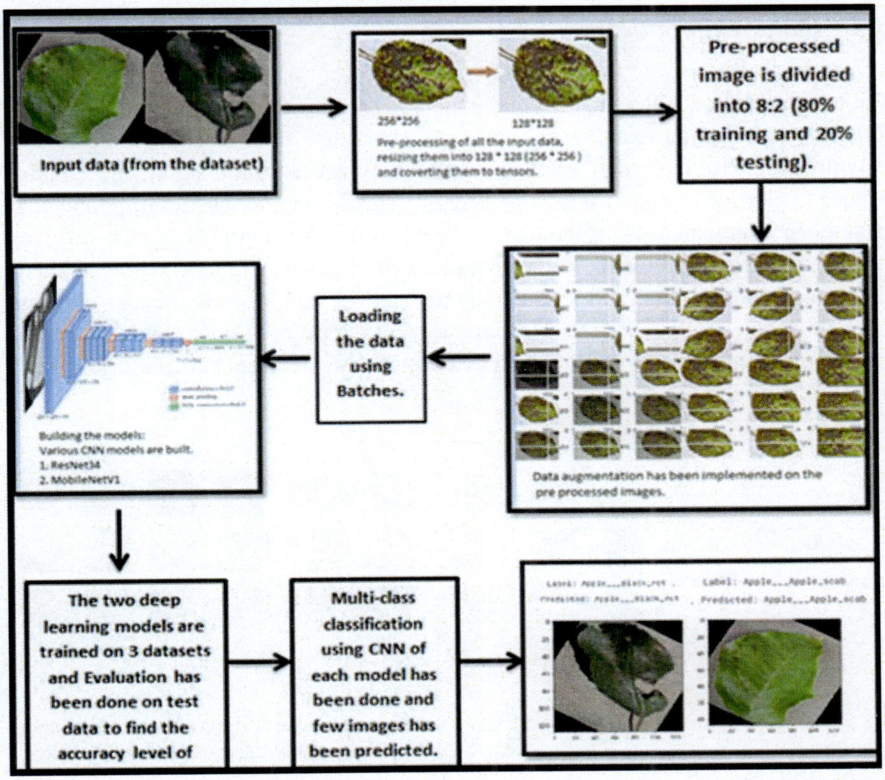

Fig. 1 Workflow of the project

Table 1 Information regarding class and label

Classification category	Label information
Presence of diseases	0—Healthy 1—Unhealthy

3.1 Data Preprocessing

One of the most significant processes in machine learning approaches is to preprocess the information or input data into a useable and relevant format. The data was collected in the form of PNG or Joint Photographic Expert Group (.jpeg) photographs, with labels given in the form of comma-separated values (CSV) files. The raw picture files were divided into categories based on the disorders. Every individual file was integrated into one consolidated dataset after the needed preprocessing method. Table 1 shows the information about the class and label.

3.2 Data Augmentation

Data augmentation is the most effective method for boosting the amount of training data by changing an existing dataset to create a new one. Because deep neural networks are extremely data-hungry models, they necessitate a significant amount of data to provide correct results. The Kaggle dataset contains a series of photos that have been augmented using the data augmentation approach [18], which involves making modest adjustments to the images such as image flipping, color augmentation, rotation, scaling, and so on. This updated fresh dataset is used to train the models. When using unseen data, the model will regard each little change as a new image, resulting in more accurate and better outcomes. The enhanced images are as seen below in Fig. 2.

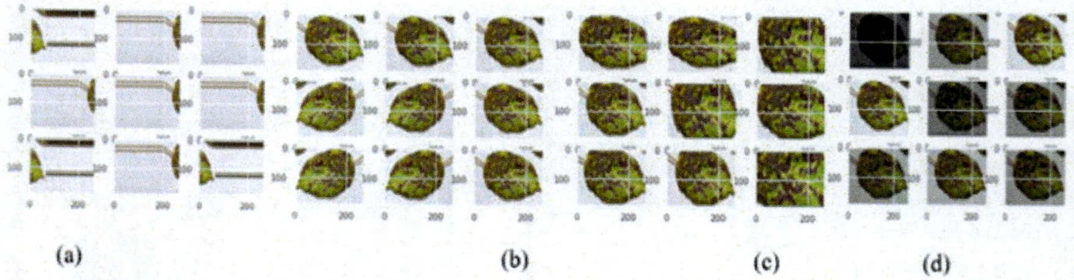

Fig. 2 Sample augmented images: **a** width shift, **b** horizontal and vertical flip, **c** zoom, and **d** color adjustment

3.3 *Proposed Approach—Transfer Learning*

In today's world, deep learning is a useful technique in which a CNN model that has previously been constructed for one job is used as a starting point for a model for a different task. A productive technique to employ pre-trained models for natural language processing (NLP) tasks, as rebuilding network models for this procedure takes a long time [19]. When dealing with predictive modeling, issues in which picture data is used as an input transfer learning are frequently used. As demonstrated in Fig. 3, this might be a prediction job with photos as well as video data as input. Deep learning concepts reduce time by eliminating the need to analyze huge amounts of data since it builds on existing knowledge [20]. We get better and more accurate outcomes as a result of this. MobileNet, ResNet, and EfficientNet are three well-known and useful designs.

MobileNetV1

TensorFlow's first mobile computer vision model, MobileNet, was created by a team of Google engineers specifically for mobile applications. Depthwise separable convolutions are preferred. As compared to a network with regular networks with the similar depth in the networks, the set of features is substantially decreased. Lightweight neural networks are built as a result [21]. When depthwise and pointwise convolutions are studied independently, a MobileNet has 28 layers. In a typical MobileNet, which is an upgraded version of other current models, the width multiplier hyperparameter may be modified or tweaked to lower the number of parameters (4.2 million). The size of the supplied picture is 224 * 224 * 3 pixels [22]. The MobileNetV1 architecture is depicted below in Fig. 4.

ResNet34

ResNet34 is a 34-layer convolutional neural network image classification model that is state of the art. The ResNet34 network's infrastructure is the network's residual building component, and it makes up the bulk of the network residual neural networks that (ResNets) are artificial neural networks (ANNs) that build networks by stacking residual blocks on top of each other [22]. A 3 × 3 max-pooling layer, an average pool

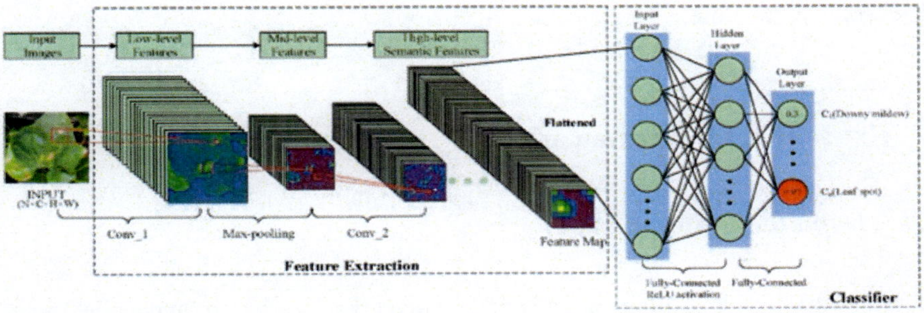

Fig. 3 Proposed plant disease classification framework

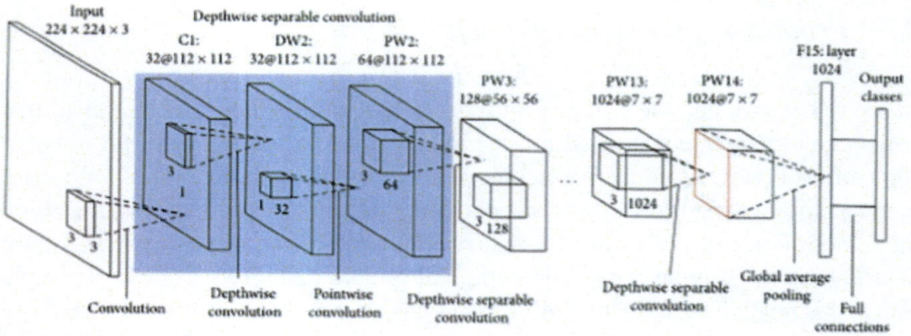

Fig. 4 Architecture of MobileNetV1 [23]

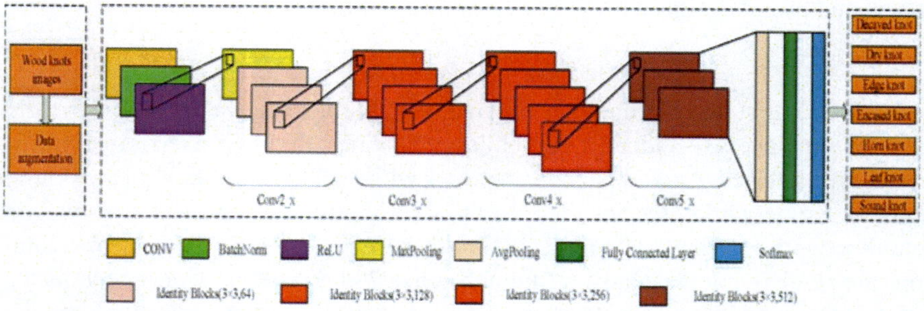

Fig. 5 Architecture of ResNet34 [24]

layer, and a fully linked layer are among ResNet's 34 layers, which are separated into 33 convolutional layers. In the "Basic Block," rectification nonlinearity (ReLU) activation and batch normalization (BN) have been provided toward the rear of all fully connected layers, and the sigmoid activation function is applied to the final layer in the typical manner. 63.5 million parameters in the ResNet34 model. The ResNet model is trained using residuals, which are the differences between the layer's inputs. The input shape for each ResNet34 model is 150 × 150 × 3. The residual building component consists of many convolutional layers (Conv), batch normalizations (BN), a rectified linear unit (ReLU) activation function, and a shortcut. Figure 5 depicts the ResNet34 architecture.

4 Experimental Framework

4.1 Implementation Details

A framework for identifying and quantifying plant diseases based on deep learning was analyzed on multiple leaf photos from diverse crops in a device with a 64 GB

RAM, Intel Xenon CPU, and a 64-bit Windows 10 operating system. Python is the programming framework in Anaconda Jupyter Notebook with TensorFlow, Keras, PyTorch, and other libraries. In this project, a learning rate of 0.001 is used. In this section, the loss and optimization functions that we used in our implementation are discussed in this section.

Loss Function

To learn, machines use a loss characteristic. It is a technique for determining how well a selected set of rules replicates the data. The loss feature will return in a large quantity if the forecasts are too far off from the actual circumstances. With the help of a few optimization characteristics, eventually, the loss function adapts to decrease prediction errors. Cross-entropy [25] has been employed as the loss function in this paper. Since the images are divided into various groups, categorical cross-entropy [26] is used. The error is computed using the loss function for each class, which ranges from 0 to 1. Categorical cross-entropy is expressed mathematically as Eq. (1)

$$\text{Loss} = -\sum_{i=1}^{\text{output size}} y_i \times \log\log \widehat{y_i} \tag{1}$$

Here, y_i is the predicted label and $\widehat{y_i}$ is the actual label [24].

Optimization Function

An optimizer is an algorithm that reduces the amount of loss in our neural network. This is accomplished by modifying our model's weights and learning rates [27]. The adaptive moment estimation (Adam) optimizer was utilized. RMSprop [28] and stochastic gradient descent [29] are combined in Adam. Adam optimizer can be represented mathematically as Eqs. (2) and (3)

$$W = W - \eta \times \left(\frac{V_{\text{dw}}}{\sqrt{S_{\text{dw}} + \varepsilon}} \right) \tag{2}$$

$$b = b - \eta \times \left(\frac{V_{\text{db}}}{\sqrt{S_{\text{db}} + \varepsilon}} \right) \tag{3}$$

W denotes weight, b denotes bias, and V is the rate at which the gradient drop is taking place. After each epoch, these equations will be utilized to adjust the weights and biases of each layer. Learning rate η and Epsilon ε (Epsilon = 10–8) are parameters that prevent zero division and are the learning rate.

Dataset Description

Three different datasets have been used for the training and testing process. Table 2 describes the dataset, and Fig. 6 illustrates the sample of each image class.

Table 2 Dataset description

Dataset name	Image type	#Images	Types
New plant disease dataset (Kaggle data repositories)	Laboratory and field images	87,900	Corn, orange, grape, peach, pepper, soybean, apple, tomato, blueberry, cherry, strawberry, and so on
Citrus leaves and fruits dataset	Laboratory and field images	609	Canker, greening, black spot, melanose, and healthy
Rice leaf diseases dataset (Kaggle data repositories)	Laboratory and field images	120	Leaf smut, bacterial leaf blight, and brown spot

Fig. 6 Sample pictures of the dataset

5 Result and Discussion

The trained dense deep learning model was evaluated using all photos from the validation set as well as unseen test images. True Positive (TP), False Positive (FP), True Negative (TN), False Negative (FN), accuracy, recall, precision, and F1-score are some of the traits that are evaluated. Proper data labels that were appropriately expected in reference to the ground truth are referred to be TP. Negative data labels that were incorrectly anticipated and categorized into a separate image label category are referred to be FP. Negative data samples that have been successfully forecasted are referred to be TN. FN stands for positive data labels that were incorrectly expected. The outcomes of training and testing the neural networks are shown below for both networks Data with both actual truth and extrapolated identities is assessed using confusion matrices.

Table 3 shows the recall, precision, specificity, and F1-score for each class. Specificity, recall, precision, and F1-score had an average value of 99.72%, 93.26%, 92.88%, and 92.98%, respectively. The trained deep learning model achieved an overall average accuracy of 98.91% on the test dataset. Table 4 shows the various performance parameters estimated on the testing set of photos, and the trained ResNet34 model surpassed the other model in comparison with its performance. Table 5 shows the time it takes for different models to compute the identical collection of photos utilized in the suggested study. Frames are processed fast by the MobileNetV1 deep learning model. In a system with GPU, the ResNet34 took

Table 3 Performance evaluation of the test dataset

Class	Specificity (%)	Recall (%)	Precision (%)	Accuracy (%)	F1_Score (%)	Class	Specificity (%)	Recall (%)	Precision (%)	Accuracy (%)	F1_Score (%)
Apple black rot	100.00	100.00	100.00	100.00	100.00	Raspberry health	100.00	67.44	87.40	74.99	62.45
Apple cedar rust	100.00	100.00	100.00	100.00	100.00	Soybean health	99.95	100.00	67.98	98.45	78.11
Apple healthy	100.00	100.00	100.00	100.00	100.00	Squash powdery_mildew	100.00	100.00	100.00	100.00	100.00
Apple scab	100.00	100.00	100.00	100.00	100.00	Strawberry leaf_scorch	100.00	99.34	99.45	67.33	74.77
Blueberry health	100.00	97.67	100.00	96.02	98.82	Strawberry health	100.00	99.45	98.45	97.56	99.34
Cherry health	99.2	100.00	100.00	93.40	98.76	Tomato bacterial spot	99.97	99.85	98.46	99.12	99.32
Cherry powdery mildew	100.00	100.00	100.00	100.00	100.00	Tomato early blight	99.94	100.00	98.88	99.74	99.81
Corn cercospora spot	99.96	100.00	97.80	100.00	98.88	Tomato late blight	99.83	78.65	89.54	89.99	99.99
Corn common rust	99.34	65.44	74.32	78.33	69.84	Tomato leaf_mold	99.96	100.00	100.00	88.88	100.00
Corn healthy	99.92	96.25	88.50	92.00	92.19	Tomato seporia leaf spot	99.56	88.91	87.69	85.99	99.45
Corn Northern leaf blight	100.00	100.00	100.00	100.00	100.00	Tomato spider_mites	99.39	87.88	88.93	88.45	85.38
Grape black rot	100.00	97.54	100.00	67.44	100.00	Tomato target spot	100.00	99.92	98.45	98.67	98.51

(continued)

Table 3 (continued)

Class	Specificity (%)	Recall (%)	Precision (%)	Accuracy (%)	F1_Score (%)	Class	Specificity (%)	Recall (%)	Precision (%)	Accuracy (%)	F1_Score (%)
Grape esca	100.00	100.00	100.00	100.00	100.00	Tomato leaf curl virus	100.00	100.00	100.00	100.00	100.00
Grape healthy	100.00	100.00	100.00	100.00	100.00	Tomato mosaic virus	100.00	99.97	98.88	99.95	99.95
Grape leaf blight	99.92	93.45	65.55	78.45	81.50	Tomato health	100.00	100.00	100.00	100.00	100.00
Orange Haunglongbing	97.56	67.13	74.34	97.44	98.45	Citrus black spot	100.00	100.00	100.00	100.00	100.00
Peach bacterial spot	99.96	100.00	98.76	100.00	98.31	Citrus canker	99.36	99.96	99.98	99.95	99.95
Peach healthy	99.90	76.44	71.35	100.00	63.53	Citrus greening	99.26	100.00	78.99	87.53	80.23
Pepper bell bacterial spot	99.96	100.00	88.93	74.62	100.00	Citrus scab	99.98	86.77	80.35	89.96	99.96
Pepper bell healthy	100.00	100.00	100.00	100.00	100.00	Rice bacterial leaf blight	99.96	99.34	99.98	90.56	99.98
Potato early blight	100.00	100.00	100.00	100.00	100.00	Rice brown spot	100.00	100.00	100.00	100.00	100.00
Potato late blight	99.92	81.00	78.31	65.78	100.00	Rice leaf smut	100.00	99.97	92.19	93.79	96.25
Potato healthy	99.96	100.00	94.76	100.00	67.34						

219.037 s to handle plant leave images with aspects of 128 × 128 pixels and a framerate of 18.932 frames per second.

The suggested approach achieved 98.91% correctness using a complex network trained on huge crop leaf picture from a variety of datasets with a wide variety of laboratory and on-field photos. Thus, these types of systems can be utilized to identify plant diseases since they provide the best accuracy with real-time performance on a dataset with intra-class and inter-class variance. The screenshots of web application which has been done using Python framework, and deep learning architecture is shown in Figs. 8, 9, and 10. Plant pathogens identification and classification approaches can be developed in the context of crop remote monitoring in order to make timely decisions and ensure healthy crop growth. The results in Fig. 7 are the average accuracy and loss of two different models.

Table 4 Evaluation of performance with various art models

Models	Specificity (%)	Recall (%)	Precision (%)	Accuracy (%)	F1-Score (%)
MobileNetV1	99.67	93.06	92.56	0.9832	92.1
ResNet34	99.78	93.47	93.21	0.9950	93.87

Table 5 Processing time to process several images of first dataset

Models	Total images	Image dimensions	Total computation time in seconds	Computational time for single image in seconds	Total frames/seconds
MobileNetV1	15,000	128 * 128	303.237	0.067	25.03
ResNet34	15,000	128 * 128	219.037	0.0492	18.932

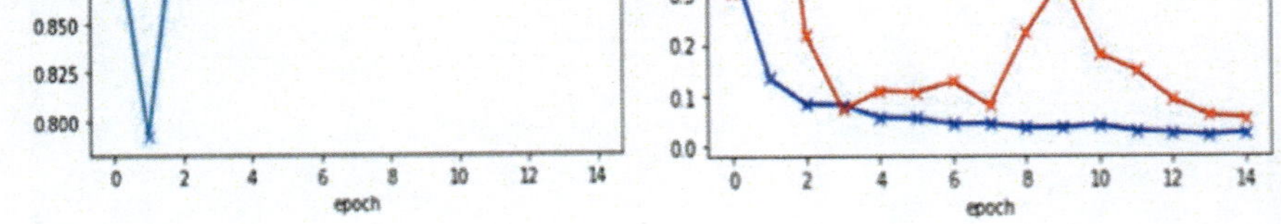

Fig. 7 Average accuracy and loss

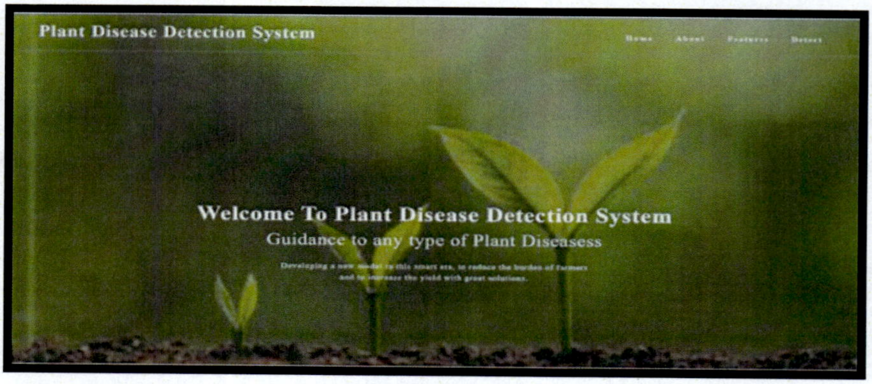

Fig. 8 Sample screenshot of the proposed system

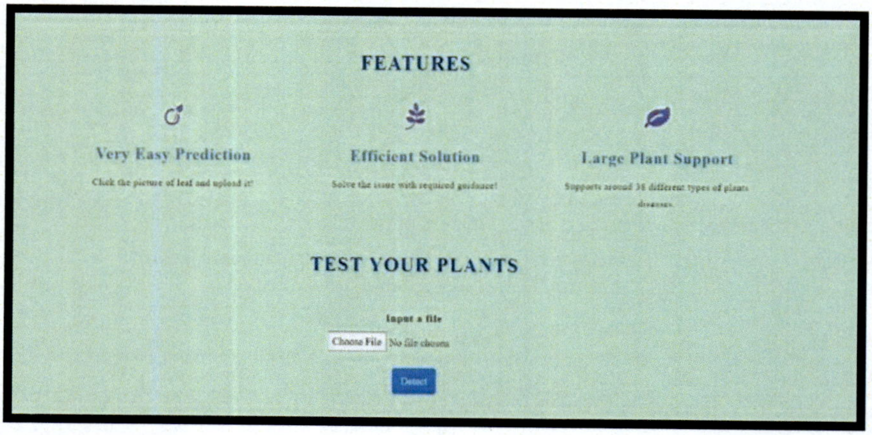

Fig. 9 Upload the image of the leaf

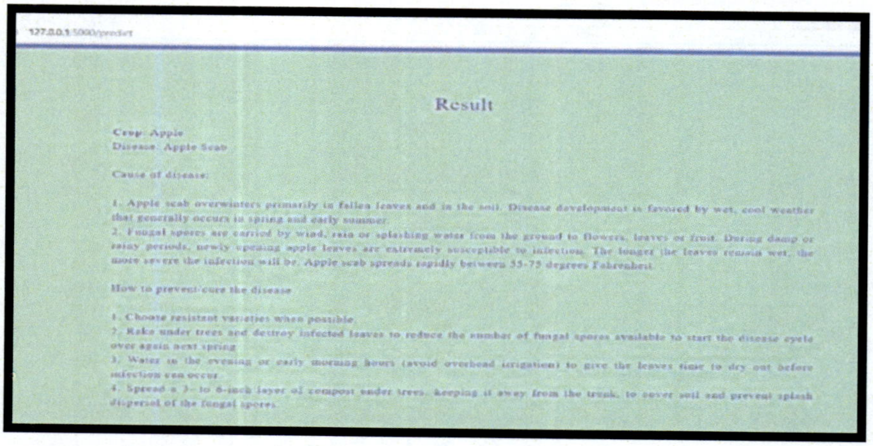

Fig. 10 Remedies of the detected diseases

6 Conclusion and Future Work

Using CNN to predict and identify agricultural illnesses is a difficult task. There have been numerous innovative ways developed for categorizing agricultural pathogens that prey on damaged crops. However, there is currently no commercially available approach for identifying illnesses that is both trustworthy and cost effective. A CNN-based plant disease prediction and analysis technique is provided in this paper. Three crop datasets are used in this study. The fully convolutional neural network is used to create the data processing model, and the data analysis approach is improved to assure the accuracy of the data analysis model. The simulation findings suggest that the method proposed is able to accurately predict and detect crop disorders, as well as having a good network model performance with an accuracy of 98.91%. The model's training time was far shorter than that of earlier machine learning approaches. According to the results of the studies, the integrated segmented and classification methods can be used well for crop disease prediction. Overall, the proposed technique has a lot of promise in terms of crop disease recognition and classification, as well as a fresh idea for the crop disease detection process. Future work could perhaps concentrate on disease and pest image analysis, also predicting the active surface of plant pathogens, and trying to judge the intensity of crop diseases and pests, required to bring out an efficient and systematic diagnosis to avoid significant economic losses, and we strategize to implement it on an integrated system to ensure and quickly detect a wide broad spectrum of biological diseases, allowing us to respond faster.

References

1. Liu Y, Zhang X, Gao Y, Qu T, Shi Y. Improved CNN method for crop pest identification based on transfer learning. https://www.hindawi.com/journals/cin/2022/9709648/
2. Hassan SM, Maji AK, Jasiński M, Leonowicz Z, Jasińska E (2021) Identification of plant-leaf diseases using CNN and transfer-learning approach. Electronics 10(12):1388
3. Sharma P, Berwal YPS, Ghai W (2020) Performance analysis of deep learning CNN models for disease detection in plants using image segmentation. Inf Process Agric 7(4):566–574
4. Food and Agriculture Organization of the United Nations, Plant Health and Food Security, International Plant Protection Convention, Rome, Italy, 2017
5. Fenu G, Malloci FM (2021) Forecasting plant and crop disease: an explorative study on current algorithms. Big Data Cogn Comput 5(1):2
6. Harvey CA, Rakotobe ZL, Rao N et al (2014) Extreme vulnerability of smallholder farmers to agricultural risks and climate change in Madagascar. Philos Trans R Soc B: Biol Sci 369(1639). Article ID 20130089
7. Tahamid A (2020) Tomato leaf disease detection using Resnet-50 and MobileNet architecture (Doctoral dissertation, Brac University)
8. Camargo A, Smith J (2009) An image-processing-based algorithm to automatically identify plant disease visual symptoms. Biosyst Eng 102:9–21
9. Singh LTP, Chouhan SS, Jain S, Jain S (2019) Multilayer convolution neural network for the classification of mango leaves infected by anthracnose disease. IEEE Access 7:4372143729
10. Mohanty SP, Hughes DP, Salathe M (2016) Using deep learning for image-based plant disease detection. Front Plant Sci 7:1419

11. Barbedo JGA (2016) A review on the main challenges in automatic plant disease identification based on visible range images. Biosyst Eng 144:52–60
12. La Y, Yi S, Zeng N, Liu Y, Zhang Y (2017) Identification of rice diseases using deep convolutional neural networks. Neurocomputing 267:378–384
13. Ma J, Du K, Zheng F, Zhang L, Gong Z, Sun Z (2018) A recognition method for cucumber diseases using leaf symptom images based on deep convolutional neural network. Comp Electron Agric 154:18–24
14. Dan B, Sun X, Liu L (2019) Diseases and pests identification of lycium barbarum using se-mobilenet v2 algorithm. In: 2019 12th International symposium on computational intelligence and design (ISCID), vol 1. IEEE, pp 121–125
15. Rangarajan K, Purushothaman R, Ramesh A (2018) Tomato crop diseases classification using pre-trained deep learning algorithmn. Procedia Comp Sci 133:1040–1047
16. Kranz J (1988) Measuring plant disease. In: Experimental techniques in plant disease epidemiology. Springer, Berlin, Germany, pp 35–50
17. Bock CH, Poole GH, Parker PE, Gottwald TR (2010) Plant disease severity estimated visually, by digital photography and image analysis, and by hyperspectral imaging. Crit Rev Plant Sci 29(2):59–107
18. Team K (2021) Keras documentation: image data preprocessing. Keras.io. https://keras.io/api/preprocessing/image/
19. Chollet F (2018) Deep learning with Python. https://livebook.manning.com/book/deep-learning-with-python/about-this-book/9
20. Brownlee J (2019, May 14) Transfer learning in keras with computer vision models. Machine Learning Mastery. https://machinelearningmastery.com/how-to-use-transfer-learning-when-developing-convolutional-neural-network-models/
21. Howard AG, Zhu M, Chen B, Kalenichenko D, Wang W, Weyand T, Andreetto M, Adam H. https://paperswithcode.com/paper/mobilenets-efficient-convolutional-neural
22. Singh N. https://iq.opengenus.org/mobilenet-v1-architecture/
23. Simonyan K, Zisserman A (2014) Very deep convolutional networks for large-scale image recognition. arXiv. https://arxiv.org/abs/1409.1556
24. Boesch G (2021, Aug 29) Deep residual networks (ResNet, ResNet50)—guide in 2021. Viso.ai. https://viso.ai/deep-learning/resnet-residual-neural-network/
25. Mannor S, Peleg D, Rubinstein R (2005) The cross entropy method for classification. In: Proceedings of the 22nd international conference on Machine learning. https://icml.cc/Conferences/2005/proceedings/papers/071_CrossEntropy_MannorEtAl.pdf
26. Parmar R (2018, Sept 2) Common loss functions in machine learning. Medium. https://towardsdatascience.com/common-loss-functions-in-machine-learning-46af0ffc4d23
27. Doshi S (2020, Aug 3) Various optimization algorithms for training neural network. Medium. https://towardsdatascience.com/optimizers-for-training-neural-network-59450d71caf6
28. Hinton G, Srivastava N, Swersky K (n.d.) Neural networks for machine learning lecture 6a Overview of mini-batch gradient descent. https://www.cs.toronto.edu/~tijmen/csc321/slides/lecture_slides_lec6.pdf
29. Robbins H, Monro S (1951) A stochastic approximation method. Ann Math Statis 22(3):400–407. https://www.jstor.org/stable/2236626

Shuffled Frog Leap and Ant Lion Optimization for Intrusion Detection in IoT-Based WSN

Dabbara Jayanayudu and A. Ch. Sudhir

Abstract In recent years, to examine the energy and security of the nodes that make up an Internet of Things (IoT) network is important. However, because of the limited resources, it is impossible to build a system that is 100% safe. An IDS is used to check all incoming traffic and identify network intrusions in WSN and IoT communication networks. It is also feasible for an attacker to steal sensors from an IoT network. To guarantee the safety of the WSN and IoT, an effective IDs system must be designed. As a result, SFLA-ALO is proposed in this study to identify intruders for WSN-IOT in order to safeguard against damaging malicious assaults. The suggested SFLA-ALO surpasses the previous systems in terms of throughput, detection rate, energy usage, and delay. MATLAB was used to assess these instances, which clearly beats existing detection systems.

Keywords Intrusion detection system (IDS) · SFLA-ALO · WSN · Ant line optimization · IoT

1 Introduction

WSNs are a kind of wireless network in which data transfer from a source to a base station is possible without the need of any infrastructure [1]. The Internet of things (IoT) has lately emerged as a superset of the previously outlined pattern of networking. Because of their distributed nature, IoT-WSN networks provide a significant security problem [2]. In addition to data transmission and reception, these IoT devices are used to connect numerous devices to the Internet [3]. An IoT network known as a WSN is what this research is committed to protecting. In the recent decade, machine learning and artificial intelligence-based IDSs have been extensively studied [4]. IoT networks link millions of sensors wirelessly, making the network resource

D. Jayanayudu (✉) · A. Ch. Sudhir
Department of EECE, GST, GITAM (Deemed to be University), Visakhapatnam, India
e-mail: d.jayanaidu@gmail.com

A. Ch. Sudhir
e-mail: camanapu@gitam.edu

K. A. Reddy et al. (eds.), *Proceedings of Fourth International Conference on Computer and Communication Technologies*, Lecture Notes in Networks and Systems 606,
https://doi.org/10.1007/978-981-19-8563-8_2

constrained [6]. Nodes in a WSN may move freely within the network due to the great mobility of the network [7]. Despite the fact that the Internet of things (IoT) offers many opportunities for creating an efficient system, power consumption seems to be a significant issue [8]. Because of the WSN's dynamic nature, routes between nodes often change, necessitating the need of an efficient routing protocol [9]. Due to the increasing mobility of network nodes, discovering the route and tracing it becomes a difficult task [10]. The major contributions of this research are analysis of WSN and IoT-based communication networks' security needs and probable harmful attacks.

- Concentrate on the protocols for IoT-WSN network intrusion detection.
- To design and implement a secure WSN-IoT application security algorithm named SFLA-ALO.

2 Literature Survey

Jain [11] presented a genetic algorithm (GA)-based biometric authentication system for IoT-WSN environment. The developed system provided both user level security and data level security using the biometric-based cryptography system. Granjal and Pedroso [12] developed a framework for intrusion detection and reaction in Internet-integrated CoAP WSN. The conventional methods in IoT for intrusion detection were failed to protect the IoT-WSN environment from the denial of service attacks. Srividya and Devi [13] presented an intrusion detection system based on the multi-strategic trust metrics for WSN. Ioannou and Vassiliou [14] developed a statistical analysis tool based on binary logistic regression (BLR) to monitor and detect the attacks in the IoT-WSN system. Here, the detection modules were generated with the help of BLR in an offline training stage by using the malicious activity and the benign local node activity from the representative two routing attacks. Haseeb et al. [15] implemented a framework to provide data protection over the malicious activities in mobile IoT devices. Shende and Sonavane [16] developed an energy and trust-aware multicast routing protocol based on the crow whale-energy trust routing (CrowWhale-ETR) optimization algorithm. The crow optimization and the whale optimization algorithms were integrated to develop the CWOA. This CWOA was utilized to find the optimum path for data transmission based on the evaluation of energy and trust of each node in the network. The secure communication in the network was achieved by updating the energy and trust of the individual notes after every data transmission. At the initial stage, the delay was low, but during the high time value, the delay was also increased.

3 Problem Statement

The following is a list of the most common WSN routing challenges.

- IoT networks have significant security challenges due to the increasing mobility of WSN nodes.
- A problem for safe routing is posed by IoT nodes' ability to self-organize and function without external infrastructure.
- The multicast routing method is computationally exhausting and does not take into consideration priority assignment criteria for the discovery of routes.

4 Proposed Method

The major goal of this work is to build SFLA-ALO for energy conscious multicast routing in WSNs. Both sorts of malicious behavior black hole attack and distributed denial of service (DDOS) attack used in the studies are discussed in this which makes the malicious node appear as if it has the shortest path.

4.1 *Computation of the Fit Factor*

DDoS assaults and black holes may both be launched from these rogue devices, which may already malicious scripts pre installed. When it comes to selecting secure nodes for secure and efficient communication, the fit factor is an important consideration. According to the trust and energy model of IoT nodes:

$$T_{ij} = T_{i,j}^{\text{direct}} T_{i,j}^{\text{indirect}} T_{i,j}^{\text{recent}} T_{i,j}^{\text{bytes}} \tag{1}$$

The trust used for the evaluation of the node trust is direct trust $T_{i,j}^{\text{direct}}$, indirect trust $T_{i,j}^{\text{indirect}}$, $T_{i,j}^{\text{recent}}$, and bytes trust $T_{i,j}^{\text{bytes}}$.

4.2 *Shuffled Frog Leap Algorithm*

Memeplexes communicate their thoughts and ideas in a restoration process that involves a series of stated stages for memetic progress. In the meantime, local inquiry and reorganization steps are carried out. Equation (2) is used to represent a frog.

$$X_i = X_{i1}, X_{i2}, \ldots, X_{\text{in}} \tag{2}$$

A frog's location can move no more than $D_{\max}$ degrees from its initial position; thus, we use random numbers between 0 and 1.

$$\text{Change in frog position } D_i = \text{rand} * (X_b - X_w) \tag{3}$$

$$\text{New position } X_W = \text{current position } X_W + D_i \tag{4}$$

$$D_{\max} > D_i > -D_{\max} \tag{5}$$

If this procedure generates an improved outcome, it substitutes the poorest frog, or else the controls in (3) and (4) are reiterated with reference to the global best frog (X_g replaces X_b). If there is no enhancement turn out to be probable in this scenario, at that time novel resolution is randomly produced to substitute that frog.

4.3 Ant Lion Optimization (ALO)

Generally, ALO primarily consists of five processes: those are random movement of ants, building, traps, catching of prays, and rebuilding of traps. The ant's position is randomly kept in the matrix M_{ant} that is given as Eq. (6):

$$M_{\text{ant}} = \begin{bmatrix} \text{Ant}_{1,1} & \text{Ant}_{1,2} & \cdots & \text{Ant}_{1,d} \\ \text{Ant}_{2,1} & \text{Ant}_{2,2} & \cdots & \text{Ant}_{2,d} \\ \vdots & \vdots & \cdots & \vdots \\ \text{Ant}_{n,1} & \text{Ant}_{n,2} & \cdots & \text{Ant}_{n,d} \end{bmatrix} \tag{6}$$

Here, $A_{i,j}$ designates the jth value of variable of ith ant, n represents the number of ants, and number of variables characterized as d. In objective function f, this is given as Eq. (7), and the fitness ant should be retained within the M_{OA} matrix.

$$M_{\text{OA}} = \begin{bmatrix} f[\text{Ant}_{1,1} & \text{Ant}_{1,2} & \cdots & \text{Ant}_{1,d}] \\ f[\text{Ant}_{2,1} & \text{Ant}_{2,2} & \cdots & \text{Ant}_{2,d}] \\ \vdots & \vdots & \cdots & \vdots \\ f[\text{Ant}_{n,1} & \text{Ant}_{n,1} & \cdots & \text{Ant}_{n,d}] \end{bmatrix} \tag{7}$$

M_{antlion} and M_{OAL} are mentioned as the location and fitness of ant lion, and the matrices are given by below Eqs. (8) and (9).

$$M_{\text{antlion}} = \begin{bmatrix} \text{AntL}_{1,1} & \text{AntL}_{1,2} & \cdots & \text{AntL}_{1,d} \\ \text{AntL}_{2,1} & \text{AntL}_{2,2} & \cdots & \text{AntL}_{2,d} \\ \vdots & \vdots & \cdots & \vdots \\ \text{AntL}_{n,1} & \text{AntL}_{n,2} & \cdots & \text{AntL}_{n,d} \end{bmatrix} \tag{8}$$

$$M_{\mathrm{OAL}} = \begin{bmatrix} f([\mathrm{AntL}_{1,1}\ \mathrm{AntL}_{1,2} \cdots \mathrm{AntL}_{1,d}]) \\ f([\mathrm{AntL}_{2,1}\ \mathrm{AntL}_{2,2} \cdots \mathrm{AntL}_{2,d}]) \\ \vdots \quad \vdots \quad \cdots \quad \vdots \\ f([\mathrm{AntL}_{n,1}\ \mathrm{AntL}_{n,1} \cdots \mathrm{AntL}_{n,d}]) \end{bmatrix} \tag{9}$$

The roulette wheel is used to estimate the high probability for the ant section that will be used to find the best ant lion. For the trapping procedure, Eq. (10) is applied.

$$C_i^t = \mathrm{Antlion}_j^t + C^t \tag{10}$$

$$d_i^t = \mathrm{Antlion}_j^t + d^t \tag{11}$$

where the random walks are designated adjacent to the ant lion via roulette wheel which is referred as R^t.

5 Results and Discussion

The performance of detection rate denotes the precise attacker detection, and delay performance mentions the interval reserved for transmitting the data between the IoT nodes in the network. The proposed SFLA-ALO technique promotes the better performances in all the metrics such as the detection rate, throughput, and highest energy with minimal delay. The result segment computes the proportional investigation of the proposed SFLA-ALO approaches and depends on the performance indexes in the existence of two types of attacks: One refers to the black hole attack and other one refers to DDoS attack. This investigation is deliberated with 50 nodes in the MATLAB/simulation setting as discussed in below.

5.1 Black Hole Attack at 50 Nodes

In this segment, the investigation is developed with 50 nodes in the existence of the black hole attack. In the beginning of the rounds, there is no delay, and with the increase in time, the delay increases. In the establishment of the node sequences, there is not much delay and formulas when the time starts to increase; it gives the increase in delay. On the other hand, from the Fig. 1, it strongly denotes that the delay of the proposedSFLA-ALO accomplished a marginal delay rate while associated with the existing crow whale method. The delay of proposed SFLA-ALO and crow whale-ETR are 0.2511 and 0.3363, respectively, while the interval is at 20 s. Figure 2 illustrates the detection rate of the proposed SFLA-ALO approaches with respect to

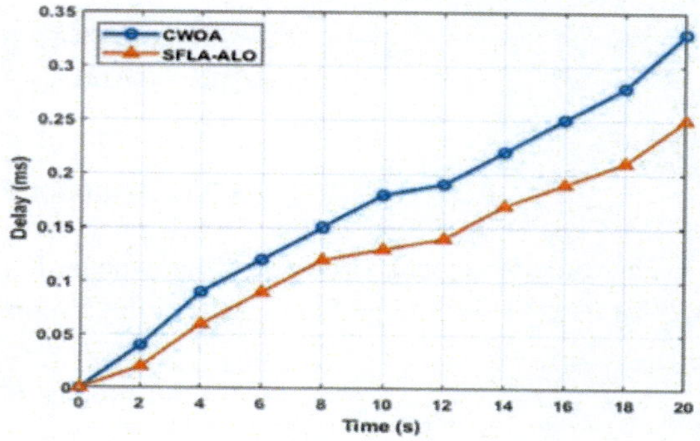

Fig. 1 Performance of delay

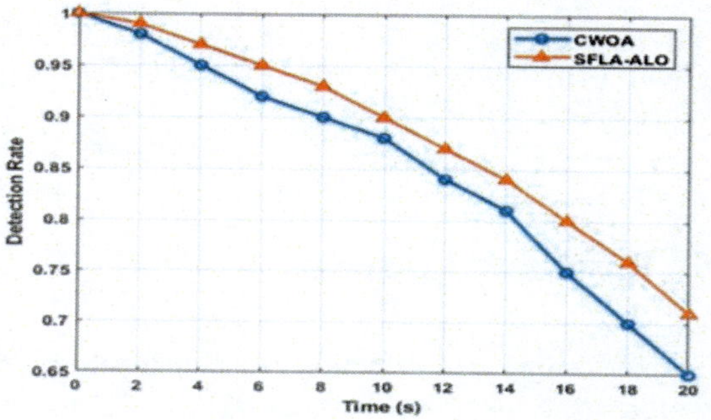

Fig. 2 Performance of detection rate

time. The detection rate obtained in the proposed SFLA-ALO and crow whale-ETR is 0.712 and 0.651, respectively, at the interval of 20 s.

Correspondingly, the examination of the energy performance is computed in Fig. 3. The results of the proposed SFLA-ALO and CWOA are 71 and 65 at the time of 20 s. Figure 4 illustrates the performance of throughput for proposed SFLA-ALO and CWOA, respectively. From the result of throughput, it clearly illustrates the proposed SFLA-ALO and CWOA that achieve the value of 0.2 and 0.05, respectively, at the time of 20 s.

5.2 *DDOS Attack at 50 Nodes*

In this section, the investigation is developed with 50 nodes in the existence of the DDoS attack. In the establishment of the node sequences, there is not much delay, when the time starts to increase; it gives the increase in delay. On the other hand, from Fig. 5, it strongly denotes that the delay of the proposed SFLA-ALO technique accomplished a marginal delay rate while associated with the existing crow whale method. The delay of proposed SFLA-ALO and crow whale-ETR is 0.2388 and 0.2890, respectively, while the interval is at 20 s.

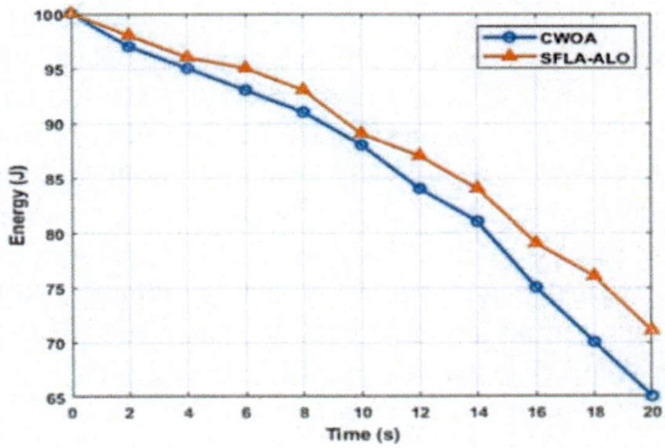

Fig. 3 Performance of energy

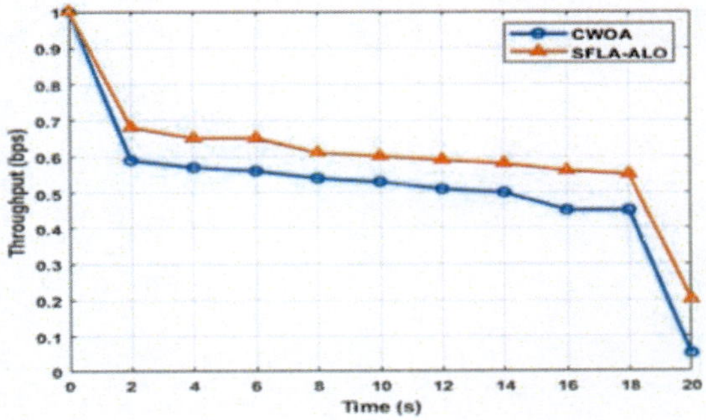

Fig. 4 Performance of throughput

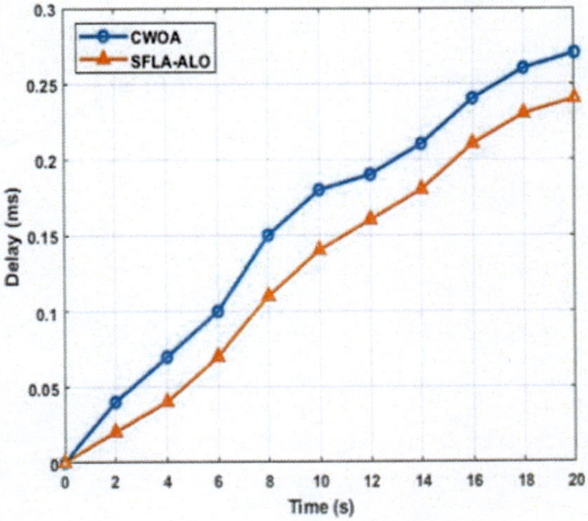

Fig. 5 Performance of delay

Figure 6 illustrates the detection rate of the proposed SFLA-ALO approaches with respect to time. The detection rate obtained in the proposed SFLA-ALO and crow whale-ETR is 0.72 and 0.67, respectively, at the interval of 20 s (Fig. 6).

Figure 8 illustrates that the performance of throughput for proposed SFLA-ALO and CWOA, respectively. From the result of throughput, it clearly illustrates the proposed SFLA-ALO and CWOA achieve the value of 0.19 and 0.05, respectively, at the time of 20 s. From the all simulation results for both the attacks (black hole and DDoS), it is determined that the proposed SFLA-ALO technique attained the improved performance in all the parameters like detection rate, delay, throughput, and energy in the existence of 50 nodes.

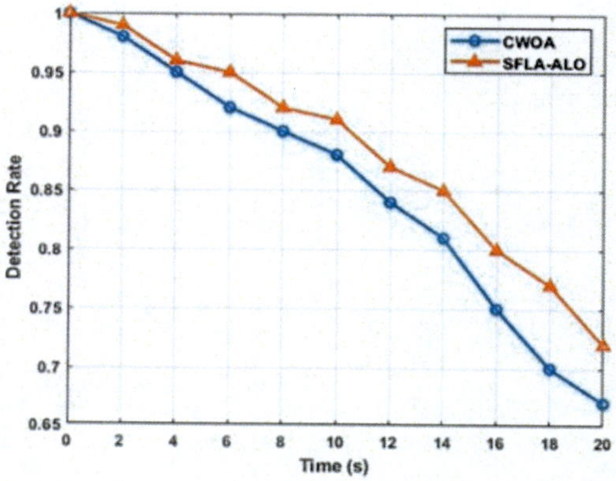

Fig. 6 Performance of detection rate

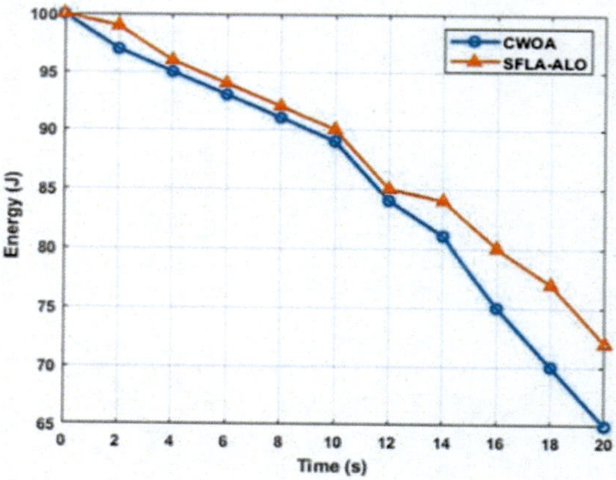

Fig. 7 Performance of energy

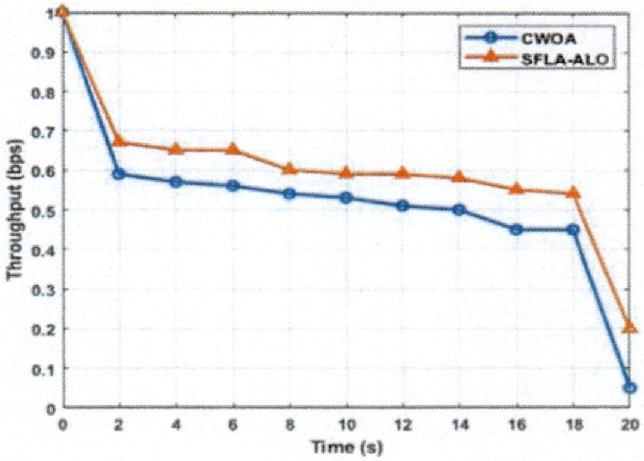

Fig. 8 Performance of throughput

6 Conclusion

One of the most difficult issues in networking is ensuring security while maximizing energy efficiency. Monitoring IDS through IoT-WSN necessitates an enhanced focus on security. In this study, we provide an IoT-WSN safe routing intrusion prevention architecture. Improved network performance and protection against malicious attacks are the primary goals. The greedy method for data routing is used by the majority of energy-efficient techniques, which rely on static sensor nodes. As a consequence, dynamic circumstances do not allow for such solutions. The preservation of IoT security and privacy is essential to IoT services, but it also presents a significant challenge to IoT security. We now have an abundance of information thanks to the Internet's many communication channels and social media platforms. The SFLA-ALO approach proposed in this research is a novel intrusion detection tool for the Internet of things. The suggested SFLA-ALO approach increases security by measuring performance metrics such as detection rate, throughput, latency, and energy consumption, according to simulation findings.

References

1. Butun I, Morgera SD, Sankar R (2013) A survey of intrusion detection systems in wireless sensor networks. IEEE Commun Surv Tut 16(1):266–282
2. Borkar GM, Patil LH, Dalgade D, Hutke A (2019) A novel clustering approach and adaptive SVM classifier for intrusion detection in WSN: a data mining concept. Sustain Comput: Inf Syst 23:120–135
3. Pundir S, Wazid M, Singh DP, Das AK, Rodrigues JJ, Park Y (2019) Intrusion detection protocols in wireless sensor networks integrated to internet of things deployment: survey and future challenges. IEEE Access 8:3343–3363
4. Amouri A, Alaparthy VT, Morgera SD (2020) A machine learning based intrusion detection system for mobile Internet of Things. Sensors 20(2):461
5. Halder S, Ghosal A, Conti M (2019) Efficient physical intrusion detection in Internet of Things: a Node deployment approach. Comput Netw 154:28–46

6. de L, Martínez M, Scheffel RM, Fröhlich AA (2020) An analysis of the gateway ıntegrity checking protocol from the perspective of ıntrusion detection systems. In: Design automation for embedded systems, pp 1–23
7. Karabiyik U, Akkaya K (2019) Digital forensics for IoT and WSNS. In: Mission-oriented sensor networks and systems: art and science, pp 171–207
8. Rani S, Maheswar R, Kanagachidambaresan GR, Jayarajan P (2020) Integration of WSN and IoT for smart cities. Springer
9. Chen H, Meng C, Shan Z, Fu Z, Bhargava BK (2019) A novel low-rate denial of service attack detection approach in ZigBee wireless sensor network by combining Hilbert-Huang transformation and trust evaluation. IEEE Access 7:32853–32866
10. Rajeswari AR, Kulothungan K, Ganapathy S, Kannan A (2021) Trusted energy aware cluster based routing using fuzzy logic for WSN in IoT. J Intell Fuzzy Syst:1–15
11. Jain JK (2019) Secure and energy-efficient route adjustment model for Internet of Things. Wirel Pers Commun 108(1):633–657
12. Granjal J, Pedroso A (2018) An intrusion detection and prevention framework for internet-integrated CoAP WSN. Security and Communication Networks
13. Srividya P, Devi LN (2021) Multi-strategic trust evaluation for intrusion detection in wireless sensor networks. Int J Intell Eng Syst 14(2):106–120
14. Ioannou C, Vassiliou V (2018) An intrusion detection system for constrained WSN and IoT nodes based on binary logistic regression. In: Proceedings of the 21st ACM ınternational conference on modeling, analysis and simulation of wireless and mobile systems, pp 259–263
15. Haseeb K, Islam N, Almogren A, Din IU (2019) Intrusion prevention framework for secure routing in WSN-based mobile Internet of Things. IEEE Access 7:185496–185505
16. Shende DK, Sonavane SS (2020) CrowWhale-ETR: CrowWhale optimization algorithm for energy and trust aware multicast routing in WSN for IoT applications. Wirel Netw:1–19

A Comprehensive Alert System Based on Social Distancing for Cautioning People Amidst the COVID-19 Pandemic Using Deep Neural Network Models

Kanna Naveen, Nagasai Mudgala, Rahul Roy, C. S. Pavan Kumar, and Mohamed Yasin

Abstract The World Health Organization (WHO) has suggested a successful social distancing strategy for reducing the COVID-19 virus spread in public places. All governments and national health bodies have mandated a 2-m physical distance between malls, schools, and congested areas. The existing algorithms proposed and developed for object detection are Simple Online and Real-time Tracking (SORT) and Convolutional Neural Networks (CNN). The YOLOv3 algorithm is used because YOLOv3 is an efficient and powerful real-time object detection algorithm in comparison with several other object detection algorithms. Video surveillance cameras are being used to implement this system. A model will be trained against the most comprehensive datasets, such as the COCO datasets, for this purpose. As a result, high-risk zones, or areas where virus spread is most likely, are identified. This may support authorities in enhancing the setup of a public space according to the precautionary measures to reduce hazardous zones. The developed framework is a comprehensive and precise solution for object detection that can be used in a variety of fields such as autonomous vehicles and human action recognition.

Keywords YOLOv3 algorithm · Dataset · Object detection · Convolutional Neural Networks (CNN) · Action recognition

K. Naveen · N. Mudgala · R. Roy · C. S. Pavan Kumar (✉)
Department of Information Technology, Sultan Qaboos University, Al Khoudh, Muscat, Oman
e-mail: pavanchitturi@vrsiddhartha.ac.in

R. Roy
e-mail: 188w1a1211@vrsiddhartha.ac.in

M. Yasin
Department of Math and Information Technology, Sultan Qaboos University, Alkhoudh, Muscat, Oman
e-mail: mohyasin@squ.edu.om

K. A. Reddy et al. (eds.), *Proceedings of Fourth International Conference on Computer and Communication Technologies*, Lecture Notes in Networks and Systems 606,
https://doi.org/10.1007/978-981-19-8563-8_3

1 Introduction

Social distancing is evidenced as a critical initiative amidst the pandemic caused by contagious COVID-19 virus. However, people have been unaccustomed to maintaining the required 2 m separation between themselves and their surroundings. A dynamic monitoring system able to detect infringements in social distance between people and immediately notify them is the necessity of the hour. The proposed system is a model for detecting and notifying people in real time that could make people who aren't following social distance to be more cautious. Using this alerting system, the people who are not adhering to the recommended distancing norms could be identified. A count of social distancing violations for each frame of the video would be displayed. Many changes were made to the object detection algorithms in order to improve fulfillment concurrently on speed and accuracy. As a result of ongoing research in the disciplines of computer vision and deep learning, the deep learning techniques have indeed been significantly associated with enhanced object recognition accuracy. An ample amount of research has been conducted on existing object detection algorithms in order to select the algorithm. Object detection algorithms are classified into two types: two-stage detectors and single-stage detectors. The algorithms covered by two-stage detectors are RCNN, Fast RCNN, and Faster RCNN, while the algorithms covered by single-stage detectors are You Only Look Once v1, You Only Look Once v2, You Only Look Once v3, and Single Shot Multibox Detector. The results of two-stage detection algorithms are maximally accurate but are generally slow. Single-stage detectors are faster than two-stage detectors, but these result in inaccurate results.

2 Objectives

This project is for cautioning the people who are violating the social distancing norms and in result the organizations could take the support of this project and could monitor whether the social-distancing norms are well followed, using a novel single-stage model approach for increasing speed without compromising much accuracy with the YOLOv3 algorithm, which improves object detection speed while maintaining accuracy in comparison with that of the similar other algorithms.

3 Literature Review

This project was created after reviewing the following literature. In each of these papers, we have studied several technological aspects, offering a basic overview of object detection schemes that combine two object detectors [1]. Although single-stage detectors are significantly faster than two-stage object detectors, two-stage sensors achieve the best prediction rates. It describes the detection method, which is based on regional suggestion and regression, as well as the system's advantages and disadvantages [2]. SSD produces more accurate results, and YOLO operates more quickly. Because of the speed with which the execution was accomplished, the proposed solution helps make the use Mobile Net SSD [3]. Using a simplified IoT paradigm would be resulting in excessive electrical energy consumption. Even minor movements, such as a strong breeze or wildlife, would be misinterpreted as human appearance suggests wearing a social distancing device that uses a microprocessor and an ultrasonic sensor to determine the distance between two persons [4]. When compared to these image processing algorithms, detection is much more accurate. This, however, does not guarantee that each individual has the detector with them. It employs a deep neural network, a mask Regions with Convolutional Neural Networks (RCNN) for identifying faces in video frames [4]. The CNN algorithm can be used with large datasets and detects without the assistance of humans, but it is slow. Although the CNN algorithm can handle large datasets and identify without human intervention, it is slower. When considering distancing in public spaces, CNN-based object detectors with a recommended social distancing algorithm produce promising results [5]. CNN models are being used in image recognition and text mining. It is important in classification. Small objects, on the other hand, are difficult to detect. A new adaptive detection methodology for effectively recognizing and monitoring people is constructed using both interior and exterior contexts [6, 7].

4 Proposed Methods in Continuous Development

The proposed model helps to identify people who violate social distance rules in crowded environments. Deep learning and computer vision techniques are used, and the people in the video frame were detected using an open-source object detection network based on the YOLOv3 algorithm, and the pedestrians were identified. In this application, these object classes are ignored. As a result, for each recognized individual in the image, the best bounding boxes with centroids will be generated, and the centroids will be used for distance measurement. The centroids of two people can be used to calculate the distance between them. Let the first person's centroid be (x_1, y_1) and the second person's centroid be (x_2, y_2), and the distance between their centroids be (x_2, y_2) is calculated by squaring difference of the respective coordinates to the whole root of the sum of their product. The model's flow diagram and the procedure conducted on the input video. The first step is to load a video frame and count the number of people in the frame. Following the detection of people, the distance between them is measured, and if the distance is higher than or equal to the social distance, a green bounding box surrounds the person. In this situation, the individual is marked with a red bounding box, and the number of infractions is increased and displayed on the screen. This is a recursive procedure that continues until the entire video frame is identified. And the social distance between people would be measured if the distance is less than 2.5 ft. They would be bounded in red

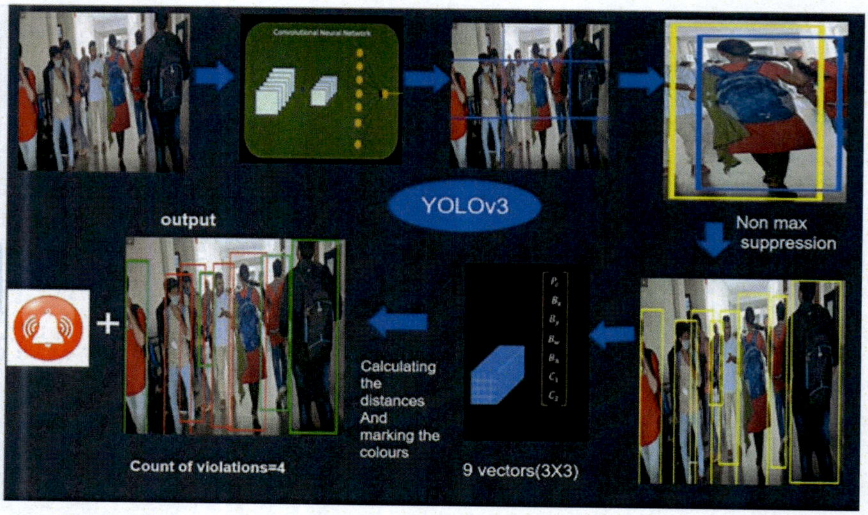

Fig. 1 Architecture diagram

boxes and if the distance is greater than 2.5 ft. they would be bounded in the green boxes. So, we could identify the people who are not following the social distancing in red boxes and the ones following in green boxes, hence could notify them accordingly with a beep sound using an alarm [8].

The first input frame from the movie is imported, as shown in Fig. 1 and previously explained, and a grid is formed using a convolution neural network. Assume that it takes 3 by 3 grids to produce 9 vectors, each with 7 variables. Probability, x-axis length, y-axis length, height, and breadth come first, followed by classes c_1 and c_2. This vector is now used to identify the picture and create boundary boxes. If we have more than one, non-max suppression occurs when the bounding box with the lowest probability is deactivated. The pictures are now bound with high probability to the boxes, and the distance between the bounding boxes is determined. In this investigation, to determine people, the YOLO algorithm was used. The YOLO method learns bounding box coordinates (t_x, t_y, t_w, t_h), object confidence, and matching class label probabilities (P_1, P_2, ..., P_c) to distinguish objects from a given input image. The YOLO was trained using the COCO dataset [9], which contains 80 labels, including human and pedestrian classes. Only box coordinates, object confidence, and pedestrian object class from the YOLO model detection result were used for person detection in this study [4] (Fig. 2).

5 Performance Metrics

On manually labeled videos, performance was evaluated. To summarize the model performance, the F1 score, accuracy, and prediction time are shown. The system's average accuracy is 91.2%, and its average F1 score is 90.79%. The overall system's

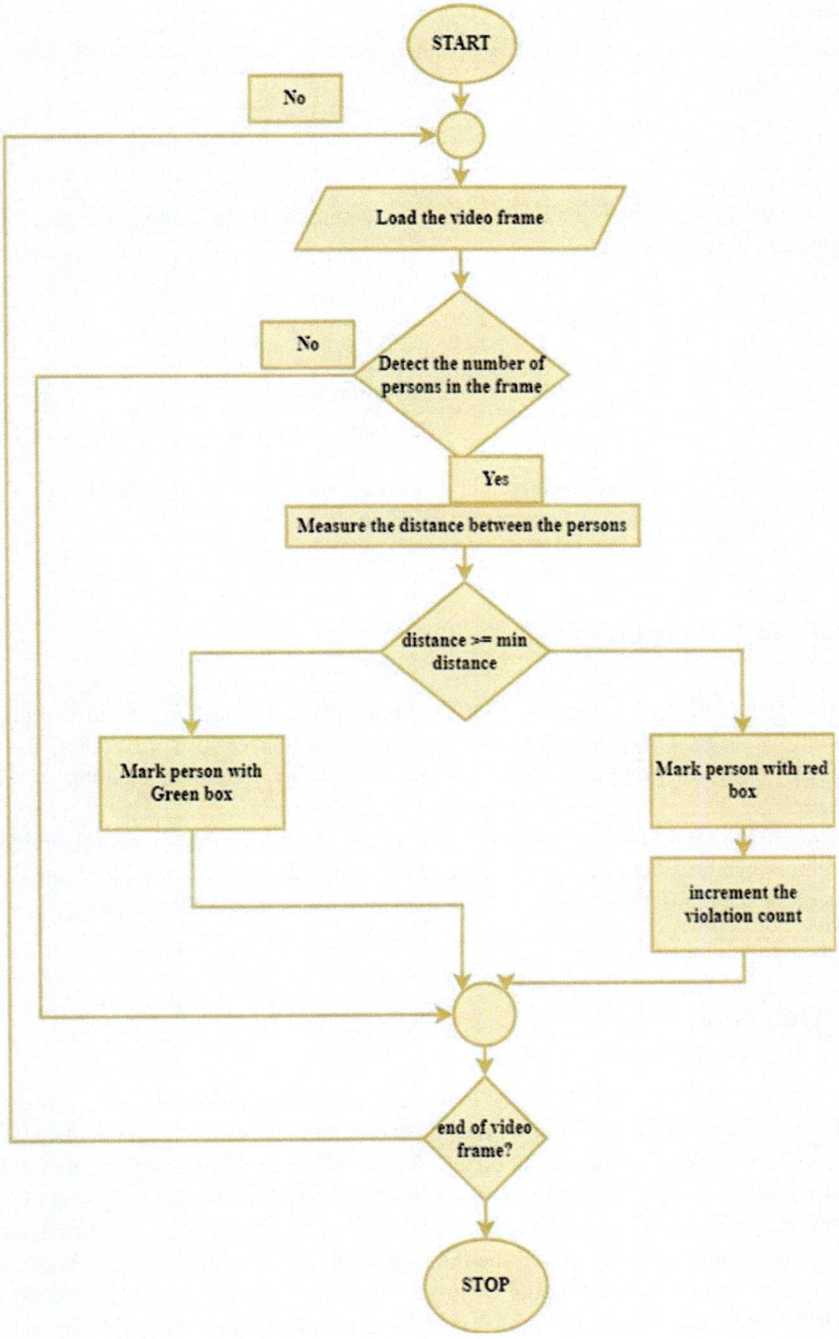

Fig. 2 Proposed work's process flow

Table 1 Showing the values of YOLOv3

Model	Accuracy %	F1 Score %	Average time in process 1 s of video (in seconds)
Social distancing	89.79	88.51	0.0042
Person detection	93.46	93.57	5.24

average prediction time is 7.12 s for a one-second video frame, with person detection taking the longest, at 5.24 s.

$$\text{Accuracy} = \frac{\text{TN} + \text{TP}}{(\text{TN} + \text{FN}) + (\text{TP} + \text{FP})} \tag{1}$$

$$F1\ \text{Score} = 2 * \frac{\text{Precision} * \text{Recall}}{(\text{Precision} + \text{Recall})} \tag{2}$$

where TN stands for true negative, TP stands for true positive, FN stands for false negative, and FP stands for false positive (Table 1).

6 Data Collection

In this proposed model, the MS COCO (Microsoft Common Objects in Context) dataset was used. This is a large-scale dataset that can be used for object recognition, segmentation, key-point detection, and captioning. The collection contains 328 K photos.

Splits: MS COCO dataset was first released in 2014, with 164 K photos divided into three groups: the training, validation, and test. Additionally, the 2017 released version of the dataset contains a new unannotated set of 123,000 Images [3].

7 Outcome

The concerns are illustrated from top to bottom, with boxes indicating each runner in terms of image processing. To acquire an accurate number, we forecast the distances by computing the centroids of the boxes. These red colored boxes represent people who are too far away from this individual [10]. These green hue boxes represent those who keep a safe distance from this. In all of the above examples, slower object detection algorithms may fail to determine the right distance since the distances between people are continuously changing owing to their motions; hence, the quicker one stage detector object detection model is utilized to suit the objective. Yolo operates at a fast speed, with an accuracy of 0.358 and a time limit of 0.8 s per every frame. The concerns are illustrated from top to bottom, with boxes indicating each runner in terms of image processing. To acquire an accurate number, we forecast the distances by computing the centroids of the boxes [11].

These red colored boxes represent people who are too far away from this individual. These green hue boxes represent those who keep a safe distance from this. In all of the above examples, slower object detection algorithms may fail to determine the right distance since the distances between people are continuously changing owing to their motions; hence, the quicker one stage detector object detection model is utilized to suit the objective. Yolo operates at a fast speed, with an accuracy of 0.358 and a time limit of 0.8 s per every frame. Similarly, the YOLO algorithm may forecast more than one bounding box for a single runner. However, only the bounding box with the highest probability will be evaluated [9].

8 Comparison Values of Various Methodologies as Determined by Various Studies

As a result, there is no issue with counting the same object too many times. Only runners who do not keep the necessary distance will be tallied in the number of infractions. Aside from that, each violation will result in a warning. So that it is simple to determine how many people are breaking.

In the test 1, test 2 shown above a video stream from Velagapudi Ramakrishna Siddhartha Engineering College's Information Technology is evaluated. The input video shows a group of individuals strolling at a steady pace. The detection procedure has been finished, and the results are depicted in Figs. 3 and 4, respectively.

In comparison between YOLOv3 and Faster RCNN, YOLOv3 can take eight frames more than Faster RCNN per second. YOLOv3 can work better in terms of speed in comparison with Faster RCNN (Figs. 5, 6 and 7). Table 2 shows that YOLOv3 has higher accuracy of 93.46% than other algorithms.

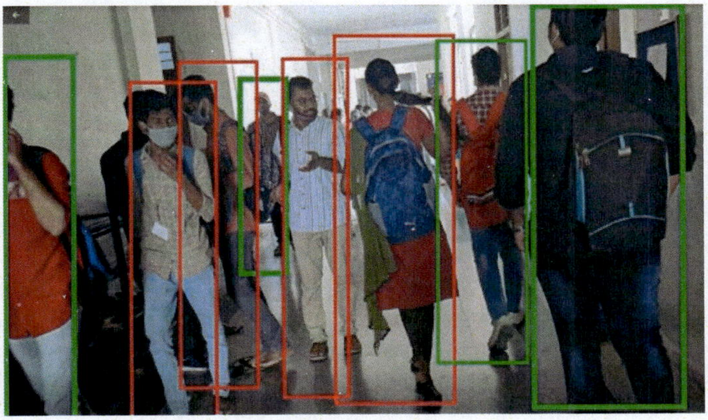

Fig. 3 Test 1's social-distancing detection results

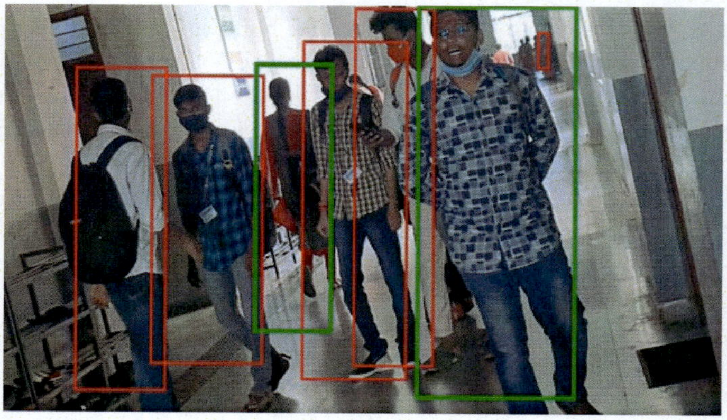

Fig. 4 Test 2's social-distancing detection results

Fig. 5 Output with Faster RCNN

9 Conclusion and Future Scope

Image processing techniques are used to detect social distancing violations. This design was validated using a video of a group of people competing in a running race. The visualization results confirmed that this method could determine the distance between people, which can also be used in public places such as bus stops, shopping malls, and hospitals. Furthermore, it can be improved by combining it with mask detection. It can also be improved by modifying the system capabilities and implementing more advanced algorithms for faster detection. For social distancing

Fig. 6 Output with YOLOv3

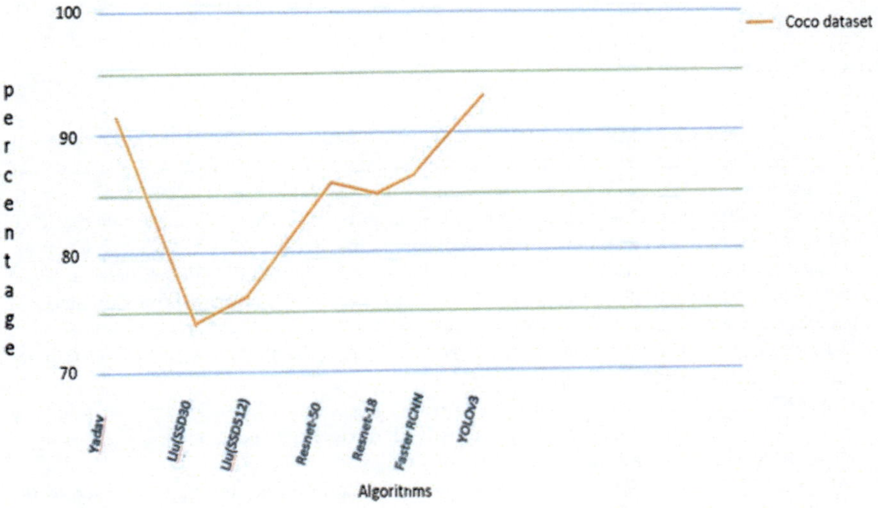

Fig. 7 Percentage graph on comparing different algorithms that outperforms with COCO dataset

detection, the flows are depicted from top to bottom as boxes denote each runner [13]. To get an accurate value, we estimate the distances by calculating the centroids of the boxes. The red boxes represent the runners who are the furthest away from the finish line. The runners in the green boxes are those who keep a safe distance from these runners. The system was successfully tested, and it was able to detect social distancing accurately. The errors are possible as a result of the runners running too

Table 2 Comparison of the values of different methodologies used by different researchers

Model	Accuracy in %
Yadav et al. [12]	91
ResNet-18	85.3
ResNet-50	86.5
YOLOv3	93.46
Liu et al. [13] (SSD512)	76.8
Liu et al. [13] (SSD300)	74.3
Faster RCNN	87.3

close to this runner. However, the obtained results have a certain number of limitations. According to the results of the system tests that have been performed, the object detection model that has been used for detecting people has difficulty in correctly detecting people outdoors and there have been issues with distant scenes too. In this case, we may not be able to determine the correct distance [14]. The YOLO algorithm can also detect the runner's half body as an object by displaying the bounding box. The visualization results confirmed that this approach was effective. Furthermore, it can be improved by combining it with mask detection [7].

References

1. Hou YC, Baharuddin MZ, Yussof S, Dzulkifly S (2020) Social distancing detection with deep learning model. In: 2020 8th International conference on information technology and multimedia (ICIMU), pp 334–338. https://doi.org/10.1109/ICIMU49871.2020.9243478
2. Wei W (2020) Small object detection based on deep learning. In: Proceedings of the IEEE international conference on power, intelligent computing and systems (ICPICS), pp 938–943
3. Gupta S, Kapil R, Kanahasabai G, Joshi SS, Joshi AS (2020) SD-measure: a social distancing detector. In: Proceedings of the IEEE 12th international conference on computational intelligence and communication networks, pp 306–311
4. Ansari MA, Singh DK (2021) Monitoring social distancing through human detection for preventing/reducing COVID spread. Springer
5. Adarsh P, Rathi P, Kumar M (2020) YOLO v3-tiny: object detection and recognition using one stage improved model. In: Proceedings of the IEEE 6th international conference on advanced computing and communication systems (ICACCS), pp 687–694
6. Madane S, Chitre D (2021) Social distancing detection and analysis through computer vision. In: 2021 6th International conference for convergence in technology (I2CT), pp 1–10. https://doi.org/10.1109/I2CT51068.2021.9418195
7. Tyagi A, Rajput D, Singh A (2021) A review on social distancing auto detection techniques in perspective of COVID' 19. In: 2021 Fifth international conference on I-SMAC (IoT in social, mobile, analytics and cloud) (I-SMAC), pp 1–6. https://doi.org/10.1109/I-SMAC52330.2021.9640663
8. Pan X, Yi Z, Tao J (2021) The research on social distance detection on the complex environment of multi-pedestrians. In: 2021 33rd Chinese control and decision conference (CCDC), pp 763–768. https://doi.org/10.1109/CCDC52312.2021.9601818
9. Saponara S, Elhanashi A, Gagliardi A (2021) Implementing a real-time, AI-based, people detection and social distancing measuring system for Covid-19. Springer

10. Hou YC, Baharuddin MZ, Yussof S, Dzulkifly S (2020) Social distancing detection with deep learning model. In: 2020 8th International conference on information technology and multimedia (ICIMU), pp 334–338
11. Melenli S, Topkaya A (2020) Real-time maintaining of social distance in Covid-19 environment using image processing and Big Data. In: 2020 Innovations in intelligent systems and applications conference (ASYU), pp 1–5. https://doi.org/10.1109/ASYU50717.2020.9259891
12. Indulkar Y (2021) Alleviation of COVID by means of social distancing and face mask detection using YOLO V4. In: 2021 International conference on communication information and computing technology (ICCICT)
13. Shao Z, Cheng G, Ma J, Wang Z, Wang J, Li D (2022) Real-time and accurate UAV pedestrian detection for social distancing monitoring in COVID-19 pandemic. IEEE Trans Multimedia 24:2069–2083. https://doi.org/10.1109/TMM.2021.3075566
14. Hossam H, Ghantous MM, Salem MA (2022) Camera-based human counting for COVID-19 capacity restriction. In: 2022 5th International conference on computing and informatics (ICCI), pp 408–415

An Improved Image Enhancing Technique for Underwater Images by Using White Balance Approach

G. Geetha and A. Sai Suneel

Abstract In recent days, a lot of study has been done on improving the visual quality of underwater and undersea imaging in submarine and military operations to find the hidden structure designs and sea excursions. This paper proposed an underwater image enhancement method based on colour constancy theory. When we compare with much of the existing research, the time and complexity of the proposed method are low, and it achieves excellent performance. Firstly, we analyse the underwater imaging model and distortion. Then, by compensating the red channel and applying local white balance, a linear transformation is performed. Finally, results are obtained by applying the histogram equalization to the RGB images. We also measure the image quality by taking the parameters PSNR, UIQM, UCIQE, and Entropy. These parameters are compared to proposed and existing approaches; however, our method produces higher image quality.

Keywords Underwater image · Image enhancement · White balance · Histogram equalization

1 Introduction

Earth is an aquatic planet, and most of its surface is covered by water. If he or she dives into the water, they have to face so many problems that they have to stay for an extended period of time in order to conduct experiments [1]. Exploration of the oceans is not an easy task. At present, most of the researches going on in these oceans but due to poor imaging environment quality of the images are bad. The low quality of underwater images leads to low efficiency when humans use these image sensors to explore the ocean. In Fig. 1, we can clearly see the distortion caused by the underwater conditions. Underwater, the quality of the image degrades and light properties differ compared to air [2]. Only one way to get clear underwater images is

G. Geetha · A. Sai Suneel (✉)
School of Engineering and Technology, Sri Padmavati Mahila Visvavidyalayam, Tirupati, India
e-mail: saisuneel.adem@gmail.com

K. A. Reddy et al. (eds.), *Proceedings of Fourth International Conference on Computer and Communication Technologies*, Lecture Notes in Networks and Systems 606,
https://doi.org/10.1007/978-981-19-8563-8_4

Fig. 1 Several types of underwater images

by applying an underwater image enhancement method. It can reduce this problem [3].

The importance of deep water image pre-processing [1] is described are as follows: Light ray properties such as scattering and absorption diminish image quality captured in deep water. Video or photograph shot from deep water, such as an unknown stiff scene, as well as the scene's depth and low light sensitivity owing to marine snow. When vehicles travel, environmental factors such as lighting differences, water turbidity, and blue complexion have varying degrees of influence.

There are mainly two aspects [4]: Firstly, as shown in Figs. 1a, b, due to the attenuation of underwater light propagation, the colour of the image degraded significantly, making underwater images seem bluish or greenish. Secondly, due to the scattering of underwater light, underwater images lose their texture details. Jerlov classifies the water into two types [5]. There are two types of water: open ocean and coastal waters. Now, looking at the problem of colour degradation and poor contrast in underwater images, this paper analyses underwater imaging models and attenuation of underwater images. Firstly, we have to eliminate that severe attenuation in that image and, by performing the linear transformation based on colour constancy theory [6], the colour degradation of underwater images recovered from the attenuated state. Finally, we adjust the brightness of the images.

Figure 1 depicts several sorts of visuals based on the different water types. The clarity of the image changes in oceans, seas, and rivers because sunlight does not pass through the water as we travel deeper into it. Depending on the water type, the images seem green or bluish.

2 Literature Survey

Ancuti et al. [7] proposed a novel method for improving underwater images and videos. It is based on the fusion principle, that is, it is influenced by the inherent qualities of the source picture. Our degraded image first undergoes white balancing. It removes the unwanted colour cast, but this method solely does not handle the problem of visibility. Here, the contrast local adaptive histogram equalization is used to obtain the second input. Since it works well and the distortion is small. Next, the weight maps of this algorithm evaluate the image qualities that identify the spatial pixel relationships. The weight measure must be designed with the intended appearance of the restored output in mind. It handles the global contrast by applying the Laplacian filter on input of the channel, but this weight is inadequate to take back the contrast. So, we have to take another measurement, which is the local contrast weight. It strengthens the contrast. Saliency weight is to highlight the selective objects that lose their fame in the underwater world, and exposedness weight estimates how well a pixel is exposed, and all these weights are summed up by using a normalised weight. Finally, the multi-scale fusion process is completed; that is, the restored output is adding up all the fused additions of all the inputs. Although this method has high contrast and excellent performance, it can easily cause noticeable distortion in images and amplify the noise in images.

Ancuti et al. [3] proposed an effective technique to improve the underwater captured images. It builds on a two-step strategy. Firstly, the original image is given to the white balancing algorithm, which removes the unwanted colour casts. The two-step strategy consists of combining local white balance and image fusion. To enhance the edges and reduce the loss of contrast, we use image fusion. First, we perform gamma correction on the white-balanced image. The second input, a sharper version of the white-balanced picture, is also used, and it reduces the degradation caused by the scattering. Next, the weight maps are used during the combining process to represent pixels with higher weight values in the final image. Here, the Laplacian contrast, saliency, and saturation weights are used. Saturation weight enables chrominance information to be adapted by highly saturated regions. The reconstructed image is created in the following naive fusion step by fusing the inputs with weight measurements at each pixel location. Finally, the output of the multi-scale fusion process is obtained by adding all the inputs fused contributions. Although this method has high contrast and excellent performance, it can easily cause noticeable distortion in images and amplify the noise in images.

Berman et al. [4] proposed a method: it reconstructs underwater sceneries from a single photograph using several spectral profiles of various water kinds. The challenge can be simplified by estimating the two global parameters, namely the attenuation ratios of the blue–red and blue–green colour channels. But we do not know the water type. Each has a set of attenuation ratios that are known and constant. To begin, we calculate the covering light. If we wanted to detect the pixels that correspond to veiling light, we had to create an edge map using the Structured Edge Detection Toolbox with a pre-trained model and default settings, then group the pixels into haze-lines to gain a first estimate of the blue channel's transmission. Finally, use a guided image filter using a contrast enhanced input picture as direction to set up the transmission. Then, the restored image is calculated and performed the white balance on that image, and then return the image to the gray world assumption on pixels. We perform the restoration several times with various attenuation ratios before deciding on the optimum one based on the gray world assumption. We discovered gray world assumption which produces the best results. This method only performs well if it meets the assumptions about the underwater environments, and this is one of the drawbacks.

Huang et al. [8] proposed a simple but very effective method. It has three major steps: contrast correction, colour correction, and quality evaluation. Firstly in contrast correction, after this RGB channel decomposition, we apply colour equalization and the relative global histogram stretching to the image. And next, relative global histogram stretching ignores the histogram distribution of distinct channels in the image and uses the same parameters for all RGB channels. By using the bilateral filter, we can eliminate the noise after the transformation, and this is given to the colour correction process. In this colour correction, we apply simple histogram stretching on the "L" component and adjust the "a" and "b" in CIE Lab colour space. Next, this CIE Lab colour space is given to the adaptive stretching of "L", "a", and "b". The channels are then combined and returned to the RGB colour model. Finally, a contrast and colour-corrected output image may be generated, and we evaluate the effectiveness of the proposed method using five quality evaluation models. But, in RGHS, the main drawback is that it lacks the ability to correct the colour casts in underwater images.

3 Existing Method

In this existing method, we are having three steps there are as follows:

1. Red channel compensation
2. Colour correction
3. Histogram stretching.

3.1 Red Channel Compensation

If we get a good-quality image, it requires equal colour values of RBG components. But unfortunately, the underwater pictures are rarely colour corrected [9]. As we know, underwater, as we go deeper into the water, colours like red, blue, and green will attenuate. The red channel has a wavelength attenuation of 600 nm, the blue channel has a wavelength attenuation of 525 nm, and the green channel has a wavelength attenuation of 475 nm. If we directly apply the white balance approach to the images, the result is unsatisfactory. So, first we have red channel compensation on the images, and then we apply white balance, and finally we get an effective result.

3.2 Colour Correction

After this red channel compensation, we still need to correct the red channel because it is having severe attenuation [10]. A traditional method like gray world also fails to correct the images. This is because of the colour degradation of images, which is not uniformity. So we divide the images into patches. The effect of distance on the extent of distortion may be ignored for each patch, and colour constancy can be used to rectify red channel distortion. For each patch, we use the gray world method to solve the weight map for the local white balance.

3.3 Histogram Stretching

We apply the method of histogram stretching to the R, G, and B channels of the images. Since the colour correction approach, we presented would cause the image contrast to decrease, and because of applying local white balance would cause the image gray scale to be concentrated [10]. But, after this histogram stretching, the image contrast and quality are poor. The below mentioned Fig. 2a shows a raw image, while Fig. 2b shows red channel correction applied to the raw image. Figure 2c shows the colour correction we used to rectify the image, and Fig. 2d shows how we performed histogram stretching to increase the image contrast.

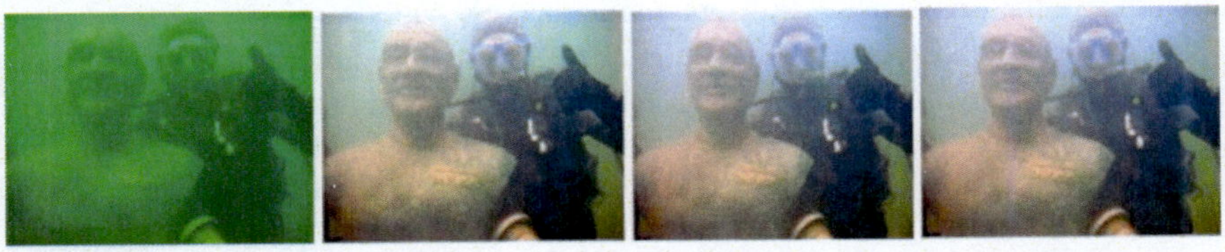

(a) (b) (c) (d)

Fig. 2 **a** Raw image, **b** after red channel compensation, **c** colour correction, and **d** histogram stretching

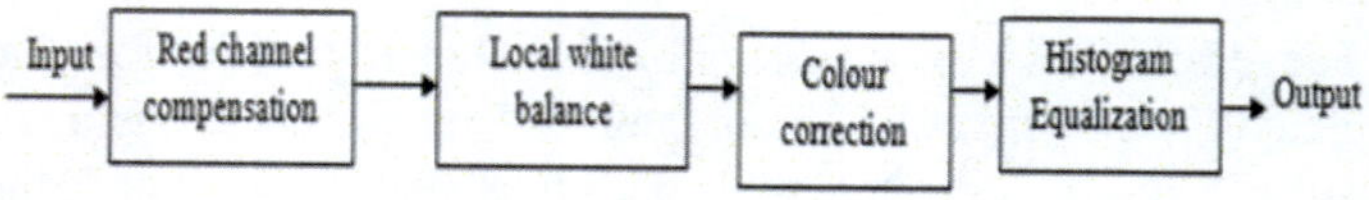

Fig. 3 Block diagram of proposed method

4 Proposed Method

At present, underwater imaging model accepted by community Jaffe-McGlamery [11, 12] model. The difference in the degree of distortion in a picture is simply linked to the distance between the image's objects and the lens. Because it does not account for the change in distortion degree with distance, the traditional gray world [6] technique is unsuitable for underwater picture augmentation. Gray world only improves images by multiplying them by a single coefficient, which result in both local over-and under-enhancement. In this proposed method, like the existing method, we also use red channel compensation and colour correction methods, but we don't apply histogram stretching because still we have to improve the image contrast. Instead of histogram stretching, we use histogram equalization on the images to improve the image contrast.

4.1 Histogram Equalizaton

The histogram equalization technique uses the histogram to modify the contrast of an image [2]. After this colour correction method, the image gets low image contrast. To eliminate this, we are using histogram equalization. When this histogram equalization is applied to the RGB channels of the image, it will improve the image contrast and give a better image quality. The block diagram is designed in such a way that we send an input to the red channel compensation and then apply local white balance for the specific channel, correct the picture using the colour correction technique, and improve the image using histogram equalization (Fig. 3).

Figure 4a depicts the raw picture, and Fig. 4b red channel correction applied to raw image. Figure 4c depicts the colour correction, and Fig. 4d depicts how we applied histogram equalization to boost image contrast and yield a high-quality image.

5 Experimental Results

We took the above images from the below mentioned references. Figure 5a contains raw images, and in Fig. 5b a fusion-based method [7] is used. In Fig. 5c, a fusion-based method [3] is used. In Fig. 5d shows relative global histogram stretching. In Fig. 5e, we first compensate the red channel and use the white balance method, then

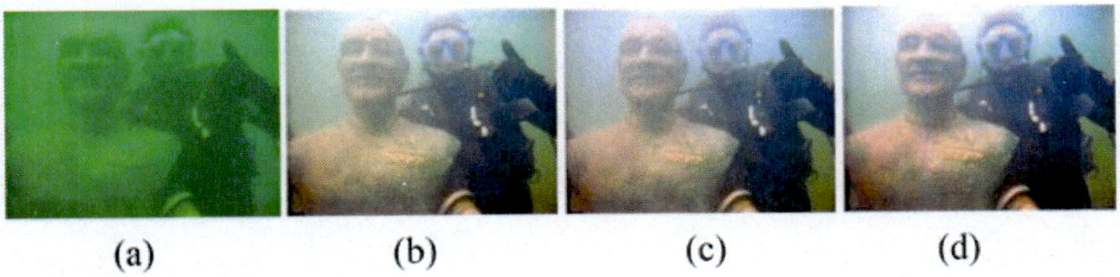

Fig. 4 **a** Raw image, **b** after red channel compensation, **c** colour correction, and **d** histogram equalization

colour correction method and finally we use histogram stretching. Figure 5f shows red channel compensation and local white balance, followed by colour correction and histogram equalization. Compared to those methods, our proposed method gives a good-quality image and improves the image contrast.

In Table 1, here we calculate some parameters that is peak signal-to-noise ratio (PSNR), underwater image quality measure (UIQM) [13], underwater colour image

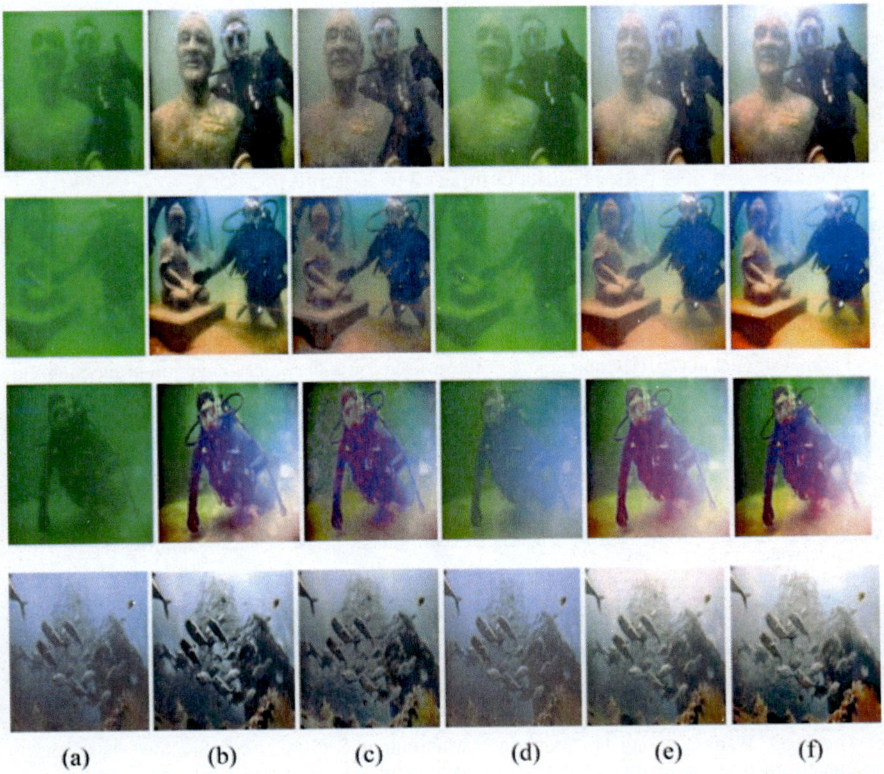

Fig. 5 **a** Raw image, **b** fusion method [7], **c** fusion method [3], **d** RGHS, **e** existing method, and **f** proposed method

Table 1 Comparison of quality metrics of the images

Quality measures	Fusion method [7]	Fusion method [3]	RGHS	Existing method	Proposed method
PSNR	54.16	54.15	54.16	54.17	54.21
UIQM	4.17	4.20	1.31	3.7	3.9
UCIQE	30.73	28.53	13.82	30.61	31.62
Entropy	7.9	7.4	7.6	7.92	7.99

quality evaluation (UCIQE) [14], and Entropy. To evaluate an image, we take these parameters, and we can strongly say that our proposed method got better results compared to other methods.

6 Conclusion

In this paper, we propose an underwater image enhancement technique based on a white balance approach. By applying white balance on the image after the red channel compensation, the colour degradation of the image can be corrected. Through the gray world method and histogram equalization, we get a good image contrast and high-quality image. We also measure the image quality by calculating the parameters PSNR, UIQM, UCIQE, and Entropy. Our proposed method has excellent performance and gets better results compared to existing methods.

References

1. Padmavati G et al (2010) Comparison of filters used for underwater image pre-processing. Int J Comp Sci Netw Security 10(1):58–65
2. Singh B, Mishra RS, Gour P. Analysis of contrast enhancement techniques for underwater image. IJCTEE 1(2):190–195
3. Ancuti CO, Ancuti C, De Vleeschouwer C, Bekaert P (2017) Color balance and fusion for underwater image enhancement. IEEE Trans Image Process 27(1):379–393
4. Berman D, Treibitz T, Avidam S (2017, Sept) Diving into haze-lines: color restoration of underwater images. BMVC 1(2)
5. Jerlov NG (1976) Marine optics. Elsevier
6. Buchsbaum G (1980) A spatial processor model for object colour perception. J Franklin Inst 310(1):1–26
7. Ancuti C, Ancuti CO, Haber T, Bekaert P (2012, June) Enhancing underwater images and videos by fusion. In: 2012 IEEE conference on computer vision and pattern recognition. IEEE, pp 81–88
8. Huang D, Wang Y, Song W, Sequeria J, Mavromatis S (2018, Feb) Shallow water image enhancement using relative global histogram stretching based on adaptive parameter acquisition. In: International conference on multimedia modeling, pp 453–465

9. Iqbal K, Odetayo M, James A, Salam RA, Talib AZH (2010, Oct) Enhancing the low quality-images using unsupervised colour correction method. In: 2010 IEEE international conference on systems, man and cybernetics. IEEE, pp 1703–1709
10. Zhang H, Li D, Sun L, Li Y (2020) An underwater image enhancement method based on local white balance. In: 5th International conference on mechanical, control and computer engineering (ICMCCE), pp 2055–2060
11. Jaffe JS (1990) Computer modeling and the design of optimal underwater imaging systems. IEEE J Ocean Eng 15(2):101–111
12. McGlamery BL (1975) Computer analysis and simulation of underwater camera system performance. SIO Ref 75(2)
13. Panetta K, Gao C, Agaians S (2015) Human-visual-system-inspired underwater image quality measures. IEEE J Ocean Eng 41(3):541–551
14. Yang M, Sowmya A (2015) An underwater color image quality evaluation metric. IEEE Trans Image Process 24(12):6062–6071

32-Bit Non-pipelined Processor Realization Using Cadence

K. Prasad Babu, K. E. Sreenivasa Murthy, and M. N. Giri Prasad

Abstract In modern electronics field, power constraint is emphasizing factor, since most of electronic devices are handheld. With the advent of IoT, processors design with high performance, low power dissipation is of primary concern. In this work, a 32-bit non-pipelined processor is implemented using Cadence tool of 0.35 μm technology library with minimal working units along with optimal performance. Static power, net power, dynamic power for the proposed design are obtained in nano-watts. Basic building blocks of the design include the CPU along with memory controller. The data size is 32-bit, and address size is 16-bit. Four-bit operational code is employed. Nine instructions are used in the design. Four-bit bidirectional IO is implemented. Memory is divided for ROM and RAM input–output ports. Design avail clock frequency of 80 MHz. RTL simulation is done along with synthesis, place and route, and LVS and DRC rule check. For the work implemented, no of cells of CPU are 1127, memory controller are 2071, the total leakage power is 4.38 nW, dynamic power is 56436431.19 nW, short-circuit power is 41474454.01 nW, and net power is 14960177.18 nW.

Keywords CPU · Dynamic power · Short-circuit power · Static power · Net power · 32-bit processor

K. Prasad Babu
15PH0426, Department of ECE, JNTUA, Anantapuramu, Andhra Pradesh, India

K. E. Sreenivasa Murthy (✉)
Department of ECE, RECW, Kurnool, Andhra Pradesh, India
e-mail: kesmurthy1969@gmail.com

M. N. Giri Prasad (✉)
Academics and Audit, JNTUA, Anantapuramu, Andhra Pradesh, India
e-mail: giriprasadmn.ece@jntua.ac.in

Department of ECE, JNTUA, Anantapuramu, Andhra Pradesh, India

K. A. Reddy et al. (eds.), *Proceedings of Fourth International Conference on Computer and Communication Technologies*, Lecture Notes in Networks and Systems 606,
https://doi.org/10.1007/978-981-19-8563-8_5

1 Introduction

The basic working principle of a processor is fetch, decode, execute, and store the instructions. ALU is heart of any processor. Arithmetic unit performs the logical and arithmetic tasks. Processors can be RISC/CISC, and most of the times RISC processors are suitable for low power applications and hence suited for portable or embedded applications. These have very few instructions of preset length, additional common registers, load–store structural design, basic addressing types for collective execution of instructions in a faster way, and the area required is less when compared to CISC Processors. For execution of instructions, different types of instruction formats are used like R-Type, B-Type, and I-Type. The PC provides the next address location to be fetched. The instruction register decodes and stores the fetched one. In the decode process destination register, source register, address of the memory location, or immediate value is assigned based on the operation to be performed. Short-circuit power dissipation occurs throughout switching of transistors. Dynamic power is due to the charge and discharge of the output load. The equation representing it is

$$P_{\text{dyn}} = C_{\text{load}} V^2 F \alpha \quad (1)$$

$$P_{\text{lek}} = I_{\text{lek}} V_D \quad (2)$$

$$P_{\text{sck}} = V_D I_{\text{sc}} \quad (3)$$

All these three power equations are the contributors for total net power dissipation in any design.

1.1 Literature Review

There are many works done and proposed for processor design. RISC processors are majorly employed for their prominent usage. In [1], Chandran Vankatesan et al. have implemented the design of 16-bit RISC processor using 45 nm. In [2], Nirmal Kumar et al. have proposed using separated LUT's for embedded system. In [3], Indu et al. had implemented the design of low power pipelined RISC processor. In [4], Chandran et al. employed rounding technique for energy efficiency of multipliers. In [5], Samiappa et al. coded for convolution applications of processor. In [6], Topiwala implemented the MIPS 32-bit processor using Cadence. In [7] Jain had implemented the 32-bit pipelined processor on spartan fpga board. In [8] Rupali et al., have analysed the Instruction fetch-decode blocks of 32-bit pipelined processor. In [9] Gautham et al., have proposed low power 5 stage MIPS-32.

2 Implementation

The basic blocks of processor implemented is shown in Fig. 1. The inputs to the CPU block are Clock and RESET; CPU block in this design is composed of InstructReg, PrgCntr, and Accmltr. The memory IO controller block is combination of separate RAM, ROM, input–output controller, and multiplexer. The data to CPU is feedback from the output of Mux to the CPU block. The CPU outputs data from CPU, address, write enable are fed to the Memory IO controller block.

The format for opcode is as shown below, where only most significant bits are used for selection 31 to 28 bits, and remaining are unused.

31	28	27 16	15 0
OPCODE		UNUSED OPERAND	

The nine type of instructions used are revealed in below table.

Commands	Operational code	Operation
Addition	1	AC ≤ AC + mem(IR[15:0])
Shift left operation	2	AC ≤ AC << mem(IR[15:0])
Shift-right operation	3	AC ≤ AC >> mem(IR[15:0])
Load immediate	4	AC ≤ IR[15:0]
Load	5	AC ≤ mem(IR[15:0])

2.1 Simulations

The proposed work simulation is started with ISE for coding in Verilog, and Fig. 2 indicates the RTL view of CPU and memory controller unit. By using Cadence Simvision, the output waveform is obtained for the code. Total number of instances and their short-circuit power consumption, leakage power consumption, dynamic power consumption are calibrated using RTL encounter. Layout is obtained as the final step of the processor work.

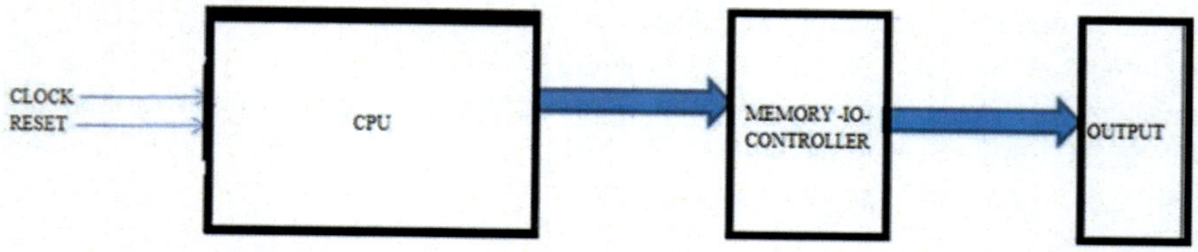

Fig. 1 Block diagram of processor

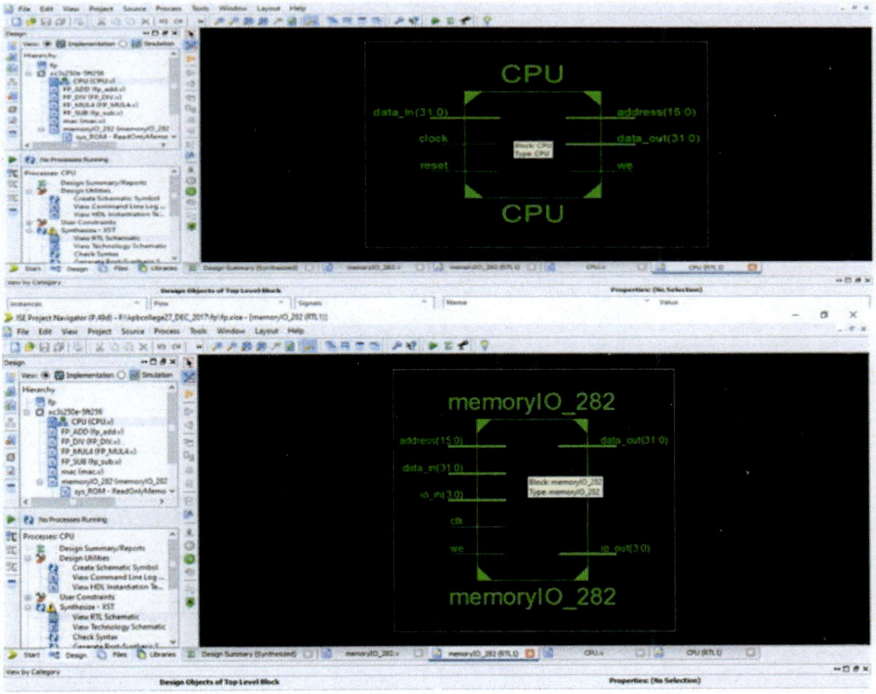

Fig. 2 ISE RTL simulation view of the CPU and memory

Figure 2 indicates the schematic view of processor and its memory unit. Data bits size is 32, and address size is 16-bit. Four input output values.

The above Fig. 3 depicts the waveform notation obtained on invoking the Simvision. All the blocks are synced w.r.t clock and accordingly to write enable IR, PC, AC Datain, and Dataout are processed.

The RTL values variation can observed in Fig. 4.

Area synthesis report gives us the no of cells used in the design, there occupancy in µm values. Figure 5 emphasizes on this. The critical part of the entire design, the usage of the various instances of the design, is clearly mentioned in Fig. 6, and the memory controller is utilizing most of the area.

An individual basic gate with total number of mapping instances, area is mentioned in Fig. 7, and it is also in µm values.

Figure 8 discloses us the power consumption of different types. Leakage power, short-circuit power, net power, dynamic power in nw. Figure 9 signifies the synthesis layout of the design. Lastly, the layout generation of the 32-bit processor design is revealed in Fig. 10.

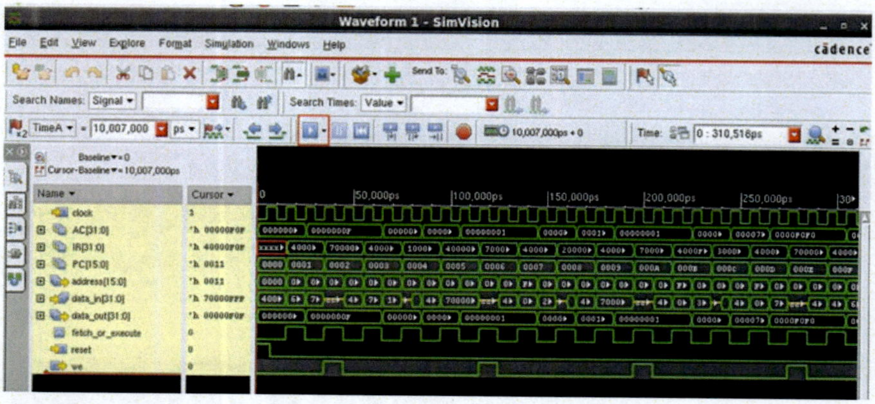

Fig. 3 Simulation waveform using Simvision

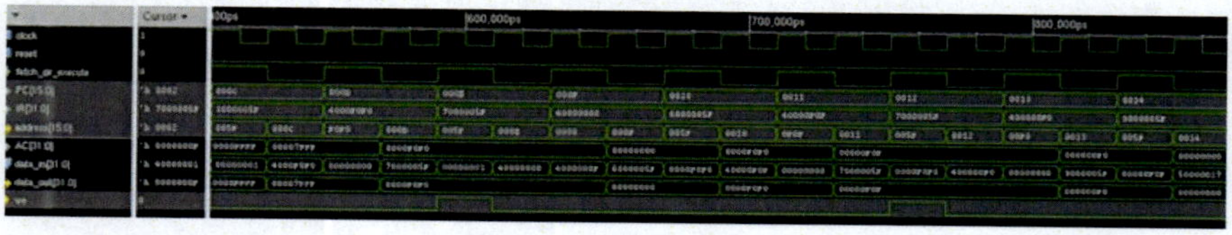

Fig. 4 RTL simulation

Instance	Cells	Cell Area	Net Area	Total Area
	3198	518973.00	224245.17	743218.17
cpu_inst	1127	97715.80	65570.61	163286.41
cpu_inst/add_34_26	16	2202.20	562.33	2764.53
cpu_inst/add_42_34	249	16962.40	11710.99	28673.39
cpu_inst/sll_47_34	193	14305.20	10615.17	24920.37
cpu_inst/srl_52_34	194	14341.60	10635.77	24977.37
memory_controller_inst	2071	421257.20	125707.48	546964.68
memory_controller_inst/data_mux_inst	76	4568.20	2804.42	7372.62
memory_controller_inst/io_controller_inst	15	3057.60	398.57	3456.17
memory_controller_inst/ram_inst	1919	409773.00	117219.04	526992.04
memory_controller_inst/rom_inst	47	2912.00	2245.18	5157.18

Fig. 5 Area synthesis report. This shows a figure consisting of different types of cells. The corresponding cell area and net area are obtained

2.2 *Program Code*

The below code is for testing basic

```
//Test the store to ROM and add to AC
5'h00:dataout<=32'h4000000f; //LI 000f
5'h01:dataout<=32'h7000005f; //STO 05f
5'h02:dataout<=32'h40000001; //LI 0001
5'h03:dataout<=32'h1000005f; //ADD 005f
//Result(AC)should be 10f
//Test the shift left(SL)
5'h04:dataout<=32'h40000001;
```

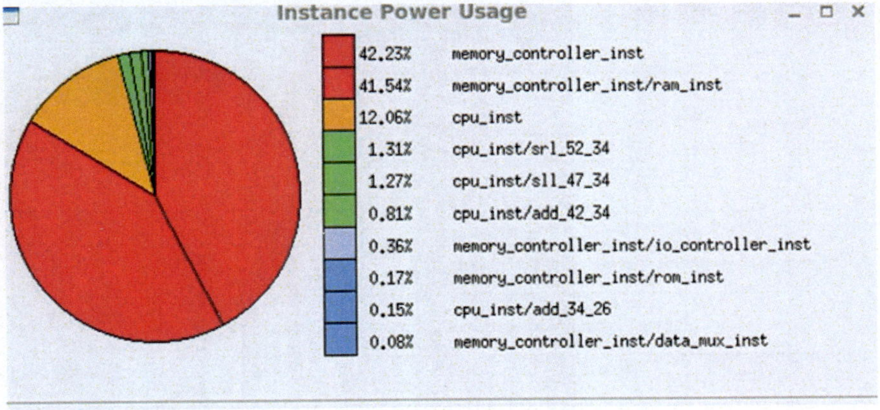

Fig. 6 Total power report. This shows a figure consisting of different types of blocks. The corresponding percentage of net power usage individually is obtained

Report Mapped Gates

Technology libraries: physical_cells
s35_CORELIB_TYP 3.02
s35_IOLIB_TYP revision 1.0

Operating conditions: typical

Interconnect mode: ple

Gate	Instances v	Area	Library
TOTAL	3230	521539.20	
DFE1	1064	348566.40	s35_CORELIB_TYP
AOI221	603	54873.00	s35_CORELIB_TYP
NAND22	338	18454.80	s35_CORELIB_TYP
INV3	127	4622.80	s35_CORELIB_TYP
NOR21	123	6715.80	s35_CORELIB_TYP
NAND41	110	10010.00	s35_CORELIB_TYP
IMUX21	104	9464.00	s35_CORELIB_TYP
INV2	83	3021.20	s35_CORELIB_TYP
AOI211	81	5896.80	s35_CORELIB_TYP
CLKIN3	62	2256.80	s35_CORELIB_TYP
NAND31	51	3712.80	s35_CORELIB_TYP

Fig. 7 No. of mapped gates

Cells	Leakage (nW)	Internal (nW)	Net (nW)	Switching (nW)
3198	4.38	41474454.01	14960177.18	56434631.19
1127	0.83	9213141.20	8640585.41	17853726.61
16	0.02	115636.04	84258.54	199894.58
249	0.14	619536.33	772880.88	1392417.21
193	0.12	971203.23	1308317.22	2279520.45
194	0.12	1002991.51	1328086.54	2331078.05
2071	3.54	32261312.81	1392456.65	33653769.46
76	0.04	64370.20	266566.78	330936.97
15	0.03	277482.57	14202.37	291684.95
1919	3.44	31732072.24	751441.19	32483513.43
47	0.02	129612.03	249327.12	378939.14

Fig. 8 Total net power consumption of cells. This shows a figure consisting of different types of cells. The corresponding values of cells and their power usage individually are obtained

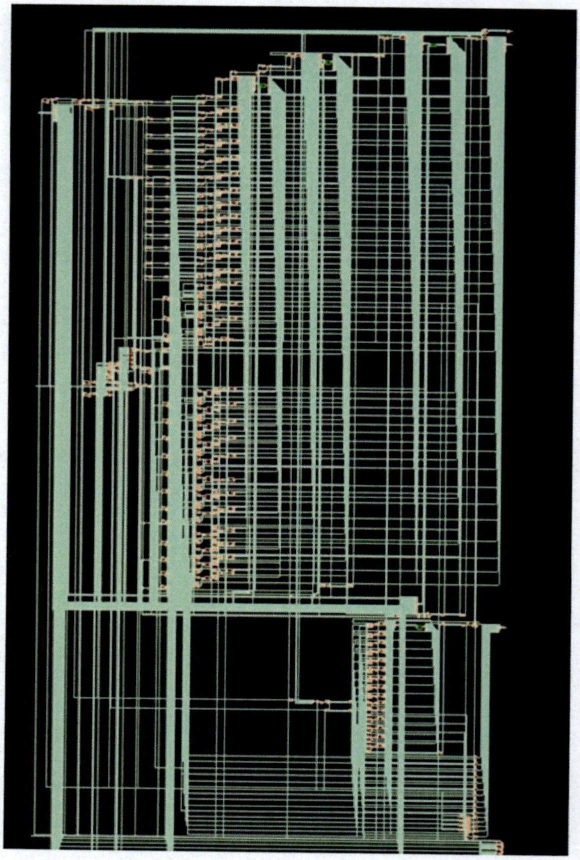

Fig. 9 Synthesis simulation of design

```
5'h05:dataout<=32'h7000005f;
5'h06:dataout<=32'h4000ffff;
5'h07:dataout<=32'h2000005f;
//Result(AC)should be fffe
//Test the shift left(SL)
5'h08:dataout<=32'h40000001;
5'h09:dataout<=32'h7000005f;
5'hOA:dataout<=32'h4000ffff;
5'hOB:dataout<=32'h3000005f;
//Test the OR
5'hOC:dataout<=32'h4000fOfO;
5'hOD:dataout<=32'h7000005f;
5'hOE:dataout<=32'h40000000;
5'hOF:dataout <=32'h6000005f;
//Test the AND
5'h10:dataout<=32'h4000OfOf;
5'h11:dataout<=32'h7000005f;
5'h12:dataout<=32'h4000OfO;
5'h13:dataout<=32'h9000005f;
//Branch
5'h14:dataout<=32'h80000000;
```

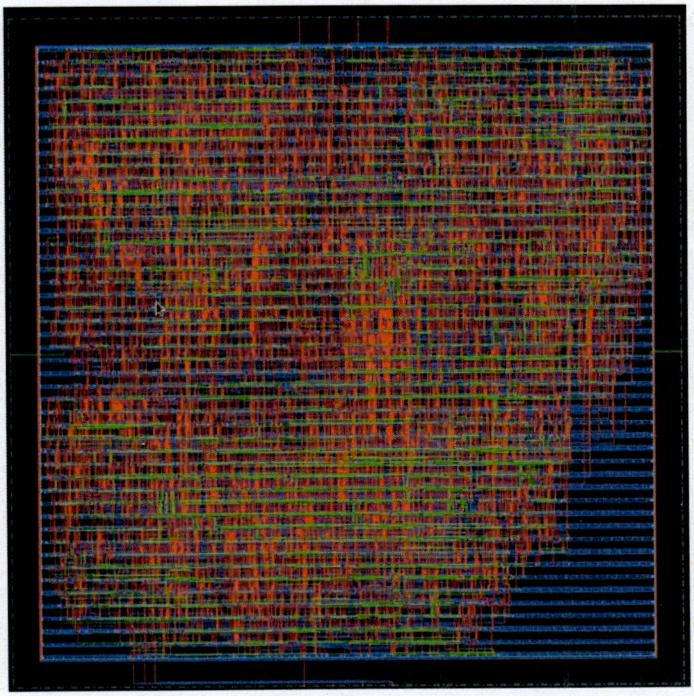

Fig. 10 Post-layout simulation of design

3 Conclusion

This work implemented the design of processor with minimum no of instructions. The short-circuit power obtained is 41,474,454.01 nw, dynamic power is 56,436,431.19 nw, and static power is 4.38 nw. Among all the instances CPU inst is consuming 17,853,726.61 nw switching power and 0.83 nw leakage power. Memory controller is consuming 33,653,769.46 nw dynamic and 3.54 nw leakage power. The future scope of this work is to upgrade the no of instructions, apply power gating, and clock gating techniques for static and dynamic power dissipation, respectively.

Acknowledgements I am very much thankful for the guide and support given by my Supervisor Dr. K. E. Sreenivasa Murthy, in doing this work and special thanks to the Co-Supervisor Dr. M. N. Giri Prasad for helping me in work.

References

1. Venkatesan C, Thabsera Sulthana M, Sumithra MG, Suriya M (2019) Design of a 16-Bit Harvard structure RISC processor in cadence 45 nm technology. In: 2019 5th international conference on advanced computing and communication systems (ICACCS), 978-1-5386-9533-3/19/$31.00 ©2019 IEEE, pp 173–178
2. Kumar RN, Chandran V, Valarmathi RS, Kumar DR. Bitstream compression for high speed embedded systems using separated split LUTs. J Comput Theor Nanosci 15(Special):1–9
3. Indu M, Arun Kumar M (2013) Design of low power pipelined RISC processor. Int J Adv Res Electr Electron Instrum Eng 2(Aug 2013):3747–3756
4. Chandran V, Elakkiya B. Energy efficient and high-speed approximate multiplier using rounding technique. J VLSI Des Sig Process 3(2, 3)
5. Sakthikumaran S, Salivahanan S, Bhaaskaran VK (2011) 16-Bit RISC processor design for convolution applications. In: IEEE international conference on recent trends in information technology, June 2011, pp 394–397
6. Topiwala MN, Saraswathi N (2014) Implementation of a 32-Bit MIPS based RISC processor using cadence. In: IEEE International conference on advanced communication control and computing technologies (ICACCCT), 2014, pp 979–983
7. Jain N (2012) VLSI design and optimized implementation of a MIPS RISC processor using XILINX tool. Int J Adv Res Comp Sci Electron Eng (IJARCSEE) 1(10), Dec 2012
8. Balpande RS, Keote RS (2011) Design of FPGA based instruction fetch & decode module of 32-bit RISC (MIPS) processor. In: 2011 IEEE. https://doi.org/10.1109/CSNT.2011.91
9. Gautham P, Parthasarathy R, Balasubramanian K (2009) Low power pipelined MIPS processor design. In: proceedings of the 2009, 12th international symposium, 2009 pp 462–465

Metaverse: The Potential Threats in the Virtual World

K. Ghamya, Chintalacheri Charan Yadav, Devarakonda Venkata Sai Pranav, K. Reddy Madhavi, and Ashok Patel

Abstract The virtual world seems like it's a bit old term, let's use the metaverse. Neal Stephenson, a science fiction writer, created the term metaverse in 1992. "The concept of a fully immersive virtual world where people assemble to socialize, play, and work," according to its most basic definition. From basic games and shopping to complex meetings, we can now be organized in the virtual world called the metaverse. In general, we are habituated to working in the physical world but metaverse is the future where we will be playing games, virtual shopping, virtual meetings, creating our virtual world, and investing in stocks. Cryptocurrencies are the fuel for metaverse to keep it working. Non-Fungible Tokens (NFTs) are the cryptocurrencies that are used to stand for real-world objects. These NFTs cannot be hacked since they use blockchain technology that allows the decentralization of data to prevent these hacks. Now comes the billion-dollar question is this metaverse safe? As we know everything cannot be 100% perfect and it's also had some shortcomings. Anyone can enter into a highly confidential meeting using our unique avatar, and this arises much dubiousness, is our data safe in the metaverse?, privacy concerns, digital boundaries, and social engineering. This makes metaverse a bit scary even though it has some dominance. This requires creating a more secure, robust metaverse, and awareness in the people.

Keywords Metaverse · Blockchain · Web 3.0 · Virtual world · Security · Privacy issues

K. Ghamya · C. C. Yadav · D. V. S. Pranav · K. Reddy Madhavi (✉)
CSE, Sree Vidyanikethan Engineering College, Tirupati, India
e-mail: kreddymadhavi@gmail.com

A. Patel
Department of Computer Science, Florida Polytechnic University, IST 2068, 4700, Research Way, Lakeland, FL 33805-8531, USA
e-mail: apatel@floridapoly.edu

K. A. Reddy et al. (eds.), *Proceedings of Fourth International Conference on Computer and Communication Technologies*, Lecture Notes in Networks and Systems 606,
https://doi.org/10.1007/978-981-19-8563-8_6

1 Introduction

Metaverse will be a world of mixed reality, where it is a combination of web 3.0, blockchain, virtual reality, and augmented reality technologies. Consider the web that we are using is two-dimensional, in simple terms metaverse can be three-dimensional. Without any secure networks or secure web [1], it will be quite difficult to reduce the privacy issues that we are going to face in the upcoming shortcomings in the metaverse. Many multinational companies are building projects taking metaverse as their major topic. Companies like Facebook, which recently changed to Meta as their parent company and Microsoft are working to develop metaverse. Now the question arises? When did this all start, the revolution started with web 3.0, where the main focus is on the decentralization of data and its maintenance by users? Similarly, blockchain uses a block of codes that are enabled such that it's hard to mutate the code or hack into it. And the final technology is called mixed reality, a hybrid of both augmented and virtual reality in the virtual world [2, 3]. Some metaverse traits can already be found in virtual video game environments. Workplace socializing tools like Gather and games like Second Life. In virtual settings, town connects various elements of our lives. These programs aren't quite as advanced as the Metaverse, but they're near. The metaverse has yet to be created. It is clear that most companies' large companies that have claimed to have developed the metaverse are targeting virtual meetings, virtual games as well virtual events as shown in Fig. 1. Although, there happen to be various use cases of the metaverse. Generally, the development of metaverse consists of three phases: A digital twin is a virtual representation of an object or system in the digital world of the metaverse. Simply put, digital natives (also known as NFTs) allow creatives to profit from their work. It may be unique for traditional creatives, but it is in the digital realm, so it should be familiar.

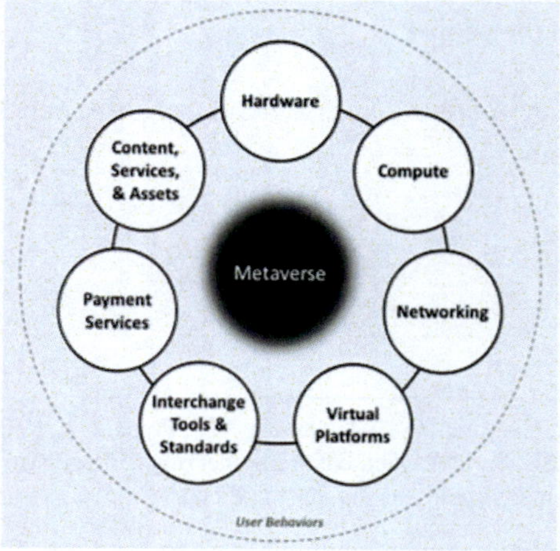

Fig. 1 Metaverse

The rest of the paper is organized as follows, Sect. 3 discusses the role of each technology in the metaverse and its issues, Sect. 4 shows the discussions on the findings of the study, and Sect. 5 presents the conclusion.

2 Various Technologies in the Metaverse

2.1 Role of Virtual Reality

Utilizing virtual reality technology, we can create a hybrid universe where the impossible is possible. You can step into different stores, explore additional items, gain knowledge, and participate in fantastic occasions. It fosters a sense of community, but it also allows us to feel the warmth of hugs and holding hands from loved ones all over the world. Using VR technology can enable metaverse to disrupt vast industries even more. Let's suppose Facebook Horizons enables you to design your ideal virtual landscape. You connect with people worldwide and establish relationships with them. Meanwhile, VR gloves are estimated to be the next big thing. Worldwide, HTC, PlayStation VR, Oculus quest, and Valve index are among the most popular VR headsets.

2.1.1 Issues in VR

Virtual reality (VR) is encountering several major difficulties on its way to broad acceptance, despite its many benefits. A lot of huge investors like Google and Facebook have thrown billions into the VR business, allowing for some incredibly powerful devices to join the market in the last year, such as the Oculus Rift. According to Riccitiello, the problem is that consumers were not prepared for everything [4]. When seen from the perspective of an organization, where a design team may require many VR machines for its process, the cost issue becomes much more acute. Even if the price barrier is solved, VR still confronts a significant challenge in the shape of a scarcity of must-have content. As a result, it becomes a significant market impediment.

2.2 Role of Augmented Reality

In the metaverse, augmented reality transforms real-life items and characters into digital visual components. Virtual reality also generates a virtual environment using computer-generated graphics. Virtual components can be embedded in the real world using augmented reality. For example, Facebook uses virtual reality headsets and augmented reality smart glasses to introduce metaverse to users on desktop and mobile devices. The increased use of AR and VR technologies implies that metaverse development is becoming more understandable [5].

2.2.1 Issues in Augmented Reality

The biggest attempt by augmented reality to catch up to virtual reality and into the public consciousness may also prove to be one of the most difficult challenges it faces. The addition of augmented reality to regular mobile devices puts augmented reality in the hands of hundreds of millions of people right away. That augmented reality experience is far from perfect. Although Apple and Google developers have done an incredible job of bringing augmented reality to devices that weren't designed specifically for it [5, 6], most consumers' first exposure to the technology will be restricted to what a mobile device can achieve.

2.3 *Role of Blockchain*

The blockchain that enables the usage of Non-Fungible Tokens or NFTs has proven to be immensely useful in the accomplishment of digital ownership, governance, value transfer, accessibility, and interoperability. Considering its infinite potential and capabilities, metaverse will furnish ample opportunities for businesses of all shapes and sizes, thereby, accelerating the growth of major economic industries such as real estate, eCommerce, entertainment, and media as shown in Fig. 3. To ensure the full functioning of the blockchain metaverse, all participants must see and interact in the same virtual landscape. A decentralized ecosystem powered by blockchain technology enables tens of thousands of independent nodes to seamlessly synchronize [7].

We've seen large corporations develop private blockchain solutions for companies who prefer to keep certain information classified. Top tech firms like IBM and Intel deploy these blockchain technology solutions to businesses looking to improve supply chain problems.

2.3.1 Issues in Blockchain

It's critical to remember that the endpoints of the majority of blockchain transactions are significantly less secure. As a result of bitcoin trading or investment, a sizeable amount of bitcoin may be deposited into a "hot wallet," or virtual savings account. The actual blocks of the blockchain cannot possibly be more secure than these wallet accounts. The absence of clear legislative standards presents another challenge to blockchain security (Fig. 2).

The blockchain sector has no homogeneity, making it challenging for developers to learn from others' mistakes. It is clear that blockchain technology isn't completely

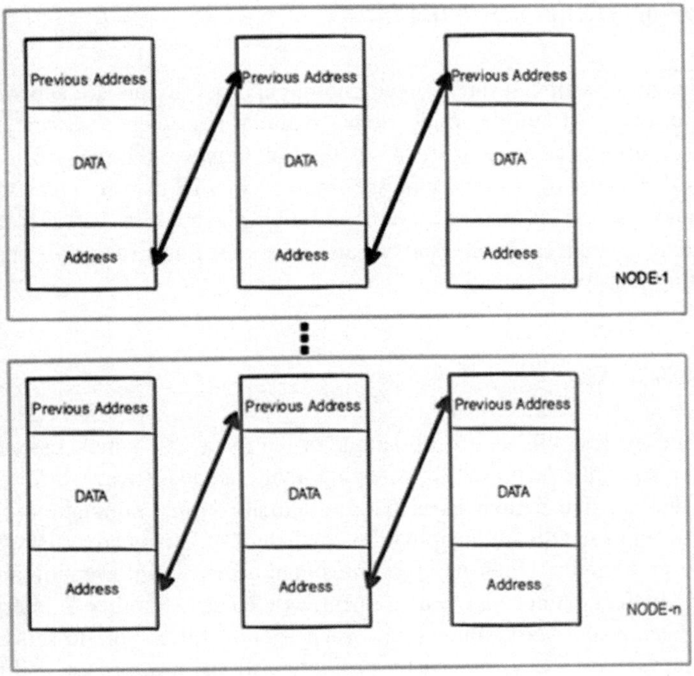

Fig. 2 Blockchain

Fig. 3 AI and blockchain

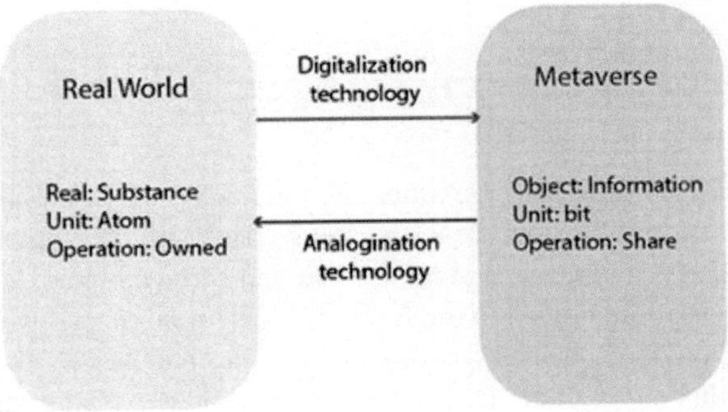

secure. It's crucial to understand every facet of blockchain security as a result, whether a single user acquired cryptocurrency in numerous transactions was a mystery to hackers. Contrarily, the potential of blockchain privacy protection has not yet been fully realized. According to a study, chaff coins and mixins are missing from about 66% of the transactions that were looked at. Chaff coins or mixins may make it more difficult for hackers to determine the relationship between the coins used in a transaction.

2.4 *Role of Artificial Intelligence*

With the fusion of artificial intelligence, mobile app developers are showing more interest in creating AI mobile applications. Artificial intelligence is critical for the metaverse experience because it improves the link between the real world and the digital world. The use of artificial intelligence in improving user contact and experience AI can be used to make more realistic and lifelike avatars, as well as to tailor the user experience to their preferences. It can also be used to improve social connections by simulating real-world interactions in virtual places.

2.4.1 Issues in AI

Artificial intelligence will soon become one of the metaverse's most essential, and potentially hazardous, features. I'm referring about agenda-driven artificial agents that seem and act like regular users but are actually virtual simulations that will engage us in "conversational manipulation" on behalf of paying advertising. This is particularly problematic when AI algorithms gain access to information about our preferences, beliefs, habits, and temperament, as well as the ability to interpret our facial expressions and verbal inflections. Such agents will be able to sell us better than any salesperson. They may easily promote political propaganda and targeted misinformation on behalf of the highest bidder, and it won't only be to offer us items and services (Fig. 3).

2.5 *Role of NFT*

Since real-world identities are connected to digital avatars, NFTs can be used to limit who has access to the metaverse. With the implementation of NFT-controlled access, the metaverse NFT token first surfaced in 2019. Guests were admitted via an NFT-based ticket to the first NFT.NYC conference, which took place in 2019. Even though no one could recognize it as the "metaverse," the conference provided a good illustration of NFT metaverse interaction [8]. NFTs have the potential to play a key role in the greater ecosystem of the metaverse.

2.5.1 Issues in NFT

Privacy-focused cryptocurrencies have attempted to address this issue in various ways, but smart contracts may exacerbate the situation. The contract details must be merged with the NFT when it is issued, allowing anyone to extract information that would otherwise be proprietary. While a few blockchain projects are attempting to remedy this, the larger networks, such as Ethereum, the most popular network for NFTs, are not.

3 Are We Ready for Metaverse?

Accenture, like most corporations, faced the dilemma of how to onboard new workers when they couldn't come to the office when COVID directed staff to work from home. The answer devised by the consulting behemoth was to transport them to the Nth Floor for orientation. The Nth Floor is a virtual office where coworkers may collaborate as if they were all in the same room using a virtual reality headset. In an interview, Yusuf Tayob, group chief executive of operations, said, "We distributed the device to new hires and then held training sessions on the Nth Floor." "I have my Oculus device on my desk across from me, and I can now go to the Nth floor and interact with peers." There is no doubt that businesses and supply chain managers are interested—just look at Facebook's name change to Meta. For example, in a digital twin, you'd be able to visualize the impact of modifications and adjustments to your operations rather than merely scenario planning by running reports. "You might link the physical space to supplier data, publicly available data like weather data, or other digital twins," he said. "The setting becomes considerably more lively." While the metaverse may arrive sooner than robotics, Tayob (chief executive of Accenture Operations) believes it will be a gradual process. He sees five stages in the evolution of digital twins, some of which are currently taking place. We've heard of the metaverse, and it appears that new world order is on the way. Various theories, multiple thoughts, possibilities, expectations, and, of course, news are all circulating to keep us guessing and imagining what it will be like [9]. As a result of our long hours of screen time, whether on laptops, PCs, mobile phones, tablets, or even smartwatches, we are living in a mini-metaverse.

3.1 Issues in Metaverse

One of the questions that arise as we consider the concept of a digital universe is whether users will be required to use a single digital identity or "avatar" across the entire metaverse, or if will there be numerous avatars for different micro-pocket communities. This is similar to logging into an iOS app using your Facebook ID, which is linked to your Google Account, as you may do now. To access the app, you're essentially utilizing three distinct IDs. In the metaverse, how will identification and transparency work? This is something that the developers must first resolve, as the wallet address is insufficient [9, 10].

4 Discussions

Some speculate that the metaverse may be the internet's future. Many businesses are investing in the development of the metaverse. It is critical to ensure that no monopoly exists in the shared virtual environment. Addictions to the Internet and smartphones are becoming commonplace. As a result, virtual world addiction may become the next big thing. Furthermore, the metaverse contains entertainment, commerce, games, and a variety of other addicting activities. Even in this day and century, not everyone has an Internet connection. Many people lack basic digital skills. Many people will be unable to benefit from the metaverse because of the digital divide. Few companies may have control over the metaverse, leaving power and influence in the hands of a few individuals [10].

5 Conclusion

The metaverse's powers are clearly depicted in the final perspective on the metaverse. The promise of a fully immersive web presence incorporating a variety of components, such as social media, entertainment, video production, and other contemporary technology, is very advantageous to the metaverse. On the other hand, worries about privacy and security, as well as the need for advanced technologies, emerge as major metaverse issues. The benefits and drawbacks of the metaverse are considered to create a balanced view of what the metaverse is and could be. Metaverse enthusiasts and organizations should study both sides of the metaverse before making a final conclusion.

References

1. Park S-M, Kim Y-G (2022) A metaverse: taxonomy, components, applications, and open challenges. IEEE Access
2. Wang Y, Su Z, Zhang N, Liu D, Xing R, Luan TH, Shen X (2022) A survey on metaverse: Fundamentals, security, and privacy. arXiv preprint arXiv:2203.02662
3. Kim T, Jung S (2021) Research on metaverse security model. J Korea Soc Digi Indus Inf Manage 17(4):95–102
4. Ning H, Wang H, Lin Y, Wang W, Dhelim S, Farha F, Ding J, Daneshmand M (2021) A survey on metaverse: the state-of-the-art, technologies, applications, and challenges. arXiv preprint arXiv:2111.09673
5. Di Pietro R, Cresci S (2021) Metaverse: security and privacy issues. In: 2021 Third IEEE international conference on trust, privacy and security in intelligent systems and applications (TPS-ISA), pp 281–288. IEEE
6. Far SB, Rad AI (2022) Applying digital twins in metaverse: user interface, security and privacy challenges. J Metaverse 2(1):8–16
7. Masadeh R (2022) Study of NFT-secured blockchain technologies for high security metaverse communication

8. Brown Sr R, Shin SI, Kim JB (2022) Will NFTS be the best dıgıtal asset for the metaverse?
9. Skalidis I, Muller O, Fournier S (2022) The metaverse in cardiovascular medicine: applications, challenges and the role of non-fungible tokens. Can J Cardiol
10. Nath K (2022) Evolution of the ınternet from web 1.0 to metaverse: the good, the bad and the ugly

A Mobile-Based Dynamic Approach to Comparative Study of Some Classification and Regression Techniques

Vijay Souri Maddila, Madipally Sai Krishna Sashank, Paleti Krishnasai, B. Vikas, and G. Karthika

Abstract With the ever-growing complex world of Machine Learning algorithms and data processing techniques, the entry requirements for beginners are steadily rising. So, to allow a much lower entry ceiling into the world of Machine Learning, an Android application is proposed in this study that informs the user about the performance of a few particular supervised algorithms for any given data set. This study informs on three main aspects, namely an Android application that can be developed on flutter as the front end, an API that handles the upload of Comma Separated Values (CSV) file, processing of data, and sending the results back to the front end, and the Machine Learning pipeline that automates the data preprocessing, model creation, and model evaluation for the classification task. The data preprocessing includes data preparation with SMOTE, missing values filling, Scaling and Transformation of data, Feature Engineering, and Feature Selection. To build models, libraries such as Scikit-Learn, XGBoost, and LightGBM are used. Finally, when the user uses the application, they only need to upload a CSV file with the label of the data set renamed to "Target." As soon as the file is uploaded, the server starts the automation process. Once the results are formed, the view of the application changes to a new screen where the results are displayed as a table.

Keywords Machine Learning · Pipeline · Automation · Flutter

V. S. Maddila · M. S. K. Sashank · B. Vikas (✉) · G. Karthika
Department of Computer Science and Engineering, GITAM School of Technology, GITAM (Deemed to be University), Visakhapatnam, Andhra Pradesh 530045, India
e-mail: vikasboddu30@gmail.com

G. Karthika
e-mail: kgidijal@gitam.edu

P. Krishnasai
Department of Computer Science and Engineering, Design and Manufacturing, Indian Institute of Information Technology, Kancheepuram, Chennai, Tamil Nadu, India
e-mail: ced18i039@iiitdm.ac.in

K. A. Reddy et al. (eds.), *Proceedings of Fourth International Conference on Computer and Communication Technologies*, Lecture Notes in Networks and Systems 606,
https://doi.org/10.1007/978-981-19-8563-8_7

1 Introduction

Beginners are perplexed and doubtful about which algorithm suits a particular problem. Furthermore, comparing the performance of several models generally entails a significant amount of time and effort. For example, we can predict which algorithm would perform best on a specific task, but we must test these algorithms on the specific use case and inspect the finer features and approach details. To reduce this overhead, our application is aimed to automate these activities and provides a comparison between different models using flutter.

The world is moving on to intelligent systems that can do a lot more than the currently existing classical systems. With the advent of Machine Learning, computer systems can now be created to learn and be taught, how to do new things. Machine Learning also gives the ability of senses to some extent to computes, such as computer vision, audio processing and sentiment analysis. This field achieved a state of maturity during the recent years and is now able to exert a tremendous amount of influence on the regular world. Machine Learning is so omnipresent now that every person in this world is somehow related to a machine learning algorithm. Machine Learning grew out of the field of artificial intelligence or AI and was strictly a branch that dealt with data and experience.

There is an abundance of data created due to the boom in information systems and Internet specifically. Almost every person in this world has an online presence and is generating data as we speak. Although this data by itself does not amount to much, its utilization thorough Machine Learning and AI can cause huge stirs in the world. Organization that collect the user data can understand what a particular user likes or dislikes and can provide this valuable insight to whoever is willing to buy. As of now Machine Learning is so potent that with the right data about a person, the machine could predict their entire day with minor altercations. Although this seems to invade privacy, with controlled usage it could be used to make human lives better. One example is allowing an AI to predict an underlying disease by understanding the daily habits of a person etc.

Another aspect responsible for boom in Machine Learning is the improvement in computational power. Right now, computational power is abundant and readily available, this makes the high requirements of Machine Learning algorithms attainable at relatively lost cost. Along with the cost, the availability of computational power as cloud resources and distributed resources makes further makes it easy to attain the power. Previously when computation time was expensive and centralized availability of computational power made, researching and experimenting with Machine Learning difficult. Now, almost anyone with a google account could use powerful CPUs, GPUs, and TUPs to arrive at a Machine Learning solution.

With all these advancements and popularity increase for Machine Learning, there is a need for programmers and developers to acclimate to the new technology. Machine Learning is already widely used and is being implemented in every industry. Software engineers and developers should now have expertise in Machine Learning to facilitate the organization in keeping up with the trend of using AI and ML in services and products. But, Machine Learning involves statistics and complex modeling paradigms that dictate how learning happens and how the results can be delivered. Learning these paradigms requires time and steady practice. Similarly, statisticians and professionals with background in other mathematics fields can also enter the ML world. But the barrier of programming and computational approach might slow them down. In such scenarios, a one stop solution that can build and compare the models for a required problem can become crucial.

Hence, with this study, the entry barrier to the world of Machine Learning from the perspective of computational approach is vastly reduced. The user will be able to comprehend the performance of a Machine Learning algorithm through a simple interface and be able to make informed decisions based on the results provided.

2 Literature Review

A programmer needs a lot of practice and knowledge about all the algorithms to know which algorithm works better for that specific problem. It is quite tedious and timetaking to conclude which algorithm works better and faster for a specific problem. So, our goal with this project is to make it easier for the programmer, newcomers, and students of Machine Learning to understand which algorithm best discovers a solution to a certain problem. Following researches show that the comparative analysis of the algorithms helps us in finding the accurate results by finding the best algorithm for the certain problem in less time.

Researches show that the comparative analysis of the Machine Learning algorithms can be used in finding the best algorithms for many problems in many fields [1]: The authors of this paper put the data sets through many Machine Learning algorithms such as Support Vector Machine, decision tree, logistic regression, and random forest to predict the presence of dementia across the data set. Comparing the results obtained by the algorithms suggested that the Support Vector Machine algorithm gives the accurate results of dementia when compared to the remaining algorithms. This emphasized the importance of SVM in actual use cases and hence is included in this project as one of the major algorithms to test.

Intrusion detection should happen whenever necessary and is one of the challenging tasks in the modern networking industry [2]. A network should be continuously monitored to detect the intrusion so we need an intrusion detection system which will monitor network for the sudden intrusion. In this paper, several classification techniques and Machine Learning algorithms have been considered to categorize the network traffic. Out of the classification techniques, we have found nine suitable classifiers like BayesNet, Logistic, IBK, J48, PART, JRip, Random Tree, Random Forest, and REPree. This study showed which algorithms are implemented in a situations that demanded speed and precision. Hence, some of the algorithms are included in the project for classification.

It is very necessary to efficiently distribute the electricity across the population for reducing the power loss [3]. Smart Grids (SG) have the potential to reduce the power loss when distribution. Many algorithms are used to analyze and predict the most suitable one that can be applied to SGs. Out of Support Vector Machines (SVM), K-Nearest Neighbor (KNN), Logistic Regression, Neural Networks, Naive Bayes, and decision tree classifier, have been deployed for predicting the stability of the SG.

Identifying the drugs have received a lot of interest so based on 443 sequence-derived protein features, we applied the algorithms to identify if the protein is druggable and also to know the superior algorithm of the chosen set of algorithms [4]. Neural Network is the best classifier, with the accuracy of 89.98%.

Identifying the stock marketing trends is quite challenging for anyone, so we can use the comparative analysis of Machine Learning algorithms to identify which ne works the best for the identification [5]. After using five techniques i.e., Naive Bayes, Random Forest, K-Nearest Neighbor (KNN), Support Vector Machine (SVM), and Softmax, results show that Random Forest algorithm performs the best for the data sets that are large, and Naive Bayesian performs the best for the data sets that are small.

3 Methodology

3.1 Application Development Approach

In this project, there are four major steps to arrive at the desired end result of a comparative analysis flutter application. They are as follows.

3.2 Building Machine Learning Pipeline

To build a Machine Learning Pipeline, three major processes need to be automated namely the data processing, the model designing and building, and model evaluation.

The first and foremost thing for an ML pipeline is to read the data correctly. To ensure that the pipeline understands the data and realizes the correct task at hand, a schema or structure of the data with clear labels must be obtained. This schema includes the column names, data type of each column, the target column in case of a supervised task, etc. This information is highly specific to the data set on which the algorithms will be trained and tested on. So, once the data set is selected, an automation script which creates the schema.

After building the schema, the pipeline should move onto preparing the data for Machine Learning models. For the data processing section, the following methods are used:

1. Data Scaling: When data has varying scales (ranges of values) and there are no distributions to which the data fits, we use normalization to bring the data to a standard scale.

2. Data Transformation: This gives an immediate sense of the relationship between features of the data set.
3. Feature Engineering: Feature Engineering is the creation or generation of new features (columns) through studying feature interactions and feature grouping.
4. Feature Selection: Feature selection is the decision-making step that decides which columns or features from the data set are relevant to the task at hand.

After preprocessing, the pipeline should move onto structure the data in a way that enables model training, testing, and validation. Since most algorithms are chosen from libraries such as Scikit-Learn, XGBoost, LightGBM, and CatBoost, which use simple fit function calls with the training portion of the data, the data must be split into training and validation sets.

After the data is prepared for the models, the model building process should start in the pipeline. Here, the pipeline manages the training of different models in parallel with its corresponding data. The same data is not used for training all the models, Feature Selection plays a critical role in understanding which features are best for a given task and ML algorithm. All the models are trained and then tested with the training and validation sets, respectively. The metrics for comparative analysis is all collected together and displayed side by side to bring out more information about each of the algorithm's performance.

3.3 Enabling Machine Learning Pipeline in a Flask Environment

To deploy the Machine Learning pipeline, it should be embedded into a server environment. This can be done with the flask web framework which creates routes for a specific task in the web. Before starting with the coding part, we need to download flask and some other libraries. Here, by using virtual environment, all the libraries are managed and made easier for development and deployment. Create script.py file in the project folder and implement the Machine Learning pipeline code as designed above. Then we import the libraries and by using app = Flask(__name__), we create an instance of flask. @app.route('/') is used to set the URL that should trigger the function index() and in the function index used render_template('index.html') to display the script index.html in the browser.

3.4 Through Flask Create an API that Takes a CSV File and Send Results

Once the Machine Learning pipeline is set into the flask environment, it should be enabled to accept and send data. A flow of data must be established from the API call to Machine Learning pipeline and again to the API call. The flask API handles the calls through HTTP and Werkzeug libraries which are responsible for HTTP calls and file handling, respectively.

3.5 *Create a Flutter Application that Can Interact with the API and Show Results from the Server*

Finally, the API should be interacting with an interface to display the results. Similarly, the user will also need an interface to upload the data and look at the results. To allow the application to develop and expand into multiple formats in the future, Flutter is used. Flutter's dart language allows easy application building for multiple platforms (Figs. 1 and 2).

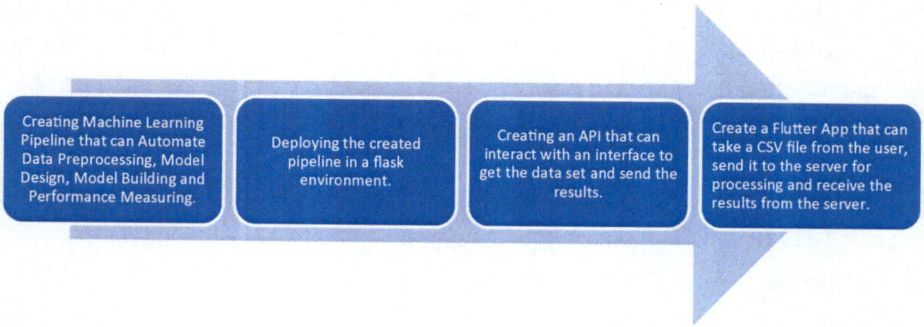

Fig. 1 Development methodology

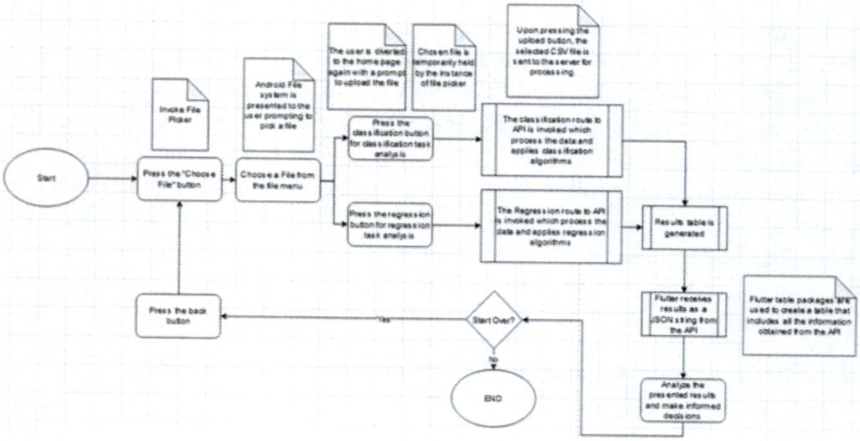

Fig. 2 User experience and underlying function calls

The flow of the application is as follows: Once the user starts the application, they are greeted with a home page that has three buttons. There is a file selection button and two file upload buttons. Initially, the user has to click on the file selection button and pick their CSV file. Once the file is selected, the application comes back to the home page with a changed text over the buttons informing the user that the file is picked. Once the file pick confirmation text is shown, the user can move onto uploading the file for either regression task or classification task. If the user clicks on the regression task, the regression path of the API is invoked and the Machine Learning pipeline processes the data accordingly. Similarly, when the user clicks on the classification upload button, the classification path of the API is invoked and the pipeline processes the data accordingly. Once the processing of the pipeline is done, a pandas table is generated with comparative results of all the tested algorithms. This table is converted into a json and sent to the application interface through an API call. Once the interface receives the data, it shows the user a view of tabulated data in a new screen.

4 Results

Let us understand the working of the application through an example. Let's assume the user wants to perform a comparative analysis on different algorithms for a classification task on a data set called boston.csv (https://www.kaggle.com/datasets/puxama/bostoncsv) (Fig. 3).

The boston.csv file has 15 columns with 1 target column and 14 features. The label column has 5 classes which describe the favorability level of living in the location. All the features in the data set are describing a particular location inside boston. The goal of this classification task is to predict the class of a particular location depending on 14 features (Fig. 4).

In the above screenshot a, we see the home page of the application, with three buttons. To continue, the user clicks on the file picking button at the bottom right of the screen which takes them into the android file space. In the above screenshot b, we see the file picker of android where the user can select the file they want to upload or perform the task on. Once the user selects the boston.csv data set, the file picker models locks onto the file location and keeps it ready for upload to the API (Fig. 5).

In the above screenshot a, we see an alert generated by the application so that the user know which algorithms which be checked on the data set. For classification,

crim	zn	indus	chas	nox	rm	age	dis	Target	tax	ptratio	black	lstat	medv
0.00632	18	2.31	0	0.538	6.575	65.2	4.09	1	296	15.3	396.9	4.98	24
0.02731	0	7.07	0	0.469	6.421	78.9	4.9671	2	242	17.8	396.9	9.14	21.6
0.02729	0	7.07	0	0.469	7.185	61.1	4.9671	2	242	17.8	392.83	4.03	34.7
0.03237	0	2.18	0	0.458	6.998	45.8	6.0622	3	222	18.7	394.63	2.94	33.4
0.06905	0	2.18	0	0.458	7.147	54.2	6.0622	3	222	18.7	396.9	5.33	36.2
0.02985	0	2.18	0	0.458	6.43	58.7	6.0622	3	222	18.7	394.12	5.21	28.7

Fig. 3 Overview of boston.csv

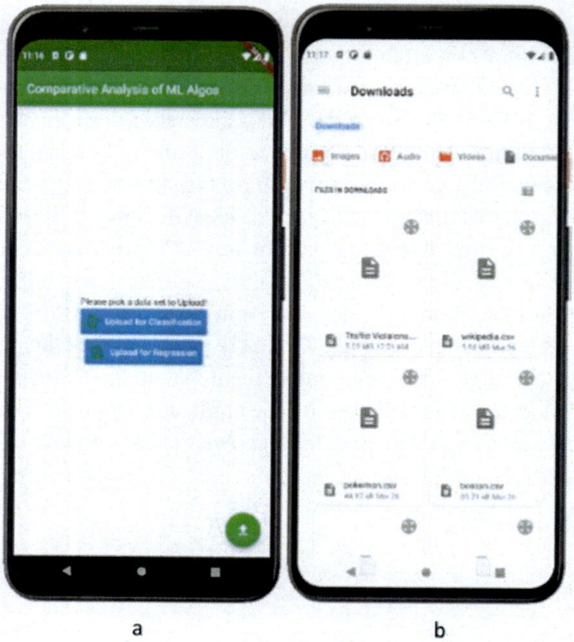

Fig. 4 Flutter application showing the home page and android file picker

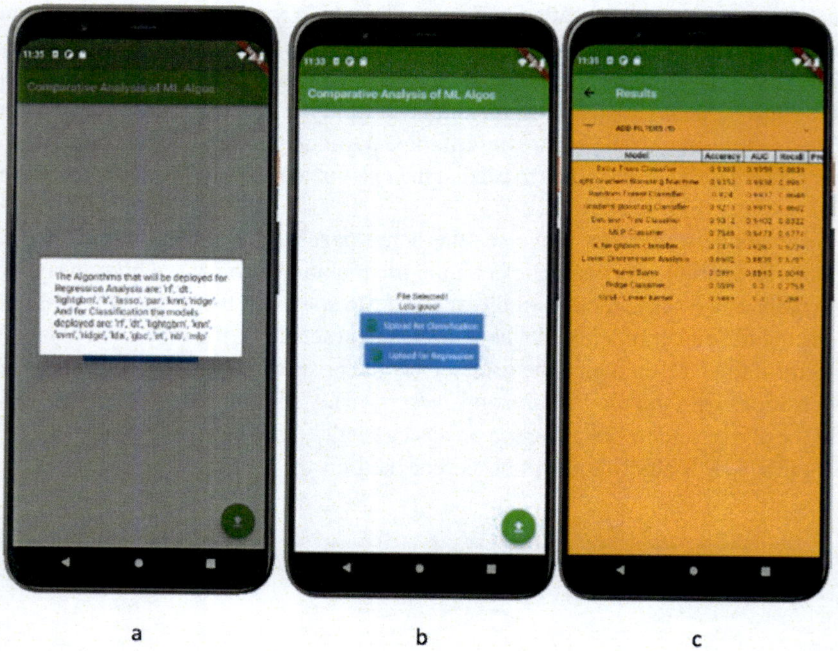

Fig. 5 Flutter application showing an alert to inform the user about the algorithms

we use Random Forest, decision trees, light gradient boosting, K-nearest neighbors, Support Vector Machine, Ridge Classifier, extra trees, multi-layer perceptron, and gradient boosting classifier. In the screenshot b, we see the next step which is after the removal of the alert, the application shows text saying "File selected" which informs the user that they can proceed with the upload of file. Since the user is using a classification data set, they will have to choose the first button which says "Upload for classification."

Once the back end is done with the processing, the user is automatically diverted to a new screen as per screenshot c, where a tabular information is presented. This table gives comparative results of all the algorithms previously mentioned. If the user wishes to restart the whole process or try the application with a new data set, they simply have to press the back button at the navigation bar or the system's back button at the bottom. The whole process can be redone.

5 Conclusion and Future Scope

This study gives an overview of how to create an automated Machine Learning workflow and embed it into a user interface. To perform the Machine Learning tasks, the pipeline must include data preparation, data processing, model building, and model evaluation. The final product obtained is an application that allows a beginner of the Machine Learning world to quickly understand how some algorithms perform on a data set for a given task. This allows the users to learn and comprehend the specificities of different algorithms easily.

The current application can be improved as follows:

1. The current application fails for extremely large data set due to constraints in available memory. This can be solved by optimizing memory usage and chunking the data wherever possible.
2. Having the ability to train and test the Machine Learning models on device can be a huge help, but currently due to available libraries and processer dependencies this is not possible. It can be explored in the future when processors get more powerful and tiny ML algorithms improve.
3. Explore more efficient automation techniques for data processing and model building.

References

1. Bansal D, Chhikara R, Khanna K, Gupta P (2018) Comparative analysis of various machine learning algorithms for detecting dementia. Procedia Comp Sci 132:1497–1502
2. Choudhury S, Bhowal A (2015) Comparative analysis of machine learning algorithms along with classifiers for network intrusion detection. In: 2015 International conference on smart

technologies and management for computing, communication, controls, energy and materials. IEEE, pp 89–95
3. Bashir AK, Khan S, Prabadevi B, Deepa N, Alnumay WS, Gadekallu TR, Maddikunta PKR (2021) Comparative analysis of machine learning algorithms for prediction of smart grid stability. Int Trans Electr Energy Syst 31(9):e12706
4. Jamali AA, Ferdousi R, Razzaghi S, Li J, Safdari R, Ebrahimie E (2016) DrugMiner: comparative analysis of machine learning algorithms for prediction of potential druggable proteins. Drug Discov Today 21(5):718–724
5. Kumar I, Dogra K, Utreja C, Yadav P (2018) A comparative study of supervised machine learning algorithms for stock market trend prediction. In: 2018 Second international conference on inventive communication and computational technologies (ICICCT), pp 1003–1007

Land Cover Mapping Using Convolutional Neural Networks

Cheekati Srilakshmi, Pappala Lokesh, Juturu Harika, and Suneetha Manne

Abstract Using deep learning, the proposed method provides a novel approach for UC Merced satellite photos. The main aim of the deep learning method is to draw out a large number of features without human interaction. Adding object-based segmentation to deep learning further improves classification accuracy. Remote sensor images are accurately classified using deep object-based feature learning with CNN. This method is based on the extraction of deep features and their application to object-based classification. The proposed machine extracts intensive capabilities the usage of predefined filter values, which improves standard overall performance when compared to randomly initialised clear out values. In complicated satellite pictures, the object-based classification technique can preserve edge statistics. Object-based totally deep studying is used to growth type accuracy and reduce complexity. The proposed object-based deep learning strategy is used to seriously enhance the type accuracy. Challenge-based category outperformed the experiments.

Keywords Deep learning · UC Merced · Convolutional Neural Network (CNN) · Image classification

1 Introduction

Land cowl mapping and detection, water useful resource detection, agricultural usage, wetland mapping, geological statistics, and concrete and nearby making plans are handiest few of the packages for classifying numerous regions of far flung sensing pix. However, due to its complexity, categorization of far off sensing photographs stays a time-eating operation. In categorization, feature extraction can be very large. The class manner loses efficiency when characteristics are selected manually with human interaction. As an end result, we use an independent function mastering

C. Srilakshmi (✉) · P. Lokesh · J. Harika · S. Manne
Department of Information Technology, VR Siddhartha Engineering College, Vijayawada, India
e-mail: srilakshmichowdarychikati@gmail.com

S. Manne
e-mail: hodit@vrsiddhartha.ac.in

K. A. Reddy et al. (eds.), *Proceedings of Fourth International Conference on Computer and Communication Technologies*, Lecture Notes in Networks and Systems 606,
https://doi.org/10.1007/978-981-19-8563-8_8

generation together with deep studying to reinforce performance. Huge geographical regions are physically impossible to display inside the beyond, aerial images turned into used to pick out land goals in order that humans may physically interact with them. Because this approach calls for a big variety of gadget and human input, gazing changes in huge regions through the years are extremely difficult. The far off sensing era has been delivered because of this. The method of obtaining statistics approximately an item or phenomena without having direct contact with it or on website online remark is referred to as faraway sensing. Land cover mapping and detection, water useful resource identification, agricultural utilisation, geological information, and concrete and nearby planning are just a few examples of what this phrase refers to.

2 Related Work

In 2021, Ali Jamali presented "Improving land use land cover mapping of a neural network using three optimizers of multi-verse optimizer, genetic algorithm, and derivative-free function," which describes how facts on land use land cover are crucial for land control and planning. To growth the accuracy of faraway sensing photo categorization, the use of a small-sized neural community, three optimizers of the multi-verse optimizer, genetic set of rules, and by-product-unfastened characteristic are designed inside the MATLAB programming language. The outcomes are compared to the ones of a medium-sized neural network created inside the MATLAB programming language based totally on the results of the assessments. Landsat-8 imagery with a spatial decision of pixel-based totally is used [1].

In 2021, Yao Li, Peng Cui, Cheng-Ming Ye, and Jose Marcato Junior presented "Accurate Prediction of Earthquake-Induced Landslides Based on Deep Learning Considering Landslide Source Area," which explains that earthquake brought on landslide (EQIL) is an unexpectedly converting system happening at the Earth's floor; this is firmly managed by way of the Earthquake in question and propensity situations. To provide an explanation for the complex link and enhance spatial prediction accuracy, they present a deep studying framework that takes into account the source location function of EQIL. To isolate the source area of an EQIL, we first employed high-decision remote sensing photographs and a virtual elevation version (DEM). For EQIL prediction, shallow machine getting to know models most effective take use of applicable parameters, consistent with this look at [2].

In 2018, "Identification of farm regions in satellite pictures using Supervised Classification Technique," proposed by Rahul Neware and Amreen Khan outlines the process. Finding land usage or area covered by reviewing prior satellite data and providing analytics is a remote sensing difficulty. This work examines the use of supervised classification to identify farm areas from satellite images. Minimum distance, maximum like hood, spectral angle mapping, parallelepiped classification, and land cover signature classification are some of the mathematical procedures used for classification [3].

In 2015, Suruliandi and Jenicka who were presented texture-based classification of remotely sensed pictures and defined how texture-based type is crucial in land use land cover packages, which are remotely sensed photos. Multivariate nearby binary sample (MLBP), multivariate neighbourhood texture pattern (MLTP), multivariate advanced neighbourhood binary pattern (MALBP), wavelet, and Gabor wavelet were used to extract texture facts on this research. On IRS-P6, LISS-III records, texture-based category became finished, and the consequences have been assessed the usage of the error matrix, class accuracy, and kappa facts. According to the effects of the assessments, MLTP outperformed different texture fashions [3].

3 Proposed Methodologies

3.1 VGG16

VGG-16 is a 16-layer deep convolutional neural community. You can use the ImageNet database to load a pretrained model of the community that has been skilled on over one million photographs. The community can categorise photographs into a thousand distinct item classes, such as keyboards, mice, pencils, and a variety of animals. As an end result, the network has learnt a spread of wealthy feature representations for a spread of pix. The network's image enter size is 224 × 224 pixels. This model takes the input image and turns it into a 1000-value vector.

$$\hat{y} = \begin{bmatrix} \widehat{y0} \\ \widehat{y1} \\ \widehat{y2} \\ \widehat{y3} \\ \vdots \\ \widehat{y999} \end{bmatrix} \quad (1)$$

The configuration "D" in the table below is referred to as VGG16. There are 16 weight layers in the configuration "C." In stacks 3, 4, and 5, however, the last convolution layer is a 1 × 1 filter. This layer was utilised to boost the decision functions' nonlinearity without compromising the layer's receptive field. Unless otherwise noted, configuration "D" shall be referred to as VGG16 throughout this discussion. Unless otherwise noted, configuration "D" shall be referred to as VGG16 throughout this discussion (Fig. 1).

The input to the cov1 layer is a 224 × 224 RGB photo with a hard and fast size. That photo is going to be processed through a stack of convolutional (conv.) layers with an extremely slim receptive area: 33 (the smallest size is to capture the notions of left/right, up/down, and centre). It also uses eleven convolution filters in one of the planned steps, which may be thought of as a linear adaptation of the enter channels

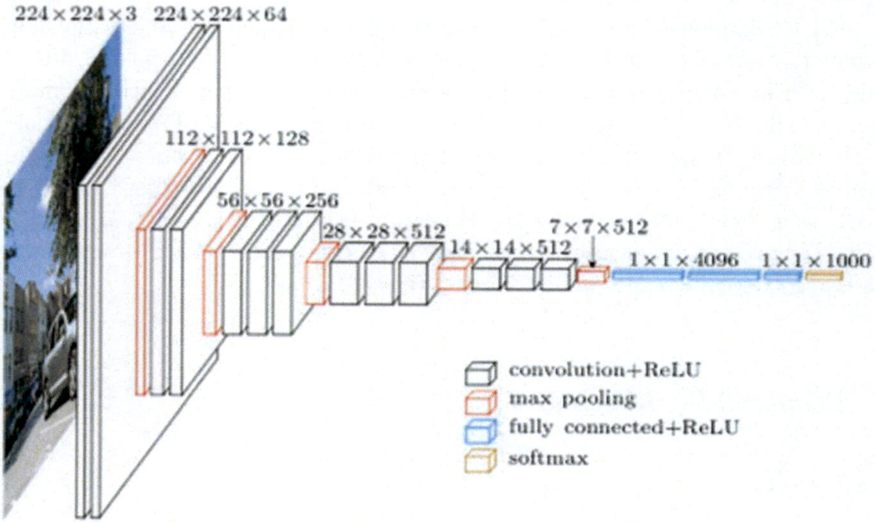

Fig. 1 VGG16 architecture diagram

(observed by the way of nonlinearity). The convolution stalk was about to 1 pixel, and the spatial padding of conv. layer enter is a set to one pixel for 33 conv. layers in order to the spatial resolution was stored after the convolution. Five max-pooling layers, which follows a part of the conv. layers, do spatial pooling (now not all of the conv. layers are travel with by means of max-pooling). Max-pooling is accomplished with stalk 2 throughout a 22 pixel body.

3.2 Convolutional Neural Network (CNN)

Because of its efficacy in spatial feature exploration, the CNN method is useful for high-resolution picture categorization. Deep features are extracted from the LISS III picture using CNN as a feature extractor. The acquired deep characteristics are merged with object-based textural information to further boost efficiency. A neural network is generally trained in two stages: The input is routed entirely through the network in this phase. Gradients are back propagated (backprop), and weights are modified at this step.

It is made up of input and output layers, as well as numerous hidden levels in between. It's far typically used to categorise photos, cluster them based on similarity, and recognise gadgets within scenes. They are algorithms that can recognise a wide variety of visible statistics. Two key elements in CNN are the locally related network and parameter sharing. If we appoint a completely connected community, a massive wide variety of parameters will be essential. As a result, locally linked

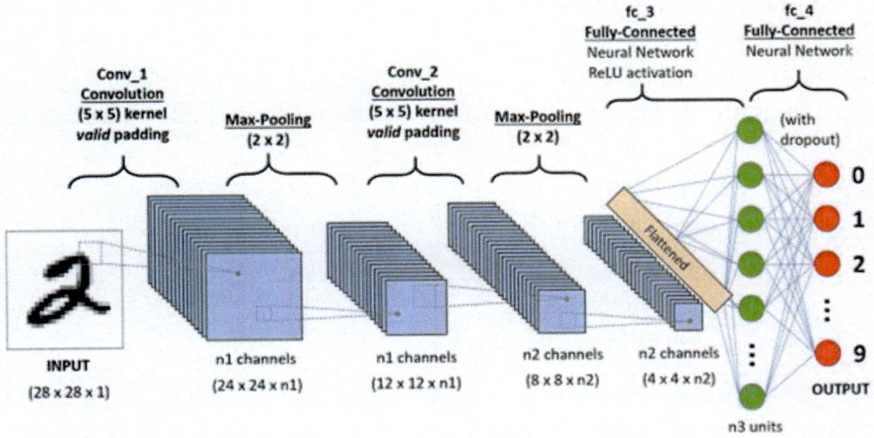

Fig. 2 CNN architecture diagram

networks are employed to decrease parameter requirements. Also, instead of utilising fresh parameters every time, parameter sharing allows you to reuse previously used parameters (Fig. 2).

3.3 Fast.AI

Fast.ai is a non-profit research organisation that focuses on deep learning and AI. the objective of making deep learning more accessible to the general public "Practical Deep Learning for Coders" is a massive open online course (MOOC) that requires only knowledge of the programming language Python as a prerequisite. Fast.ai is a deep modern-day library that offers practitioners with excessive-degree components which can offer effects in conventional deep today's domain names speedy and without problems, in addition to teachers with low-stage components that can be mixed and coupled to create novel techniques. It strives to attain each dreams without sacrificing usability, flexibility, or normal performance. This is made viable through a layered layout that represents the not unusual underlying styles extremely present day numerous deep modern and information processing techniques in language which are easy to apprehend.

4 Experimental Analysis

We utilised Google Colaboratory, which is a free service provided by Google Inc., to create this project. Using the WGET package and the following URL, we can easily retrieve the data in Jupyter Notebook/Google Colab.

```
!wget
https://www.dropbox.com/s/u83ae1efaah2w9o/UC
MercedLanduse.zip
!unzip UCMercedLanduse.zip
```

We're ready to study the data once we've downloaded it and unzipped it. First, let's look at the labels. Pandas helped us read the labels. The information is stored in One-Hot Encoded format. Each image includes 17 labels, with "0" indicating that the label is not present in the image and "1" indicating that it is there. We have 2100 photographs in total (Fig. 3).

Checking the dataset's distribution for data imbalances is a critical phase in the process. To store the classes and their numbers, we first establish a new data frame. The following visualisation depicts the dataset's class imbalances. There are almost 1200 photographs in the pavement class and just 100 shots in the aeroplane class (Fig. 4).

We'll need to prepare the data for the training. Our data labels are in One-Hot Encoded format, which I expected to be difficult. Fortunately, a quick search of the Fast.ai Forum revealed that Fast.ai has a native method for multiple-labels in the One-Hot Encoding format. When labeling the dataset, we must provide the column names as well as the fact that it is a multi-category dataset. After we've created the data source, we can use Fast.ai's data bunch API to feed it through. In addition, we make certain data enhancements. After we've created the data source, we can use Fast.ai's data bunch API to feed it through. In addition, we make certain data modifications.

Next, we build a learner and provide it the data bunch we made, the model we want to use (in this case, ResNet34), and the metrics we want to use (accuracy thresh and F Score) (Figs. 5, 6, 7, 8, 9 and 10).

	IMAGE\LABEL	airplane	bare-soil	buildings	cars	chaparral	court	dock	field	grass	mobile-home	pavement	sand	sea	ship	tanks	trees	water
0	agricultural00	0	0	0	0	0	0	0	1	0	0	0	0	0	0	0	1	0
1	agricultural01	0	0	0	0	0	0	0	1	0	0	0	0	0	0	0	0	0
2	agricultural02	0	0	0	0	0	0	0	1	0	0	0	0	0	0	0	0	0
3	agricultural03	0	0	0	0	0	0	0	1	0	0	0	0	0	0	0	0	0
4	agricultural04	0	0	0	0	0	0	0	0	0	0	0	0	0	0	0	1	0

Fig. 3 Dataset distribution diagram

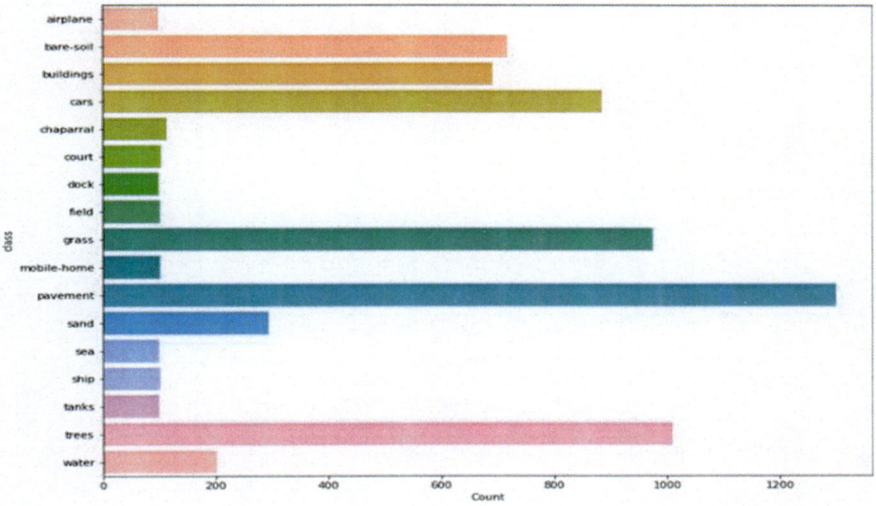

Fig. 4 Dataset distribution diagram

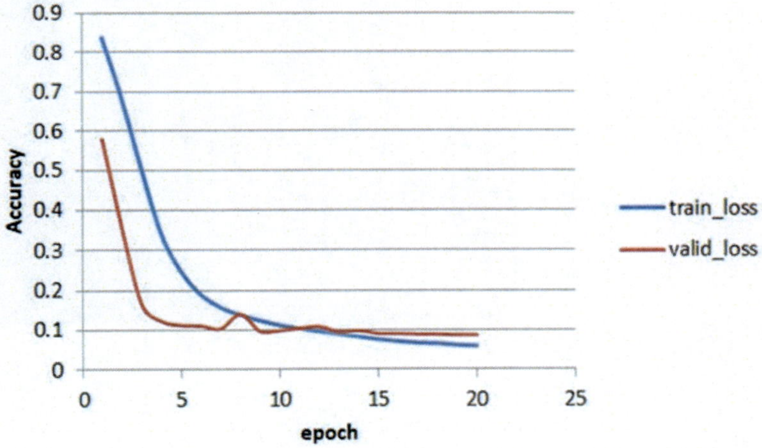

Fig. 5 Accuracy metrics diagram

5 Architecture

In our architecture diagram it states that how to do perform the process of UC Merced to land use mapping. This project use various techniques for bringing the land use mapping (Fig. 11).

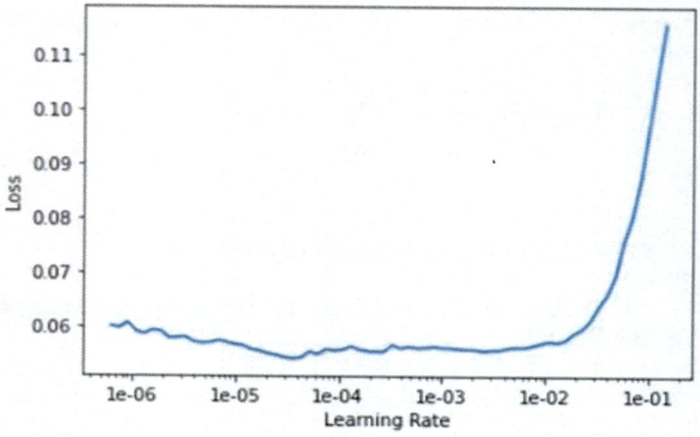

Fig. 6 Learning rate diagram

Fig. 7 Input image1 for object detection

$$\text{Mean } \overline{x} = \frac{1}{N^2}\sum_{i=0}^{N}\sum_{j=1}^{N} x_{ij} \tag{2}$$

$$\text{Variance } V = \frac{1}{N^2}\sum_{i=0}^{N}\sum_{j=1}^{N}(x_{ij} - \overline{x}) \tag{3}$$

$$\text{Entropy} = \sum_{i=1}^{N}\sum_{j=1}^{N}(C(i, j))\log(C(i, j)) \tag{4}$$

$$\text{Contrast} = \sum_{i,j=0}^{N}(i - j)^2 C(i, j) \tag{5}$$

Fig. 8 Input image2 for object detection

Fig. 9 Output for input image1

```
MultiCategory tensor([0., 0.,
1., 1., 0., 0., 0., 0., 1., 0.,
1., 0., 0., 0., 1., 1., 0.])
```

Fig. 10 Output for input image2

```
MultiCategory tensor([1., 0., 0.,
1., 0., 0., 0., 0., 0., 0., 1.,
0., 0., 0., 0., 0., 0.])
```

$$\text{Energy} = \sum_{i=1}^{N} \sum_{i=1}^{N} C(i - j)^2 \tag{6}$$

$$\text{Local consistency} = \sum_{i,j=0}^{n} 1/\left(1 + (i - j)^2\right)C(i, j) \tag{7}$$

$$\text{Cluster Shade} = \sum_{i,j=0}^{n} \left(i - M_x + j - M_y\right)^3 C(i, j) \tag{8}$$

$$\text{Cluster prominence} = \sum_{i,j=0}^{n} \left(i - M_x + j - M_y\right)^4 C(i, j) \tag{9}$$

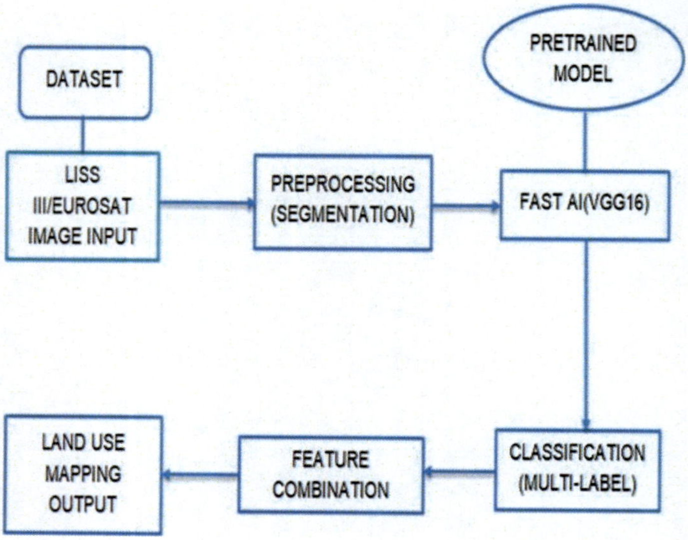

Fig. 11 Project flow diagram

where $M_x = \sum_{i,j=0}^{n} iC(i, j)$ and $M_y = \sum_{i,j=0}^{n} jC(i, j)$

$$\text{Correlation} = \frac{\sum_{i=1}^{N}\sum_{j=1}^{N}[ijC(i, j)] - \mu_x\mu_y}{\sigma_x\sigma_y} \tag{10}$$

where $\mu_{x=\sum_i^N i \sum_j^N C(i,j)}$ $\mu_{y=\sum_i^N j \sum_j^N C(i,j)}$

$$\sigma_x^2 = \sum_i^N (a - \mu_x)^2 \sum_j^N C(i, j)\sigma_y^2 = \sum_i^N (b - \mu_x)^2 \sum_j^N C(i, j)$$

$$\text{NDVI} = (\text{Near IR band} - \text{Red band})/(\text{Near IR band} + \text{Red band}). \tag{11}$$

6 Observations

By using this project, we can identify the changes in the area done from previous years till now. And also we can identify multiple objects in a satellite image which we are using in the project from the dataset we loaded.

7 Conclusion

Using deep neural networks, we trained a multi-label category classification model. We also used additional photos to make assumptions about the model.

Acknowledgements Authors would like to thank Department of IT, VR Siddhartha Engineering College and Siddhartha Academy of General and Technical Education (SAGTE) for giving permission to carry out this work on Deep Learning Server.

References

1. Li Y, Cui P, Ye C, Marcato J Jr, Zhang Z, Guo J, Li J (2021) Accurate prediction of earthquake-induced landslides based on deep learning considering landslide source area. Remote Sens 13:3436. https://www.mdpi.com/2072-4292/13/17/3436
2. Roshandel S, Liu W, Wang C, Li J (2021) 3D ocean water wave surface analysis on airborne LiDAR bathymetric point clouds.Remote Sens 13(19):3918. https://doi.org/10.3390/rs13193918
3. Liu X, Chen Y, Wei M, Wang C, Goncalves, WN, Marcato J Jr, Li J (2021) A building instance extraction method based on improved hybrid task cascade. IEEE Geosci Remote Sens Lett. https://doi.org/10.1109/LGRS.2021.3060960

Establishing Communication Between Neural Network Models

Sanyam Jain and Vamsi Krishna Bunga

Abstract Artificial intelligence is transforming every industry and revolutionizing the planet earth. For years, machines have been able to accomplish feats that would seem magical to us: collect data, remember information, and compute fast. Now they are getting close to reaching a level of intelligence that can devise answers to some questions just like a person. Furthermore, neural networks are scattered throughout today's popular apps and programming languages, allowing artificial intelligence to help us make smarter decisions. Eventually, these AI algorithms could also transform our fields of art and science into an era where creativity arises from advanced deep learning systems rather than human brain power—a time where humans hold the primary role in an advisory capacity only, directing AI processes on very difficult tasks. This paper explores training sets of neural networks to probe how communication can happen between them. We find that communication across these networks is volatile by running a range of tests in a variety of conditions, and we also learned that defining a common "language" of communication between various networks can be difficult.

Keywords Neural networks · Adversarial networks · Multi-agent communication · Deductive situation · Modeling interpretation · Communication

1 Introduction

To what extent can neural network models communicate with each other and discover each other's identity? How would they use this information in a competitive setting? For example, in a social deduction game, players attempt to uncover each other's hidden allegiance—typically with one "good" team and one "bad" team. Players

S. Jain
Maharaja Agrasen Institute of Technology, Delhi, India
e-mail: sanyam.04514804919@eee.mait.ac.in

V. K. Bunga (✉)
Andhra University, Visakhapatnam, India
e-mail: vamsikrishh0099@gmail.com

K. A. Reddy et al. (eds.), *Proceedings of Fourth International Conference on Computer and Communication Technologies*, Lecture Notes in Networks and Systems 606,
https://doi.org/10.1007/978-981-19-8563-8_9

must utilize deductive reasoning to find the truth or instead lie to keep their role hidden. In this paper, we explore if neural networks can be successfully trained to compete in a scenario such as this, and how would the opposing parties interact during the period of debate.

1.1 Among Us

Among Us is a currently popular social deduction game, where the "imposters" attempt to sabotage and kill all of the "crewmates." Crewmates have to complete tasks and figure out who the imposters are and eliminate them before the imposters win. At certain points in the game, after periods of no direct communication, players debate the roles of each individual based on information previously acquired through their personal experience. At the end of this discussion, every player votes on a single player to be eliminated. The player with the most votes is eliminated, and if there is a tie, no one is voted out. We chose to emulate this game based on the overall simplicity of the two roles and the requirement of communication for either party to succeed. If the crew do not exchange information and all vote the same person, the vote could result in a tie or a crew being eliminated. If the imposters do not bluff, the crew can easily spot the liars among the group. This provides ample room to explore and experiment with the communication between the two opposing parties.

1.2 Adversarial Networks

Within this design space, there are adversarial parties working against each other. In the deep learning realm, adversarial situations appear in adversarial examples [1] and within generative adversarial networks (GANs) [2]. In particular, the latter often designs a contest between two neural networks, in the form of a zero-sum game. We build upon these concepts and foundations in our work.

1.3 Multi-agent Communication

Inherently, a social deduction game requires multiple agents to be trained and contested. This has been explored within the deep learning problem space with multi-agent sub problems. Both cooperative [3, 4] and adversarial [5] communication has been experimented with, showing that models can effective share and also selectively protect information. We reference these approaches and we generate active adversarial communication between neural network models.

2 Approach

In order to examine the interactions between multiple agents—some of who are secretly imposters who must "lie," we chose to construct an "Among Us"-esque scenario for our models to partake in [6]. On a high level, we wanted each agent to first collect information independent of the others. Then, there should be some phase of communication between the models—qualitatively, it is during this stage that the agents will attempt to deduce who the imposters are, and the imposters will attempt to blend in. Finally, the agents all vote on who they think is a likely imposter. The upcoming sections explain how we chose to model each of these phases in greater detail.

2.1 *Deductive Situation*

During the period of communication in Among Us, the crew must do tasks and gain information while the imposters must sabotage and kill the crew. We decided to simplify the "game" by both removing the tasks and killing and making the entire perception of each player predetermined [7]. Specifically, each agent is given as input a matrix consisting of N events. During each event, the agent "sees" some subset of the other players (sight is reflexive and symmetric, but not transitive) [8]. They also may experience a "sabotage," which means they are in the prescience of an imposter who is sabotaging. They cannot see the imposter, but the imposter can see them during this event. Ultimately, each agent will receive a $N\ (P + 1)$ matrix, where P is the total number of players—the first P values of a row indicate who the agent is seeing, and the final value indicates a sabotage.

These event matrices are generated randomly using four key parameters: the total number of players (P), the number of those players who are imposters (I), the chance that any given pair of players will see one another during an event (view chance), and the chance that any given imposter will sabotage during an event (sabotage chance) [9].

2.2 *Modeling Interpretation and Communication*

In order to process the input events, our agent model has an LSTM, which generates c_N and h_N, which are used as the initial inputs h_0 and c_0 to the next phase: communication. Communication is also modeled with an LSTM. During each "round" of communication, every agent contributes a message vector of size M via a small MLP using h_t as input. These messages are collected into a matrix of size $M\ P$, which is fed as input into each agent's LSTM so that their memory can be updated before the next round of communication—there are R rounds in total. We chose to model

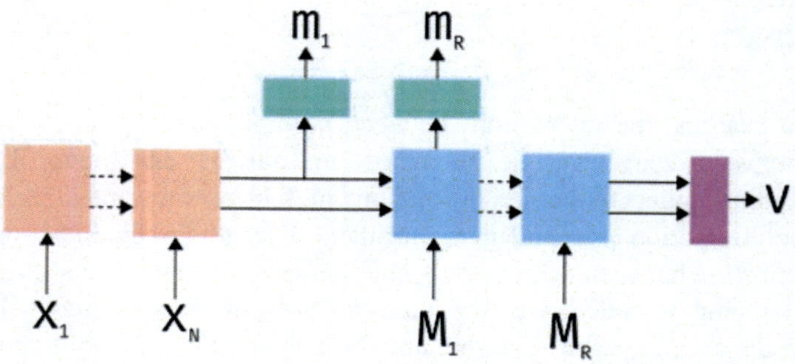

Fig. 1 Diagram of the agent model. The red section is the perception LSTM, which takes in a sequence of events. The blue section is the communication LSTM, which receives messages and generates messages using the green MLP. Finally, the purple section is the voting MLP, which produces the agent's vote vector

communication this way because it is a simple and symmetric way for the agents to pass information between each other in multiple rounds. Figure 1 shows a simplified diagram of the complete architecture.

2.3 *Zero-Sum Target*

The last stage is the most simple one—voting. The model simple takes the h_R and c_R from the end of the communication LSTM and feeds them through a small MLP finalized with a softmax layer. This results in a probability vector of a confidence that a specific player should be "voted out."

At the end of voting, we calculate a "crew score." This score is simply the maximum vote-off score that any imposter received, where votes are averaged across all agents. Clearly, the imposters would like to minimize the votes on themselves, so their loss function for training purposes is simply the crew score [10]. Inversely, the crew's loss function is the negation of the crew score. This takes the form of a zero-sum game with similar application as in generative adversarial networks [11].

2.4 *Training Scheme*

There were multiple decisions we had to make when attempting to train the models. Similar to GANs, we decided the best approach to train the two adversarial models was to use alternating training to help find eventual convergence and juggle two different optimizations.

2.5 *Challenges*

The main challenge in building and running the model came mainly in the form of training time, gradient overlap, and hyperparameter search. As discussed previously, to help speed up training time and reduce gradient overlap between multiple different models training over the same dataset, we had to only train either one imposter or crew, and keep the rest constant. This limits the quick adaptivity and may lead to the models attempting to train against the constant cooperative models rather than against the adversarial model.

Furthermore, the immense number of combinations of hyperparameters and input and outputs sizes, along with no precedent for recommended values, lead us to generalize our model significantly. Although this offers the extended ability for customizability and exploration, it severely reduces reproducibility over small changes.

Another important challenge to recognize is the extreme "black box" architecture of the model. With our current model iteration, we have no current approach to visualizing exactly what the model is doing, especially with communication. Therefore, there are assumptions and estimations when creating conclusions about the model.

2.6 *Situation Hyperparameters*

To list the variable hyperparameters and input and outputs sizes for the player model, overall, we have batch size, number of epochs, epoch length, the global LSTM hidden layer size, and the amount of players and imposters. For the perception phase, there is the chance a player will view another player, the chance a sabotage occurs, and the number of N events. The communication stage has variables the message size M and the number of communication rounds R (Fig. 2).

3 Results

Due to the unprecedented architecture, we decided our overall goals were to find the balance between imposter and crew score, locate trends within the variable changes, and attempt to understand how the models communicate, to a degree. We conducted many training sessions over multiple different combinations of variables.

A typical training session started with the first two alternating epochs, which were considered as the "initialization" of the two models. Training the imposters first gave them time to learn to not vote for each other. Then, the crew had learned the same thing on the next epoch. After this initial training, then the models started to significantly train against each other.

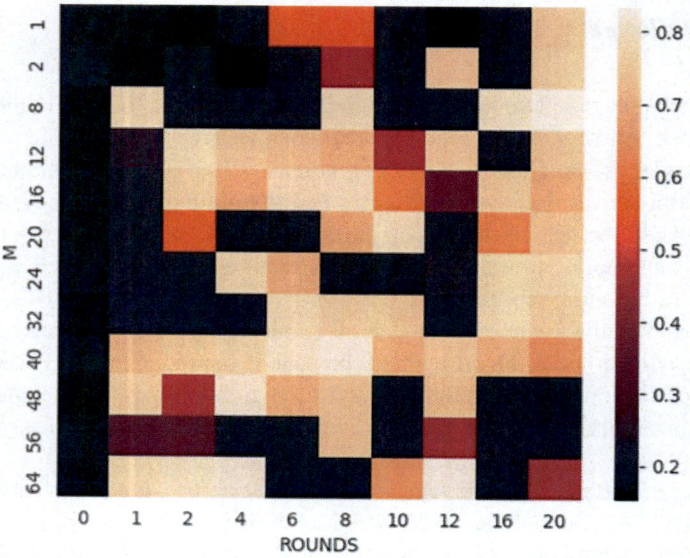

Fig. 2 Grid search heat map over the size parameters M (message vector) and rounds (the number of communications) displaying the ending crew score after 6 epochs. Figure 3 contains the other parameters used

Through our search, we found a trend to three different types of results: an imposter 'win,' a crew 'win,' and a convergence. An imposter win happens under certain circumstances where the crew are unable to employ a better strategy than the crew's equally voting all other candidates. A crew win occurs with the opposite—the crew can determine exactly who the imposters are, and crew vote for one imposter. A convergence happens to place itself in the middle, for what we assume to be a fair, balanced game between the two opposing parties.

Due to resource constraints, we were not able to get as much data as we had hoped for. This lead to a consistency issue, where the same variable combination would yield different results because of certain randomness within training the models.

3.1 Oscillating Scores

The hopeful goal was to find the set of variables that lead to not only a convergence, but an almost "tug-of-war" battle when one model trains against the other and vice versa. This is graphically shown via an oscillating crew score around the eventual convergence crew score. This would mean that the two models are successfully balanced and able to improve their strategy with training time against the opposition. Figure 3 shows an example of a single run which converges to a crew score of around 0.6.

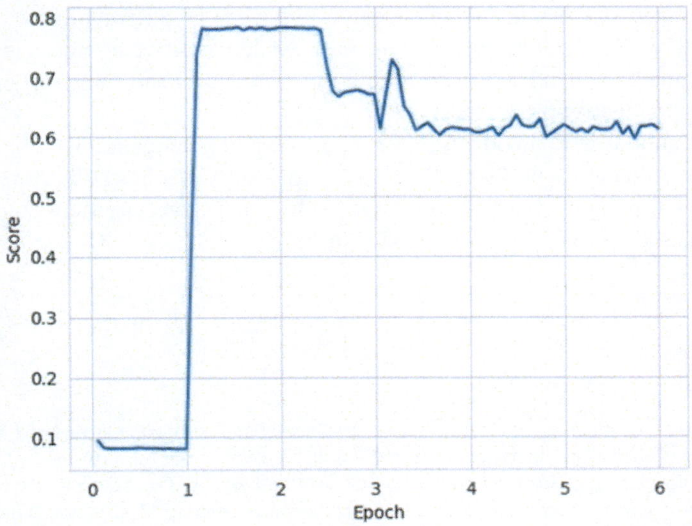

Fig. 3 Crew score over time for a particular training run. The crew model is being trained in odd epochs, and imposters in even epochs

3.2 Interpreting Communication Vectors

Unfortunately, we were not able to interpret the communication vectors due to the disconnect between representation of information. However, we found an interesting result when the players were given no identifying information (e.g., there were no sabotages whatsoever in the perception stage), the crew could still win the situation. At first glance, this seems almost impossible because the crew would be unable to identify the imposters via the necessary events. Further thought lead us to make an estimated conclusion that the crew model was actual creating a hidden tag or key to identify themselves within the communication stage, therefore identifying the imposters immediately. The imposters would then have to train to figure out this key or become helpless [10]. This situation resembles almost a cryptographic adversarial problem, unlocking more future directions this work could potentially travel.

4 Conclusion

This approach is just a start into understanding how neural networks communicate with one another. Although assumptions can be made, understanding the actual information being transferred between each model is a dark area. Innovative techniques must be found to analyze this information flow in order to fully comprehend the strategy within the game.

Furthermore, the results could help to explore more cryptographic use cases for neural networks. It was observed that the crew could identify the imposters with no prior information which leads us to believe the crew model had some sort of hidden code or tag to identify themselves.

Finally, similar work in adversarial networks could potentially give rise to more techniques utilizing them. Generative adversarial networks are at the forefront of adversarial techniques, but possibly large multi-agent adversarial communication networks could give heed to more good results.

References

1. Goodfellow IJ, Shlens J, Szegedy C (2015) Explaining and harnessing adversarial examples, 2015. Published as a conference paper at ICLR 2015
2. Goodfellow I, Pouget-Abadie J, Mirza M, Xu B, Warde-Farley D, Ozair S, Courville A, Bengio Y (2014) Generative adversarial nets. In: Ghahramani Z, Welling M, Cortes C, Lawrence N, Weinberger KQ (eds) Advances in neural information processing systems, vol 27, pp 2672–2680. Curran Associates, Inc.
3. Foerster JN, Assael YM, de Freitas N, Whiteson S (2016) Learning to communicate to solve riddles with deep distributed recurrent q-networks. CoRR, abs/1602.02672
4. Foerster JN, Assael YM, de Freitas N, Whiteson S (2015) Learning to communicate with deep multi-agent reinforcement learning. CoRR, abs/1605.06676
5. Abadi M, Andersen DG (2016) Learning to protect communications with adversarial neural cryptography. CoRR, abs/1610.06918
6. Fekri MN, Grolinger K, Mir S (2022) Distributed load forecasting using smart meter data: federated learning with recurrent neural networks. Int J Electr Power Energy Syst 137:107669
7. Ni Y, Li X, Zhao H, Yang J, Xia W, Gui G (2022) An effective hybrid v2v/v2i transmission latency method based on LSTM neural network. Phys Commun 51:101562
8. Jing Y, Ye X, Li H (2022) A high precision intrusion detection system for network security communication based on multi-scale convolutional neural network. Futur Gener Comput Syst 129:399–406
9. Liu F, Meng W, Lu R (2022) Anti-synchronization of discrete-time fuzzy memristive neural networks via impulse sampled-data communication. IEEE Trans Cybern
10. Bhushan M, Nalavade A, Bai A (2020) Deep learning techniques and models for improving machine reading comprehension system. IJAST 29
11. Kim SH, Moon SW, Kim DG, Ko M, Choi YH (2022) A neural network-based path loss model for bluetooth transceivers. In: 2022 International conference on information networking (ICOIN), pp 446–449. IEEE

Hardware Implementation of Cascaded Integrator-Comb Filter Using Canonical Signed-Digit Number System

Satyam Nigam

Abstract Among the existing implementations of high-order decimation filters, canonical signed digit (CSD) number system implementation in a cascaded integrator-comb (CIC) filter has proven to be one of the most effective filtering schemes. Since CIC filter is multiplier-less the complexity had reduced yet there is still some room for improvement. The implantation of CSD number system is aimed to improve the time response of a traditional CIC Filters. In this work, we propose to use canonical signed digits—a special manner for encoding a value in a signed-digit representation, which itself is non-unique representation and allows one number to be represented in many ways—in a cascaded integrator-comb filter—consists of one or more integrator and comb filter pairs. We compare the performance of CSD implemented CIC filter with standard CIC filter. Our empirical analysis shows that CSD implemented CIC filter can basically outperform traditional CIC filters in terms of high frequency applications, noticeably with a reduced computational effort.

Keywords Decimation filters · Number systems · Canonical signed digit · CIC filters

1 Introduction

In the modern era of digital signal processing, the requirement of more robust and accurate filters has increased. Computerized signal processors are utilized for facilitating activities of separating in high-transfer speed applications. Computerized channels are undeniably utilized because of the reality they put off various issues connected with simple channels. Filters are commonly used to reduce noise and improve quality of information. The presence of interference can veil the resultant sign, or mediate with its investigation. Notwithstanding, if signal and noise involve restrictive spectral locales, it tends to be plausible to improve the signal-to-noise proportion (SNR) with the guide of utilizing advanced channels with digital filters

S. Nigam (✉)
Electronics and Communication Department, Netaji Subhas University of Technology, Dwarka Sector-3, Dwarka, Delhi, India
e-mail: satyamn.sp20@nsut.ac.in

K. A. Reddy et al. (eds.), *Proceedings of Fourth International Conference on Computer and Communication Technologies*, Lecture Notes in Networks and Systems 606,
https://doi.org/10.1007/978-981-19-8563-8_10

to the information. Filters usually perform multiply and accumulate technique to manipulate the spectral information of the input signal. The overall all speed of the hardware required to realize a CIC filter can be improved by carrying addition in canonical signed-digit (CSD) representation. This paper sheds light on how we can implement CSD arithmetic into CIC filter structure, and optimize it for better performance.

Over the course of time, multiple attempts were made to optimize the computational complexity of digital filters. From Crochiere and Rabiner [1], multistage stage decimators and interpolators which emphasize on minimizing the number of multiplications per second to Goodman and Carey [2] half-band hardware efficient decimators and interpolators and from Peled and Liu [3] coefficient slicing approach to Hogenauer et al. multiplier-less filters [4]. CIC filters evidently outperform traditional FIR filters on FPGA implementation [5]. Recently, two new structures of CIC filters are designed to minimize pass-band droop and enhance alias rejection [6]. Previous works have solved various problems using different novel approaches. As there are much better designs of multi-rate filters available, still pondering upon the new implications of CSD numbers a Hogenauer's CIC filter design is modified in this paper. Due to the fact that it has a simple structure and relatively less complex to implement on an FPGA. The main purpose of using this structure is to achieve optimum filter performance at high frequencies.

Improving computational speed using canonical signed-digit number system is an elegant approach for less complex high frequency filtering applications [2]. CIC filters have three major blocks named as rate converter, integrator and comb. The integrator and comb section have adder units which adds samples. In order to make the filter CSD compatible upgrading, the adder block to the CSD adder block and comparing it with the responses of standard CIC filter is the core agenda of this work.

Three bit number representations are shown in Table 1. Nonzero bit occurrence probability for the CSD numbers is nearly 1/3 and for 2's complement numbers it is exactly 1/2. Acknowledging these details about the CSD number system, it is quite obvious that on addition if there are almost zero number of adjacent nonzero bits, there will be no carry generation and we can be able to add two number without using pervious carry. More details are in Sect. 4.

Table 1 Three bit CSD numbers

Number	2's complement	Canonical signed digit
−3	101	$\bar{1}01$
−2	110	$0\bar{1}0$
−1	111	$00\bar{1}$
0	000	000
1	001	001
2	010	010
3	011	$10\bar{1}$

2 Method

The filter simulations are based on FPGA Nexys A7-100T (xc7a100tcsg324-1) board. All the simulation results are calculated by Xilinx VIVADO and MATLAB R2017a.

3 CIC Decimation Filter

The cascaded integrator-comb (CIC) decimation filters are multi-rate digital filters. These filters have several advantages over traditional moving average filters such as there is no need for coefficient data storage as CIC filters do not have filter coefficients and narrow band-pass filters can be made using CIC filters, which are less complex than their FIR counterpart. The noise is reduced due to averaging. Perfect precision can be achieved using fixed-point numbers only.

The principle on which these filter work is recursive running sum. Traditional digital FIR filters require total $(D - 1)$ summations to calculate a single filter output. Moreover, there are also (D) multiplications with filter coefficients.

The recursive running sum filters shown in Fig. 1 subtract the oldest sample $x(n - D)$ from the output $y(n - 1)$ and simultaneously add present input sample $x(n)$ to obtain the present output $y(n)$. The number of computations per output sample reduces drastically by implying this methodology.

Equations 1 and 2, as in [4], will express the complete picture of recursive running sum filters

$$y(n) = \frac{1}{D}[x(n) - x(n - D)] + y(n - 1) \tag{1}$$

And the transfer function for a second order CIC filter can be expressed as

$$H(z) = \frac{Y(z)}{X(z)} = \frac{1}{D}\left[\frac{1 - z^{-D}}{1 - z^{-1}}\right]^2 \tag{2}$$

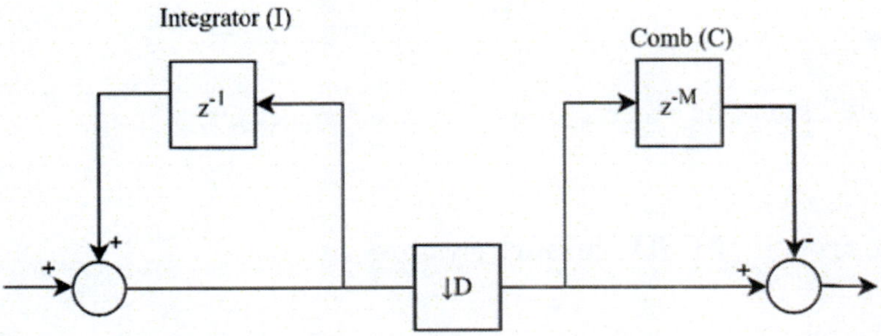

Fig. 1 CIC decimation filter

Table 2 CIC decimation filter parameters

Parameter name	Value
Decimation factor (D)	2
Filter order (N)	2
Differential delay (M)	1
Input bit width (bit)	16
Output bit width (bit)	18
Output rounding option	Truncation

The application of filter is endless as they support high data-rate filtering [3]. Modern technologies such as quadrature amplitude modulation, delta-sigma ADC and DAC use CIC filters. CIC filters are best to design anti-aliasing filters prior to decimation. CIC filters provide BIBO stability, are linear phase, and have a finite length impulse response (Table 2).

Frequency characteristics of CIC filers are defined in Eq. 3, as mentioned in [4].

$$H_{\text{CIC}}\left(e^{j2\pi f}\right) = e^{-j2\pi f(D-1)/2}.\frac{\sin\left(\frac{2\pi f D}{2}\right)}{\sin\left(\frac{2\pi f}{2}\right)} \tag{3}$$

CIC filters possess low pass frequency characteristics.

3.1 Design Example

The Hogenauer decimation filter design [3] is simulated using SIMULINK. A discrete-time multi-rate CIC decimation filter having decimation factor $D = 8$, differential delay $D = 8$ and having three sections is designed to work with 48 kHz sampling frequency. Total implementation cost consists of total of six adders and six delay elements. The above-implemented filter has the following transfer function:

$$H(z) = \frac{Y(z)}{X(z)} = \frac{1}{8}\left[\frac{1 - z^{-8}}{1 - z^{-1}}\right]^3 \tag{4}$$

Figure 2 shows the frequency response of the designed filter.

4 Approach of CSD Number System

A CSD number is a vector of digits. These digits involve $\{1, 0, \overline{1}\}$. Where two digits 1 and 0 represent standard binary and $\overline{1}$ represents -1 in decimal format. We can

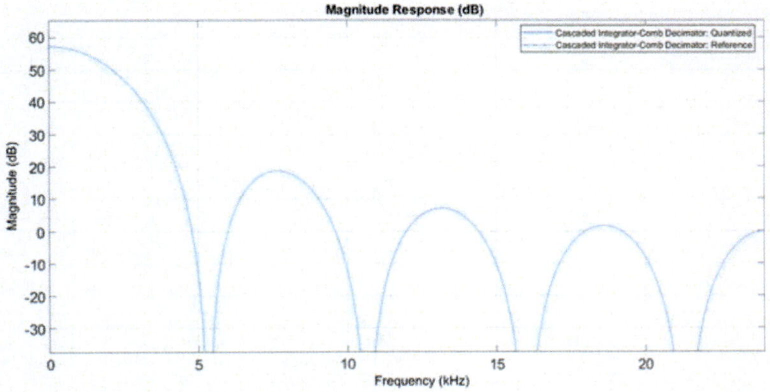

Fig. 2 Frequency response (linear) of the designed CIC filter

use these digits to for avoiding the occurrence of consecutive 1's. To achieve this, we first need to convert the binary number to CSD number with the help of recursive algorithms. These algorithms works on the principle that if the numbers of 1's in a number is k having the value $2^k - 1$, then we can represent the number using only two 1's and $k - 1$ zeros. Other than this some higher radix CSD representations are also discussed in [7]. Higher radix CSD numbers can further reduce the complexity of the systems.

Now focusing on simple CSD representation, we have designed an adder circuit, which saves carry propagation and gives out result in single iteration. This can be achieved once we complete the conversion process from CSD numbers to encoded CSD formats. The requirement of encoding is compulsory to achieve optimized addition operation. Negative and positive encoding [8] is one such technique we use to convert a CSD number to an encoded CSD format. As a part of this encoding technique, an algorithm converts $\overline{1}$, 0 and 1 to $\langle 10 \rangle$, $\langle 00 \rangle$ and $\langle 01 \rangle$, respectively. Rest one combination will be treated as do not care on hardware implementation.

To further elaborate how we can convert a binary number to CSD, we can look into the algorithm explained in [9].

The CSD adder consists of combinational hardware logic gates based on the truth table in Table 3 and use iterations to calculate sum and carry simultaneously. From Table 3, it is evident that the addition module has no dependency over previous carry. On hardware level modeling, we can get all these values in a single clock cycle ready to recalculate next operation. The adder is modeled in MATLAB also to inculcate the behavior in CIC filter (Fig. 3). As the filter consists of adder the step responses are similar and the robustness of the system increases. The real novelty in the design lies under the CSD adder modules. As we have introduced a direct decimal to encoded CSD number conversion algorithm that checks the stored decimal number in a variable and directly records a code accordingly Fig. 4.

This implementation improves path latency by using carry free addition, and the whole system can run at higher clock frequencies. Filter responses are simulated

Table 3 Truth table for encoded CSD number addition

X_B	X_A	Y_B	Y_A	S_B	S_A	C_B	C_A
0	0	0	0	0	0	0	0
0	0	0	1	0	1	0	0
0	0	1	0	1	0	0	0
0	1	0	0	0	1	0	0
0	1	0	1	0	0	0	1
0	1	1	0	0	0	0	0
1	0	0	0	1	0	0	0
1	0	0	1	0	0	0	0
1	0	1	0	0	0	1	0

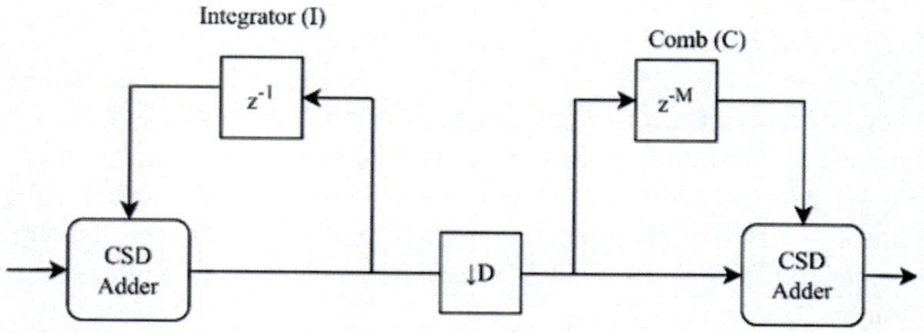

Fig. 3 CIC filter with CSD adder modules

using MATLAB, and the rest implementation is doing using VIVADO software. The software provides the interface to convert hardware description of the filter to a schematic.

5 Hardware Implementation Results

The hardware implementation of CSD adder is done using Verilog HDL. This will help us analyze the power and delay optimization of the system. The parameters of CIC filter that are incorporated on the FPGA implementation are shown in Table 2. Simulations are done using Xilinx Simulator (XSIM) on Intel Core i5 sixth generation processor. Simulation results are shown in Fig. 5.

The path delay (delay experienced by a sample from input to output of the circuit) is known as data-path delay. The comparison in Table 4 shown the timing improvement of the CSD number system implementation in CIC decimation filters.

The filter consumes 0.022 W with CSD adder at frequency 50 MHz and normal CIC filter consumes 0.028 at 100 MHz. Filter input has 16 bit width (B_{in}), while output having 18 bits (B_{out}) by using formula mentioned in [4] for full resolution

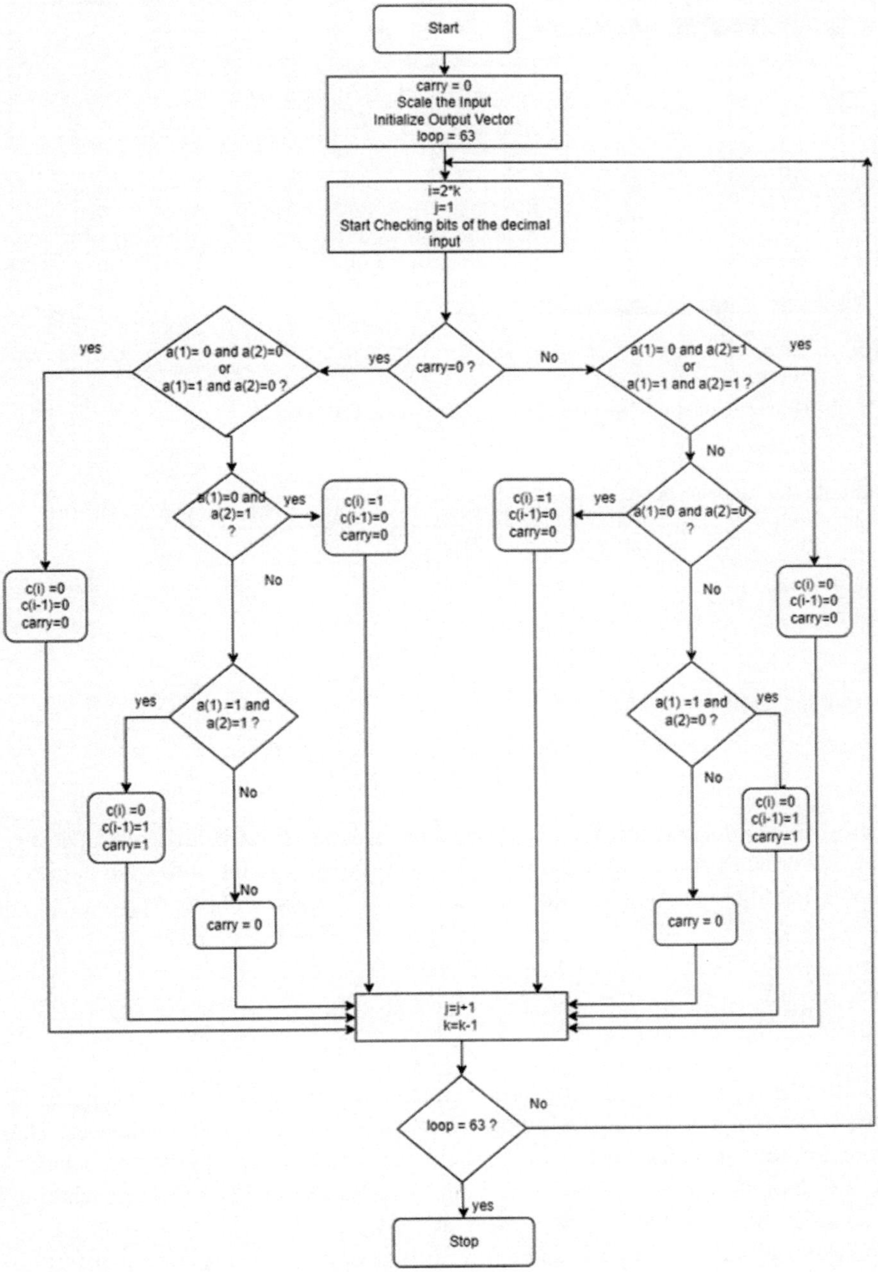

Fig. 4 Proposed algorithm for decimal to encoded CSD numbers

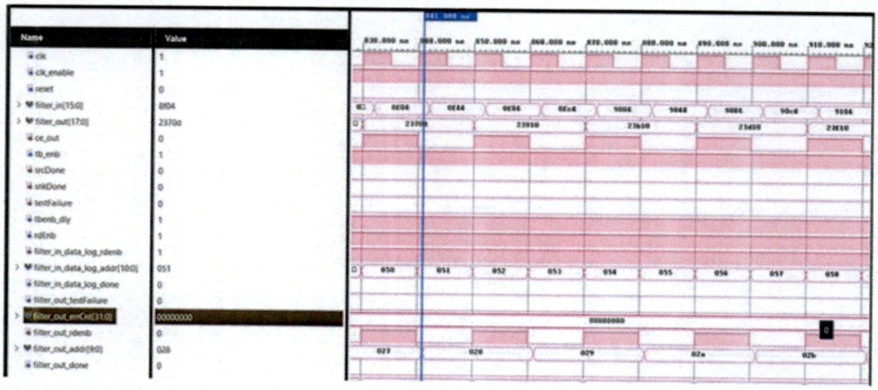

Fig. 5 Xilinx simulation results for 16 bit CIC filter with CSD adder

Table 4 Net data-path delay comparison

Delay	CIC (ns)	CIC with CSD adder (ns)
Data-path delay	5.203	5.111
Logic delay	3.123	3.244
Route delay	2.080	1.867

output bit width.

$$B_{\text{out}} = \lceil B_{\text{in}} + N \log_2 DM \rceil \tag{5}$$

Here, for the implemented filter, D (conversion factor) is 2 and M (differential delay) is 1. The number of sections (N) are two. The post implementation utilization statistics shows that filter having CSD adder requires 802 LUTs and 108 Flip-flops.

6 Conclusion

A different approach for adding two samples in a filter is presented. Using CSD number system has their own advantages in making high performance devices. The design methodology used in this work require further analysis of this implementation yet results look promising. The algorithm as well as the CSD adder are working seamlessly both in hardware as well as in software implementation. With all the timing constraints met, there is an improvement in data-path delay. We can therefore conclude that the CSD arithmetic have caused some improvement in time domain while maintaining the integrity of the signal. CSD number while reducing switching activity also prevent carry to propagate, thus overall reducing computational effort of the system.

References

1. Crochiere R, Rabiner L (1975) Optimum FIR digital filter implementations for decimation, interpolation, and narrow-band filtering. IEEE Trans Acoust Speech Signal Process 23(5):444–456
2. Goodman D, Carey M (1977) Nine digital filters for decimation and interpolation. IEEE Trans Acoust Speech Signal Process 25(2):121–126
3. Peled A, Liu B (1974) A new hardware realization of digital filters. IEEE Trans Acoust Speech Signal Process 22(6):456–462
4. Hogenauer E (1981) An economical class of digital filters for decimation and interpolation. IEEE Trans Acoust Speech Signal Process 29(2):155–162
5. Jing Q, Li Y, Tong J (2019) Performance analysis of resample signal processing digital filters on FPGA. EURASIP J Wirel Commun Netw 31:1–9
6. Aggarwal S, Meher PK (2022) Enhanced sharpening of CIC decimation filters, implementation and applications. Circuits, Syst Signal Process:1–23
7. Coleman JO (2001, Aug) Express coefficients in 13-ary, radix-4 CSD to create computationally efficient multiplierless FIR filters. In: Proceedings European conference on circuit theory and design
8. Parhami B (1988) Carry-free addition of recoded binary signed-digit numbers. IEEE Trans Comput 37(11):1470–1476
9. Hewlitt RM, Swartzlantler ES (2000, Oct) Canonical signed digit representation for FIR digital filters. In: 2000 IEEE workshop on signal processing systems, SiPS 2000. Design and implementation (Cat. No. 00TH8528). IEEE, pp 416–426

Study of Security for Edge Detection Based Image Steganography

Nidhi Jani, Dhaval Vasava, Priyanshi Shah, Debabrata Swain, and Amitava Choudhury

Abstract Due to the increase in the speed of the computer networks, the advantage in information communication has also increased and thus, the importance of information security cannot be overstated. The method of steganography has the purpose of making the communication hidden by wrapping the data into some other form. Many steganography formats of files are available, but still images in digital form are considered most famous due to their internet frequency. There are plenty of algorithms for data hiding but might have a compromise in quality of image. In this paper, a new technique is proposed for performing image steganography by utilizing edge detection techniques for grayscale images. In this proposed method, edges are detected by converting the image into grayscale and then, text is embedded into the digital image. Different methods like Canny, Sobel, Prewitt and Laplacian are applied here for having better secrecy and also for enhancing the stego image as well for getting correct data that was embedding.

Keywords Steganography · Edge detection · Canny · Prewitt · Sobel · Laplacian · LSB

1 Introduction

A transaction of information has a very important part in everyone's day-to-day life and so it is utmost important to secure the information from getting into an unauthorized person's hand. Steganography method is a way through which one can hide information in order to prevent detection of messages. The motive of steganography closely resembles cryptography. Cryptography does the encryption of a message while in steganography one can hide information behind any image. So in steganography, we can say that the communication happening between any two individuals will not be visible and understandable simply [1]. The process of using this method through which the message is not visible will not be dubious on the recipient's part

N. Jani · D. Vasava · P. Shah · D. Swain (✉) · A. Choudhury
Pandit Deendayal Energy University, Gandhinagar, Gujarat, India
e-mail: debabrata.swain7@yahoo.com

K. A. Reddy et al. (eds.), *Proceedings of Fourth International Conference on Computer and Communication Technologies*, Lecture Notes in Networks and Systems 606,
https://doi.org/10.1007/978-981-19-8563-8_11

[2]. Steganography use of image, video, audio or text files is done for information embedding.

Steganography methods are mainly of five different types depending on the cover object, which can be text, image, video, audio and network. So of all the methods most frequently used are image steganography in which an image is used as a cover object for embedding the secret information. So in image steganography for various images, there will be several file formats and mostly all of them are there for particular application.

In the image steganography, lossy and lossless are the compression types [3]. These two types are not similar but both of them will be helpful in saving storage. In lossy type, smaller files are created and image data that are excess are discarded from the original image. In lossless technique, information is hidden in important areas. So more preferred formats of image are lossless one for image steganography [4].

Steganography on the whole image can sometimes distort the image to a level at which the modifications in the image are perceptible to a human eye. Also, several techniques have been evolved over time through which an adversary can easily extract the concealed message from the image. Much research work has been carried out to overcome this limitation with image steganography, and it involves embedding the confidential data in some specific regions of an image called region of interest (ROI). One such region is the edges present in an image. Any changes made in the smooth area of the image is easily detected by the human vision, but when the same changes are done on the edges of an image, there is a high probability of going unnoticed or undetected [1].

An image's basic characteristic is that the image edge is a combination of values, which are gray, having huge pixel change in the image [5]. The initial step of detecting an edge is processing the image, and once the detection is done, the results will have influence on analysis of image and direct recognition [6]. Thus, detecting edges has important significance.

In this paper, the contribution of the author includes collection of information from various sources and designing the work. The contribution also consists of analyzing the performance of the proposed system and drafting the article regarding the same.

This section includes an introduction about steganography, why it is used, how it is used and why image steganography is used despite the different steganography methods available and also an introduction about edge detection methods. Section 2 consists of a literature survey about various edge detection methods. In Sect. 3, an algorithm developed for the proposed system is mentioned. In Sect. 4, performance analysis about the proposed work is done, and lastly, conclusion and future work is included in Sect. 5.

2 Literature Survey

Simrat et al. mentioned in their paper proposed a scheme by utilizing steganography with the method 2k correction and also method for edge detection. They proved their technique to be a better one and that can be accomplished by embedding large amounts of data in the edges. The reason for this was because the possibility of detecting the hidden data by the human eye is quite difficult [6].

Nitin et al. mentioned in their paper about a new method proposed which says that for embedding the data into the grayscale image, edges were utilized. They used a method of Canny edge detector for detection of edge position from the image, and then, they used LSB method, which is basically an insertion method, which will be used for embedding data in the image [7].

Shahzad et al. proposed a scheme that has better security for image steganography. The enhanced method has a secret key built on LSB method random. Also, they have advantage in edge detection that gives more capacity for data embedding. They also mentioned a high PSNR value that gives no difference that can be noticed between stego and cover image [8].

Saurabh et al. suggested an approach that can hide the message in an effective manner. The scheme mentioned method by change in some pixel value of color. The pixel that is selected by the author that does not represent color value instead of that characters are presented. The resultant image will look similar to the original image [9].

Setiadi et al. proposed a new steganographic method based on edge detection that uses a combination of Canny and Sobel edge detections. It then uses LSB substitution technique to hide the secret message. The scheme appends a special character at the end of the message to indicate the end of the message while extracting the hidden message. This means that the hidden message cannot deliver strings with special characters, thus limiting it to be just alphanumeric. Also, as edge detection is performed on the entire image, there are chances that we may not get the exact embedded message because the changes in the LSB of the original image may impact the edge detection process at the receiver end. Thus, the integrity of the message is not preserved in some cases [10].

There are various edge detection methods already available. Here, in this paper, we have used four different edge detection methods and they are Canny edge detection, Prewitt edge detection, Laplacian edge detection and Sobel edge detection.

3 Proposed Methodology

The current mentioned scheme uses edge detection techniques and least significant bit (LSB) technique to conceal the secret message in the cover object. The cover image is first pre-processed by converting the RGB image to grayscale image, and then, the last bit of every 8-bit pixel value is set to zero. For edge detection, the image is passed through five stages namely image noise reduction phase, gradient calculation phase (where we apply different kernels like Sobel, Prewitt and Laplacian), non-maximum suppression phase, double thresholding phase and edge tracking by hysteresis phase. For double thresholding, the following conditions were used for classifying the edges as strong and weak edges

$$(x, y) = \text{strong edge}, \ \text{if } I(x, y) > 200 \tag{1}$$

$$(x, y) = \text{heterogenous edge}, \ \text{if } 50 \leq I(x, y) \leq 200 \tag{2}$$

$$(x, y) = \text{weak edge}, \ \text{if } I(x, y) < 50 \tag{3}$$

where $I(x, y)$ indicates the magnitude of the pixel at position (x, y).

The heterogeneous edge is then classified as either strong or weak edge, depending on the connectivity with the strong edge in the edge tracking by hysteresis phase. After this phase, we will get an edge-detected image of the original image. We store the secret data in those detected edges in the original image. The image formed after this stage, will be our stego image. The initial steps (i.e., Step 1, 2, 3 and 4) are same for both the encoding and the decoding phase of the proposed method as mentioned below:

Step 1: Convert the cover image into grayscale image
Step 2: Change the rightmost (LSB) bit of every pixel to zero.
Step 3: Perform edge detection on that image using the below mentioned steps

(a) Apply Gaussian blur to smooth the image
(b) Detect edge direction and intensity by computing the gradient of the smoothen image.
(c) Carry out non-maximum suppression on the modified image to thin out the edges
(d) Implement double threshold detection on non-maximized suppressed image.
(e) Perform edge tracking by hysteresis and an edge-detected image will be generated.

Step 4: From the edge detection image in an array.

Encoding Phase:

Step 5: Convert the secret message from ASCII to binary format.
Step 6: Compute the length of the message in 8-bit binary format and prepend it to the binary format of the secret message.
Step 7: Perform LSB substitution of secret data in the original image at the stored edge positions.

Decoding Phase:

Step 5: With the help of stored edge positions, extract the first 8-bits of the message, which will determine the length of the message.
Step 6: Then continue extracting the message bit by bit until the length of the message.
Step 7: Convert the extracted message from binary format to ASCII format.

4 Results and Performance Analysis

Image steganography using edge detection techniques like Canny, Prewitt, Laplacian and Sobel were used to detect edges, and further, the secret message was concealed in those edges. The results obtained by using the Sobel edge detection method are shown in Figs. 1 and 2, and the concealed image obtained as a result of the proposed method is shown along with the original image.

In order to perform comparative analysis of the different techniques, the following performance metrics were considered: mean squared error (MSE), root mean square error (RMSE), peak signal-to-noise ratio (PSNR), embedding capacity (EC), structural similarity index (SSMI) and image quality index (Q) as discussed in [11–13].

For "lena.png" and "lena.jpeg" image, the measurements obtained from the proposed algorithm are shown in Tables 1 and 2, respectively.

As seen from Table 1, Prewitt and Sobel methods perform almost similarly for a PNG image. The Laplacian method achieves better results than other methods in

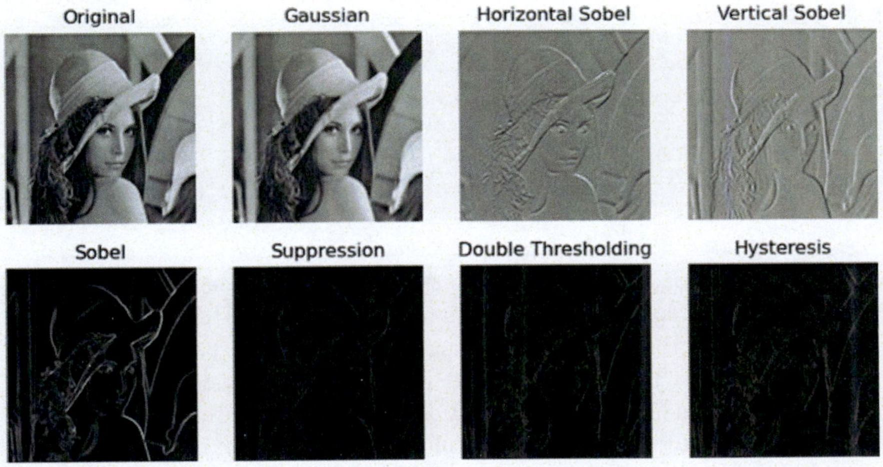

Fig. 1 Images obtained after each phase of the Sobel edge detection method **a** Original (cover) image. Image obtained after **b** applying Gaussian blur, **c** applying horizontal Sobel filter, **d** applying vertical Sobel filter, **e** calculating the magnitude, **f** non-maximum suppression phase, **g** double thresholding phase and **h** edge tracking by hysteresis phase

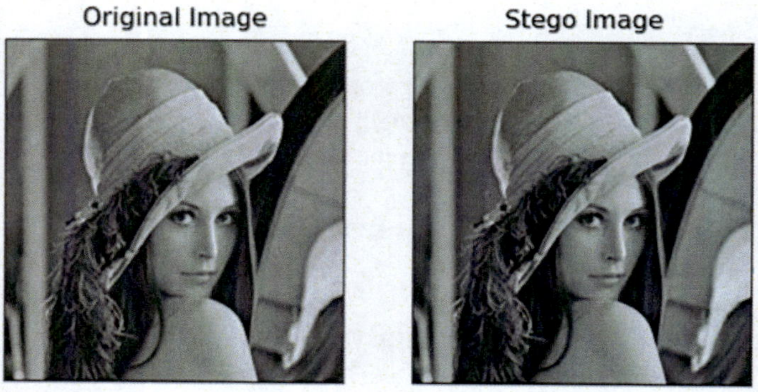

Fig. 2 **a** Cover image **b** stego image

Table 1 Evaluation parameter's value for a cover image of PNG format

PNG	Canny	Prewitt	Sobel	Laplacian
MSE	3.319×10^{-4}	3.433×10^{-4}	3.433×10^{-4}	3.052×10^{-4}
RMSE	1.823×10^{-2}	1.853×10^{-2}	1.853×10^{-2}	1.747×10^{-2}
PSNR	82.921	82.774	82.774	83.285
EC	1.819×10^{-1}	2.721×10^{-2}	3.604×10^{-2}	2.247×10^{-3}
Q	0.9999999176978239	0.9999999148576904	0.9999999148588887	0.9999999243181324
SSMI	0.9999999188751874	0.9999999160756837	0.9999999160768654	0.9999999254007945

Table 2 Evaluation parameter's value for a cover image of JPEG format

JPEG	Canny	Prewitt	Sobel	Laplacian
MSE	3.378×10^{-3}	4.4×10^{-3}	3.911×10^{-3}	4.089×10^{-3}
RMSE	5.812×10^{-2}	6.633×10^{-2}	6.254×10^{-2}	6.394×10^{-2}
PSNR	72.844	71.696	72.208	72.015
EC	2.388×10^{-1}	7.036×10^{-2}	8.156×10^{-2}	2.013×10^{-2}
Q	0.9999992501429389	0.9999990233526488	0.9999991317968545	0.9999990922817996
SSMI	0.9999992597597294	0.9999990358777614	0.9999991429313567	0.9999991039230585

terms of MSE, RMSE and PSNR but at the cost of EC. The Canny method has the highest embedding capacity among the four methods, and values for other parameters are close enough. For JPEG images, the Canny method performs better than others as evident from Table 2. Thus, we can say that the Canny method is ideal for performing edge detection based image steganography. As this scheme focuses more on the area of secrecy, the proposed scheme achieves much better results but at the cost of better payload capacity and is also practically feasible.

5 Conclusion and Future Work

Though there are several steganography methods available, in this paper, we have discussed image steganography, which is the most prominent method for information hiding. The main file formats of image have several ways of hiding information with several weak as well as strong points. This paper consists of methods like Canny, Sobel, Prewitt and Laplacian for edge detection. This helps one to understand the methods in a better way with their capability as well as robustness. Canny edge detection is considered to be the best among the all four edge detection methods. For different image formats and images, Canny edge detection works best because it is time efficient as well as simple to implement. Along with that edge detection using Canny edge detection method is less noisy when compared with other methods. Thus, it depends on the user as well as the application on which algorithm to use. In this paper, we have converted the image into grayscale for better edge detection. So instead of this RGB channel can be used through which embedding capacity can be increased but while doing the same take care of mean square error might be increased. Also, the use of key can be done for encoding and decoding for more security. When normal strings are used as keys high security will not be provided. Use of cryptographic functions in order to implement keys will give us more security but at the same time embedding capacity might get compromised. Therefore, it is important to check all the parameters while performing the task.

References

1. Luo W, Huang F, Huang J (2009) A more secure steganography based on adaptive pixel-value differencing scheme. Springer Science Business Media, LLC
2. Johnson NF, Jajodia S (1998) Exploring steganography: seeing and unseen, 0018-9162/98/$10.00 © 1998 IEEE
3. Moerland T. Steganography and steganalysis. Leiden Institute of Advanced Computing Science
4. Lal M, Singh J (2008) A novel approach for message security using steganography. In: 3rd International conference of advance computing and communication technologies, 08-09 Nov, 2008, APIIT, Panipat, India
5. Xiaofeng Z, Yu Z, Ran Z (2011) Image edge detection method of combining wavelet lift with canny operator. Proc Eng 15:1335–1339
6. Kaur SP, Singh S. A new image steganography based on 2k correction method and Canny edge detection. Int J Comput Bus Res. ISSN: 2229-6166
7. Jain N, Meshram S, Dubey S (2012) Image steganography using LSB and edge-detection technique. Int J Soft Comput Eng (IJSCE) 2(3). ISSN: 2231-2307, July 2012
8. Alam S, Kumar V, Siddiqui WA, Ahmad M (2014) Key dependent image steganography using edge detection. In: Fourth international conference on advanced computing & communication technologies
9. Singh S, Agarwal G (2010) Use of image to secure text message with the help of LSB replacement. Int J Appl Eng Res 1
10. Setiadi DRIM, Jumanto J (2018) An enhanced LSB-image steganography using the hybrid canny-Sobel edge detection. Cybern Inf Technol 18(2):74–88. https://doi.org/10.2478/cait-2018-0029

11. Pradhan A, Sahu AK, Swain G, Sekhar KR (2016) Performance evaluation parameters of image steganography techniques. In: 2016 International conference on research advances in integrated navigation systems (RAINS), pp 1–8
12. Asamoah D, Oppong E, Oppong S, Danso J (2018) Measuring the performance of image contrast enhancement technique. Int J Comp Appl 181:6–13. https://doi.org/10.5120/ijca2018917899
13. Gaurav K, Ghanekar U (2018) Image steganography based on Canny edge detection, dilation operator and hybrid coding. J Inf Secur Appl 41:41–51. ISSN 2214-2126

Fake Face Image Classification by Blending the Scalable Convolution Network and Hierarchical Vision Transformer

Sudarshana Kerenalli, Vamsidhar Yendapalli, and C. Mylarareddy

Abstract A face has been used as a primary and unique attribute to authenticate individual users in emerging security approaches. Cybercriminals use the double-edged sword "image processing" capabilities to deceive innocent users. The underlying technology is based on advanced machine learning and deep learning algorithms. The intentions of cyber criminals range from simple mimicking or trolling to creating violent situations in society. Hence, it is necessary to resolve such problems by identifying the fake face images generated by expert humans or artificial intelligent algorithms. Machine learning and artificial neural networks are used to resolve the issue. In this work, we have designed an approach for detecting deep learning-generated fake face images by combining the capabilities of the scalable convolutional neural networks (CNN) "EfficientNet" and hierarchical vision transformer (ViT) "shifted window transformer". The proposed method accurately classifies the fake face images with a 98.04% accuracy and a validation loss of 0.1656 on the 140 k_real_fake_faces image dataset.

Keywords Fake face images · Generative adversarial networks (GAN) · Vision transformers (ViT) · EfficientNet · Classification · Convolutional neural network (CNN)

1 Introduction

Computers can distinguish humans based on their distinct physical or biometric characteristics. Such systems have the potential to be employed in a variety of near real-world applications, including security and telemedicine systems. Preventing illegal

S. Kerenalli (✉) · V. Yendapalli · C. Mylarareddy
Department of CSE, SoT, GITAM University, Bengaluru, Karnataka 562163, India
e-mail: ksudarsh@gitam.in

V. Yendapalli
e-mail: vyendapa@gitam.edu

C. Mylarareddy
e-mail: mylarareddyc@gmail.com

K. A. Reddy et al. (eds.), *Proceedings of Fourth International Conference on Computer and Communication Technologies*, Lecture Notes in Networks and Systems 606,
https://doi.org/10.1007/978-981-19-8563-8_12

access to physical infrastructures such as networks, databases, and other facilities is a problem of the context to be resolved urgently. Face recognition systems allow a computer to recognize a person based on facial traits. Face recognition compares the face's structure, geometry, and proportions [1]. Face recognition systems are often used for facial authenticity rather than identification because it is simple to swap someone's face, obscure the masking of the particular face, and modify the environment in digital photos for conveniently executing cybercrime. Face pictures are often published in databases on social media sites like Facebook, YouTube, and Twitter without the subject's permission. In exchange for profits, governments and businesses get access to data from these companies. Another issue is that the leak of stored biometric data is irreparable, and the face image databases are vulnerable in and of themselves. Fake news and the spread of disinformation on social networking sites can weaken faith in digital media. Manipulated photos and films have been used to incite conflicts.

Many nations view the technology that allows fake images, videos, or audio generation as a danger [2]. The various approaches for identifying phony image manipulation come along with their own merits and limitations. These methods include one-class learning, ensemble learning, multi-modal learning, identity-aware detection, and securing the sources by tracking. The proposed work attempted to build an ensembled model that blends the salient features of an EfficientNet [3] and shifted window vision transformer [4] models to classify fake face images. The proposed method achieved a 98.04% accuracy with a validation loss of 0.1656 on the 140 k_real_fake_faces image dataset.

The article is organized as follows: In Sect. 2, the literature survey is discussed. Section 3 discusses the architecture of the proposed method. In Sect. 4, the experimental results are presented and discussed. Section 5 concludes the given work and describes the future avenues.

2 Literature Review

This section presents brief information on face manipulation and generation techniques along with the detection strategies. Face modification methods have substantially advanced over time. Several technologies, including deepfakes and face morphs, have been presented in the literature to achieve realistic face manipulations. Manipulated photos and videos might be nearly indistinguishable from genuine content to the untrained eye.

2.1 Fake Image Generation

The commonly used algorithms for fake image generation range from simple cuts and pastes to generating visually appealing images using deep learning techniques.

Primary deep learning methods include autoencoders (AE) and generative adversarial networks (GAN).

Encoder-decoder [5] pairs are used in the AE-based generator. They are taught how to dismantle and recreate one of the two faces that will be swapped. The decoder is then switched to reassemble the target picture. DeepFaceLab, DFaker1, and DeepFaketf2 are among the examples. GANs [6] use two neural networks, one for generation and one for discrimination for image generation. The generator network draws the random noise to generate the sample to deceive the discriminator. The discriminator learns to differentiate the generated sample against the actual sample. The discriminator score is iteratively fed into the generator network to learn a better approximation of the real sample. The iteration stopped when the discriminator could not differentiate between the real and generated images. WGAN, StarGAN, DiscoGAN, and StyleGAN-V2 are some examples.

A style transfer-inspired swapping generator architecture for generative adversarial networks was developed by Karras et al. [7]. The new scheme offers intuitive, scale-specific artificial feature generation by automatic learning and unsupervised separation of high-level features—such as posture and identity. The resulting pictures also generate random variations (e.g., freckles, hair). This generator scheme has state-of-the-art capability in terms of distribution quality, resulting in superior interpolation features, and better disentangles the varying latent characteristics. In the current work, a data set authored by XHLULU [8] is used for training and evaluation purposes. This dataset contains a set of high-quality human faces with diverse styles.

2.2 *Fake Image Detection Techniques*

The various approaches for identifying fake image manipulation come with their merits and limitations. Abdulreda and Obaid [9] studied the earlier work to examine deepfakes, principles, and counterfeiting strategies. ImageNet was used by Touvron et al. [10] to train a robust convolution-free transformer. On ImageNet, the vision transformer obtained top-1 accuracy of 83.1% on single-crop evaluation. More importantly, they offered a transformer-specific teacher-student strategy. It is based on a distillation token, which guarantees that the student pays attention and learns from the teacher. They showed how beneficial this token-based distillation might be, especially when using a convent as a teacher. As a result, they outperformed the ConvNets for ImageNet and when transferring to other tasks.

The fused facial region feature descriptor (FFRFD) is presented as an alternative to mining more subtle and broad characteristics of deepfakes. It is a discriminative feature description vector for practical and quick detection. DeepFake faces contain more minor feature points in facial areas than actual faces, according to their study. To improve the generalizability, FFRFD takes advantage of such crucial insights. FFRFD to be trained with a random forest classifier to accomplish efficient detection. Tests on six large-scale Deepfake datasets show that this lightweight strategy successfully has an AUC of 0.706, outperforming most state-of-the-art approaches [11].

To detect deepfake videos, Kolagati et al. [12] created a deep hybrid neural network model. They gathered data about numerous face features from the videos using facial landmark recognition. This information is fed into a multilayer perceptron, which is used to understand the distinctions between actual and deepfake videos. They utilize a convolutional neural network to extract features and train on the videos simultaneously. These two models are used to create a multi-input deepfake detector. The model is trained using a subset of the Deepfake Detection Challenge Dataset and the Dessa Dataset. The suggested model produces good classification results with an accuracy of 84%, an AUC score of 0.87, accuracy of 84%, and an AUC score of 0.87.

A transfer learning based on the ResNet50v2 architecture was described for detecting manipulated images, especially spliced images. The image splicing approach was employed with the pre-trained weights of a YOLO CNN model to see whether they generated the photos had been intentionally tampered with. Vision transformer-based models and self-attention processes have piqued researcher curiosity to acquire visual representation successfully. Convolution layer injection and the construction of local or hierarchical structures are among them. Several solutions add substantial architectural complexity [13].

A self-attention mechanism is implemented into CNNs to mimic long-range interactions. It was pretty tricky due to the issue of convolutional kernel locality. Recent research has discovered that a self-attention-only structure with no convolution works effectively [14]. The original ViT beats convolutional networks, which need hundreds of millions of images to train; however, such a data demand is not always practical. Data-efficient ViT (DeiT) solves this problem by combining neural network-based teacher distillation. Despite its promise, it adds to the supervised training complexity, and current reported performance on data-efficient benchmarks still falls short of convolutional networks [10].

3 Proposed Method

The detailed workflow for the EfficientNet hierarchical vision transformer approach is illustrated in Fig. 1. The given set of input images is preprocessed using various image augmentations. The preprocessed image set is used to extract the features from the EfficientNet, and then the hierarchical vision transformer block classifies the images as real or fake.

3.1 Preprocessing

Image augmentation operations are carried out on the dataset to generate more data to train the model. They include:

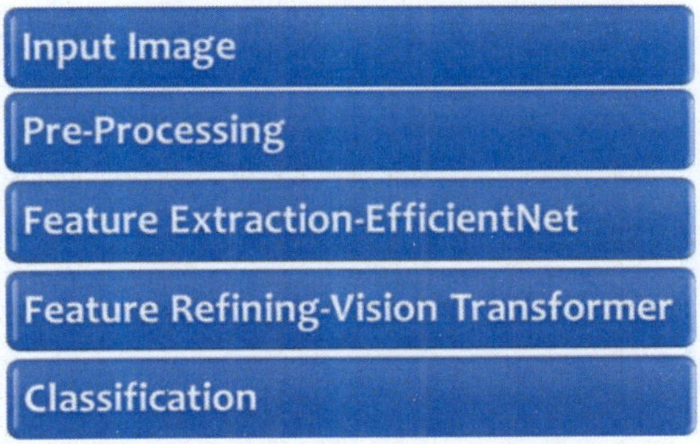

Fig. 1 Workflow for the EfficientNet hierarchical vision transformer

1. **Resize**: Each color image is resized into 384 × 384 pixels.
2. **Random Flip**: Each color image is flipped randomly via the horizontal and vertical axis.
3. **Random Zoom**: Each color image is randomly zoomed 20% over the horizontal axis and 30% over the vertical axis.
4. **Random Rotation**: Each color image is randomly rotated 20% along the horizontal axis and 30% over the vertical axis with reflection.

3.2 *EfficientNet*

Randomly selecting the network depth, width, or image resolution is a common tactic to scale a CNN network for training and validation. This method involves tuning the network manually and frequently produces sub-optimal results. EfficientNet [3] is a systematic compound scaling method. This method appropriately resizes the network width, depth, and resolution correctly. A compound scaling coefficient p is a hyper-parameter used to scale coefficients of the network width w_1, depth d_1, and image resolution r_1 for available computational resources.

$$\text{Depth} = d = d_1^p \tag{1}$$

$$\text{Width} = w = w_1^p \tag{2}$$

$$\text{Resolution} = r = r_1^p \tag{3}$$

The values w_1, d_1, r_1, are scaled using a small grid search over the network parameters with a restriction of $w_1^1 \times d_1^2 \times r_1^2 \approx 2$. This restriction imposes that, for any new chosen value of “p”, the sum of all the floating-point operations should increase “2p” times approximately.

Stage i	Operator $\hat{\mathcal{F}}_i$	Resolution $\hat{H}_i \times \hat{W}_i$	#Channels $\hat{C}_i$	#Layers $\hat{L}_i$
1	Conv3x3	224 × 224	32	1
2	MBConv1, k3x3	112 × 112	16	1
3	MBConv6, k3x3	112 × 112	24	2
4	MBConv6, k5x5	56 × 56	40	2
5	MBConv6, k3x3	28 × 28	80	3
6	MBConv6, k5x5	14 × 14	112	3
7	MBConv6, k5x5	14 × 14	192	4
8	MBConv6, k3x3	7 × 7	320	1
9	Conv1x1 & Pooling & FC	7 × 7	1280	1

Fig. 2 EfficientNetB0

Randomly selecting the network depth, width, or image resolution is a common tactic to scale a CNN network for training and validation. The baseline network of the EfficientNet dictates the model scaling effectiveness. It uses a neural architecture search to optimize accuracy and efficiency. In EfficientNet B0, an inverted bottleneck convolution network architecture similar to MobileNetV2 and MnasNet is used. This architecture is quite prominent than the MobileNetV2. It compensates FLOP budget. The architecture of EfficientNetB0 is as shown in Fig. 2.

3.3 SWIN Transformer

A shifted window transformer [4] is an efficient general-purpose backbone network proposed for computer vision tasks. It is shown in Fig. 3. The transformer representations are computed hierarchically by moving the windows. It limits the self-attention calculation to non-overlapping local windows and keeps the connections across the windows. The famous transformer structure and its variant for image classifications employ global self-attention leading to quadratic complexity in identifying the relationship between a token and all other tokens. The self-attention is computed inside local windows for efficient modeling, but global connections are lost. A window shifting and partitioning approach switches between partitioning configurations among the sequential SWIN blocks to retain the global connections. Transformer blocks for cross-window interactions are included, as is the efficient computation of non-overlapping home windows. An input image is partitioned into a set of non-overlapping patches. They are evenly arranged. Each window contains $M \times M$ patches. The SWIN transformer block computes the global multi-headed self-attention over an $h \times w$ patches window image.

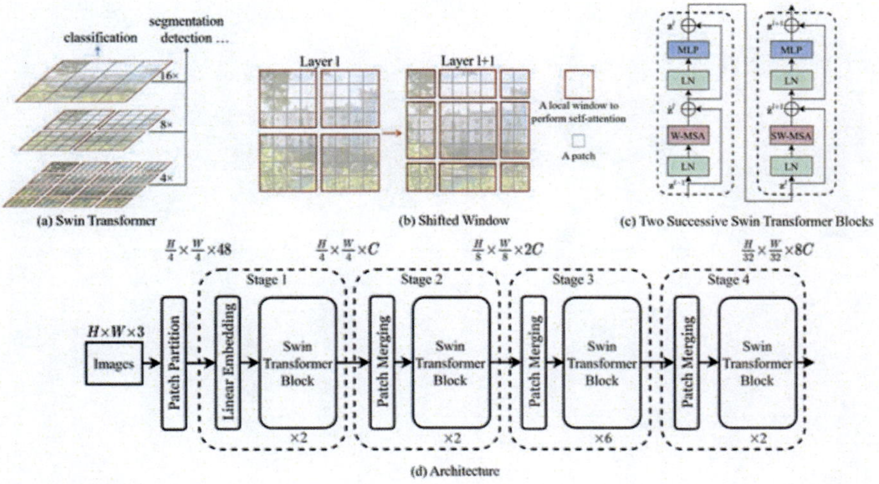

Fig. 3 SWIN transformer

4 Results and Discussions

The details of the experimental design, dataset, and results are presented in this section.

4.1 140 K Real and Fake Faces

The "140 K Real and Fake Faces" dataset consists of 70,000 StyleGAN2 generated counterfeit images and 70,000 real images collected from Flickr.com by Nvidia Incorp. This dataset consists of many high-quality face images with subjects of different sex, age, and even real-world fake faces [8]. The dataset is split into train, test, and validation subsets. The train set has 135,000 images, and test and validation groups have 5000 images each. An input dataset has an equal number of labeled images and belongs to two classes, "real and fake".

4.2 Experimental Set-Up

The proposed approach is executed on a Dell Precision 5820 Tower Workstation running at a 4.1 GHz Intel Xeon W-2225 Quad-Core processor having 64 GB installed RAM on Windows 11 Pro operating system. The algorithm is programmed using the Keras framework with python language. The model is trained over ten epochs on the "140 K Real and Fake Faces" dataset. Augmentation results are as shown in Fig. 4.

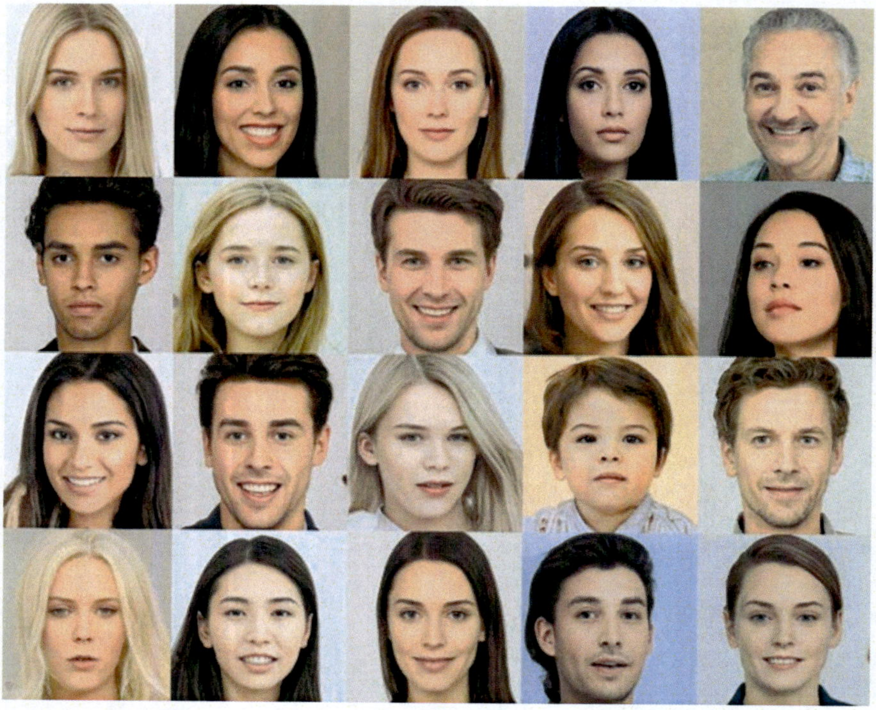

Fig. 4 Data augmentation

4.3 Discussions

We used validation accuracy to monitor the model's performance in the proposed method. Figure 6 presents the training and validation accuracy, and Fig. 5 presents the training and validation losses for ten epochs. The results show that the training and validation accuracy increase as the number of epochs increases. The training loss and validation loss decrease significantly in the initial epochs and gradually over the subsequent epochs. It is because of the model learns better during the initial epochs and improves over the successive epochs.

5 Conclusion

In this work, we combined the capabilities of the EfficientNet, the scalable convolutional neural networks (CNN), an algorithm for feature extraction, and the hierarchical SWIN vision transformer for classification. On the validation set, the proposed model obtains an accuracy of 98.04% and a validation loss of 0.1656. The above method is applied to process the real-time image data in our future work.

Fig. 5 Model loss

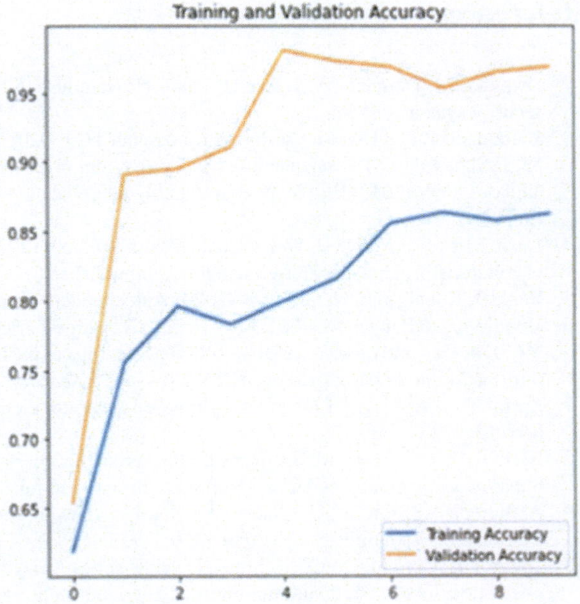

Fig. 6 Model accuracy

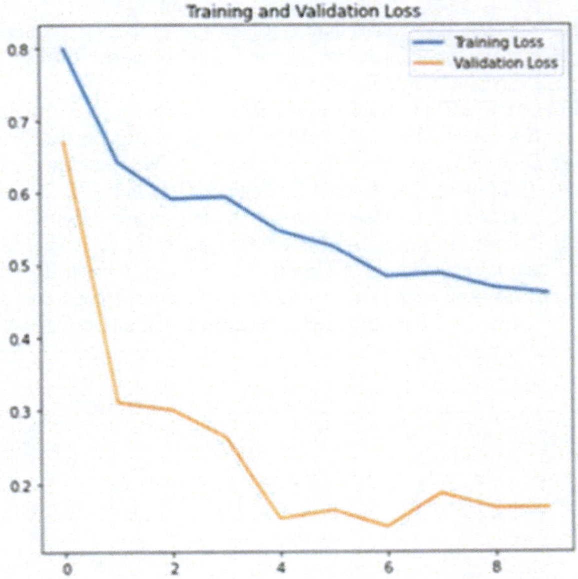

References

1. Owayjan M, Dergham A, Haber G, Fakih N, Hamoush A, Abdo E (2013) Face recognition security system
2. Kietzmann J, Lee LW, McCarthy IP, Kietzmann TC (2020) Trick or treat? Business Horizons 63. ISSN: 0007-6813, Sciencedirect Ltd. https://doi.org/10.1016/j.bushor.2019.11.006
3. Tan M, Le QV (2019) EfficientNet: rethinking model scaling for convolutional neural networks (in press)
4. Liu Z, Lin Y, Cao Y, Hu H, Wei Y, Zhang Z, Lin S, Guo B (2021) Swin transformer: hierarchical vision transformer using shifted windows, unpublished
5. Bank D, Koenigstein N, Giryes R (2020) Autoencoders (in press)
6. Goodfellow I, Pouget-Abadie J, Mirza M, Xu B, Warde-Farley D, Ozair S, Courville A, Bengio Y (2014) Generative adversarial nets. In: Proceedings of the international conference on neural information processing systems, NIPS 2014, pp 2672–2680
7. Karras T, Laine S, Aila T (2019) A style-based generator architecture for generative adversarial networks (in press)
8. XHLULU, 140k real and fake faces. Published on February 2020, accessed on 10 May 2022, https://www.kaggle.com/datasets/xhlulu/140k-real-and-fake-faces
9. Abdulreda AS, Obaid AJ (2022) A landscape view of deepfake techniques and detection methods. Int J Nonlinear Anal Appl 13:745–755
10. Touvron H, Cord M, Douze M, Massa F, Sablayrolles A, Jegou H (2020) Training data efficient image transformers & distillation through attention (in press)
11. Wang G, Jiang Q, Jin X, Cui X (2022) FFR FD: effective and fast detection of DeepFakes via feature point defects. Inf Sci 596:472–488
12. Kolagati S, Priyadharshini T, Rajam VMA (2022) Exposing deepfakes using a deep multilayer perceptron—convolutional neural network model. Int J Inf Manage Data Insights 2:100054. ISSN: 26670968. Elsevier BV
13. Qazi EH, Zia T, Almorjan A (2022) Deep learning-based digital image forgery detection system. Appl Sci 12:2851. ISSN: 2076-3417. https://doi.org/10.3390/app12062851
14. Dosovitskiy A, Beyer L, Kolesnikov A, Weissenborn D, Zhai X, Unterthiner T, Dehghani M, Minderer M, Heigold G, Gelly S, Uszkoreit J (2020) An image is worth 16 x 16 words: transformers for image recognition at scale, unpublished
15. Tolosana R, Rathgeb C, Vera-Rodriguez R, Busch C, Verdoliva L, Lyu S, Nguyen HH, Yamagishi J, Echizen I, Rot P, Grm K (2022) Future trends in digital face manipulation and detection. In: Advances in computer vision and pattern recognition, vol 2. ISSN: 21916594. Springer Science and Business Media Deutschland GmbH, pp 463–482. https://doi.org/10.1007/978-3-030-87664-7_21

Performance Analysis of Osteoarthritis from Knee Radiographs Using Convolutional Neural Networks

Sivaprasad Lebaka and D. G. Anand

Abstract Osteoarthritis (OA) is listed among the chronic disease which shows a lot of impact on the health-related issues among the people. The ancient or the earlier scoring method and the physical diagnosis processes would be requiring man's higher involvement and time. This article will involve in developing the automatic OA diagnosis by keeping the base of convolutional neural architecture (CNN) for implementing the rheumatologist to diagnose and the planning the treatments. Our article will be including the various CNN architecture like DenseNet121, VGG16, ResNet50 with InceptionV3 this may or may not have the argumentations of the data carried out from the RA diagnosis. By the end of the 50th epoch, InceptionV3 accuracies have reached to the 98.91% with no data augmentations with least errors of 1.65%; DenseNet121 reaches to the 96.57% (training set). InceptionV3 it reaches 96.7% to the validations that will indicate the InceptionV3 with lower variance.

Keywords Convolutional Neural Network (CNN) · RA · Inceptionv3 · VGG16 · ResNet50 · X-ray

1 Introduction

Most of the patients who is suffering from the OA will be classified as the different types of OA like knees, hips, with the spine osteoarthritis. OA can be easily diagnosed by doctors from physical testing with medical image of the OA patient that is accumulated in the hospitals. The accuracy of the OA in the patients is very time talking process. Many of the articles have been considered the automatic detection of the OA from the images that are dependent on the deep learning algorithms [1–3]. But, the medical image will use the health behaviors that will collect data from

S. Lebaka (✉)
Department of Electronics and Communication Engineering, Visvesvaraya Technological University, Belagavi, Karnataka, India
e-mail: sivaprasadlebaka@gmail.com

D. G. Anand
Rajiv Gandhi Institute of Technology, Bengaluru, India

K. A. Reddy et al. (eds.), *Proceedings of Fourth International Conference on Computer and Communication Technologies*, Lecture Notes in Networks and Systems 606,
https://doi.org/10.1007/978-981-19-8563-8_13

the statistical data that are easy for collecting and using the images of the medicine [4]. By taking into account the medical image parameters for predicting the occur of the various types of OA which will signifies the impacts on the pro-active with prevention of the medical care. Here, we will be using the deep neural network in detection of occurring the OA by taking the data got from the statistics to use the medical and health behavior data [5, 6]. The analysis of the component along the scale of the quantile transformers was used in generating the features by using the background of the patient through the record that is identifying the occurrences of the OA.

OA is led by the disability and considered as the great social costs of the aged citizens. According as the age is increased and the increase in the obese, the OA will be much widespread than the earlier years [7, 8]. As the time passes, there is a vast increase in the insight to originate pathogenesis in the pains OA. Preventing and diseases modifications were the areas to be targeted from the many research endeavor that indicate the huge potentials [9]. The prevalent and incapacitating disorders that are significant and rising in the health burden along the notable implication of individual affected, medical system, wider socioeconomic cost. A convolutional neural network (CNN) the part of the machine learning technique utilized the automatic OA [10, 11] detecting multilayer neural network structures. CNN application will be including the automatic OA complexity predications, detecting early and various stage of management of gout arthritis. Deep learning application used for the analyzing medical image. DenseNet has 121 layer CNN; CheXNet network trained among 10,000 X-ray images with ten different diseases. The Xception model and VGG based model used for image classifications. These models includes the performance analysis of the CNN architectures like ResNet50, DenseNet121, VGG16, and InceptionV3 for the RA classification purpose A correct predictions OA that is very essential step to effectively diagnose and preventing severes OA.

2 Literature Survey

2.1 Classification of the Knee Joints Using Pre-trained CNNs

Mora et al. [11] and Mandl et al. [12] have represented the collected some of the knee joint image that are spliced to train (~60%) tested (~25%) according to the KL grade. To classify the images of the knee joints, they also tried to get the features of the fully connected, the pooling, and the convolution layer of the VGG16, VGG-M-128, and BVLC CaffeNet. The binary classification along with the multi-classification, linear SVM is trained separately along the gained results. These results of the classifications were achieving the CNNs that were comparing the knee classifications of OA image used Wndchrm. Classification accuracy is computed which utilizes the convolutional (conv4, conv5), and (pool5) layers are high in comparison are connected layer feature.

There were minimum variation of the classification accuracies got from the features of VGG-M-128 net and BVLC reference CaffeNet by comparing to VGG16.

2.2 Classifications of Knee Joint from Fine-Tuned CNN Numbering

Antony et al. [13] present the multi-classes the classification has resulted in the fine-tuning of BVLC CaffeNet with the VGG-M-128 network. These authors eliminated the VGG16 networks while experimenting this because of the differentiation in the accuracies in the already trained CNN were too less; the fine-tuned VGG16 had greater computational expenses. These data is then divided to train (60%), validate (10%) and testing (30%) dataset used in fine-tune. For increasing, the total numbers of the datasets of these samples can be including the right to left-flipped images of the knees to the trained dataset. Network is fine-tuned to 20 epochs by utilizing the learning rate of 0.001 to transfer layer, boost it in novel introduced layer by the factor of 10. This performance of the fine-tuned BVLC CaffeNet is significantly better compared to the VGG-M-128.

3 Material and Methods

CNN architecture will have the hidden layers, input layers along with the output layers. Input image represents by input layer, with all the feature that were leant on the hidden layer, result gained from output layers. This architecture makes us utilize the multiple convolution layer, pooling layers, and activation layers. The convolution layer of CNN is one of the important properties in the extraction. These filters are utilized in finding the various properties in various level through applications of multiple filter along the various kernel size the input image.

a. **Image dataset**: The X-ray data is collected through ChanRe Rheumatology along with Immunology Center, Bengaluru, Karnataka, India. Data includes normal and RA affect image. Data contains 398 radiographs image that has 180 usual and 168 OA image. The 398-dataset divided to 278 images (89%) to train and 35 image (30%) to validate. Images were of the dimension of 256 * 256. Figure 1 has samples of normal and OA dataset.

b. **Data augmentation**: Deep learning requires the large datasets for producing the accurate productions on the testing phases. Many datasets for accurate predictions need the testing images. Data augmentation necessary steps to increase overall performance of the network, and it avoids overfitting, irrelevant pattern reorganization, and memorizing condition occurred during training phase of the network. All the algorithms trained with both original dataset as well as augmented dataset.

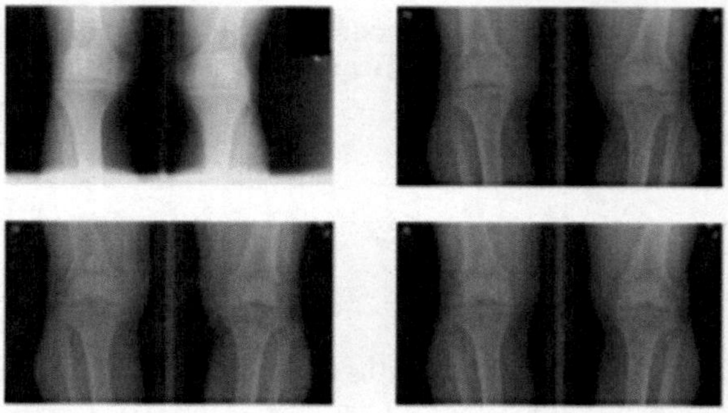

Fig. 1 OA images and normal images

Table 1 Description of data augmentation operations

Data augmentation operations	Values
Wide shift range	0.1
Height shift range	0.1
Shear range	0.01
Zoom range	0.15
Horizontal flip	True
Flip mode	Reflect
Rotation range	5

Initially, images are resized to 256 * 256 sizes. Data augmentation operations include width shift and height shift by 0.2, shear transformations by 0.02, zoom range by 0.17, and rotation by 7 degree. All the images are normalized to 256 * 256 sizes (Table 1).

c. **CNN architectures**: The work includes CNN architecture ResNet50, DenseNet121, VGG16, and InceptionV3. TensorFlow machine learning platform used to create different CNN architecture. The algorithms were given the training through the actual datasets along with the augmented sets. Image were resized to 256 * 256 sizes.

Binary cross-entropy is given by,

$$\mathrm{CE} = -\sum_{i=1}^{C'=2} t_i \log(f(s_i)) = -t_1 \log(f(s_1)) - (1 - t_1) \log(1 - f(s_1)) \quad (1)$$

where $f(s_i)$ represents sigmoid function given by,

$$f(s_i) = \frac{1}{1 + e^{-s_i}} \quad (2)$$

4 Results and Discussion

Various CNN architectures such as ResNet50, DenseNet121, and VGG16 along with InceptionV3 trained with normal RA X-ray image. Each of them was trained to nearly 60 epochs. Training was performed to both original and augmented dataset by resizing the images to 256 * 256.

4.1 Evaluation Metrics

CNN architectures like DenseNet121, VGG16, ResNet50, and InceptionV3 may or may not have data argumentation trained. The performance of the each model is evaluated through different parameters like accuracies, losses, sensitivity, specificities, recall, precisions that would F1 score. Table 2 explained various verities of the performances were analyzed formula. Here, the metrics TP,FP,FN and TN represents numerical values of the true negative, false negative, false positive, true positive. The performance metrics change over various CNN architectures is analyzed, and the values are shown in Table 3. Performance matrices values changes with different CNN architectures are shown in Fig. 2.

Accuracy plot indicates that InceptionV3 performance in train and validate set of augment and non-augment compares with various model. On the other hand, losses were plotted shows the variations the category crosses-entropies loss in various network. InceptionV3 results minimum categorical cross-entropy loss in augmented and non-augmented data were comparing through different models. This kind of conclusion is given as InceptionV3 performed better comparative to plot of F1-scores, precisions, recalls, specificities. By 50th epoch, the accuracies of InceptionV3 and DensNet121 reached 95% and above. InceptionV3 performs well on validation set and validation accuracy more compared to all other model. Network with good validation accuracy indicates network has low variance. Also, it is observed that wide gap between training and validation set in F1-scores, accuracies, precisions, specificities along with recalls in VGG16, ResNet50 (VGG16) indicates those networks

Table 2 Description of performance analysis formulas

Performance parameter	Formula
Accuracy	(TP + TN)/(TP + FN + TN + FP)
Loss	(FP + FN)/(TP + FN + TN + FP)
Sensitivity	(TP)/(TP + FN)
Specificity	(TN)/(TN + FP)
Precision	(TP)/(TP + FP)
Recall	(TP)/(TP + FN)
F1-score	2 *(Precision * Recall)/(Precision + Recall)

Table 3 Performance analysis of ResNet 50, DenseNet121, VGG16, and InceptionV3

ResNet	ResNet	VGG16	VGG16	DenseNet121	DenseNet121	Inception	Architecture
Yes	No	Yes	No	No	Yes	No	Augmentation
0.2029	0.1179	0.3986	0.2666	0.072	0.0778	0.0187	Loss
0.9348	0.9456	0.8307	0.9127	0.9728	0.9673	0.989	Accuracy
0.9337	0.9487	0.7908	0.9103	0.9701	0.9673	0.998	F1
0.943	0.9669	0.9587	0.8711	0.9903	0.9807	1	Precision
0.9256	0.9339	0.688	0.9598	0.9519	0.9581	0.9828	Recall
0.9390	0.9706	0.9598	0.8757	0.9882	0.9789	1	Specificity
34.6734	0.6663	2.5059	25.4664	9.1435	0.8949	0.0242	Val_loss
0.5264	0.6845	0.4218	0.5265	0.5267	0.6318	1	Val_accuracy
0	0.79	0.3528	0	0	0.4618	1	Val_F1
0	0.7	0.377	0	0	0.74	1	Val_precision
0	1	0.3336	0	0	0.3336	1	Val_recall
1	0.5	0.5	1	1	0.9	1	Val_specificity

non-general to InceptionV3 and DenseNet121. From recall plot, InceptionV3 recall value is 0.9825 on training set in case of non-augmented and even performs well validations. InceptionV3 values were indicated are better robust toward predicting the disease by comparing other network. Through specificity plot shows that, specificity of InceptionV3 and DenseNet performs well on validation set. InceptionV3 and DenseNet network validation specificity values are 0.9 for non-augmented and 1 augmented case, respectively. Validation specificity value indicates predication of non-disease cases. InceptionV3 and DenseNet both networks good in predication of non-disease cases. It is observed that the size of InceptionV3 has high VGG16 and ResNet50. ResNet, the inception layers receive o/p through second or the third layers though its false in DenseNet; therefore, the higher memory would be required are represented in DenseNet for prediction.

5 Conclusion

Our study was the comparative work of various CNN architectures of the RA disease diagnose. Various CNN architectures uses DenseNet121, VGG16, ResNet50 with InceptionV3. The network trained the original and data arguments dataset of RA detections. Finally, 50th epoch, InceptionV3 accuracies reach 96.1% zero data augmentation at lower error at 1.65%, DenseNet121 reached 96.57% (training set). InceptionV3 it reached 96.1% on validation set indicates InceptionV3 which is less variances by comparing. Accurately, the plots will be indicating the InceptionV3 performing the training and validations set of augmentation and non-augmentation data by comparing different models. Simultaneously, the loss plot will be showing

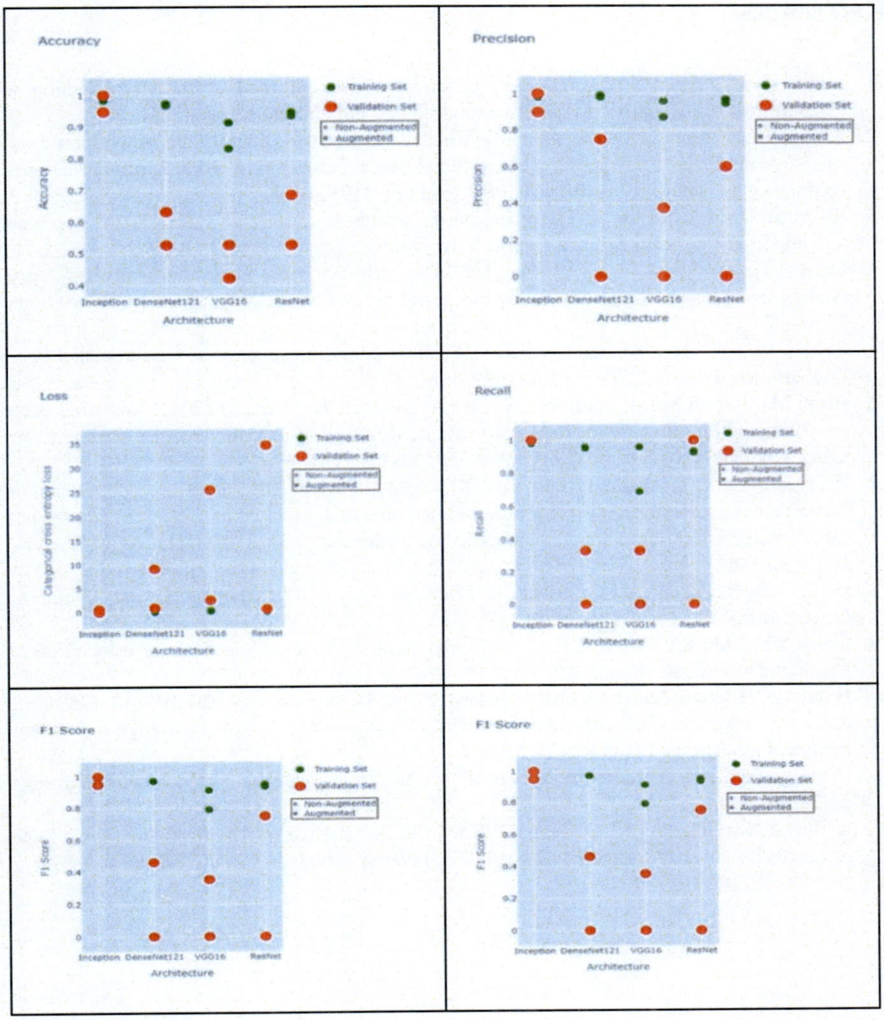

Fig. 2 Performance analysis of ResNet50, DenseNet121, VGG16, and InceptionV3

variations of the categorical cross-entropy's losses with the many networks. InceptionV3 results less categorical cross-entropy losses in augmentation and non-augmentation data comparing to different models. The conclusion is drawn that InceptionV3 performs well on the different plot of F1-scores, precisions, recalls, specificity.

References

1. Lim J, Kim J, Cheon S (2019) A deep neural network-based method for early detection of osteoarthritis using statistical data. Int J Environ Res Public Health 16(7):1281
2. Antony J, McGuinness K, O'Connor NE, Moran K (2016, Dec) Quantifying radiographic knee osteoarthritis severity using deep convolutional neural networks. In: 2016 23rd International conference on pattern recognition (ICPR). IEEE, pp 1195–1200
3. Kokkotis C, Moustakidis S, Papageorgiou E, Giakas G, Tsaopoulos DE (2020) Machine learning in knee osteoarthritis: a review. Osteoarthr Cartilage Open 2(3):100069
4. Chen P, Gao L, Shi X, Allen K, Yang L (2019) Fully automatic knee osteoarthritis severity grading using deep neural networks with a novel ordinal loss. Comput Med Imaging Graph 75:84–92
5. Saleem M, Farid MS, Saleem S, Khan MH (2020) X-ray image analysis for automated knee osteoarthritis detection. SIViP 14(6):1079–1087
6. Awan MJ, Rahim MSM, Salim N, Rehman A, Nobanee H, Shabir H (2021) Improved deep convolutional neural network to classify osteoarthritis from anterior cruciate ligament tear using magnetic resonance imaging. J Pers Med 11(11):1163
7. Wahyuningrum RT, Anifah L, Purnama IKE, Purnomo MH (2019, Oct) A new approach to classify knee osteoarthritis severity from radiographic images based on CNN-LSTM method. In: 2019 IEEE 10th ınternational conference on awareness science and technology (iCAST). IEEE, pp 1–6
8. Glyn-Jones S, Palmer AJR, Agricola R, Price AJ, Vincent TL, Weinans H, Carr AJ (2015) Osteoarthritis. The Lancet 386(9991):376–387
9. Chow YY, Chin KY (2020) The role of inflammation in the pathogenesis of osteoarthritis. Mediators of inflammation, 2020
10. Hunter DJ, Bierma-Zeinstra S (2019) Osteoarthritis. The Lancet 393(10182):1745–1759
11. Mora JC, Przkora R, Cruz-Almeida Y (2018) Knee osteoarthritis: pathophysiology and current treatment modalities. J Pain Res 11:2189
12. Mandl LA (2019) Osteoarthritis year in review 2018: clinical. Osteoarthritis Cartilage 27(3):359–364
13. Antony J, McGuinness K, O'Connor NE, Moran K (2016) Quantifying radiographic knee osteoarthritis severity using deep convolutional neural networks. 1195–1200. https://doi.org/10.1109/ICPR.2016.7899799

Efficient Motion Detection and Compensation Using FPGA

N. Sridevi and M. Meenakshi

Abstract Moving target detection plays a vital role in computer vision applications, which requires rigorous algorithms and high computation; also, realization of these algorithm in real time is difficult. Hence, in this paper, FPGA implementation of moving object detection by correcting the unwanted motion is proposed. The sub-components of the architecture are optimized to get the optimization of the entire module. Also, the performance of the proposed technique is validated by calculating the different performance metrics, and the method is compared with some of the method from literature. The experimental results illustrate the approach is excellent in detecting the moving objects even when the camera is moving and less hardware is utilized to detect the moving targets efficiently. The designed architecture is implemented on Xilinx 14.5 Zynq Z7-10 series FPGA development board.

Keywords FPGA architecture · Object detection · Motion detection · Compensation

1 Introduction

Visual moving object detection and tracking has attained pronounced advances in the past decades and has been used in many applications such as traffic monitoring and control, security purpose, surveillance, military, sports, and much more. Real-time computer vision applications require rigorous algorithms and high computation; due to the limited input and output capabilities and processing power, it is difficult to run these algorithms on general purpose computers. Hence, a high-speed dedicated hardware development is essential. There are many methods to determine the motion and compensate the unwanted motion that hinders the efficient detection of moving parts in the video sequences. Traditionally, many methods exist to determine the motion in the video sequence. VLSI realization of these methods degrades the detection accuracy therefore not suitable for hardware implementation. However, the full

N. Sridevi (✉) · M. Meenakshi
Department of Electronics and Instrumentation Engineering, Dr. Ambedkar Institute of Technology, Bangalore, India
e-mail: sridevee@gmail.com

K. A. Reddy et al. (eds.), *Proceedings of Fourth International Conference on Computer and Communication Technologies*, Lecture Notes in Networks and Systems 606,
https://doi.org/10.1007/978-981-19-8563-8_14

search (FS) method illustrates consistent data flow hence appropriate for hardware realization. The FS block matching method of motion estimation is most popular technique. Here, the block is divided into many sub-blocks of finite number; from this, the best match is detected to estimate the motion vector.

Contribution: This paper focuses mainly on architecture level, and the contributions are (i) The "Controller" architecture is optimized by using simple counter-architecture. (ii) The regular adder is supplanted with high-speed Kogge–Stone adder architecture. (iii) The uses of data reuse technique reduce the overall hardware utilizations.

1.1 Related Work

Numerous methods are available in the literature to detect and compensate the motion; some of them are discussed below. Authors in paper [1] used two steps to obtain the motion vectors. A rough motion vector was obtained in the first step, and in the second step, search area is reduced to compute the vectors. Block matching with RANSAC approach to estimate the movement is proposed in [2] to compensate for ego-motion. Also, the author of the paper developed a prototype and implemented on FPGA. FPGA-based motion estimation is explained in paper [3]. Here, correlation between pixels in the reference and current frames is used for estimating the motion. Further, to reduce the lighting issues, the author of the paper used normalized correlation. In reference [4], two steps are used for searching the blocks to estimate the motion initially; using reference frame, a partial measure for distortion was constructed and is extended further to find the motion vectors. A review on various motion estimation was explained in paper [5]. To reduce the zonal improvement range the authors of paper [6] used Wavelet transform to get the starting point from current frame and reference frame. This improved the TZ search by removing the data dependency. The authors of the paper [7] claim that the designed architecture for processing uses pipeline technique, reduced latency, highest output, complete utilization of hardware.

The rest of the paper is organized as follows; the brief insight about the methodology of designing the architecture and their sub-components are presented in Sect. 2 which follows the results of FPGA implementation and simulation in Sect. 3. Finally, Sect. 4 concludes with some remarks.

2 Architecture Framework

Figure 1 illustrates the architecture proposed to estimate the motion in the video sequences. The current block having a block size of 16×16 is stored in the external memory from which the motion vector is estimated. However, the pixel data of the computational zone and current block are directed to DEMUX unit, which further

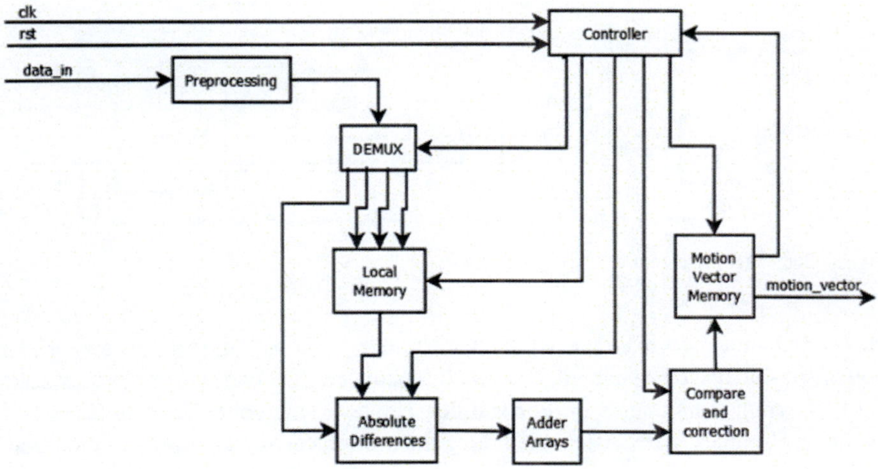

Fig. 1 Architecture framework

distributes these data pixels to three different memory units called SUBM1, SUBM2 and SUBM3. The pixel values stored in these memory units are used for finding the motion using successive approximation difference (SAD). Further, motion correction and compensation is accomplished using the compare and correct module. However, for the next frame, the motion vector stored in the memory as a motion topographies is taken as reference. Moreover, the data path for the complete system framework is controlled by the controller module.

The raw video data are not compatible to process in the hardware because of various formats of video frames. Hence, it necessitates to convert the raw video sequences in to fixed number of frames having standard size. Here, in this work, MATLAB and system generator is used for conversion. After preprocessing, the description of each module is given below.

2.1 Controller Unit

The controller in this architecture is used to regulate the data flow by enabling or disabling a suitable module. The operation starts by setting the select line of DEMUX to zero. This in turn directs the pixels of the first sub-matrix to move to SM1. Absolute difference calculation from current zone will begin after 16 clock cycles. Now, the SM2 block is selected by controller through DEMUX. Until 49th clock pulse, the absolute differences are estimated for the two submatrices, and the final absolute value starts at 50th clock pulse. The rationality is realized using the state machine counter-rationalities. Further, the corresponding architecture is shown in Fig. 2 which comprises of counter-unit, decision unit, and encoder unit. The operation of each unit is discussed below. When the reset signal goes high, for every positive clock cycle,

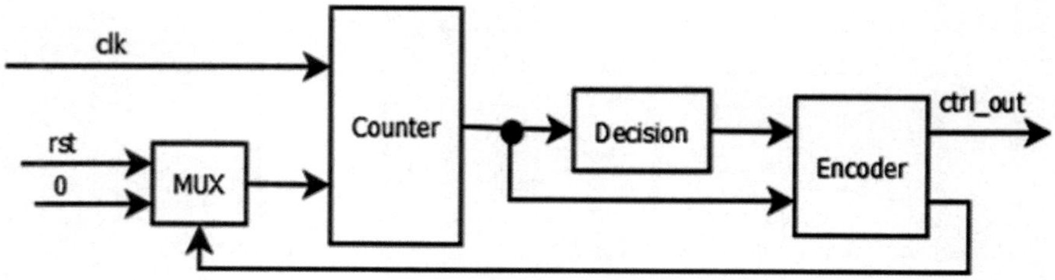

Fig. 2 Controller architecture

the counter-unit starts to count. The decision unit then chooses the proper categorization of sub-blocks to be enabled. Finally, the encoder unit encodes the units to enable the processing components in proper order. However, the controller unit is initialized to preliminary state. After calculating the motion vector corresponding to three successive 16 × 16 sub-matrices, then the process again repeats from the begining.

By using the parallel architecture of Kogge–Stone adder, the absolute difference in terms of binary arithmetic is achieved in this work. There by the speed of processing is improved.

2.2 *Motion Recognition*

The motion recognition module comprises of absolute difference, adder array, and decision unit, through which the motion vectors are estimated to find the actual movements present in the video sequences. The method adopted for absolute difference calculation is illustrated in Fig. 3. Here, due to the Kogge–Stone adder the entire architecture is optimized.

Subtraction is performed through addition using binary arithmetic from which the sign of the value is determined using concatenation module. Based on these value, the MUX will send either the data or data in 2 s complement form. Here, in this work, Kogge–Stone adder is used to compute the 2 s complement there by the architecture is optimized. However, all the matched motion vectors stored in the memory are motion vectors obtained from overlapping blocks using the comparator. The memory

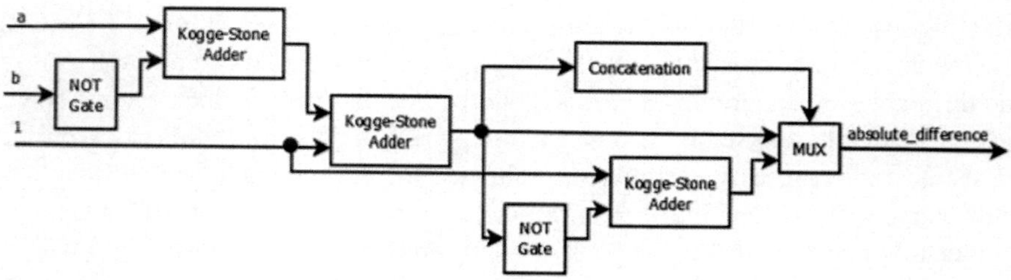

Fig. 3 Architecture for absolute difference calculation

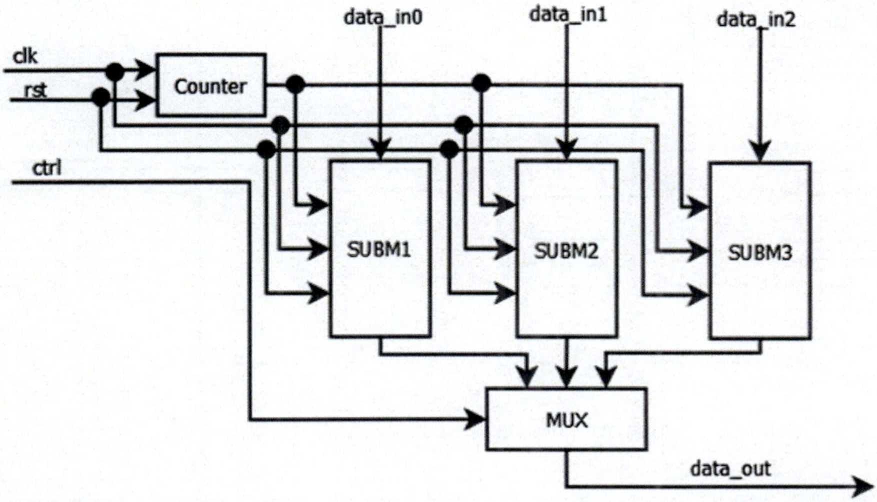

Fig. 4 Data flow architecture

is composed of three sub-parts, namely SUBM1, SUBM2 and SUBM3. Through the select line of the DEMUX module from top to bottom way row by row, the pixels enter in to the memory in 16 bits. The architecture of local memory is given in Fig. 4.

2.3 Motion Correction

The estimated motion vectors stored in the memory are compared to determine the vectors that represent the motion of the object in the video. However, the false detection from this technique is removed by using correction module and is given Fig. 5. Block of memory is used for storing the motion vectors obtained from SAD. After comparing the motion vectors using comparator, the vectors are then stored in separate modules which are then interpolated using interpolation array for detecting the correct motion vector. Moreover, the controller module will control the entire data flow of the architecture.

3 Results and Discussion

In this work, the accuracy of detection is demonstrated with two different scenario of traffic videos. The proposed approach is synthesized using Xilinx Zynq-Z7-10 series FPGA board and is coded using VHDL language. Figure 6 demonstrates the detection of moving object from the normal traffic flow. Similarly, Fig. 7 shows the detection of moving vehicles from moving camera. The unwanted motion instigated by the motion

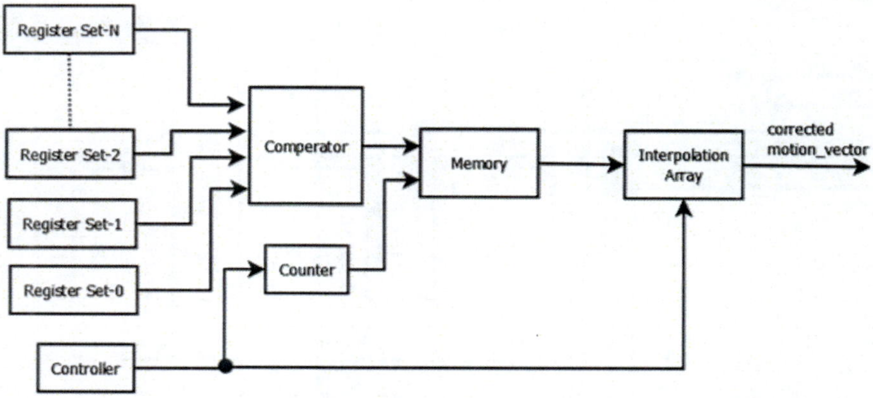

Fig. 5 Compare and correction module

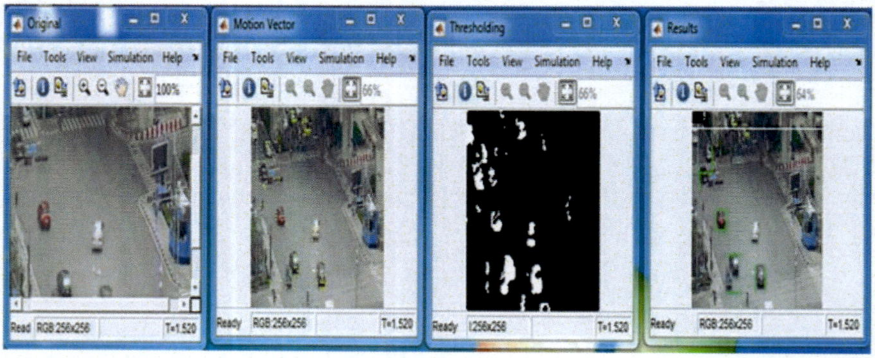

Fig. 6 Moving object detection with static camera

of the camera which may be placed on the moving platform is corrected using the compare and correction unit, which is given in Fig. 5. However, the performance metrics of rate of true detection (TR), rate of false detection (FR), and moving objects not detected (NR) are calculated from randomly selected frames and are given in Table 1.

3.1 FPGA Implementation

The designed system framework is synthesized on Xilinx FPGA development board. The resources utilized in implementing the technique are shown in Table 2.

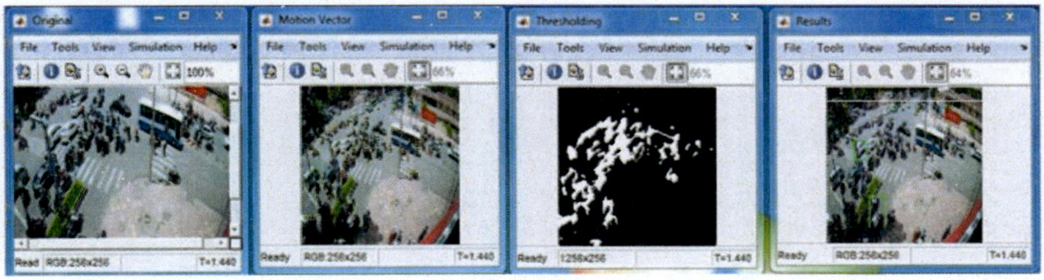

Fig. 7 Moving object detection with dynamic camera

Table 1 Illustration of performance metrics

Scenarios	TR (%)	FR (%)	NR (%)
Normal traffic	93.78	3.97	1.5
Moving camera	92.93	4.714	2.14

Table 2 Resource used for implementing the technique

Parameters	FPGA	No. of slices	Slice LUTs	LUT-FF pairs	Maximum frequency (MHz)
SAD	Zynq-Z7-10 (xc7z010-1clg400c)	6144	6299	2910	1432.665
Comparator		135	146	86	321.404
Controller		21	460	12	650.347
Memory		6154	2184	1850	707.214
Complete unit		12,450	908	517	321.404

3.2 *Performance Evaluation*

The experimental results obtained are compared with the methods from literature and are listed in Table 3 in terms of hardware used (Table 4).

Table 3 Comparison of accuracy of detection with existing

Authors	Chih-Yang et al. [8]	Yu and Fen [9]	Sridevi and Meenakshi [10]	Zhu et al. [11]
Methodology	Bit plane with co-occurrence matrix	Optical flow method	Modified Gaussian mixture	You Only Look Once
Accuracy of detection in %	86.3	89	93.18	88.59

Table 4 Comparison of hardware resources used with existing

Parameters	Platform	Slice register	Slice LUTs	LUT-FF pairs	Maximum frequency (MHz)
Proposed	Zynq-7	12,450	908	517	321.404
She and Peng [12]	Virtex-5	13,245	–	–	200
Cho et al. [13]	Zynq-7	12,767	37,761	27,875	128

4 Conclusion

This paper proposes a hardware framework design based on FPGA to estimate and compensate the motion. In order to detect only the moving objects from the video frame, the method has to confiscate the motion induced by the camera. The proposed method detects and compensates the unwanted movements. Also, to achieve optimized hardware utilization, different techniques are used such as Sum of Absolute Difference (SAD), controller and memory block. The Kogge–Stone adder is used to optimize the sum of absolute difference calculation and modified absolute difference block. The basic logical elements are used to optimize the comparator and compensation block. Further, the controller module is optimized by adopting the counter-operation using FSM modeling. Finally, the entire framework is tested and synthesized on Xilinx FPGA development board.

References

1. Chatterjee SK, Vittapu SK (2019) An efficient motion estimation algorithm for mobile video applications. In: 2019 Second ınternational conference on advanced computational and communication paradigms (ICACCP)
2. Tang JW, Shaikh-Husin N, Sheikh UU, Marsono MN (2016) FPGA based real-time moving target detection system for unmanned aerial vehicle application. Int J Reconfig Comput
3. Viorela Ila V, Garcia R, Charot F, Batlle J (2004) FPGA implementation of a vision-based motion estimation algorithm for an underwater robot. In: Becker J, Platzner M, Vernalde S (eds) Field programmable logic and application. FPL 2004. Lecture notes in computer science, vol 3203. Springer, Berlin, Heidelberg
4. Paramkusam V, Reddy VSK (2014) A novel block-matching motion estimation algorithm based on multilayer concept. In: 2014 IEEE ınternational conference on multimedia and expo (ICME)
5. Bnadou R, Hiramori M, Iwade S, Makino H, Yoshimura T, Matsuda Y (2016) A study on motion estimation algorithm for moving pictures. In: 5th IEEE global conference on consumer electronics, pp 1–3
6. Pakdaman F, Hashemi MR, Ghanbari M (2020) A low complexity and computationally scalable fast motion estimation algorithm for HEVC. Multimed Tools Appl, Springer, vol 79: 11639–11666
7. Singh K, Shaik RA (2015) A new motion estimation algorithm for high efficient video coding standard. In: Annual IEEE India conference, 2015

8. Chih-Yang L, Zhi-Yao J, Wei-Yang L (2016) Image bit-planes representation for moving object detection in real-time video surveillance. In: IEEE International conference on consumer electronics, Taiwan, pp 1–2, May 2016
9. Yu Z, Fen Fen W (2016) Improved optical flow algorithm of moving object detection. In: IEEE International conference on ınstrumentation and measurement, computer, communication and control, pp 196–199, Feb 2016
10. Sridevi N, Meenakshi M (2020) Efficient movement compensation and detection algorithm using Blob detection and modified Kalman filter. In: 2020 5th International conference on communication and electronics systems (ICCES), pp 264–268
11. Zhu H, Wei H, Li B, Yuan X, Kehtarnavaz N (2020) Real-time moving object detection in high-resolution video sensing. Sensors
12. She J, Peng D (2020) FPGA-based motion estimation algorithm optimization. In: Microprocessors and microsystems, vol 20. Springer, pp 1–13
13. Cho J, Jung Y, Kim DS, Lee S, Jung Y (2019) Moving object detection based on optical flow estimation and a Gaussian mixture model for advanced driver assistance systems. Sensors:1–14

Discover Crypto-Jacker from Blockchain Using AFS Method

T. Subburaj, K. Shilpa, Saba Sultana, K. Suthendran, M. Karuppasamy, S. Arun Kumar, and A. Jyothi Babu

Abstract The blockchain technology is used throughout the world for digital ledgers of transactions. To maintain all participant transactions on a blockchain, distributed ledger technology (DLT) is used. Blockchain technology has become a popular method of transferring huge amounts due to the pandemic situation. A hacker was able to exploit some part of a chain, smart contract, exchange, or stealing cryptocurrency, that is, a hack or a theft. Such attacks are referred to as crypto-Jack attacks. The Wormhole cryptocurrency platform was hacked in Feb 2022, resulting in a loss of $326 Millions. Many cryptocurrency Web pages are being hacked every day. Most of the attacks are focusing the financial purpose, so banking attackers are globally raised. To identify crypto-jackers and to trace out the hijackers, we proposed the new technology.

T. Subburaj
Department of Computer Applications, Rajarajeswari College of Engineering, Bangalore, India
e-mail: shubhurajo@gmail.com

K. Shilpa (✉) · S. Sultana
Department of CSE, CMR Technical Campus, Hyderabad, Telangana, India
e-mail: shilpamtech555@gmail.com

S. Sultana
e-mail: sabasultana014@gmail.com

K. Suthendran
Department of Information Technology, Kalasalingam Academy of Research and Education, Krishnan Koil, Srivilliputhur, Tamilnadu, India
e-mail: k.suthendran@klu.ac.in

M. Karuppasamy
Department of Computer Applications, Kalasalingam Academy of Research and Education, Krishnan Koil, Srivilliputhur, TamilNadu, India

S. Arun Kumar
Department of Computer Science and Engineering, Bethesda Institute of Technology and Science, Gwalior, India

A. Jyothi Babu
Department of MCA, Sree Vidyanikethan Engineering College, Tirupati, India

K. A. Reddy et al. (eds.), *Proceedings of Fourth International Conference on Computer and Communication Technologies*, Lecture Notes in Networks and Systems 606,
https://doi.org/10.1007/978-981-19-8563-8_15

Keywords Block chain · Crypto-jack · AFS · DLT · Attack model

1 Introduction

Data integrity is one of the most important characteristics of blockchains, as it is decentralized and verifiable. By using blockchain technology, certain positions of malfunction can be avoided. In financial transactions and information exchange, blockchain technology has demonstrated how security can be transformed. In addition, it employs distributed ledger technology to store records in a decentralized way. Blockchain technology stores all our transaction information, hash algorithm, and user identification number for every user associated with a transaction in its blocks. The blockchain technology allows cryptocurrency transfers without the traditional middleman, thanks to its open-source nature. In particular, blockchain is utilized to keep a particular place from failing. The fault in any one of the blocks in a network system affects the entire network. To solve the above problem, blockchain uses state machine replication (SMR) technology. Cryptocurrency Bitcoin is the first to use blockchain technology.

Distributed ledger technology (DLT) is at the core of blockchain technology. It is immutable, secure, programmable, anonymous, and distributed. DLT uses cryptography based on digital signatures. The most popular blockchain-based digital currencies are Bitcoin and Ethereum.

Blockchain is considered the most secure technology by the majority of people. There is, however, some vulnerability. Business and industrial sectors benefit greatly from blockchain technology, without a doubt [1]. Cryptocurrencies are normally used during pandemics (COVID-19) to make money transfers. In some cases, these attacks were caused by wrong deployments, weak networks, or inappropriate tuning to secure the systems. The situation is being exploited by cybercriminals to steal money. Blockchain has been used by crypto criminals to launch attacks. Security threats are evolving, with new types posing large-scale, irreversible threats. The threats can be divided into client-oriented or core-oriented attacks. Common attack methods used by cybercriminals include reverse proxy phishing, crypto-jacking, dusting, and clipping [2].

One type of spoofing attack is reverse proxy phishing. The attacker uses man-in-the-middle (MitM) techniques to listen in on the conversation between two users. This technique is primarily used to crack two-step passwords. Crypto-jacking involves stealing sensitive information from a PC or mobile without the victim's knowledge. Dusting is a new type of cybercrime. A blockchain transaction requires decrypted secured data to be collected. Transaction amounts are also stolen. When the transaction is done online, clipping is a way to steal money.

Last year, cybercriminals stole $4.25 billion in cryptocurrency assets, nearly three times as much as they stole in 2020. Several of the top ten most expensive crypto breaches occurred in 2021, according to statistics collected by Comparitech.

2 Crypto-Jacking Attack

Crypto-jacking has become increasingly common across the world today. Criminals are constantly learning new methods. On September 2017, cybercriminals began launching crypto-jacking attacks [3].

2.1 How Crypto-Jacking Characters Are Extend

The most common methods cybercriminals use to steal currency from cryptocurrencies are file-based, browser-based, and cloud-based crypto-jacking. Crypto-jacking attacks are unique types of attacks in the crypto sphere. An attack of this kind involves having hackers create a fake environment around a particular block in a blockchain so that they can manipulate the artificial node in order to commit crimes.

2.2 Attack Methods

(a) **Bonding method**

Bonding with a node starts with the client checking if the node is available. A client trying to add a node to its table will attempt all the above checks if all are successful. If there is still space in the table, an additional node may be added. A client checks whether the node is alive by sending a ping to it. If a pong response is received by the node, the bonding has been made. Bonding success causes the client to update the entry for the node in both its database and table simultaneously.

(b) **Impulsive pings**

Unsolicited pings are received by clients, which are responded to with pong messages, and then successfully bond with other clients.

(c) **Lookup Table**

In this technique, it may be used for customers to find out the new nodes. This method is predicting the closeness node. In this lookup table, it finds the closeness using bitwise exclusive OR calculation between nodes. The client sends findnode with target node message to nearest 16 nodes in the lookup table.

3 Crypto-jacking Attack Model

To launch a crypto-jacking attack on the Ethereum network, there are generally two methods. By establishing TCP connections to an attacker's malicious node before the victim can establish TCP connections to itself, an attacker can crypto-jacking an Ethereum client, and the other is to own the table and crypto-jacking. A victim-centric crypto-jacking attack framework is designed in light of the attack methods. The crypto-jacking framework defines four states for nodes, and based on the change in state, we can determine if a node is currently being attacked by crypto-jack [4]. Figure 1 shows the crypto-jacking model.

3.1 Running

The running state means a node which has already running on at least last 24 h. Every node was maintained the database and table. The database of the every node is maintained in the ping message and also pong message. In table, every node automatically filled the information about the SHA3 features.

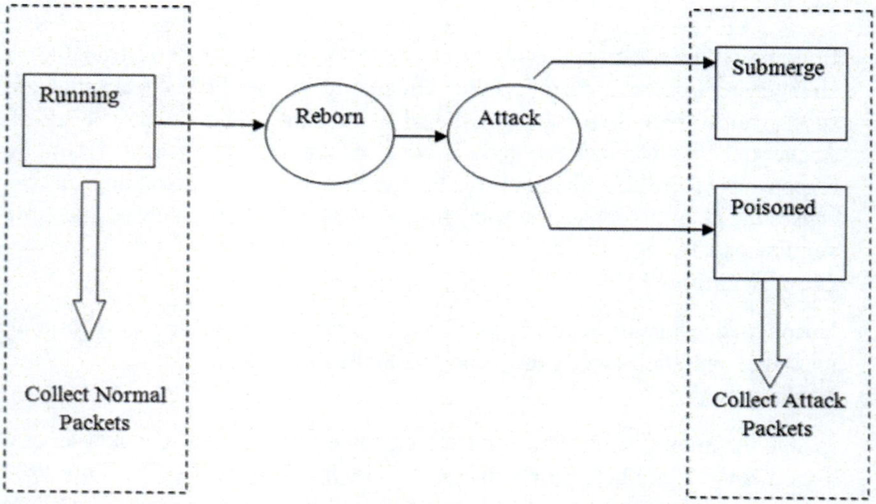

Fig. 1 Crypto-jacking attack model

3.2 *Reborn*

When the crash or recover occur in node, the node will be reboot. After rebooting the node, node's information from the table is deleted. As a result, an attacker is now able to initiate connections or carefully-crafted packets to a rebooting victim once the node has been accessed. It may be best to collect malicious packets at this time [5].

3.3 *Submerge*

A state submerge occurs when an attacker establishes Maxpeers from its own adversaries to the victim's TCP connections[6]. The victim is forced to set all connections to incoming at this point.

3.4 *Poisoned*

A table poisoned when the attacker inserts a crafted nodeID into the victim table. There is a high probability that the victim forms all outgoing connections to the attacker's nodes.

An active crypto-jacking attack shows the states changing. Node A is in the running state when it has been active for a while. A node A needs many ping requests to launch an attack by an attacker B. After victim A becomes reborn for some reason (as we would say, it reboots), its attack probability increases. Depending on the configuration of A's connection, it will either be submerged or poisoned. A may enter the submerge state if it does not create outgoing TCP connections. By contrast, a crafted message from attacker B will poison the table of A.

The ground data, which is needed for studying our detection models, must be collected beforehand [7]. Geth V 1.6.6 client is designed as the victim. First, we gather packets from normal access connections, then we make an attack script that attempts to send ping repeatedly at the target. After a victim reboots, packet collection begins and continues until all connections from that victim is occupied or the victim's table is filled with node entries from our node.

4 Proposed Approaches

Main objective of the proposed system is to identify the attacks from cybercriminals in cryptocurrency.

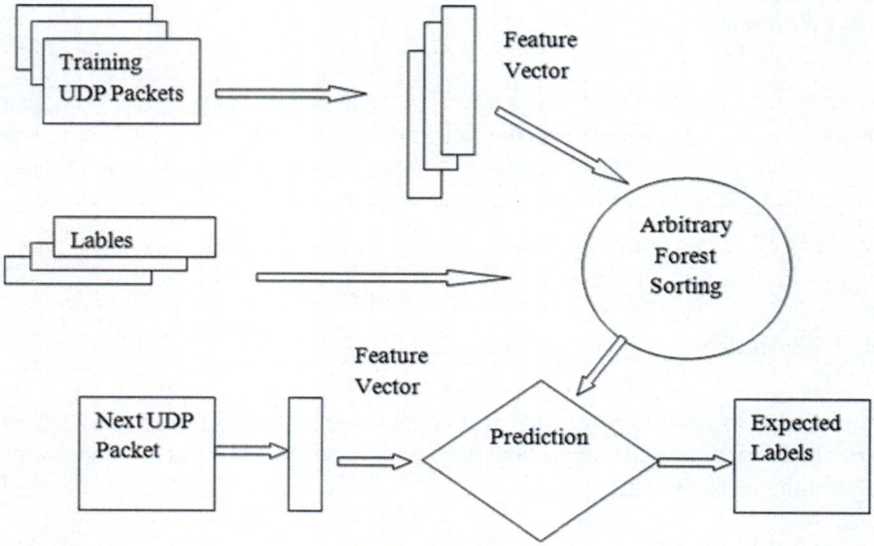

Fig. 2 Proposed model training process system

To detect the crypto-jacking attack, we are developed the tool based on the arbitrary forest decision sorting. Attackers are continually sending the requests to victim nodes. Victim nodes are collecting the all UDP packets from the spontaneous nodes. We are analyzing the incoming packets of crypto-jacking attack.

4.1 Proposed Model Training Process System

The process of the training model is show in Fig. 2.

a. **Selection Process**

In this paper, we explore the characteristics of crypto-jacking attacks based on the information entropy of data streams. Entropy calculation is used to describe crypto-jacking attacks [8]. Information entropy calculation via the following formula:

$$H(X) = -\sum_{i=1}^{n} p(x_i)\log_2(p(x_i)) \quad (1)$$

Equation 1 is used to find the entropy value of X, where $p(x_i)$ is the probability of X. Entropy value of X about Y is

$$H(X/Y) = -\sum_{j=1}^{m} p(x_j)\sum_{i=1}^{n} p(y_j/x_i)\log_2(p(y_{i/}x_j))) \quad (2)$$

In our proposed method, entropy value of source IP address, destination IP address, and destination port number is used to category the features of normal flow and the attacked flow. We find the flow of the data from the source to destination using Eq. 3.

b. **Data flow calculation**

$$\text{Flow of Data} = -\sum_{j=1}^{m} p\left(di_j\right) \sum_{i=1}^{n} p(Si_i/di_j)\log_2(p(Si_i/di_j)))$$
$$= -\sum_{j=1}^{m} \frac{A_j}{S} \sum_{i=1}^{n} \frac{B_{ij}}{A_j} \log_2\left(\frac{B_{ij}}{A_j}\right). \tag{3}$$

The following features are used to easily identify the attacks: packets_size, access_frequencies, and access_time [9, 10].

c. **Arbitrary Forest Sorting Method**

Machine learning algorithms such as the arbitrary forest sorting improve detection accuracy without causing significant computational complexity. In addition to reducing over fitting and variance, arbitrary forest can resolve several problems associated with decision trees.

I. **AFS Model Training process**

Algorithm 1: AFS model training process

```
Input: T, p, k Output: DT
      for i 1 to n
         T′ = withResample_Sample(T, p);
         Att = get_Attributes(T′);
         Att′ = withoutResample_ Sample(Att, k);
         T″ = remain_Attributes(T′, Att′);
         DT[i] = create_DecisionTree(T″);
      return DT
```

II. **AFS Model Classification**

After the training process, a certain leaf node is reached by passing through each leaf in the test sample. Then calculate the probability of the sample. The AFS model relies on majority voting for classification. E = [e1, e2, ***, em], DT = [dt1, dt2, ***, dtn], CIR = [n], C = [c1, c2, ***, cn], and CR = [CR1, CR2, ***, CRm], respectively, define the test sample set, the other set is C = [c1, c2, ***, cn], and classification results are CR = [CR1, CR2, ***, CRm].

Algorithm 2: The AFS model classification process

Input: DTE

Output: CR

for i = 1;i ≤ m;i ++ do

```
    for j = i; j ≤ n; j ++ do
        CIR[j] = 0;

for j = 1; j ≤ n; j ++ do
        ClassifyResultIndex = classify (E[i], DT[j]) CIR [classifyResultIndex]++;
            maxIndex = getMaxAppeared(CIR);
            CR[i] = C[maxIndex];
        return CR;
```

5 Our Experiment

Throughout this section, we illustrate how accurate and effective our crypto-jack detector is at detecting crypto-jack attacks on Ethereum networks. We followed four basic steps in our experiment.

a. **Collection**

By sending ping repeatedly, we attempt to crypto-jack the victim by collecting the normal incoming packets during the running state. As soon as victim reboots, we begin collecting crypto-jack attack packets until our nodes have occupied all of the victim's incoming connections, and its table is filled. Wireshark is used at this stage to collect the UDP packets from the victim. Figure 3 illustrates how to capture sample data with Wireshark.

Additionally, we added the Ethereum devp2p protocol dissector plug-in to Wireshark for analysis of UDP packets collected.

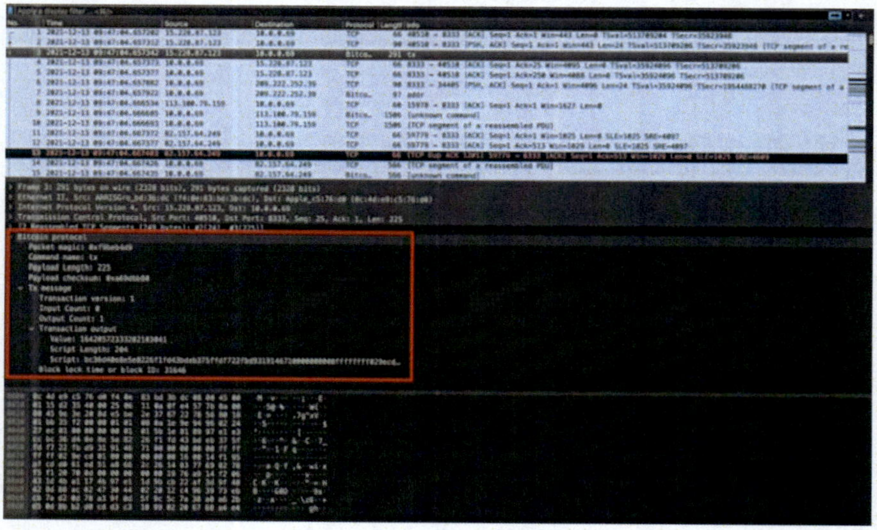

Fig. 3 Sample data with Wireshark

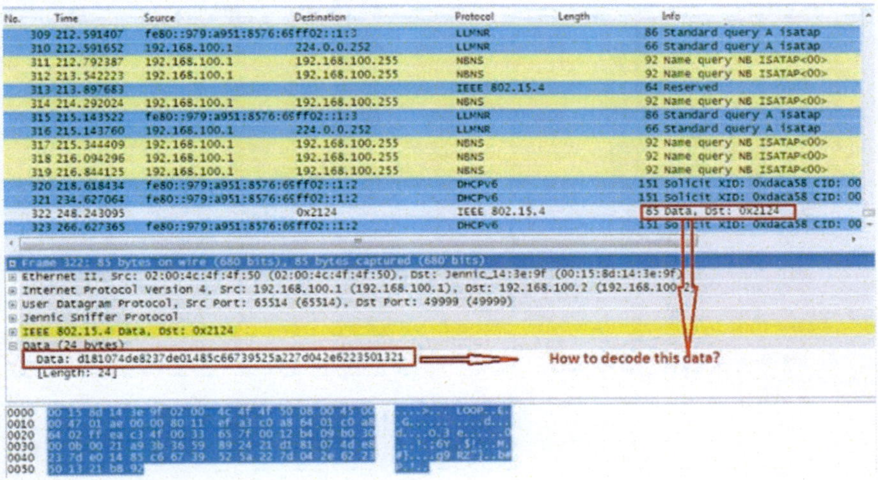

Fig. 4 Sample decoded data

b. **Preprocessing**

Data collection is using for ping, pong, findnode, and neighbor. We are currently decoding a pcap file of Ethereum packets captured into a readable format using an Ethereum UDP packet dissector for discovery protocol v4. The decoded data are shown in Fig. 4. There is a packet type, destination IP address, and source IP address for each packet. Attack traffic samples are sampled every 5 ms for 25 ms, with a single increase by 5 ms, while background data samples are sampled every 5 s. Using the data obtained from the five sets of data, a sample sequence consists of 100 samples continuously, so there are 20 sequences total.

c. **Training Model**

Initially, we analyze the distribution of UDP packets in two states using a statistical analysis. Here is the analysis. Figure 5 illustrates this. Malicious packets have a different size distribution than honest packets. An attacker must ping the victim many times to eclipse an honest node. In comparison with packets with the types of findnode, neighbors, ping, and pong will contain less data information. As a result, their sizes are distributed differently.

Figure 6 shows a higher complexity of attack access. Normally, short connection access is built for shorter connections. The attacker may wait longer for the victim to respond with a pong when the victim cannot do so on time.

In Figure 7, the chart shows that a node under eclipse attacks experiences a much higher visit frequency. Eclipse works by repeatedly sending ping requests to a victim. This is an indicator that a victim is being eclipsed. Our data are classified using random forest using these features.

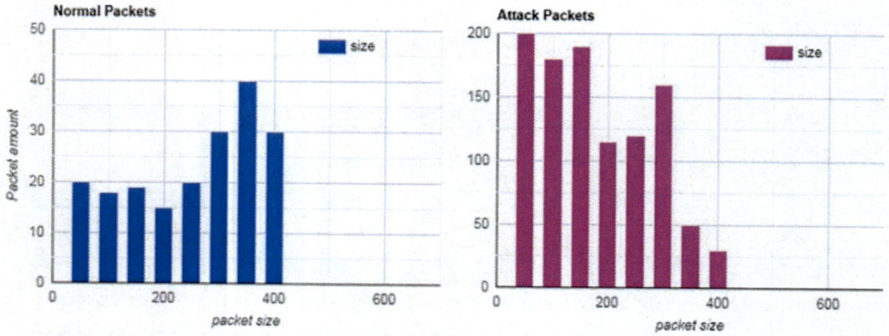

Fig. 5 Distribution of different packet size

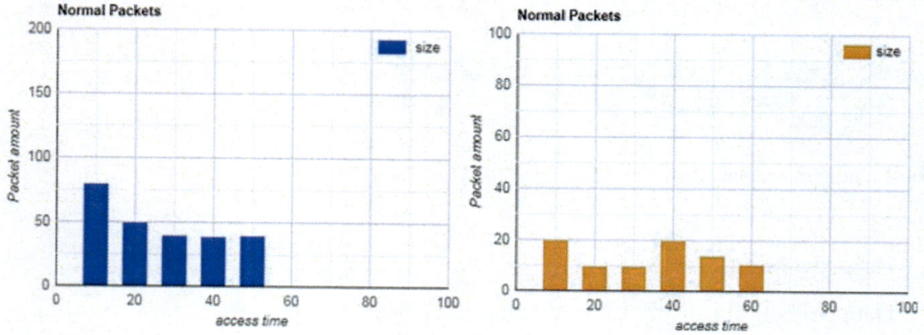

Fig. 6 Distribution of different request time

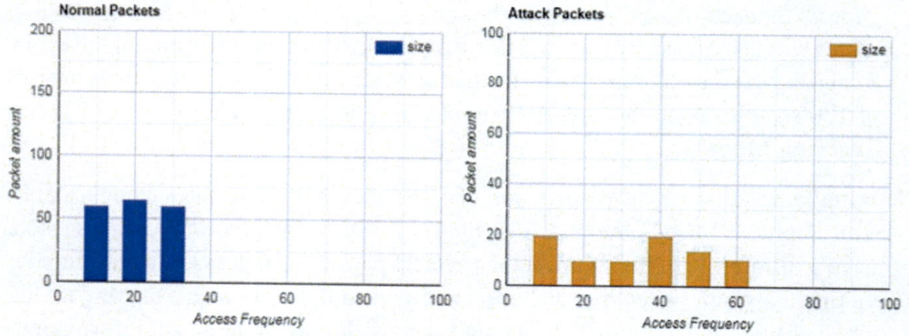

Fig. 7 Distribution of different access frequency

d. **Detection**

This data were prepared using a statistical distribution of the UDP data. Sklearn is used to build our detection model, and the test data and collected data are split 3:7. As adversarial nodes connect to our node through UDP, it will reboot several

Table 1 Result of AFS detection

S. No.	Functions	Precision	Recall	F1-Score	Support
1	Normal visit	0.67	0.95	0.82	152
2	Attack visit	0.74	0.93	0.79	131
3	Avg/Total	0.71	0.94	0.80	283

times. Detecting eclipse attacks with high probability can identify adversary connection requests.

Our detection rate is quite high in practice, with a precision rate of 72% and a recall rate of 93% (Table 1). A third of the attacked data can hit its ground label, according to the experimental results. Currently, most of the attack packets are able to be blocked by our detection model as more than 90% of malicious data can be correctly identified.

6 Conclusion

A novel crypto-jacking attack detection system is proposed to protect blockchain nodes against crypto-jacking attacks. In this paper, we define attack connection flow features and describe a novel crypto-jacking attack detection system. Our model defines the changes in the sate of the node during attacks. Detecting malicious connections is accomplished by leveraging the fusion context of the model and the stability of fitting degree as traffic increases. The characteristics of UDP data packets include packets_size, access_frequency, and access_time. With a high detection rate and low false alarm rate, our model is able to differentiate normal traffic from an attack one correctly.

7 Future Works

In future, to enhance scalability, we will be collecting larger amounts of data flow and defining more features. To determine the classification label for new incoming packets, our current model has to gather and store related data in advance. In the context of runtime data flow, a real-time detection model may provide more practical results. Future plans include leveraging other classifiers to make detection comparisons when it comes to accuracy and effectiveness. These are the open problems that we are considering in this work.

References

1. https://101blockchains.com/blockchain-security-issues/
2. https://www.digitalshadows.com/blog-and-research/cryptocurrency-attacks-to-be-aware-of-2021/
3. https://www.varonis.com/blog/cryptojacking/
4. Locher T, Mysicka D, Schmid S, Wattenhofer R (2010) Poisoning the Kad network. Lecture notes in computer science book series (LNCS) distributed computing and networking, vol 5935, pp 195–206
5. Xu G, Liu J, Lu Y, Zeng X, Zhang Y, Li X (2018) A novel efficient MAKA protocol with desynchronization for anonymous roaming service in global mobility networks. J Netw Comput Appl 107:83–92
6. Marcus Y, Heilman E, Goldberg S (2018) Low-resource eclipse attacks on Ethereum's peer-to-peer network. Cryptology ePrint Archive, 236
7. Chen S, Xue M, Fan L, Hao S, Xu L, Zhu H, Li B (2018) Automated poisoning attacks and defenses in malware detection systems: an adversarial machine learning approach. Comp Secur 73:326–344
8. Subburaj T, Suthendran K, Arumugam S (2017) Statistical approach to trace the source of attack based on the variability in data flows. In: ICTCSDM 2016, Lecture notes in computer science, LNCS 10398. Springer, pp 392–400
9. Qiang Z, Wang Y, Song K, Zhao Z (2021) Mine consortium blockchain: the application research of coal mine safety production based on blockchain. Secur Commun Netw 2021, Article ID 5553874. https://doi.org/10.1155/2021/5553874
10. Wu D, Xiang Y, Wang C (2018) Data protection technology for information systems based on blockchain. J Command Control 4(3)

Automated Detection for Muscle Disease Using EMG Signal

Richa Tengshe, Anubhav Sharma, Harshbardhan Pandey, G. S. Jayant, Laveesh Pant, and Binish Fatimah

Abstract Muscle disease is a term used to describe illnesses that affect the human muscle system. To diagnose muscle diseases such as myopathy and amyotrophic lateral sclerosis (ALS) specialists examine EMG signals. This manual method is a time-consuming procedure and needs specialized skills. In this paper, we propose an automated detection technique for the same. Proposed algorithm uses frequency decomposition method (FDM) and classifiers such as ensemble subspace k-nearest neighbour(KNN) to distinguish ALS and myopathy EMGs from normal EMG signals and obtain 92.3% accuracy for ALS versus myopathy vs normal case.

Keywords Neuromuscular disorders · ALS · Myopathy · Machine learning · Electromyogram · Frequency decomposition method

1 Introduction

Muscle diseases or neuromuscular disease (NMD) such as trophic lateral sclerosis (ALS) can be caused by a lack of mobility in a particular muscle, autoimmune disorders, ingested toxins or extended exposure to heavy metals or certain medications, or

R. Tengshe (✉) · A. Sharma · H. Pandey · G. S. Jayant · L. Pant · B. Fatimah
CMR Institute of Technology, Bengaluru, Karnataka 560037, India
e-mail: richa.t@cmrit.ac.in

A. Sharma
e-mail: ansh18ec@cmrit.ac.in

H. Pandey
e-mail: hapa18ec@cmrit.ac.in

G. S. Jayant
e-mail: gsja18ec@cmrit.ac.in

L. Pant
e-mail: lapa18ec@cmrit.ac.in

B. Fatimah
e-mail: binish.f@cmrit.ac.in

K. A. Reddy et al. (eds.), *Proceedings of Fourth International Conference on Computer and Communication Technologies*, Lecture Notes in Networks and Systems 606,
https://doi.org/10.1007/978-981-19-8563-8_16

the cause may even be genetic/hereditary in some cases. Muscle weakness, palsy, and loss of brain control over muscles are potential consequences of ALS. Another condition related to muscle fibre skin is myopathy. Myopathy is a term used to describe disorders that affect muscular tissue which leads to muscle weakness, inflammation, spasms, and cramping. If ignored, all these conditions may deteriorate over time. Hence prompt diagnosis, medical intervention, and care are strongly advised.

A common diagnostic tool for these diseases is the electromyography (EMG). EMG measures the cumulative effect of the action potentials generated by the contraction and expansion of skeletal muscles and is a good source of information regarding the muscle activity and hence very useful for the diagnosis of conditions related to muscle.

EMG is made up of a number of motor unit action potentials (MUAPs). Pertaining to ALS shows an overall reduction in the EMG amplitude associated with fasciculation and fibrillation potentials. There is also prolonged distal motor latency and slowed conduction velocity. There may also be sharp wave potentials, as shown in Fig. 1a. In muscular dystrophy or myopathy, the EMG shows motor unit potentials that are prolonged and associated with polyphasia. There is also a reduction the amplitude of EMG waves, as shown in Fig. 1b, whereas as shown in Fig. 1c, normal muscle EMG has no fasciculations or fibrillations, and continuous muscle activity can be seen with normal contraction and relaxation.

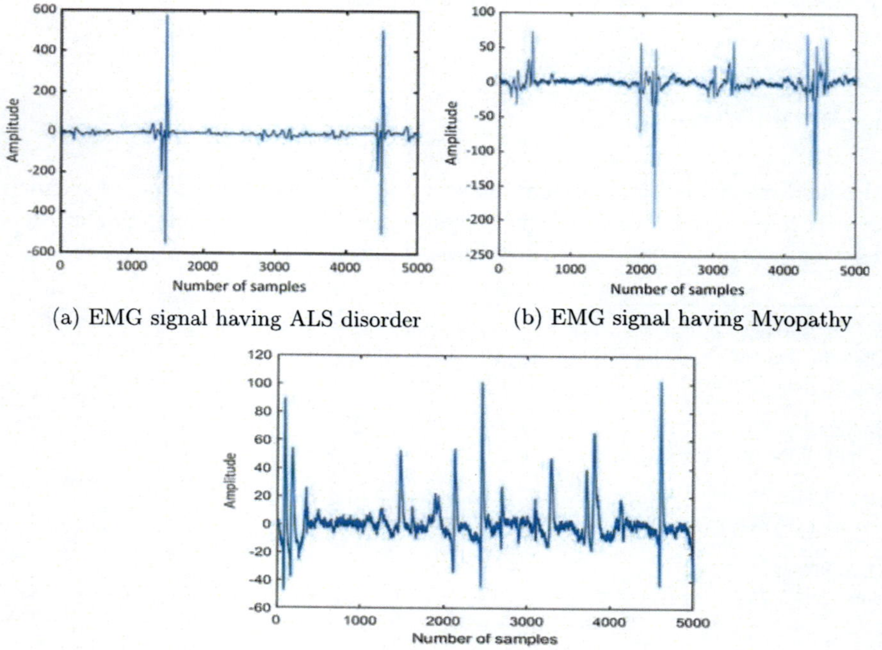

(a) EMG signal having ALS disorder (b) EMG signal having Myopathy

(c) Control object EMG

Fig. 1 EMG signals for ALS, CO, and myopathy

Various automated approaches for investigating EMG signals are proposed in the literature focusing on signal decomposition methods and different classifiers. In [3], the authors created a model based on template matching decomposition method on the mel-frequency cepstral coefficient (MFCC) of MUAPs as feature extraction technique. The suggested accuracy, sensitivity, and specificity were 92.5%, 76%, and 98%, respectively. Author [16] proposed a binary classifier for ALS and normal EMGs. Here, enhanced Hankel matrix eigenvalue decomposition (IEVDHM) method was used on MUAPs, and features like correntropy (CORR) and cross-information potential (CIP) were calculated. Comparison of tertiary classifiers based on FFNN, SVM, DT, and empirical mode decomposition (EMD)-based method was proposed in [4]. Tunable Q wavelet transform (TQWT) on MUAPs was used as feature extraction method along with some time domain feature selection in [11]. ANN-based classifier to differentiate normal and myopathic EMG signals from Biceps Brachii was proposed by [17]. Statistical features extracted from EMGLABs data are examined for this classification. EMG signal characteristics were analysed using the short time Fourier transform (STFT) approach to reduce the computational complexity in [2]. The power level of the spectrogram of ALS patients was higher than that of the normal group, and this characteristic difference was exploited in this analytical work. The continuous wavelet transform (CWT) was used for feature extraction in [1]. Authors [10] proposed a novel approach based on SEMG (Surface electrogram) and support vector classification. Time domain parameters extracted from EMG signals were used for categorization in an ANN-based classifier in [15]. The authors in [12] extracted bi- spectrum-based features with mean plosion index and fractal dimension features from EMG signals. Short time Fourier transform and statistical features were used to categorize ALS signals and normal EMG signals.

As per the above discussion, various authors have proposed muscle disease detection algorithms using artificial intelligence (AI) including deep learning or machine learning. While deep learning is more data hungry, performance of machine learning algorithms highly depends on the relevance of features computed. Since the publicly available EMG data sets are not very large, machine learning is a reasonable choice. In order to extract the best possible features, various authors have used multiscale decomposition techniques like Wavelet decomposition or empirical mode decomposition (EMD). In the recent studies, Fourier-based decomposition techniques are shown to provide better signal representation as compared to wavelet or EMD [5–9, 19]. In this work, we use the Fourier decomposition method (FDM) [18] to obtain the multiresolution signal analysis. The sub-bands so obtained are then used for computing features. Statistical features such as variance, kurtosis, entropy, and the minimum and maximum values of the sub-band signals are used to classify ALS, myopathy, and CO data using machine learning classifiers.

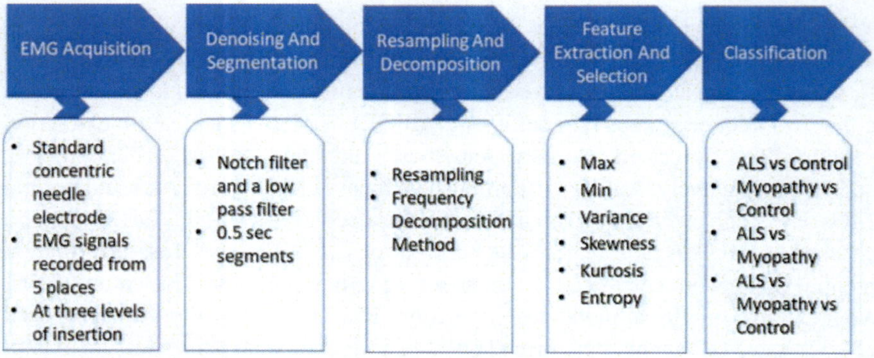

Fig. 2 Block diagram of the proposed MDD detection scheme

2 Proposed Methodology

Machine learning-based disease detection algorithms proposed here consist of the following steps, as shown in Fig. 2. First step is the biomedical signal acquisition which could be invasive or non-invasive or bio-radar-based depending on the application in hand. Second step involves data curation, removing unwanted signal, and noise and segmentation. This step requires the knowledge of data acquisition process and equipment. Third step is feature extraction and selection, here time domain or frequency domain features can be computed from the signal or from the narrowband multiscale components of this signal. Finally, the relevant features are used to train a machine learning model.

2.1 Data Set

Data set used for training the classifiers is acquired from online repository of EMGlabs N2001 at http://www.emglab.net [14]. This is a repository of clinical signals which are divided into three major subsets: normal, ALS, and myopathy. The ALS group consisted of eight participants, four male and four female aged from 35 to 67 years. The myopathy group contained seven subjects, two men and five women within the age bracket 19–63 years. The normal group consisted of ten subjects, six men, and four women aged 21–37 years, with no history or signs of neuromuscular illness. A conventional concentric needle electrode is employed. EMG signals obtained vary in location from where they are taken and level of needle insertion (low, medium deep insertion).

2.2 Pre-processing

The EMG measuring equipment's electrical network and electronic components may introduce noise which degrades the quality of the digitized EMG signal. To nullify these artefacts, a pre-processing step is done which consisted of processing the signal using a cascaded filter bank designed suitable to remove each type of noise. AC power line interference is dealt with a notch filter with cut-off frequency 50 Hz. Another noise which usually might get introduced is noise due to involuntary movement of the subject. This type of noise introduces oscillations in the form of low-frequency base line. Suppression of these baseline oscillations is done before decomposition. The denoised EMG data is then segmented to create more sample and enhance the size of the data set, as also done in [20]. Here, we have considered 0.5 s non-overlapping window to obtain the required segments.

2.3 Feature Extraction

Signal is decomposed using FDM into orthogonal intrinsic band functions (FIBFs). For detailed discussion of FDM, refer [18]. The following features are computed from each FIBF:

1. Maximum amplitude of the EMG signal in one frame.
2. Minimum amplitude of the EMG signal in one frame.
3. Variance = $\left(\frac{1}{N}\sum_{n=0}^{N-1}\left(s\left[n\right]-\mu\right)^2\right)^{1/2}$.
4. Kurtosis = $\sum_{n=0}^{N-1}\left(\frac{s[n]-\mu}{\sigma}\right)^4$ where μ and σ denotes mean and variance of $s(n)$, respectively.
5. Entropy = $-\sum_{n=0}^{N-1} p\left(s\left[n\right]\right)\log_2\left(p\left(s\left[n\right]\right)\right)$ where $p\left(s\left[n\right]\right)$ is the discrete probability of signal $s\left[n\right]$.

2.4 Feature Selection and Classification

Here, we have used different machine learning algorithms including support vector machines (SVMs) with linear, quadratic, cubic and Gaussian kernal, k-nearest neighbour (kNN), ensemble methods including ensemble bagged trees, ensemble subspace kNN. For detail discussion on classifiers refer [13]. The performance of these algorithms is compared to select the best classifier for the proposed methodology. Here, we have used tenfold cross-validation method as the data set used here is small.

3 Results and Discussions

In this section, we present the results obtained for muscle disease detection using the proposed algorithm. The data set discussed in Sect. 2 has been used here. The simulations have been carried on MATLAB 2021b to obtain the results presented in this section. We have developed muscle disease detection models for four classification tasks, namely ALS versus CO, myopathy versus CO, ALS versus myopathy, and ALS versus myopathy versus CO. In Table 1, the performance of different machine learning algorithms has been compared for the four classification tasks as mentioned above.

The data set used in this work included EMG signals collected from different muscles. Among these for four of the muscles, namely Vastus Medalis, Tibialis Anterior, Deltoideus, and Biceps Branchii, data is available for both ALS and myopathy. In order to select the most discriminative signal, we compare the classification performance of ALS versus myopathy for each of these muscles, as given in data set [14]. It seems from the result that Vastus Medialis muscle group which is part of quadricep located in front thigh is showing the best performance. However, it is pertinent to mention here that since the data present for each muscle is very small in size, our results are not conclusive (Table 2). We now present the performance of ESKNN classifier for each feature in Table 3. As it can be observed from the table, maximum

Table 1 Performance comparison of different classifiers

Classifier	Accuracy (%)			
	ALS versus CO	Myopathy versus CO	Myopathy versus ALS	CO versus ALS versus myopathy
ESKNN	95.2	93.7	96.4	92.3
EBT	92.3	84.6	94.4	87.5
KNN ($k = 10$)	88.7	89.1	93.8	82.8
SVM linear	87.7	79.6	92.3	77.7
SVM quad	90.6	84.9	93.4	82.3
SVM cubic	91.2	88.3	93.9	64.9
SVM gaussian	90.4	88.5	93.8	84.3

Table 2 Performance of model with respect to difference muscle group

Muscle	ALS versus myopathy		
	Sensitivity (%)	Selectivity (%)	Accuracy (%)
Vastus medialis	99.1	98.4	98.2
Tibialis anterior	98.2	99.1	98.1
Deltoideus	96.3	94.2	96.7
Biceps brachii	86.2	93.3	95.1

Table 3 Feature-wise results obtained ESkNN

Feature	Accuracy (%)			
	Myopathy versus CO	ALS versus CO	Myopathy versus ALS	Myopathy versus ALS versus CO
Maximum value	81.8	83.7	92.2	76.5
Minimum value	81.2	84	92.1	76.1
Variance	87	85.2	93.2	80.8
Kurtosis	55.6	75.3	72.5	52.6
Entropy	78.8	74.6	89.1	68.9

Table 4 Comparison with state of the art

Author	Classes	Accuracy (%)
Sing et al. [17]	Myopathy versus ALS	87
Belkhou et al. [1]	Myopathy versus ALS	91.11
Mishra et al. [12]	Myopathy versus ALS versus CO	88
Sengar et al. [15]	ALS versus CO	92.5
Istenic et al. [10]	Myopathy versus neuropathy versus CO	81.5
Joshi et al. [11]	ALS versus myopathy	94.42
	ALS versus CO	89.16
	Myopathy versus CO	82.41
Proposed work	ALS versus Myopathy	96.4
	ALS versus CO	95.2
	Myopathy versus CO	93.7
	ALS versus myopathy versus CO	92.3

and minimumm amplitude values are relevant features for Myopathy versus CO and ALS versus CO. This may be due to the fact that in ALS, the ampliude of the MUAPs increases by a large margin as shown in Fig. 1a, whereas for myopathy, these values decrease as shown in Fig. 1b. Since the ALS MUAPs have both high positive and negative amplitude values as compared to myopathy and control, variance as a feature works better for distinguishing these the muscle diseases. It can also be noted that the performance of kurtosis is not good as compared to other features. Finally, we compare our results for the four classification tasks, namely ALS versus CO, ALS versus myopathy, myopathy versus CO, and ALS versus CO versus myopathy, with the literature in Table 4. The proposed algorithm performs better for all these tasks as can be seen in the table.

4 Conclusions

We proposed a machine learning model for NMD detection using FDM and ensemble subspace KNN. Results are shown for binary (ALS vs. CO, ALS vs. myopathy and myopathy vs. CO) and tertiary classes (ALS vs. CO vs. Myopathy). Our model classifies NMD with accuracies: 95.2% for ALS versus CO, 96.4% for ALS versus myopathy, 93.7%, myopathy versus CO, and 92.3% ALS versus myopathy versus CO. In future, we would like to develop a subject-independent methodology where the subjects involved in training the model will not be used for testing. Also, we would like to use control subjects' data collected from different muscles and develop a muscle-independent data set. We will aim to obtain an improved algorithm with better performance metrics.

References

1. Belkhou A, Achmamad A, Jbari A (2019) Classification and diagnosis of myopathy emg signals using the continuous wavelet transform. In: 2019 scientific meeting on electrical-electronics & biomedical engineering and computer science (EBBT). IEEE, pp 1–4
2. Doulah ASU, Iqbal MA, Jumana MA (2012) Als disease detection in emg using time-frequency method. In: 2012 international conference on informatics, electronics & vision (ICIEV). IEEE, pp 648–651
3. Doulah A, Fattah S (2014) Neuromuscular disease classification based on mel frequency cepstrum of motor unit action potential. In: 2014 international conference on electrical engineering and information & communication technology. IEEE, pp 1–4
4. Dubey R, Kumar M, Upadhyay A, Pachori RB (2022) Automated diagnosis of muscle diseases from emg signals using empirical mode decomposition based method. Biomed Signal Process Control 71:103098
5. Fatimah B, Javali A, Ansar H, Harshitha B, Kumar H (2020) Mental arithmetic task classification using Fourier decomposition method. In: 2020 international conference on communication and signal processing (ICCSP). IEEE, pp 0046–0050
6. Fatimah B, Preethi A, Hrushikesh V, Singh BA, Kotion HR (2020) An automatic siren detection algorithm using Fourier decomposition method and MFCC. In: 2020 11th international conference on computing, communication and networking technologies (ICCCNT), pp 1–6. https://doi.org/10.1109/ICCCNT49239.2020.9225414
7. Fatimah B, Singh P, Singhal A, Pachori RB (2020) Detection of apnea events from ecg segments using Fourier decomposition method. Biomed Signal Process Control 61:102005
8. Fatimah B, Singh P, Singhal A, Pachori RB (2021) Hand movement recognition from semg signals using Fourier decomposition method. Biocybern Biomed Eng 41(2):690–703
9. Fatimah B, Singh P, Singhal A, Pramanick D, Pranav S, Pachori RB (2021) Efficient detection of myocardial infarction from single lead ecg signal. Biomed Signal Process Control 68:102678
10. Istenič R, Kaplanis PA, Pattichis CS, Zazula D (2010) Multiscale entropy-based approach to automated surface emg classification of neuromuscular disorders. Med Biol Eng Comput 48(8):773–781
11. Joshi D, Tripathi A, Sharma R, Pachori RB (2017) Computer aided detection of abnormal emg signals based on tunable-q wavelet transform. In: 2017 4th international conference on signal processing and integrated networks (SPIN). IEEE, pp 544–549
12. Mishra VK, Bajaj V, Kumar A (2016) Classification of normal, als, and myopathy emg signals using elm classifier. In: 2016 2nd international conference on advances in electrical, electronics, information, communication and bio-informatics (AEEICB). IEEE, pp 455–459

13. Mitchell TM (1997) Machine learning. McGraw-Hill, New York
14. Nikolic M (2001) Detailed analysis of clinical electromyography signals: EMG decomposition, findings and firing pattern analysis in controls and patients with myopathy and amytrophic lateral sclerosis. PhD thesis
15. Sengar N, Dutta MK, Travieso CM (2017) Identification of amyotrophic lateral sclerosis using emg signals. In: 2017 4th IEEE Uttar Pradesh section international conference on electrical, computer and electronics (UPCON). IEEE, pp 468–471
16. Sharma RR, Chandra P, Pachori RB (2019) Electromyogram signal analysis using eigenvalue decomposition of the Hankel matrix. In: Machine intelligence and signal analysis. Springer, pp 671–682
17. Singh A, Dutta MK, Travieso CM (2017) Analysis of emg signals for automated diagnosis of myopathy. In: 2017 4th IEEE Uttar Pradesh section international conference on electrical, computer and electronics (UPCON). IEEE, pp 628–631
18. Singh P, Joshi SD, Patney RK, Saha K (2017) The Fourier decomposition method for nonlinear and non-stationary time series analysis. Proc R Soc A Math Phys Eng Sci 473(2199):20160871
19. Singhal A, Singh P, Fatimah B, Pachori RB (2020) An efficient removal of power-line interference and baseline wander from ecg signals by employing Fourier decomposition technique. Biomed Signal Process Control 57:101741
20. Torres-Castillo JR, López-López CO, Padilla-Castañeda MA (2022) Neuromuscular disorders detection through time-frequency analysis and classification of multi-muscular emg signals using hilbert-huang transform. Biomed Signal Process Control 71:103037

Drowsiness Detection for Automotive Drivers in Real-Time

R. Chandana and J. Sangeetha

Abstract This paper presents IoT-based monitoring system for drowsiness detection for automotive drivers in real-time. The proposed system undergoes three levels of drowsiness detection system to monitor the driver drowsiness and alert him as and when required. The process begins with alcohol detection as a safety precaution, if alcohol is not sensed, then the system proceeds further to detect the face else the engine turns off. Initially, the driver's face is captured and trained using Haar cascade classifier and AdaBoost algorithm is used to select the meta-data in Haar like features. The proposed system detects only the authorised driver's face and estimates the eye closure rate, which is captured through the live streaming video from the pi camera. In level 1, if the eye-aspect ratio is below the threshold value, then a sound alerting system is generated. In level 2, if sound alert is prolonged for more than two times, a human voice alerting system is enabled and in the final level, a notification with the GPS location is sent to the driver's owner or any concerned person. The continuous retrieved data will be stored in the log file. The system uses infrared light to detect driver's drowsiness at night-time.

Keywords AdaBoost algorithm · Drowsiness detection · Eye closure rate · Haar cascade classifier · Sound alerting system · Voice alerting system

Abbreviations

EAR	Eye-aspect ratio
PERCLOS	Percentage of eyes closed

R. Chandana (✉) · J. Sangeetha
Computer Science and Engineering, MSRIT, Bengaluru, India
e-mail: chandana.r15101995@gmail.com

J. Sangeetha
e-mail: sangeethakirank@msrit.edu

K. A. Reddy et al. (eds.), *Proceedings of Fourth International Conference on Computer and Communication Technologies*, Lecture Notes in Networks and Systems 606,
https://doi.org/10.1007/978-981-19-8563-8_17

1 Introduction

Internet of Things (IoT) is a system of interrelated computing device capable of transmitting data across the wide network without any human-interference. There are various applications in IoT that exists in healthcare, smart farming, smart energy, smart city, smart transportation, etc. Also, challenges relating to IoT are security, privacy and reliability. One realistic challenge in smart transportation is safe driving. Driver drowsiness often leads to road accidents and loss of exogenous cues leads to critical events. The driver's level of concentration degrades because of less sleep and long monotonous driving. 40% of road accidents in the world are caused by driver fatigue as given by Central Road Research Institute (CRRI) [1].

The objective of this research work is to monitor the driver drowsiness in order to prevent road accidents by saving many lives and alert the driver when it is required [2]. Detection of drowsiness during day as well as during night-time is essential. The major challenges concerning our paper are detecting the face of the driver in a live video frame. However, detection of face is a challenging task here and though we detect face, the next challenge is to detect the eye and eye-state classification to determine if the driver is drowsy or not. Accurate results are provided based on different head orientations.

In [3], Dasgupta et al. mentioned the "face detection" followed by the "eye detection" and "eye-classification" based on the open or closed eye conditions, and in [4], Gou et al. stated that recognition of iris, visual interaction and driver fatigue detection has been done. Visual cues are used to characterise the level of alertness of the driver and to determine the driver fatigue. In [5], Gupta, et al. have discussed about the visual indications include movement of the eyelid, head movements, gaze movements and different facial expression. In [6], Gou et al. have stated that the Haar classifier identifies the class of face and non-face based on the knowledge learned during the training. In [7], Avizzano et al. have presented an "embedded computer vision system" is used to monitor the face of the driver and eyelid blinking in the train. The system detects the face of the driver in various variable illuminations and detects the presence of more than one face in the train's cabin. The system also detects the driver's watchfulness correctly without any false positives. The system requires low computational load and runs on embedded system. To achieve accurate facial land marking, identification of correct face location is necessary. Also, eliminating background patterns are also essential.

In [8], Chiou et al. have explained the driver monitoring system (DMSs), which is used to reduce the road accidents occurred due to drowsy driving. DMSs focus on determining any abnormal behaviour while driving. The researcher has proposed a personal-based hierarchical DMS (HDMS). In the first level, the HDMS detects the behaviour of the driver. If abnormal behaviour is found, then the second layer of the HDMS detects whether the driver is drowsy or distracted.

In [9], Mandal et al. have discussed a vision-based bus monitoring system to detect drowsiness of bus drivers. The system includes "head-shoulder detection", "face detection", "eye detection" and percentage of eyelid closure (PERCLOS). By using spectral regression, the continuous level of eyes openness is estimated. To determine the eye state, fusion algorithm is used. Based on the PERCLOS values obtained, the system analyses whether the driver is drowsy or not.

The contribution of this paper is to monitor the driver drowsiness and alert him when required. The proposed system detects only the authorised driver's face, and then, the eyes are located which is captured through the live streaming video from the camera. Face is detected involving many variations of image appearance, which involves pose variation, image orientation, illuminating conditions, occlusion and various facial expressions. After detecting the face, eyes are detected where eye closure is set and compared to the reference values. If the value descends below the threshold value, then the driver is alerted from a sound alerting system else the process continues in identifying face, eye and drowsy state continuously. If the driver continues to be drowsy for a prolonged time, then a human voice alerting system over the speaker is enabled, and a notification message via short message service (SMS) and an e-mail is sent to the owner of the vehicle. The proposed system is also used to prevent vehicle theft where only the authorised person can drive the vehicle. Any other unauthorised person face is recognised, then an alert sound gets generated and an SMS, e-mail with the GPS location is sent to the vehicle's owner. Infrared light is used to detect face at night times. The various levels of drowsiness detected are stored in the log file as a future reference for the vehicle's owner, in order to analyse the performance of the driver.

The methodology of the system is explained in Sect. 2. Section 3 determines the experimental setup, and Sect. 4 explains the results attained in the paper.

2 The Methodology

The system overview comprises of three-level drowsiness detection. "Face detection" is carried out initially, followed by "eye detection" and "eye-state classification" of eyes. In Fig. 1, the system makes use of three-level driver drowsiness detection approach. The process begins with alcohol detection as a safety precaution, if alcohol is not sensed then the engine starts else the engine will remain in off state. After the alcohol detection is done, the system detects if the driver is authorised or not. If the drowsiness is detected, then the driver is alerted from a sound alerting system as the first level of drowsiness detection. In the second level, voice alerting system is generated when the threshold reaches a certain range. Later, in the third level of drowsiness detection, the system sends an e-mail and SMS will be sent to the vehicle's owner.

Initially, the driver's face is identified and trained using Haar cascade classifier. The system proceeds by detecting the driver's face through the camera [10, 11]. The detected face will be verified across the trained faces, which are captured and only the authorised driver's face will be detected else the system generates a sound alert. If the drowsiness is prolonged for more than twice, then a voice alert is generated. If drowsiness continues, further then short message service (SMS) and an e-mail will

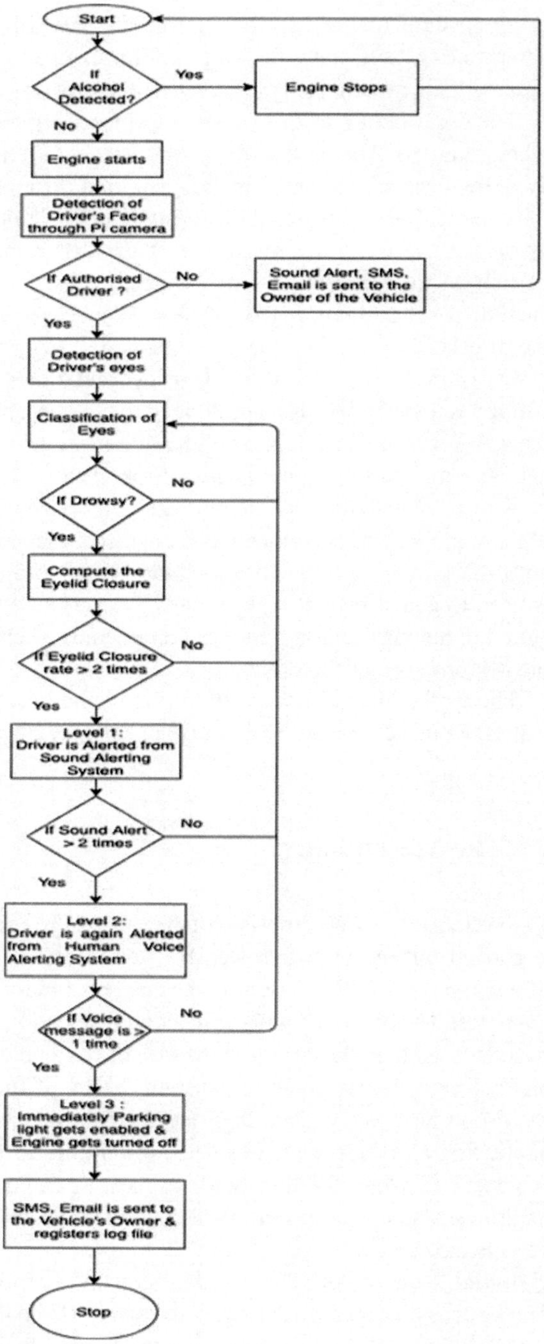

Fig. 1 Overview of the system

be sent to the owner of the vehicle. Detection of the driver's eyes is determined by the Haar cascade frontal eye detection classifier [12]. Classification of eye is determined as open, close and drowsy based on eye co-ordinates. Drowsiness of the driver is determined by computing the eyelid closure, which is based on the "eye-aspect ratio" (EAR) of both the eyes [13].

If EAR value descends below the threshold value and if eyelid closure [14, 15] occurs more than twice, then Level 1 drowsiness gets detected and driver is alerted from the sound alerting system. Subsequently in Level 2, if the alerting system is enabled more than twice, then a human voice message is generated. In Level 3, if voice message is generated more than once, and then SMS and E-mail will be sent to the vehicle's owner. Simultaneously, parking light gets enabled and engine gets turned off. The continuous retrieved data will be stored in the log file containing various levels of drowsiness detected with current date and time. The system is used for vehicles theft detection, if any other person apart from the concerned person tries to drive the vehicle, then an alerting sound, e-mail and a message notification will be sent to vehicle's owner.

2.1 Alcohol Detection

Once the driver gets seated in the vehicle, MQ-3 (Mĭngăn Qĭlai) alcohol sensor, which is placed near the driver's seat detects the existence of alcohol gases. If the value ranges from 0.05 to 10 mg/L, then alcohol will be detected else alcohol will not be sensed. Based on the alcohol sensor readings, the system allows the driver to start the vehicle's engine else it will remain in idle state.

2.2 Face Detection

Face is detected, and images are captured through camera. The detected face will be verified across the trained faces, which are captured and only the authorised driver's face will be detected.

Haar cascade classification is a machine learning algorithm where positive and negative images are used to train the classification classifier [15]. The positive images contain the images that have to be detected, and the negative images contain images other than positive images [16]. It is evident to use fast and precise detection of face. Haar cascade classifier is used based on the technique of "Haar wavelet" in order to determine the pixels in image [17]. Initially, Haar features are obtained by considering the corresponding rectangular regions in the sliding window at specific locations. This measures the intensities of the pixels in each area and assesses the difference between these quantities. It uses the concept of "integral image" to analyse the "features" [18]. Integral image is an image where in both horizontal and vertical axis we get cumulative addition of intensities on subsequent pixels. Haar cascade uses AdaBoost learning algorithm to extract the appropriate features from a large collection to produce an effective classification result [19]. Cascading technique is used to identify the face in an image and remove images that are insignificant. This reduces the number of weak classifiers and increases the detection speed [20].

2.3 Face Detection of Authorised Driver and Unauthorised Face

The system gets trained with the authorised driver's face; the training module helps to collect the pictures of authorised driver with various face orientations through camera and trains the system using Haar cascade classifier.

In Fig. 2, image samples are collected to train the face with different orientations using Haar cascade. Adaptive boosting learning ("AdaBoost") algorithm is used to select the important features from the dataset to produce an effective result of classifiers. Several weak classifiers are trained based on same training set. The strong classifier is made up of previous weak classifiers that will be jointly boosted. The efficient classifier has a greater capacity to identify the face [21, 22]. In order to process the data using AdaBoost algorithm, we need

Quality data: this tries to correct misclassifications in the training data.
Outliers: to rule out any unrealistic observations.
Noisy data: to isolate the required data from the unwanted data [23].

This technique of cascading detects the face in an image and discards irrelevant images. These images help the system to detect the faces in real-time even with

Fig. 2 Sample of images with different orientations

different head orientations. Eye is detected from upper half region of the face [24]. Here, the driver's face is captured through the live streaming video with various face orientations and training of the face is done by using Haar cascade classifier. Cascade confidence is measured across the image. The confidence level of the detected face can be calculated by the system where it checks for the similar features with the trained images [25]. This confidence level is found to improve the accuracy in detecting authorised driver's face. If the confidence level is greater than 80%, then the detected face will be authorised driver's face. Detection of unauthorised face is done in order to prevent vehicle theft. Training has been carried out by capturing driver's face with various face orientations using Haar cascade classifier [26]. Cascade confidence is measured across the image. If the confidence level is lesser than 80%, then the detected face is unauthorised driver's face.

2.4 Eye Detection

Eye is detected from the upper half region of the face [10, 27]. Haar classifier is used to train the eye images by using edge detection. Different eye images are collected and trained to detect eye from live streaming video frame [28]. Here, eye images are considered as positive images and the rest of the images, which does not have eyes are considered as negative images [29]. AdaBoost learning algorithm is used to select the important features from the dataset to produce an effective result of classifiers. This technique of cascading detects the eye in an image and discards irrelevant images.

2.5 Eye-State Classification

Eye-state classification is obtained through percentage of eyelid closure [30]. On an average, percentage of eyelid closure of a normal person per minute is calculated as follows

$$\begin{aligned} P &= E_c/(E_o + E_c) \times 100\% \\ P &= 19/(41 + 19) \times 100\% \\ P &= 0.31 \end{aligned} \tag{1}$$

where P represents percentage of eye closure, E_c and E_o represents the count of closed eyes and open eyes, respectively. Higher the P value, greater the drowsiness value.

In Eq. (1), percentage of eyelid closure is computed by calculating the ratio of count of closed eye to the sum of both the open and closed eyes. The average count of closed, open eyes for a normal human per minute is around 19 and 41, respectively. Hence, the percentage of the eye closure is computed as 0.31. Based on the threshold

Fig. 3 Eye landmarks

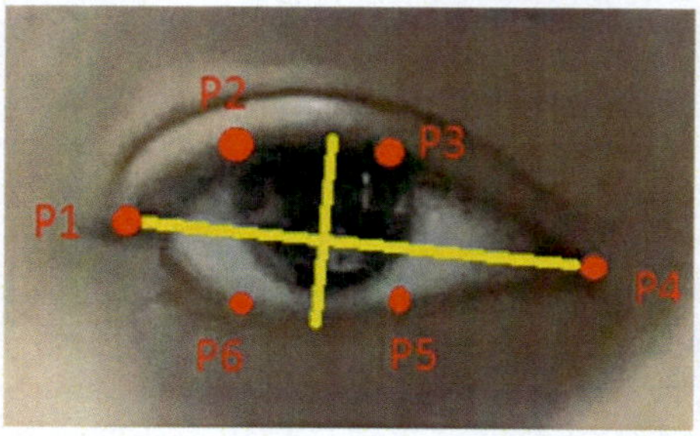

value, if the eyelid closure is lesser than 31%, then the eye state is classified as "close" else it is classified as "open" state. Eye-aspect ratio (EAR) of both the eyes is computed, and average of EAR is computed as shown below.

$$\text{EAR} = (A + B)/(2.0 * C) \tag{2}$$

where

$A = \text{Euclidean\_distance}(\text{eye}[P_2], \text{eye}[P_6])$
$B = \text{Euclidean\_distance}(\text{eye}[P_3], \text{eye}[P_5])$
$C = \text{Euclidean\_distance}(\text{eye}[P_1], \text{eye}[P_4])$.

Threshold of eye is set to a certain range (0.31) as illustrated in Eq. 1. If the EAR value falls below the specified range, then drowsiness will be detected [23, 24]. The drowsiness detection is captured for every five frames per sec. This way eye-state classifier reduces false alarm in the eye detection level [25].

In Fig. 3, six landmarks of eye are located as $P_1, P_2, P_3, P_4, P_5, P_6$ co-ordinates, respectively, where the Euclidean distances are calculated based on the distances between two points. The numerator in Eq. (2) computes the sum of distances between the vertical eye landmarks ($P_2, P_6; P_3, P_5$), while the denominator computes twice the distances between the horizontal eye landmarks (P_1, P_4).

In general

$$\begin{aligned}\text{EAR} &= \frac{||P_2 - P_6|| + ||P_3 - P_5||}{2||P_1 - P_4||} \\ &= \frac{(0.91 - 0.71) + (0.85 - 0.74)}{2(0.38 - 0.22)} \\ &= \frac{0.31}{0.32} = 0.95\end{aligned} \tag{3}$$

Here, EAR value 0.95 concludes that the eye is open.

2.6 *Levels of Drowsiness Detection*

The various levels of detection are listed as follows:

Level 1 Drowsiness: Sound Alerting System
Level 2 Drowsiness: Voice Alerting System
Level 3 Drowsiness: Alert Notification.

Level 1 Drowsiness: Sound Alerting System

In level 1, if the value of EAR drops below the threshold value of 0.31 and if eyelid closure occurs more than twice, then Level 1 drowsiness gets detected and driver is alerted from the sound alerting system.

Level 2 Drowsiness: Voice Alerting System

In level 2, when the sound alert gets generated more than twice, then the driver is again alerted from the human voice alerting system. If driver gets alerted after the voice alert, then the system again checks for the classification of eyes. The sound alert is attained when the eye-aspect ratio descends below the threshold value [26]. External speakers are connected to the Raspberry Pi which gets triggered when the second level of drowsiness is detected.

Level 3 Drowsiness: Alert Notification

In level 3, we will receive an alert notification by two ways simultaneously when drowsiness is detected. Firstly, by an SMS alert and later through an e-mail notification with GPS location which is sent to the vehicle's owner. In this level, parking light gets enabled and the engine stops. Various levels of drowsiness detection will continuously register in log file.

3 Experimental Setup

The experimental setup of the hardware is as shown in Fig. 4.

Hardware Components

The hardware components comprises of a Raspberry Pi board with a Pi camera, LCD display, 4 way relay module, alcohol sensor, buzzer, switch, parking light, vehicle's engine, which are defined in Table 1.

Software framework includes Python 2.7 with windows $\times$ 64 installer is open-source software. "Open-Source Computer Vision" library is an open-source machine learning and computer vision software focuses mainly on image processing.

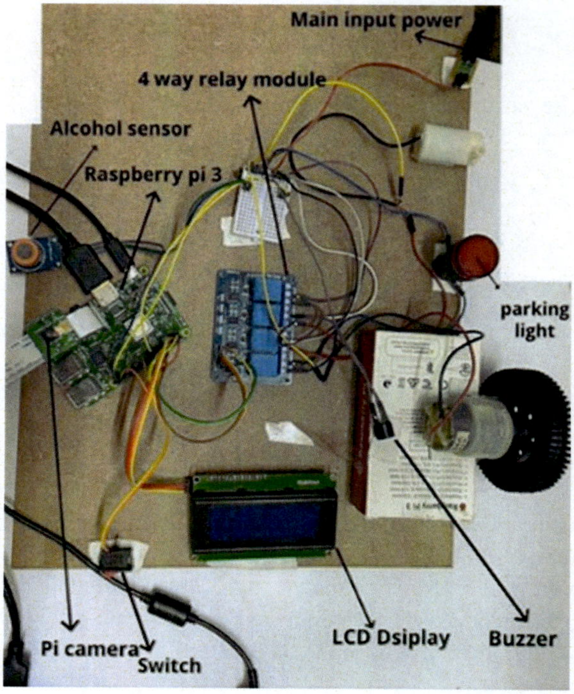

Fig. 4 Hardware components of the experiment

4 Results and Discussion

The following sections give the results attained at various stages.

4.1 Alcohol Detection

Initially, the engine will be in off state and the engine starts only when the alcohol is not sensed. The alcohol sensor is attached to the system, which checks if the driver has drunk or not. This helps in safe driving.

In Fig. 5, the alcohol is not sensed. The alcohol MQ3 gas sensor concentration ranges from 0.05 to 10 mg/L. If the value lies in between the specified range, then the engine will remain in off state. Once the alcohol detection is done, and if alcohol is not sensed, then the engine starts.

4.2 Face Detection

The system starts detecting the face and checks if the face is authorised or not.

Table 1 Hardware components

Sl. No.	Components	Functionality
1	Raspberry Pi board	Raspberry Pi 3 B+ is a device of credit size that can be plugged into a laptop. It has 1.4 GHz 64 bit quad processor, wireless LAN, power over Ethernet support. An SD card is inserted that acts as Raspberry Pi's hard-drive. It is operated by HDMI port and USB is used to connect video output to the laptop
2	Pi camera	Pi camera is basically a lightweight camera which supports Raspberry Pi. It communicates with Raspberry Pi by using MIPI camera serial interface-protocol. It has 5 MP resolution, which is capable of capturing videos and images. Dimensions having 25 mm × 20 mm × 9 mm
3	LCD display	Silicon LCD flat panel display with blue backlight is used. Blue backlight operates with 5 V DC with size 20 × 4
4	4-way relay module	The 4 channel relay module is used for high voltage and current charge control. It is integrated with Raspberry Pi, which is the microcontroller to control the actions of LED display, vehicle's engine, parking light present in the model. It also comes with LED to show the status of the relay
5	Alcohol sensor	MQ3 alcohol gas sensor detects the ethanol in the driver's breathe and provides the alcohol concentration as an output. The sensing concentration is in the range of 0.04 mg/L to 4 mg/L. It has good sensitivity and responds fast
6	Buzzer	5 V passive buzzer audio signalling device is used to produce the buzzer sound
7	Switch	Small mini boat switches with on and off button

Fig. 5 Alcohol not sensed

The system further checks if the face is authorised or not. Figure 6 represents known face detection, and Fig. 7 represents unknown face detection.

If an unknown face is detected, then an e-mail alert is sent to the vehicle's owner. The sent e-mail contains an image of the unknown driver's face with Global Positioning System (GPS) location of the vehicle as shown in Fig. 8.

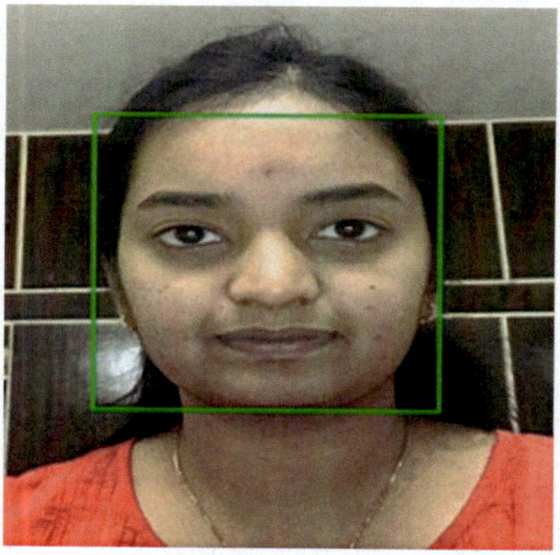

Fig. 6 Detection of known face

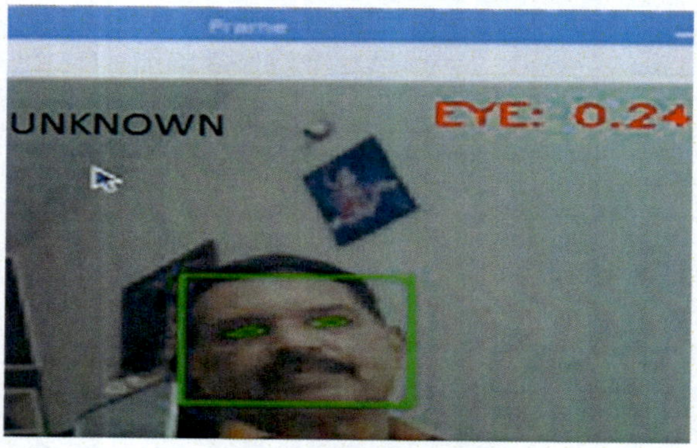

Fig. 7 Detection of unknown face

This way it helps to identify any theft activity. Also, SMS alert is sent to vehicle's owner when unknown face is identified as shown in Fig. 9.

In Fig. 9, SMS alert is received immediately by the vehicle's owner stated as "Hello owner, unknown driver found, Take action!!!" when an unknown face is recognised. This helps the vehicle's owner to take immediate action accordingly.

4.3 Eye Detection

Once the authorised face is recognised, then the system identifies the eyes as in Fig. 10.

Fig. 8 E-mail alert received by the vehicle's owner when an unknown driver is identified

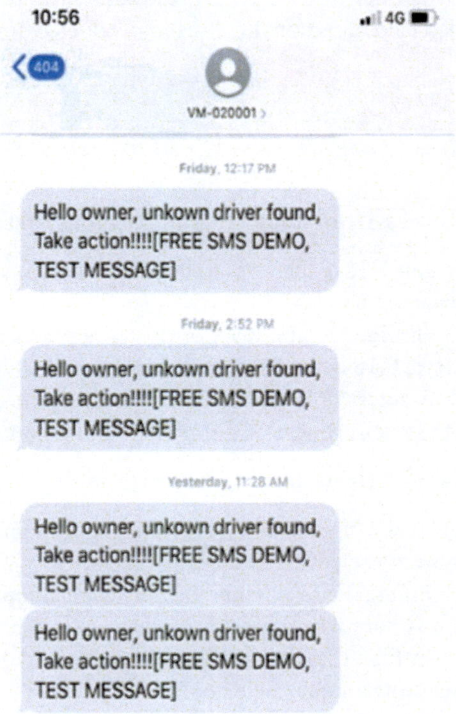

Fig. 9 SMS alert received by the vehicle's owner when an unknown driver is identified

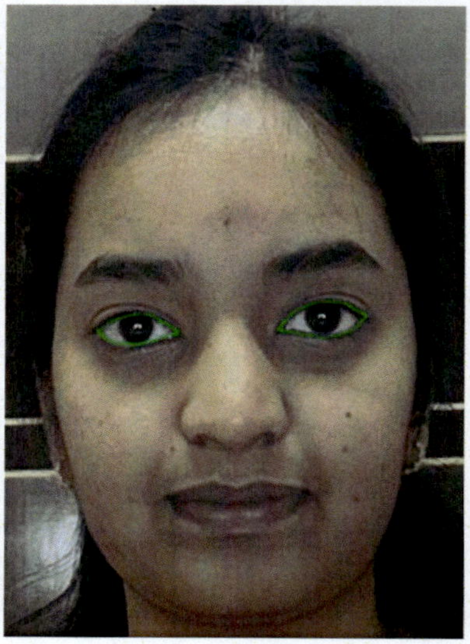

Fig. 10 Detection of eyes

On the basis of EAR, drowsiness is detected. Three levels of drowsiness are detected based on the threshold values below 0.31.

4.4 Various Levels of Drowsiness

Level 1 Drowsiness: Sound Alerting System

In level 1, if the eye-aspect ratio is below 0.31, then a sound alerting system is generated.

In Fig. 11, first level of drowsiness is detected when the EAR falls below 0.31. Here, 0.18 is the EAR, which is lesser than the threshold value (0.31). Thus, level 1 drowsiness detection is detected. Once the drowsiness is detected more than twice, the system generates a sound alarm to alert the driver.

Level 2 Drowsiness: Voice Alerting System

In level 2, if sound alert is prolonged for more than two times, a human voice alerting system is enabled through speakers.

Figure 12 represents the hardware setup of the speakers, which is connected to the system where a voice message gets enabled when drowsiness is detected.

In Fig. 13, the EAR value is 0.22, which is less than the threshold value. Thus, the drowsiness is detected.

Level 3 Drowsiness: Alert Notification

Third level of drowsiness is detected if voice alert is exceeded more than once as shown in Fig. 14.

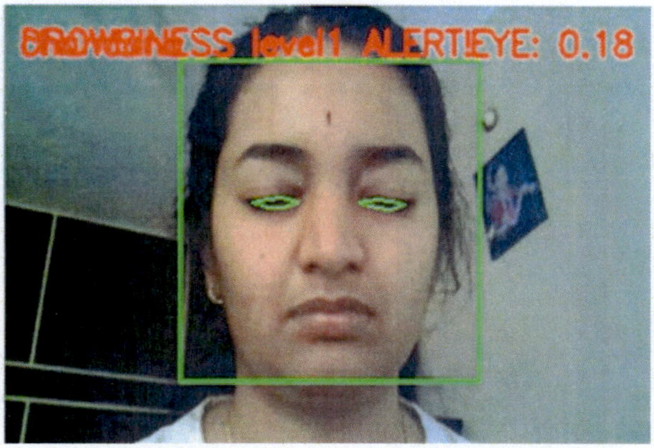

Fig. 11 Level 1 drowsiness detected

Fig. 12 External speakers connected to the system

Fig. 13 Level 2 drowsiness detected

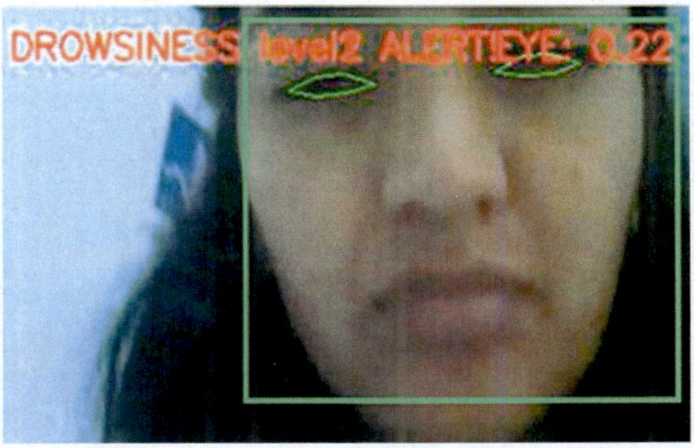

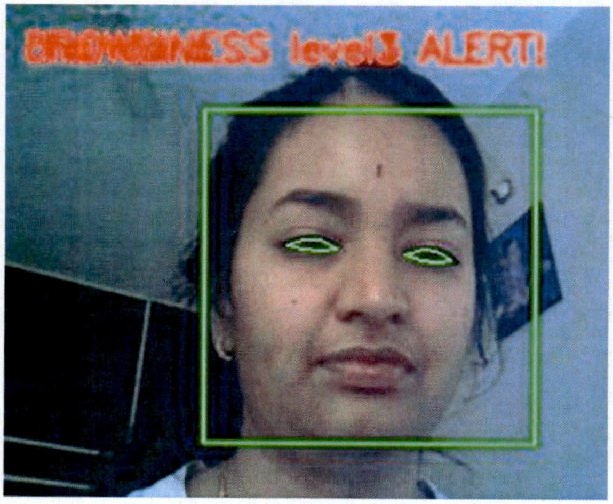

Fig. 14 Level 3 drowsiness detected

In third level of drowsiness detection, a notification with the GPS location is sent to the driver's owner or any concerned person. The drowsiness can be clearly detected during both day and night times as shown in Fig. 15a, b, respectively.

Here, EAR in daytime is 0.32 which is greater than the threshold value 0.31, hence drowsiness is not detected. The EAR during night-time driving is 0.33 where the value is greater than threshold value where the drowsiness is not detected.

In Fig. 16, E-mail alert is received immediately after the third level of drowsiness detection. Vehicle's owner receives an e-mail alert with an attachment containing the driver's face along with the GPS location. SMS Alert is sent to vehicle's owner as in Fig. 17.

Figure 17 represents a SMS, which is sent to the vehicle's owner after the third level of drowsiness detection. Here, Vonage's SMS API is used to send and receive

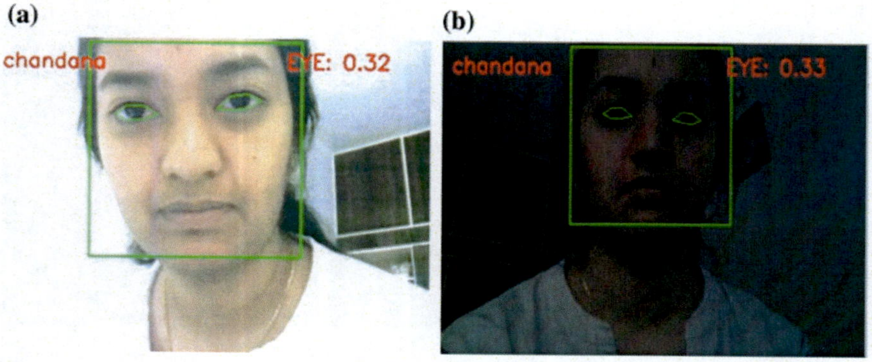

Fig. 15 **a** Detection of face and eye during daytime driving. **b** Detection of face and eye during night-time driving

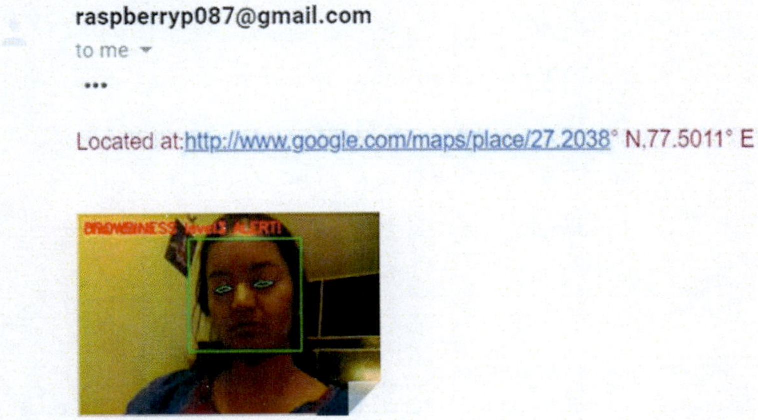

Fig. 16 E-mail alert

Fig. 17 SMS alert received by the vehicle's owner

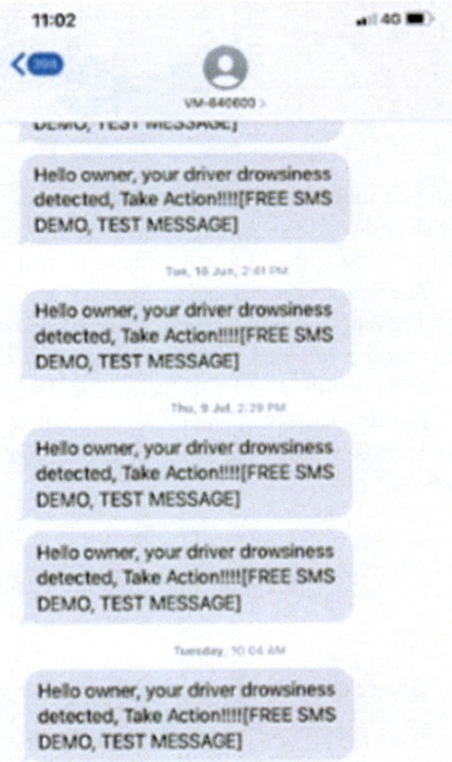

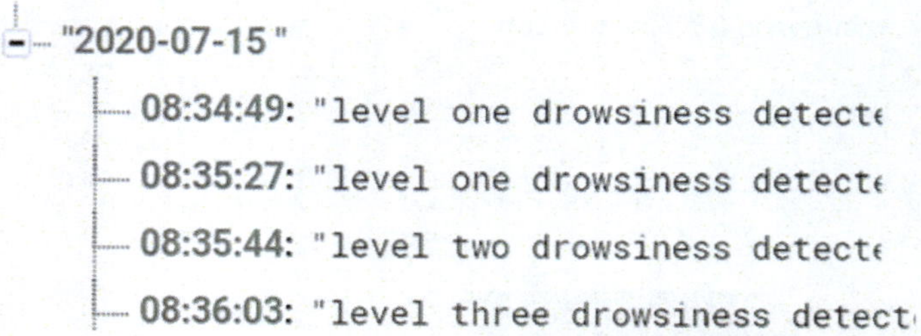

Fig. 18 Log file

Table 2 Comparisons of threshold values

Set	Total drowsy cases threshold = 0.21	Predicted drowsy	Total drowsy cases threshold = 0.31	Predicted drowsy	Accuracy (%)
I	20	18	20	19	92.5
II	20	17	20	18	87.5
III	20	19	20	19	95
IV	20	16	20	18	85

the text messages. Once the drowsiness detected, Vonage API gets triggered and a SMS is sent to the vehicle's owner using the local number. The continuous retrieved data will be stored in the log file as shown in Fig. 18.

The various drowsiness levels are stored in the log file for future reference with the current date and time. This helps the driver owner to identify the various drowsiness detection levels and take appropriate actions. Here, the first level of drowsiness is detected as followed by the second and third level of drowsiness detection.

The drowsiness of the driver is predicted by the threshold value of 0.31. When the threshold value descends below certain range, drowsiness is detected as shown in Table 2.

$$\begin{aligned} \text{Mean Accuracy} &= (92.5 + 87.5 + 95 + 85)/4 \\ &= 90\% \end{aligned} \tag{4}$$

Table 2 consists of four sets, and each set consists of 20 drowsy cases;

In set I, 18 cases are predicted as drowsy when threshold value is set to 0.21 and 19 cases are predicted as drowsy when threshold value is set to 0.31.
In set II, 17 cases are predicted as drowsy when threshold value is set to 0.21 and 18 cases are predicted as drowsy when threshold value is set to 0.31.
In set III, 19 cases are predicted as drowsy when threshold value is set to 0.21 and 19 cases are predicted as drowsy when threshold value is set to 0.31.

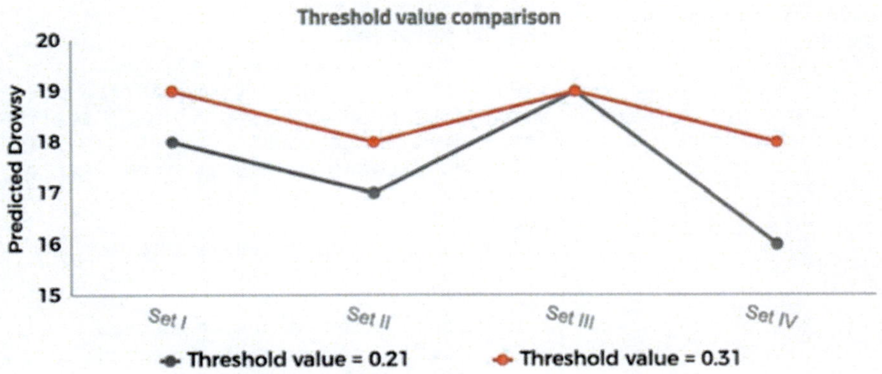

Fig. 19 Graph of threshold value comparison

In set IV, 16 cases are predicted as drowsy when threshold value is set to 0.21 and 18 cases are predicted as drowsy when threshold value is set to 0.31.

The accuracy of the drowsiness predicted is higher when the threshold value is 0.31. The mean accuracy is calculated as shown in Eq. (4) by considering four sets.

In Fig. 19, drowsiness is predicted for four sets; here, the red-line indicates the threshold value 0.31 and grey-line indicates the threshold value 0.21. The chances of predicting drowsiness are more accurate when the threshold value is 0.31.

In Fig. 20, we can see various drowsiness levels detected with date and time. At the initial level, the system checks if the driver has taken alcohol or not through the alcohol sensor. If alcohol is not sensed, then the engine starts. Subsequently, camera turns on and captures the images from the live steaming video. Preceding the system recognises the face, if known face is recognised, then eye detection will be done else unknown face will be detected and a sound alarm will be raised along with SMS and an e-mail alert. When the known face is detected and driver drowsiness is recognised, then Level 1 drowsiness will be detected and the driver gets alerted through a sound alert system. If prolonged for more than twice, Level 2 drowsiness will be detected and the driver gets alerted through human voice alerting system. Sequentially, if the drowsiness is prolonged more than once, then Level III drowsiness will be detected by enabling the parking light, which blinks the light thrice and engine gets turned off automatically.

The illustration of three-level verification is as shown in Table 3.

Here, the verification of drowsiness is carried out in three levels. The system is validated in order to check the accuracy. These results are accurate as false predictions can be eliminated in earlier levels.

The overall framework is illustrated in Table 3 showing sample data of 15 cases. Here, "*A*" and "*D*" denote the user is "alert" and "drowsy", respectively.

$$P(A) = 7/15 = 0.466667 \tag{5}$$

```
date is 2020-07-14
time is 23:26:05
loading facial landmark predictor...
[INFO] starting video stream thread...
[info] Engine On starting alcohol test
[info] Test for alcohol
[info] No alcohol sensed. engine started
buzzer on
buzzer off
[info] level 1 drowsiness detected
buzzer on
buzzer off
[info] level 1 drowsiness detected
[info] level 2 drowsiness detected
[info] level 3 drowsiness detected
Indicator Lights Blink
Indicator Lights Blink
Indicator Lights Blink
Engine is OFF
switch engine off manually
```

Fig. 20 Screen snapshot of all results

Table 3 Illustration of three-level verification

Actual	*A*	*A*	*A*	*A*	*A*	*A*	*A*	*D*	*D*	*D*	*D*	*D*	*D*	*D*	*D*
Predicted by "Level 1"	*A*	*A*	*D*	*A*	*A*	*A*	*D*	*D*	*D*	*D*	*D*	*D*	*D*	*D*	*A*
Predicted by "Level 2"			*D*				*A*	*D*	*D*	*A*	*D*	*D*	*D*	*D*	
Predicted by "Level 3"			*A*					*D*	*A*		*A*	*D*	*D*	*D*	

$$P(D) = 1 - P(A) = 1 - (7/15) = 8/15 = 0.533333 \tag{6}$$

Here, out of 15 cases, seven cases are alert and eight cases are drowsy.

The probability of alert $P(A)$ and drowsy $P(D)$ is calculated, respectively, as shown in Eqs. (5 and 6).

The probability of eyelid closure predicting as "drowsy" given when driver is actually drowsy in level 1 is obtained as

$$P(\text{Level } 1 = D/D) = 7/8 = 0.875 \tag{7}$$

The probability of eyelid closure predicting as "drowsy" given when driver is actually drowsy in level 2 is obtained as

$$P(\text{Level } 2 = D/D) = 6/8 = 0.75 \tag{8}$$

The probability of eyelid closure predicting as "drowsy" given when driver is actually drowsy in level 3 is obtained as

Table 4 System accuracy levels

Sets	Predicted	Correct prediction	Accuracy in %
I	$A = 15, D = 10$	23	92
II	$A = 13, D = 12$	21	84
III	$A = 16, D = 9$	22	88
IV	$A = 12, D = 13$	21	84

$$P(\text{Level } 3 = D/D) = 4/8 = 0.5 \tag{9}$$

The probability of eyelid closure predicting as alert given when driver is actually alert in level 1 is obtained as

$$P(A/A) = 5/7 = 0.714 \tag{10}$$

The probability of eyelid closure predicting as alert given when driver is actually alert in level 3 is obtained as

$$P(\text{Level } 3 = A/A) = 1/7 = 0.142 \tag{11}$$

This indicates that three levels of drowsiness has improved the prediction indicating the alert state *P(A/A)* = 0.714, with false positive probability as 2/7 = 0.285.

The system accuracy levels can be determined as shown in Table 4.

In Table 4, system accuracy is calculated by considering four sets of 25 samples each.

In set I, alert cases are 15, drowsy cases are 10. Out of these 25 cases, our system predicted 23 cases correctly. Therefore, the accuracy is 92%.

In set II, alert cases are 13, drowsy cases are 12. Out of these 25 cases, our system predicted 21 cases correctly. Therefore, the accuracy is 84%.

In set III, alert cases are 16, drowsy cases are 9. Out of these 25 cases, our system predicted 22 cases correctly. Therefore, the accuracy is 88%.

In set IV, alert cases are 12, drowsy cases are 13. Out of these 25 cases, our system predicted 21 cases correctly. Therefore, the accuracy is 84%.

5 Conclusion

We have proposed a three-level driver drowsiness detection system. It is vital to detect the driver's drowsiness at earlier levels and to alert him as the number of road accidents is increasing due to drowsy driving in the recent days. If the driver is detected being drowsy, the system alerts him with a sound alerting system. If prolonged for more than twice, a human voice alerting system from speaker is enabled and an alert message is sent to the owner. These continuous retrieved data gets stored in the log file for future reference by the vehicle's owner to track the driver. Infrared light is used to detect drowsiness of the driver during night-time. The proposed system is more reliable because the system has three levels of drowsiness detection of the driver. The system is also advantageous for using SMS service to inform the vehicle owner's or the concerned person concerning the loss of attention of the driver. The system is also used for vehicles theft detection, if any other person apart from the concerned person tries to drive the vehicle, then an alerting sound, e-mail and a message will be sent to the vehicle's owner. This application is applicable to monitor the face's in ATM centres, lifts and sending an alarm if any consequences occur for child and women safety.

6 Future Scope

Extension of the work can be to exploit driver health conditions in order to improvise the driver safety in driver drowsiness detection. Drowsiness can be detected based on speech signals, when a system generates a question through the speaker, then the driver has to respond through speech. Seat vibration can be implemented as a part of physical alert.

Acknowledgements We would thank the funding agency "Karnataka State Council for Science and Technology" (KSCST) for accepting the project proposed and sponsoring our project. Also our college Ramaiah Institute of Technology, CSE dept. for helping us to complete the project with favourable outcome.

References

1. Dinges DF (1995) An overview of sleepiness and accidents. J Sleep Res 4(s2):4–14
2. Lee Y-C, Lee JD, Boyle LN (2007) Visual attention in driving: the effects of cognitive load and visual disruption. Hum Factors, J Hum Factors Ergonom Soc 49(4):721–733
3. Dasgupta A, George A, Happy SL, Routray A (2013) A vision-based system for monitoring the loss of attention in automotive drivers. IEEE Trans Intell Transp Syst 14(4):1825–1838
4. Viola P, Jones MM (2004) Robust real-time face detection. Int J Comp Vis 57(2):137–154
5. Gupta S, Dasgupta A, Routray A (2011) Analysis of training parameters for classifiers based on Haar-like features to detect human faces. In: International conference on image information processing (ICIIP), Nov 2011, pp 1–4
6. Gou C, Wu Y, Wang K, Wang K, Wang F-Y, Ji Q (2017) A joint cascaded framework for simultaneous eye detection and eye state estimation. Patt Recogn 67:23–31
7. Avizzano CA, Tripicchio P, Ruffaldi E, Filippeschi A, Jacinto-Villegas JM (2019) Real-time embedded vision system for the watchfulness analysis of train drivers. IEEE Trans Intell Transp Syst
8. Chiou CY, Wang WC, Lu SC, Huang CR, Chung PC, Lai YY (2019) Driver monitoring using sparse representation with part-based temporal face descriptors. IEEE Trans Intell Transp Syst
9. Mandal B, Li L, Wang GS, Lin J (2017) Towards detection of bus driver fatigue based on robust visual analysis of eye state. IEEE Trans Intell Transp Syst 18(3), March 2017

10. Singh S, Prasad SVAV (2018) Techniques and challenges of face recognition: a critical review. Procedia Comp Sci 143; In: 8th International conference on advances in computing and communication (ICACC-2018) 2018
11. Ji Q, Zhu Z, Lan P (2004) Real-time nonintrusive monitoring and prediction of driver fatigue. IEEE Trans Veh Technol 53(4):1052–1068
12. Hong T, Qin H (2007) Drivers drowsiness detection in embedded system. In: Proceedings of international conference on vehicular electronics and safety (ICVES), Dec 2007, pp 1–5
13. Lang L, Qi H (2008) The study of driver fatigue monitor algorithm combined PERCLOS and AECS. In: Proceedings of international conference computer science software engineering, vol 1, Dec 2008, pp 349–352
14. Zhang Y et al (2019) Research and application of AdaBoost algorithm based on SVM. In: 2019 IEEE 8th joint international information technology and artificial intelligence conference (ITAIC), Chongqing, China, pp 662–666. https://doi.org/10.1109/ITAIC.2019.8785556
15. Chandana R, Sangeetha J (2021) Review on drowsiness detection for automotive drivers in real-time. Nat Volatiles Essen Oils 8(6), Jan 2021
16. Vinay A, Joshi A, Surana HM, Garg H, Murthy KB, Natarajan S (2018) Unconstrained face recognition using ASURF and cloud-forest classifier optimized with VLAD. In: 8th International conference on advances in computing and communication ICACC-2018
17. Kakade SD (2016) A review paper on face recognition techniques. Int J Res Eng Appl Manage (IJREAM) 2(2), May 2016
18. Parte RS, Mundkar G, Karande N, Nain S, Bhosale N (2015) A survey on eye tracking and detection. Int J Inno Res Sci Eng Technol 4(10), Oct 2015
19. Jin L, Niu Q, Jiang Y, Xian H, Qin Y, Xu M (2013) Driver sleepiness detection system based on eye movements variables. Hindawi Publishing Corporation, Article ID 648431
20. Fitriyani NL, Yang CK, Syafrudin M (2016) Real-time eye state detection system using Haar cascade classifier and circular Hough transform. In: IEEE 5th global conference on consumer electronics
21. Sahu M, Nagwani NK, Verma S, Shirke S (2015) Performance evaluation of different classifier for eye state prediction using EEG signal. Int J Knowl Eng 1(2), Sept 2015
22. Chan TK, Chin CS, Chen H, Zhong X (2019) A comprehensive review of driver behavior analysis utilizing smartphones. IEEE Trans Intell Transp Syst
23. Pratama BG, Ardiyanto I, Adji TB (2017) A review on driver drowsiness based on image, bio-signal, and driver behaviour. In: 3rd International conference on science and technology—computer (ICST) 2017
24. Kusuma Kumari BM, Ramakanth Kumar P (2017) A survey on drowsy driver detection system. 78-1-5090-6399-4/17/$31.00 c. IEEE
25. Ramzan M, Khan HU, Awan SM, Ismail A, Ilyas M, Mahmood A (2019) A survey on state-of-the-art drowsiness detection techniques. IEEE Access 7
26. Dhupati LS, Kar S, Rajaguru A, Routray A (2010) A novel drowsiness detection scheme based on speech analysis with validation using simultaneous EEG recordings. In: Proceedings on IEEE international conference on automation science and engineering. (CASE), Aug 2010, pp 917–921
27. Song F, Tan X, Liu X, Chen S (2014) Eyes closeness detection from still images with multi-scale histograms of principal oriented gradients. Pattern Recognit 47(9):2825–2838
28. Manjutha M, Gracy J, Subashini P Dr, Krishnaveni M Dr (2017) Automated speech recognition system—a literature review. Int J Eng Trends Appl (IJETA) 4(2), Mar–Apr 2017
29. Kashevnik A, Lashkov I, Gurtov A (2019) Methodology and mobile application for driver behavior analysis and accident prevention. IEEE Trans Intell Transp Syst
30. Tran D, Du J, Sheng W, Osipychev D, Sun Y, Bai H (2018) Human-vehicle collaborative driving framework for driver assistance. IEEE Trans Intell Transp Syst

Prediction of Dementia Using Deep Learning

Tushar Baliyan, Tarun Singh, Vedant Pandey, and G. C. R. Kartheek

Abstract Artificial intelligence and its sub-field machine learning are continuously evolving and being applied in medicine and healthcare amongst other important fields. Machine learning and deep learning are frequently used to aid dementia prediction and diagnosis. Deep learning models are better than other machine learning models for dementia detection and prediction, but they are more computationally very expensive. The objective of the work is to build a deep learning model to predict dementia. This model is designed to predict dementia from brain MRI images and is based on the concepts of deep learning and convolutional neural network (CNN). The developed model is able to identify demented and non-demented MRI images with an accuracy of 99.35%, better than existing models.

Keywords Dementia · Deep learning · Machine learning · CNN

1 Introduction

Dementia is a global health problem. Dementia is a brain condition in which a few groups of brain cells stop functioning properly, resulting in cognitive impairments. Dementia affects one's ability to think, recall, and communicate, affecting one's daily life and activities. One of the common forms of dementia is Alzheimer's disease. As per the World Alzheimer Report 2019, the number of individuals worldwide suffering from dementia is estimated to be over 50 million in 2019, with that figure anticipated

T. Baliyan · T. Singh · V. Pandey · G. C. R. Kartheek (✉)
Department of Computer Science and Engineering, CMR Institute of Technology, Bengaluru, India
e-mail: kartheek.gcr@cmrit.ac.in

T. Baliyan
e-mail: tuba18cs@cmrit.ac.in

T. Singh
e-mail: tasi18cs@cmrit.ac.in

V. Pandey
e-mail: vepa18cs@cmrit.ac.in

K. A. Reddy et al. (eds.), *Proceedings of Fourth International Conference on Computer and Communication Technologies*, Lecture Notes in Networks and Systems 606,
https://doi.org/10.1007/978-981-19-8563-8_18

to rise to 152 million by 2050 [1]. However, scientists are yet to discover a cure for Alzheimer's disease that can treat and prevent the disease precisely. Based on clinical dementia rating (CDR) value, the dementia is categorized into four stages: very mild, mild, moderate, and severe dementia. Because the treatment costs for very mild dementia patients differ a lot from that of the severe dementia patients, it is important to diagnose dementia illnesses early in order to maximize patient recovery and reduce treatment costs [2].

A major issue is incorrect diagnosis, as the majority of dementia patients are initially seen by general physicians, who often fail to recognize dementia and hence diagnose it incorrectly. Due to such late diagnosis, physicians are often unable to slow the progression of dementia and reduce debilitating behavioural changes. A simple way for diagnosing dementia early in its development might lead people to seek diagnosis and treatment sooner than later.

Recent advances in deep neural network approaches have showed a lot of promise in terms of combining massive administrative claims and electronic health record databases power, also powerful computation to generate good predictive models for healthcare. Many deep learning techniques have been used to detect and diagnose dementia along with other neurological diseases. Deep learning, unlike typical machine learning algorithms, incorporates all three fundamental processes in neural network modelling: feature extraction, feature dimension reduction, and classification. CNN and RNN have become predominant mechanisms in deep learning. In computer vision and image analysis, CNN is currently the most successful deep learning model. CNN model architectures are typically made of several layers such as convolutional layer, pooling layer, and activations. The model uses these layers to extract features from images gradually.

2 Related Work

This section of the paper explains about the various existing works on the prediction of dementia and other neurological disorders like Alzheimer's disease and Parkinson's disease. Some of these work on neuroimaging MRI dataset, whilst others review the parameters from the clinical data of patients. A thorough review of various literatures reveals that dementia is a degenerative brain condition that eventually leads to memory loss. Exploratory data analysis on longitudinal MRI dataset resulting in a technique called 'CapNet' which emphasizes the use of classification methods by which an image retrieval system is fed with images as query inputs [3]. Investigative analysis on the use of deep learning models to predict dementia using the longitudinal health information of patients reveals that the deep learning models provide a significant boost in the performance of models. Dementia is not the only neurological disease for which predictive models have been built. Analysis of algorithms like linear discriminant analysis, K-nearest neighbours, and support vector machines in identifying Parkinson's disease revealed that SVM provides better accuracy [4]. An 'ALL-PAIRS' technique developed to investigate the progression of Alzheimer's was effective when trained on patient data [5]. A predictive and preventive CNN model to predict Alzheimer's disease in the early stages along with a system that displays the preventive measures to be taken along with suggestion of medication outperforms traditional ML algorithms, when trained on both cross-sectional and longitudinal MRI scans [6]. Apart from these, a deep learning model validated on MRI scans to predict the progression towards Alzheimer's disease ranging from 6 to 18 months with a follow-up duration of 18–54 months finds a clear advantage of using 'hippocampal' features for improved prediction [7]. The use of fluorodeoxyglucose (FDG) PET and structural MRI in a 3D DenseNet brain age prediction model to see how the brain age gap relates to degenerative cognitive disorders showed an age-dependent saliency pattern of brain areas, and CNN-based age prediction provided good accuracy is proposed in [8]. A study with the goal of developing a machine learning model for predicting occurrence of Alzheimer's disease, mild cognitive impairment, and similar dementias using structured data obtained from electronic health record and administrative sources, developed a 'label-learning approach' using a cohort of patients and controls using data obtained within two years of the patient's incident diagnosis date [9]. The model achieved an accuracy of over 80% and AUC and sensitivity over 40% and thus has the utility to pre-screen patients for further diagnosis or evaluation for clinical trials.

3 Proposed Work

We propose a CNN-based model to predict dementia using MRI scan images of the patient. The MRI scan is given to the proposed model as input. In the pre-processing, the input image is resized to 128 * 128 and normalized. This image is then sent to the CNN classifier which predicts the presence of dementia as depicted in Fig. 1.

Convolutional neural networks (CNNs) are made up of various layers, which are typically the input, hidden, and output layers. The convolution takes place in the hidden layer. Computers view images as pixels, and convolution uses this ability to classify images. Features are extracted from the images in convolutional layer. The kernels in the convolutional layers scan through images to extract the feature map. Convolution layer is followed by a pooling layer. The function of the pooling layer is to reduce the feature map to prevent over fitting. Activation functions are used to activate a neuron when needed. The pooling method selected here is maxpooling2d.

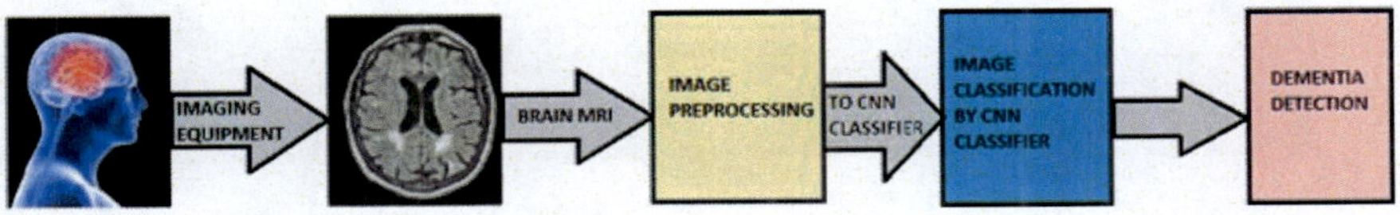

Fig. 1 Proposed working

4 Experiment

In this paper, we have used the dementia MRI dataset [10], publically available in Kaggle. The dataset is hand collected from various Websites with labels verified. This dataset has 6199 brain MRI images of size 176 × 208 pixels, and we have labelled the images with 'yes' for patients with dementia and 'no' for patients without dementia. The dataset comprised of 3190 images with 'no' labels and 3009 images with 'yes' label. The dataset is further divided into training and validation sets with training set as 80% and the validation set as 20% of the images in the dataset. The sample images of both non-demented and demented from the dataset are given in Fig. 2.

The proposed CNN model architecture is given in Fig. 3. The model is made up of convolutional layer along with a max pool layer stacked up sequentially in 3 layers. The last max pool layer is followed by two fully connected layers. Each convolutional layer of size (3 × 3) is followed by a pooling layer of size (2 × 2). The final pooling layer is flattened. This is followed by 1 fully connected layer and 1 output layer. We have used sigmoid function for activation in the output layer as shown in Fig. 4. We have used ReLU function for activation in rest of the other layers. Binary cross entropy function is used as the loss function along with Adam optimization algorithm. We trained the model for fifty epochs. We have used early stopping to stop the training when the validation loss is smaller than 0.003 for 5 epochs.

The main difference between other predictive models and our model is the use of binary cross entropy as the loss function. We used binary cross entropy since we only wanted two possibilities for our prediction model: either presence or absence of dementia. Other predictive models prefer using categorical cross entropy as the loss function.

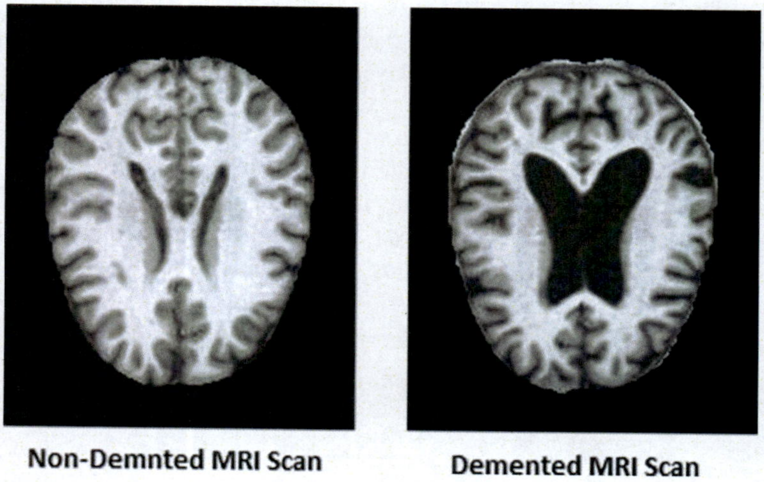

Fig. 2 Sample dataset

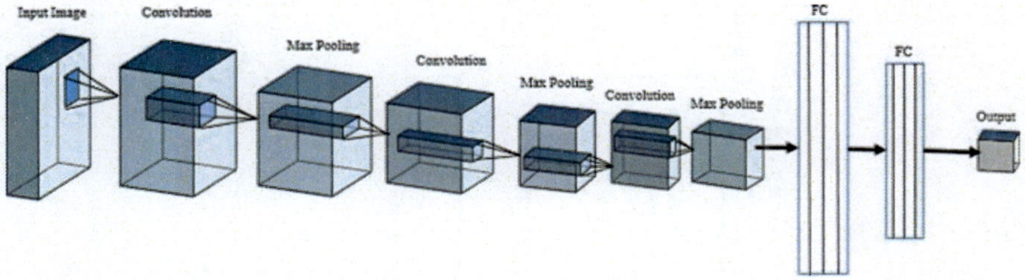

Fig. 3 CNN model architecture

5 Results

Our model has given a 99.35% training accuracy and 96.21% as validation accuracy. The training stopped at 31 epochs as change in validation loss for last 5 epochs is less than 0.003. The accuracy graph for training and validation data is given in Fig. 5. The loss graph for training and validation data is given in Fig. 6.

The comparison of the results of our model with different models is as shown in Table 1. Classification by application of query using 'CapNet' [3] has produced 92.39% accuracy, predictive and preventive-based CNN model [6] give an accuracy of 85%, and 'label-learning' approach [9] model gives an accuracy of 80%. Our model comparatively gives better accuracy of 99.335%.

6 Conclusion

In this paper, we proposed a CNN-based model for prediction of dementia. We have successfully trained our model with 3009 demented and 3190 non-demented MRI images. Sample output of our work is given in Fig. 7. Our proposed model gave us a 99.35% accuracy. The results show that by using a CNN model using 3 layers of convolutional followed by max pool layers, two fully connected layer, and an output layer using sigmoid activation, we obtained a model that has higher accuracy compared to other methods.

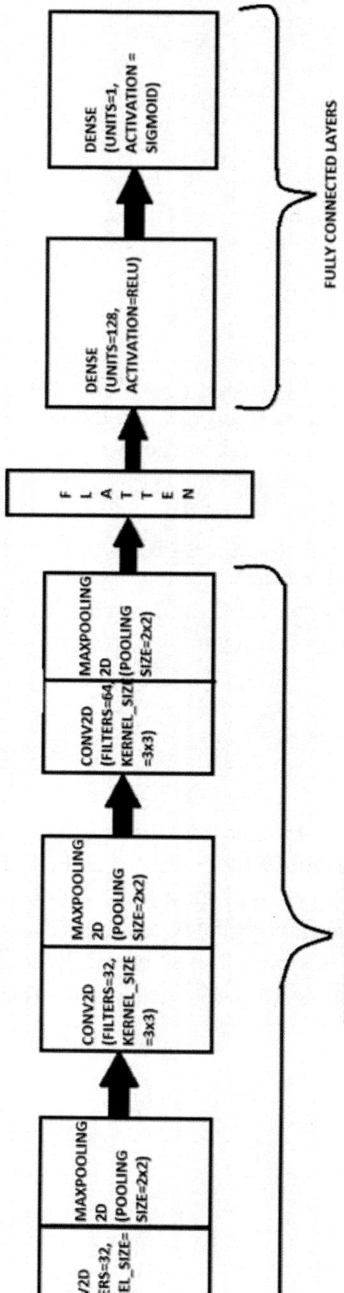

Fig. 4 CNN model working

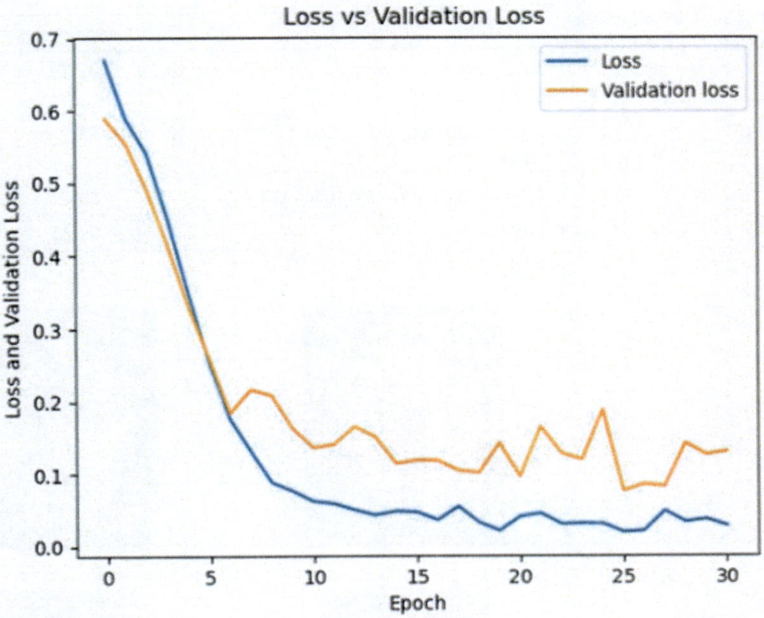

Fig. 5 Training and validation loss

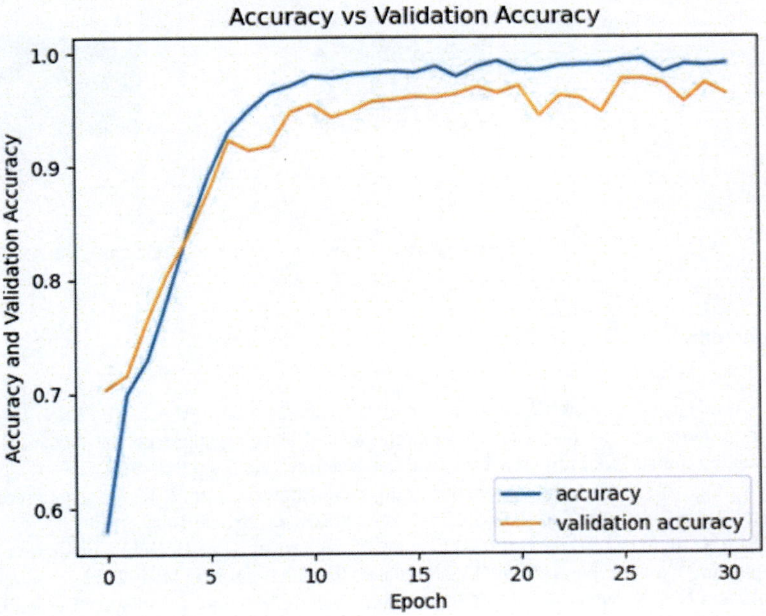

Fig. 6 Training and validation accuracy

Table 1 Comparison of our model results with different models

Model	Accuracy (%)
Classification by application of query using 'CapNet' [3]	92.39
Predictive and preventive CNN-based model [6]	85
'Label-learning' approach [9]	80
Our model	99.35

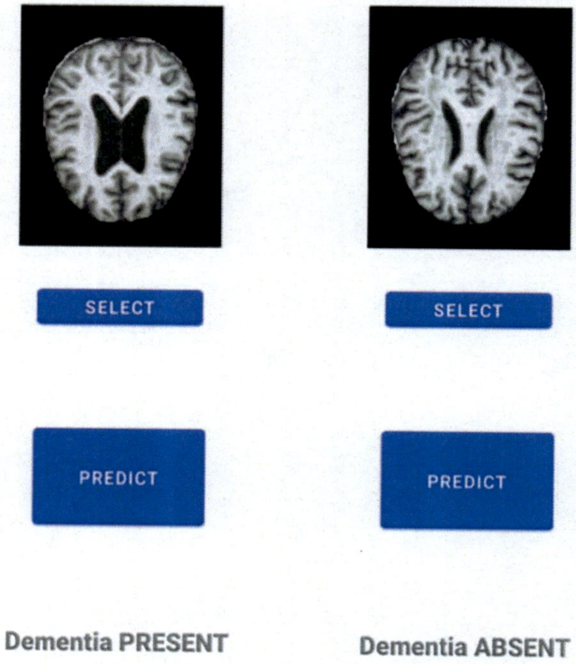

Fig. 7 Sample output

References

1. Nori VS, Hane CA, Crown WH, Au R, Burke WJ, Sanghavi DM, Bleicher P (2019) Machine learning models to predict onset of dementia: a label learning approach. In: Alzheimer's & Dementia: translational research & clinical ınterventions, vol 5, pp 918–925
2. Isik Z, Yiğit A (2019) Applying deep learning models on structural MRI for stage prediction of Alzheimer's disease. Turk J Electr Eng Comp Sci 28(1), Article 14
3. Basheer S, Bhatia S, Sakri SB (2021) Computational modeling of Dementia prediction using deep neural network: analysis on OASIS dataset. IEEE Access 9:42449–42462
4. Mathkunti NM, Rangaswamy S (2020) Machine learning techniques to ıdentify Dementia. SN Comput Sci 1:118
5. Albright J (2019) Forecasting the progression of Alzheimer's disease using neural networks and a novel preprocessing algorithm. In: Alzheimer's & Dementia: translational research & clinical ınterventions, vol 5, pp 483–491

6. Singhania U, Tripathy B, Hasan MK, Anumbe NC, Alboaneen D, Ahmed FR, Ahmed TE, Nour MM (2021) A predictive and preventive model for onset of alzheimer's disease. Front Public Health 9
7. Li H, Habes M, Wolk DA, Fan Y (2019) A deep learning model for early prediction of Alzheimer's disease dementia based on hippocampal magnetic resonance imaging data. Alzheimers Dement 15(8):1059–1070
8. Lee J, Burkett BJ, Min HK et al (2022) Deep learning-based brain age prediction in normal aging and dementia. Nature Aging 2:412–424
9. Nori VS, Hane CA, Sun Y, Crown WH, Bleicher PA (2020) Deep neural network models for identifying incident dementia using claims and EHR datasets. PLoS One 15(9)
10. Alzheimer's dataset. https://www.kaggle.com/datasets/tourist55/alzheimers-dataset-4-class-of-images

Performance Analysis of Universal Filtered Multicarrier Waveform with Various Design Parameters for 5G and Beyond Wireless Networks

Smita Jolania, Ravi Sindal, and Ankit Saxena

Abstract The next-generation cellular networks have the challenges to achieve higher data rates, low latency, and higher spectral efficiency to support the usage scenarios like massive machine type communication (mMTC) to support a high density of devices, ultra-reliable low-latency communication (URLLC) to provide high-speed mobility, and enhanced mobile broadband (eMBB) to handle larger traffic. Universal filtered multicarrier (UFMC) will be one of the possible solutions for a fifth-generation (5G) wireless network. This research paper provides a comprehensive parameterized UFMC waveform with higher-order quadrature amplitude modulation (QAM). The performance criteria such as Fast Fourier Transform (FFT) length, sub-band size, and higher-order QAM techniques with various prototype filter design constraints are analyzed and the system performance is presented. It has been concluded that UFMC is much flexible and efficient modulation technique to fulfil the dynamic requirements of 5G and beyond wireless networks.

Keywords UFMC · 5G · OFDM · MIMO · Spectral efficiency

1 Introduction

The massive deployment of wireless systems and Internet devices with new application scenarios has created demands for ubiquitous connectivity with extreme data traffic. To fulfill these needs, 5G technology has emerged to cope with challenges like increase in user density, seamless connectivity, traffic density, data rate, and extensive applications. In the current cellular network, increasing bandwidth or increasing cell density is the major factors considered to meet the requirement of peak data rate and increased capacity. The primary challenge in this approach is that the limited resources are reaching their saturation and also increasing the cost of the hardware [1].

S. Jolania (✉) · R. Sindal
IET-DAVV, Indore, India
e-mail: sprajapati2911@gmail.com

A. Saxena
Medicaps University, Indore, India

K. A. Reddy et al. (eds.), *Proceedings of Fourth International Conference on Computer and Communication Technologies*, Lecture Notes in Networks and Systems 606,
https://doi.org/10.1007/978-981-19-8563-8_19

To improve spectrum usage, the evolution of a new air interface and novel approaches to radio resources with multiple access management are needed. The design of a novel multicarrier waveform at the physical layer is to fulfill the needs of next-generation wireless networks with low peak-to-average power ratio (PAPR), high throughput, improved spectral efficiency, and reduced interchannel interference (ICI). OFDM is a widely used multicarrier modulation air interface in 4G LTE, WiMAX, optical communication, etc., but fails to meet the requirements in future scenarios of the physical layer. Due to high-frequency offset, PAPR and spectral leakage, OFDM multicarrier technique is not suitable for the next generation wireless networks [2]. For efficient utilization of spectrum with high data rate transmission and to cope with ICI, a new multicarrier modulation technique needs to be designed to bring a faster and better user experience. Various multicarrier techniques are available to meet the requirements of 5G and to improve the spectrum efficiency as discussed in the following section.

2 Concepts of Multicarrier Waveform

In filter bank multicarrier (FBMC), the spectrum is divided into multiple sub-bands which are orthogonal to each other and applies subcarrier filtering. Adaptable filters are applied at the subcarrier level to perform according to the channel conditions and use cases [3]. Although it has various advantages—time-frequency efficiency, fewer OOB emissions, and ICI proving suitable for 5G. But FBMC has very high computational complexity and incompatibility with multiple-input multiple-output (MIMO). In generalized frequency division multiplexing (GFDM), the modulated data symbols are transmitted in frequency and time, two-dimensional blocks divided into sub-symbols and subcarriers. Subcarriers are filtered with non-orthogonal pulse shaping prototype filters [4]. In this method, the major drawbacks are higher latency, incompatibility with MIMO, and complex pilot design [5]. In FBMC and GFDM, subcarrier-wise filtering is applied, but it requires a new transceiver design. Also, there are major problems with channel equalization and backward incompatibility with 4G. So we will prefer sub-band wise filtering. In universal filtered multicarrier (UFMC), sub-band filtering is applied where the total bandwidth is divided into N number of sub-bands, and filtering is applied in the frequency domain to reduce OOB emissions. Due to fine frequency filtration, shorter filter length, and compatibility with MIMO, UFMC is the best multicarrier waveform for 5G and beyond wireless networks. The remaining paper is focused on the UFMC system model and its performance.

3 Mathematical Model for UFMC Waveform

1. Let us consider a multicarrier system with the total bandwidth with C number of subcarriers indexing from [0, 1, 2, ..., $C - 1$]
2. All the subcarriers are broken down into smaller sub-bands indexing from [i = 1, 2, ..., B].
3. Each ith sub-band comprises K subsequent subcarriers, where $K = C/B$ subcarriers.
4. For the ith sub-band, where $1 \leq i \leq B$, the data blocks are represented with $x_{i,k}$, {ith sub-band and kth subcarrier, where $1 \leq k \leq K$.
5. Generating the random bit stream of data and mapped to M-QAM.
6. The QAM modulated symbol is represented as S_i (i = 1, 2, 3, ..., B) for the ith sub-band, including k_i subcarriers ($\left(\sum_{i=1}^{B} k_i = C\right)$). The QAM symbols in the frequency domain are assigned to each sub-band with length k_i [6].
7. To overcome the problem of sub-band carrier interference, the signal processing tool inverse fast Fourier transform (IFFT) is applied.
8. N point IFFT converts symbols from frequency domain (S_i) to time-domain (y_i) as shown in Eq. (2).

$$y_i = \text{IFFT}\{s_i\} \tag{1}$$

$$y_i(l) = \frac{1}{\sqrt{N}} \sum_{i=0}^{k-1} s_i(k) e^{j2\pi kl/N} \tag{2}$$

k is subcarrier index in ith sub-band.

9. Then, sub-band filtering is performed by a prototype filter of length 'L'. and generally, each filter power is normalized to unity and mathematically expressed by Eq. (3),

$$\sum_{l=0}^{L-1} |f_i(l)|^2 = 1 \tag{3}$$

10. The prototype filter should have a constant response in signal spectral range and must be suitable for communications in dispersive channels [7].
11. The summation of outputs from band filters is passed through the channel. The superposition of filtered sub-band symbols is the signal X_{UFMC} expressed as

$$X_{\text{UFMC}} = \sum_{(i=1)}^{B} y_i(l) * f_i(l), \quad \text{where } l = 0, 1, \ldots, N + L - 2 \tag{4}$$

Here, '∗' symbolizes linear convolution. Finally, the UFMC signal can be represented by Eq. (5),

$$X_K = \sum_{(i=0)}^{(K-1)} \sum_{(l=0)}^{(L-1)} \sum_{(n=0)}^{(N-1)} y_{(i,k)} f_{(i,k)}(l) e^{j2\pi k(n-1)/N} \quad (5)$$

The time-domain composite vector for a specific UFMC-based multicarrier symbol is given by Eq. (6):

$$X_K = \sum_{i=1}^{B} F_{i,k} V_{i,k} y_{i,k} \quad (6)$$

where $F_{i,k}$ is a Toeplitz matrix, comprising filter impulse response with dimension $(N + \mathrm{L} - 1) \times N$; $V_{i,k}$ is an IFFT matrix that includes the relevant columns as per the sub-band position within the available frequency range with dimension $N \times n_i$, $n_i =$ number of QAM symbols in each resource block; $y_{i,k}$ is a time-domain symbol with dimension $n_i \times 1$.

4 Mathematical Model at Receiver Side

1. $Y(k)$ is the received signal given by:

$$Y(k) = H(k).X(k) + W(k) \quad (7)$$

   $H(k)$ is channel response, $X(k)$ is the UFMC modulated transmitted signal, and $W(k)$ is the AWGN noise.
2. Then, further signal processing is applied like filtering and synchronization and applied to the fast Fourier transform (FFT) unit.

The UFMC waveform achieves better spectrum utilization than OFDM as UFMC has no cyclic prefix (CP), and sub-band filtering reduces side lobes. Although the peak-to-average power ratio (PAPR) in UFMC is better than OFDM bit it still needs to be improved in future system implementations [8]. The complete UFMC transceiver system is shown in Fig. 1.

5 UFMC System Specifications and Simulation Results

To analyze the performance of UFMC and to investigate its influence, we developed the system model in MATLAB 2021b. The system is designed for 200 subcarriers with signal to noise (SNR) of 25 dB and transmitted through the AWGN channel. The length of the received symbol in UFMC is $N + L - 1$. At the receiver, $N - L - 1$ zeros are appended, and data symbols from UFMC are recovered by applying $2N$ point FFT and QAM demodulated to get back the original signal.

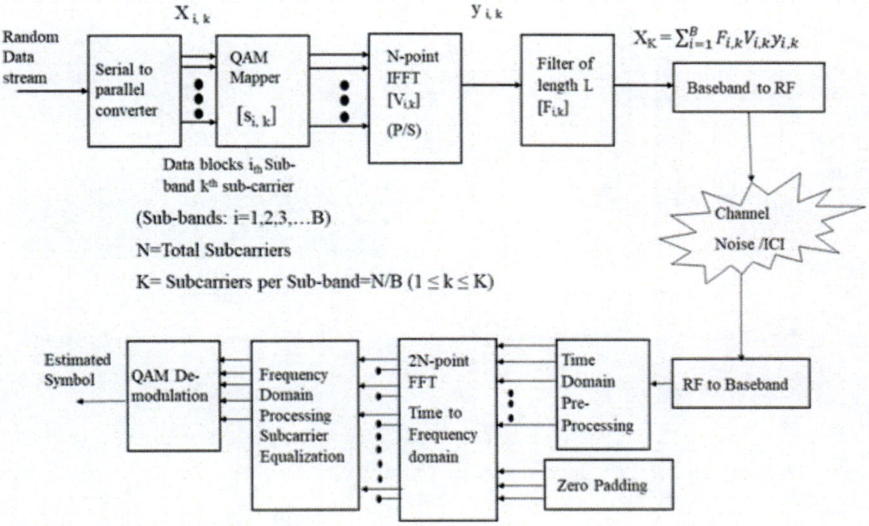

Fig. 1 UFMC transceiver systems

UFMC using different sub-band schemes: Different schemes of the sub-band ($B_{\text{sub-band}}$ = Number of Sub-bands) affect the UFMC system performance. So the allocation of the number of subcarriers per sub-band needs to be done to optimize the performance [9]. UFMC model is simulated with different system parameters, and the performance parameters such as power spectral density (PSD) and BER are analyzed. Figure 2 shows the normalized spectrum of the UFMC system with varying the number of subcarriers in each sub-band, and from Fig. 3, we conclude that as the sub-band size increases BER performance increases. The more sub-band and the smaller the sub-band size, the better performance.

Also while simulating the PSD for UFMC symbol for higher-order QAM (256 QAM) modulation and 512-FFT, from Fig. 4, we observed spectral efficiency increases as number of subcarriers in each sub-band are increased.

UFMC filter with different side lobe attenuation and filter length: Simulation is done for 256QAM, 1024 FFT, and BER is simulated. From Fig. 5a, we can conclude that the transmission efficiency and BER performance increase as the filter length decreases. From Fig. 5b, it is noted that as the side lobe attenuation is increasing, the BER performance improves.

Different FFT Size: Different FFT lengths affect the complementary cumulative distribution function (CCDF) values and BER performance [6]. The FFT size and sub-band offset size must follow the below criteria:

$$N_{\text{FFT}} \geq B_{\text{sub-band}} * K_{\text{Sub-band}} \tag{8}$$

$$\text{sub-band Offset} = \frac{N_{\text{FFT}}}{2} - \frac{B_{\text{sub-band}} * K_{\text{Sub-band}}}{2} \tag{9}$$

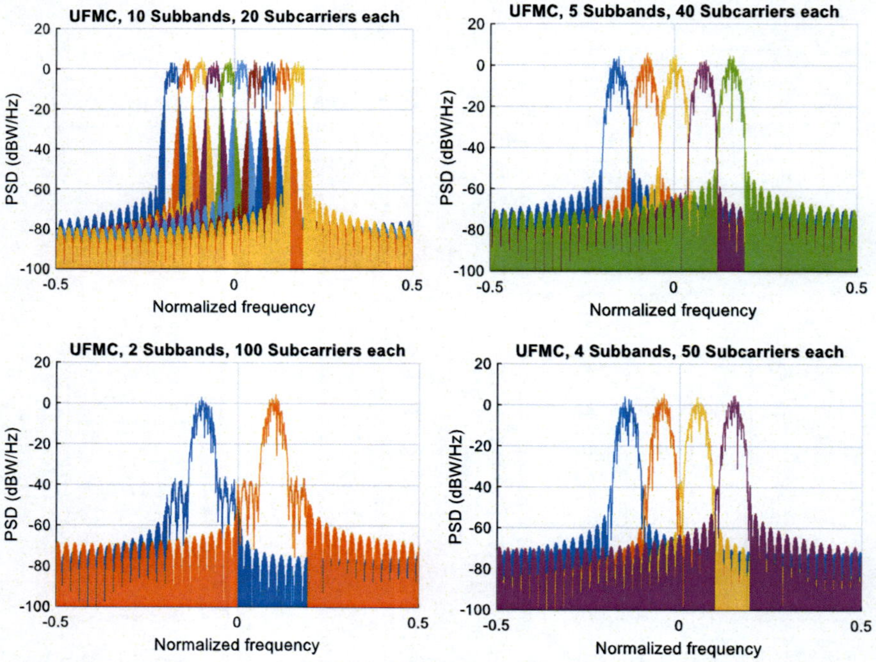

Fig. 2 PSD of UFMC with different sub-band sizes

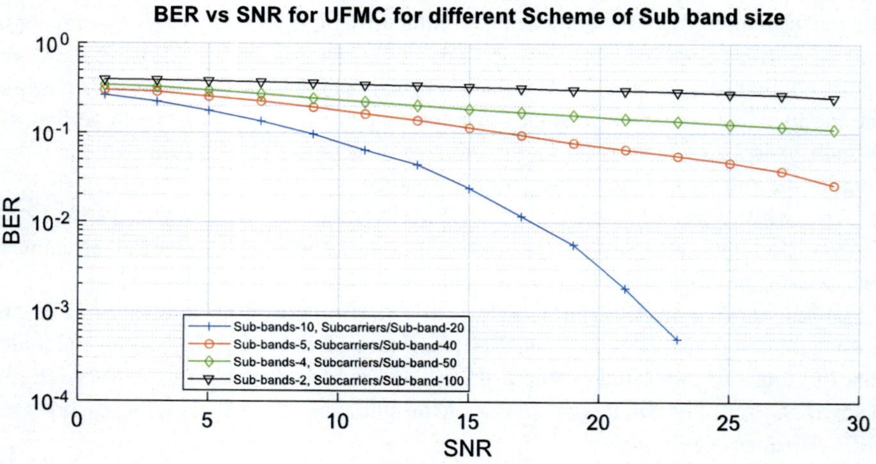

Fig. 3 BER performance of UFMC with varying sub-band size

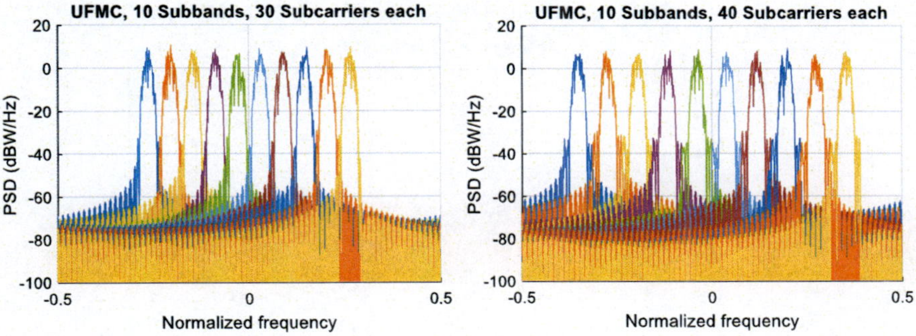

Fig. 4 PSD for given sub-band size and higher-order QAM

where $K_{\text{Sub-band}}$ = Number of subcarriers in each sub-band

Simulation results from Fig. 6a clearly show that the BER performance of the UFMC waveform scheme is improved when the FFT size increases at 256 QAM. Also, it is seen from Fig. 6b that different FFT lengths affect the system performance in terms of CCDF, and PAPR reduces with the increase in FFT size.

Different modulation: Efficient frequency spectrum utilization depends on QAM mapping and greatly affects the system capacity, so choosing higher-order QAM for UFMC performance analysis is simulated [2, 10]. To achieve good frequency spectrum utilization, we need to use the higher-order QAM mapping technique. Figure 7 shows UFMC system performance and concludes that BER increases as QAM order increases.

6 Conclusion

The multicarrier waveform UFMC is most suitable for the existing 4G as well as for future 5G and beyond systems. Good spectral efficiency due to the absence of cyclic prefix and reduced PAPR makes UFMC better than other multicarrier techniques. Sub-band filtering helps in reducing OOB emissions with the flexibility to choose sub-band size, filter length, stop-band attenuation, FFT size, and prototype window. Higher-order QAM modulation makes it best suitable with massive MIMO transmission. This research paper concluded that when the FFT size and sub-band size increase, with side lobe attenuation and filter length decreasing, the BER performance becomes better. The most important result is that for the larger FFT sizes (1024 or 2048), UFMC BER performance becomes independent of FIR filter lengths. The UFMC waveform may be the best choice for the 5G and beyond wireless networks. In future, this waveform can be implemented with a massive MIMO scenario to enhance the system capacity and spectrum utilization.

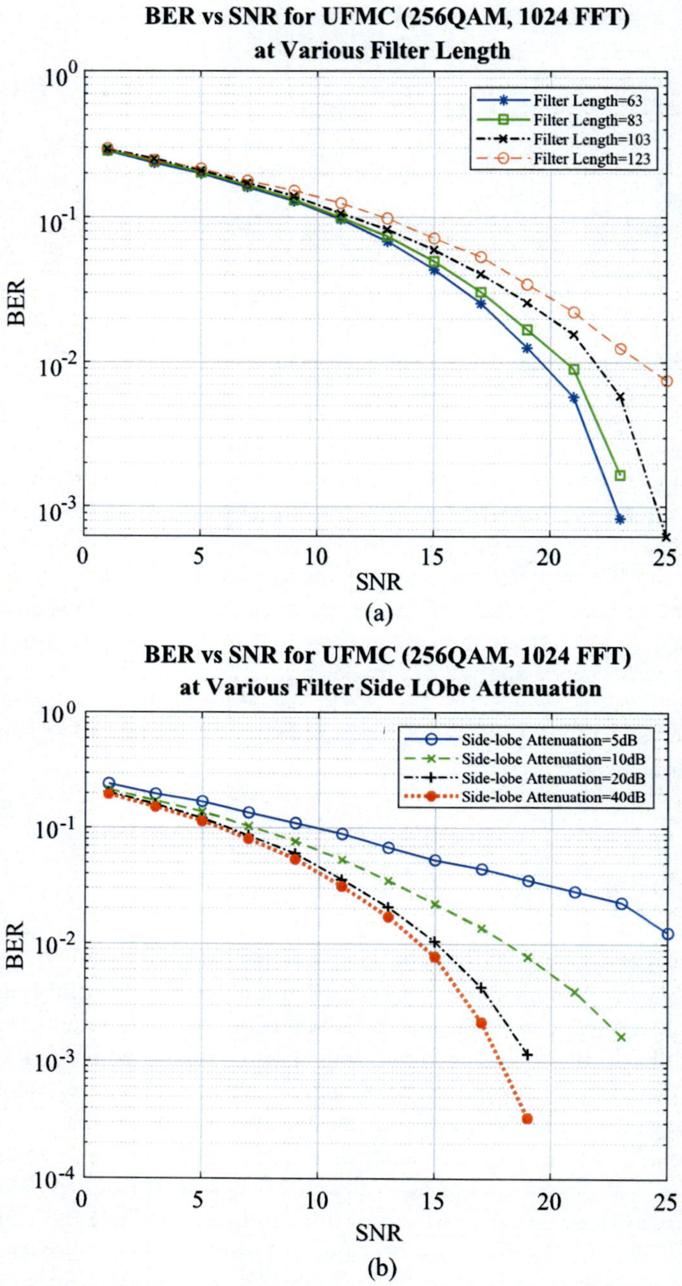

Fig. 5 **a** UFMC with different filter length, **b** BER versus SNR at different side lobe attenuation

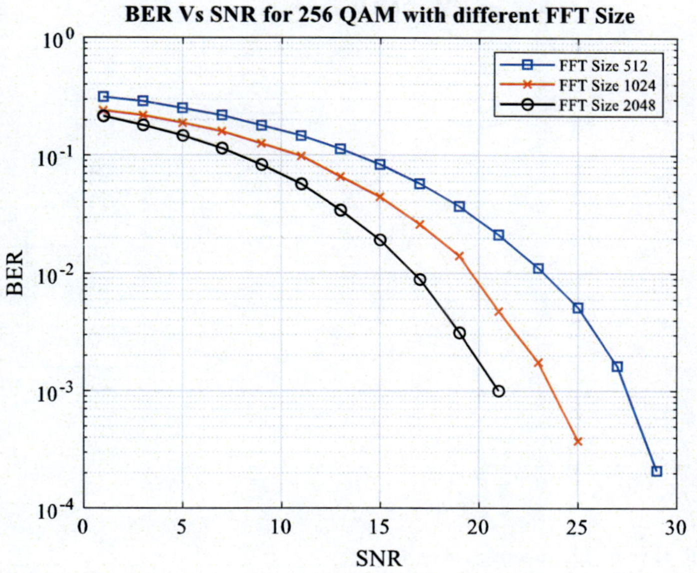

(a)

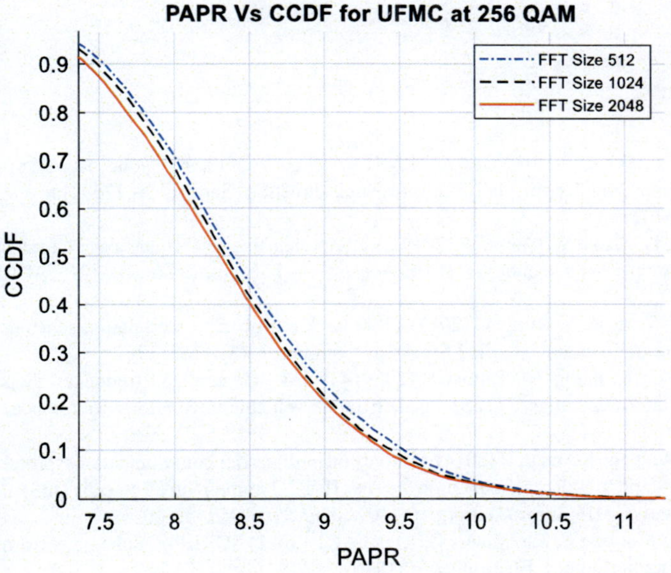

(b)

Fig. 6 CCDF and BER curve for 256QAM at different FFT size

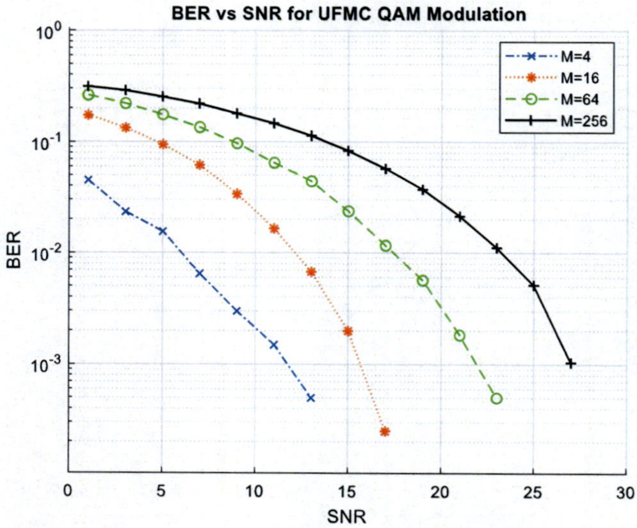

QAM Mapping	PAPR
4 QAM	9.04 dB
16 QAM	8.2379 dB
64 QAM	8.6229 dB
256 QAM	8.0416 dB

Fig. 7 BER curve for higher-order QAM

References

1. Chataut R, Akl R (2020) Massive MIMO systems for 5G and beyond networks—overview, recent trends, challenges, and future research direction. Sensors 20:2753. https://doi.org/10.3390/s20102753
2. Wei S, Li H, Zhang W, Cheng W (2019) A comprehensive performance evaluation of universal filtered multi-carrier technique. IEEE Access 7:1–1. https://doi.org/10.1109/ACCESS.2019.2923774
3. Nissel R, Schwarz S, Rupp M (2017) Filter bank multicarrier modulation schemes for future mobile communications. IEEE J Sel Areas Commun 35(8):1768–1782
4. Fettweis G, Krondorf M, Bittner S (2009) GFDM—Generalized frequency division multiplexing. In: Proceedings VTC Spring-IEEE 69th vehicular technology conference, Apr 2009, pp 1–4
5. Sahin A, Guvenc I, Arslan H (2014) A survey on multicarrier communications: prototype filters, lattice structures, and ımplementation aspects. IEEE Commun Surv Tutorials 16(3):1312–1338, Third Quarter 2014. https://doi.org/10.1109/SURV.2013.121213.00263
6. Sakkas L, Stergiou E, Tsoumanis G, Angelis CT (2021) 5G UFMC scheme performance with different numerologies. Electronics 10:1915
7. Shawqi FS, Audah L, Hammoodi AT, Hamdi MM, Mohammed AH (2020) A review of PAPR reduction techniques for UFMC waveform. In: 2020 4th International symposium on multidisciplinary studies and ınnovative technologies (ISMSIT), pp 1–6. https://doi.org/10.1109/ISMSIT50672.2020.9255246
8. Baig I, Farooq U, Hasan NU, Zghaibeh M, Jeoti V (2020) A multi-carrier waveform design for 5G and beyond Communication Systems. Mathematics 8:1466. https://doi.org/10.3390/math8091466

9. Yongxue W, Sunan W, Weiqiang W (2018) Performance analysis of the universal filtered multi-carrier (UFMC) waveform for 5G system. J Phys: Conf Ser 1169; 2018 3rd international conference on communication, image and signal processing 16–18 Nov 2018, Sanya, China
10. Kamurthi RT, Chopra SR, Gupta A (2020) Higher-order QAM schemes in 5G UFMC system. In: 2020 International conference on emerging smart computing and informatics (ESCI), pp 198–202. https://doi.org/10.1109/ESCI48226.2020.9167619

Diabetic Retinopathy (DR) Detection Using Deep Learning

Shital Dongre, Aditya Wanjari, Mohit Lalwani, Anushka Wankhade, Nitesh Sonawane, and Shresthi Yadav

Abstract Diabetic retinopathy (DR) is a condition that causes vision loss and blindness in those who have diabetes. It directly affects the blood vessels of the retina which leads to visual deficiency. Diabetic retinopathy may not have any symptoms at first, but it is early diagnosis can help to take further steps to protect your vision. Screening DR is a time-consuming procedure and requires experts like ophthalmologist. The proposed work tries to solve this problem with the help of deep learning. The ResNet34 model is trained on a dataset of fundus eye images. There are five DR stages such as 0, 1, 2, 3, and 4. Features are extracted from fundus images, and further, activation function is used to get the output. The model successfully achieves an accuracy of 0.82.

Keywords Diabetic retinopathy (DR) · Deep learning · Fundus eye images · ResNet34 architecture

S. Dongre (✉) · A. Wanjari · M. Lalwani · A. Wankhade · N. Sonawane · S. Yadav
Vishwakarma Institute of Technology, Pune, India
e-mail: shital.dongre@vit.edu

A. Wanjari
e-mail: aditya.wanjari19@vit.edu

M. Lalwani
e-mail: mohit.lalwani19@vit.edu

A. Wankhade
e-mail: anushka.wankhade19@vit.edu

N. Sonawane
e-mail: nitesh.sonawane20@vit.edu

S. Yadav
e-mail: shresthi.yadav19@vit.edu

K. A. Reddy et al. (eds.), *Proceedings of Fourth International Conference on Computer and Communication Technologies*, Lecture Notes in Networks and Systems 606,
https://doi.org/10.1007/978-981-19-8563-8_20

1 Introduction

Diabetic retinopathy (DR) is a form of diabetes which damages the retina of the eye. It is caused due to high blood sugar level. If it is not diagnosed and treated at appropriate time, then it can cause blindness. The retina is severely damaged and vision impairments result. It affects the veins that pass through the retinal tissue, causing them to leak fluid and distort vision. DR is among the most persistent diseases, alongside disorders that cause visual impairment such as waterfalls and glaucoma. DR is divided into the following stages: 0, 1, 2, 3, and 4.

The table below summarizes the various stages of DR: Each stage has its unique symptoms and characteristics, and doctors can no longer distinguish between the DR phases only on normal imaging. Furthermore, conventional diagnostic approaches are ineffective since they take a long time, causing therapy to proceed in the incorrect direction. Doctors utilized a fundus camera to diagnose retinopathy, which captures pictures of veins and nerves behind the retina. Because there are no indications of DR in the early stages of this disease, identifying it as such can be difficult. We employed several CNN algorithms for early detection so that doctors could begin therapy at the appropriate moment.

The dataset for this research was obtained from "Aravind Eye Hospital". The two CNN designs, such as VGG16 and DenseNet121, were compared, and the outcomes of both the architectures were illustrated. In recent research and projects, "deep learning" in AI has shown good results in identifying the hidden layers in different tasks, especially in the domain of medical picture analysis [1–3]. These models help to categorize illnesses, aid in medical decision-making, as well as improve persistent consideration [4]. The work is divided as follows: Section 2 contains the DR image categorization literature reviews. Section 3 has detailed description of the dataset and methodology of DL architecture. The primary outcome of this study is described in Sect. 4. Finally, Sect. 6 brings the paper to a close.

2 Literature Review

It gives a review of existing methodologies that used "deep learning" for DR automated early detection in a certain field.

Deep learning was used to develop a system for detecting DR. Because deep learning is a computational approach, it may design an algorithm by itself, and it learns from a huge number of examples and exhibits the desired behavior. Clinical imaging employs these approaches. The Messidor-2 contained 1748 pictures from 874 patients, whereas the EyePACS-1 had 963 photos from 4997 patients [5].

The automated identification of DR is critical, because it is one of the major causes of permanent vision loss in people in their working or early retirement years across the world. The licensed use is restricted to the University of Gothenburg. IEEE Xplore was used to get this document on December 19, 2020 at 22:41:58 UTC.

There are various drawbacks: Even expert medics find it challenging to classify DR pictures. As a result, deep convolutional neural network (DCNN) was used to classify DR with a 94.5% accuracy [6].

Currently, a unique DCNN is being developed that performs the initial temporal detection by identifying all microaneurysms (MAs), which is the first sign of DR. It also reliably assigns names to retinal fundus images and divides them into five groups. The architecture was evaluated on the Kaggle dataset, and it yields a QWK score of 0.851 and an AUC score of 0.844. The model has a sensitivity of 98% and a specificity of 94% in the early stage of detection, which demonstrates the efficacy of the used technique [7, 8].

On ImageNet models, transfer learning improves classification accuracy by 74.5%, 68.8%, and 57.2%, respectively [7].

With proper therapy at the early stages of DR, this form of sickness can be avoided. For the diagnosis of DR condition, a novel feature extraction approach called modified Xception architecture has been displayed in the image [8].

The objective is to utilize a universal approach to identify DR and quantify its severity with great efficiency. The use of various CNN architectures is investigated. The end outcomes of the training process describe VGG16 which had a 71.7% accuracy, VGG19 had a 76.9% accuracy, and Inception v3 had a 70.2% accuracy [9].

Unfortunately, determining the DR stage is notoriously difficult and needs experienced human interpretation of fundus images. Individual imaging of the human fundus is now being used to build an autonomous approach for DR stage detection. The technique may be utilized for early-stage detection on the APTOS dataset since it has a sensitivity and specificity of 0.99 and a QWK score of 0.925466 [10].

3 Methodology

The concept of transfer learning is implemented in this research. Transfer learning is a technique in which an already trained machine learning or deep learning model is put to use for a completely new problem statement. One of the key advantages of this technique is that even small amount of data can be utilized sufficiently to train the model. In simple terms, knowledge acquired by a pretrained model is passed on to a similar problem statement. Further, advantages of implementing transfer learning technique are the training time which is reduced, and comparatively improved performance is achieved. Figures 1 and 2 illustrate the sample fundus images of the eyes that were used to develop the model. The flowchart in Fig. 3 illustrates the workflow.

A. **Dataset**

The dataset consists of high-quality eye fundus images. The dataset for this research was obtained from Kaggle. There are a total of 5593 images. These images are of left and right eye, and clinicians have divided them into 5 classes as per the stage of DR (Fig. 4).

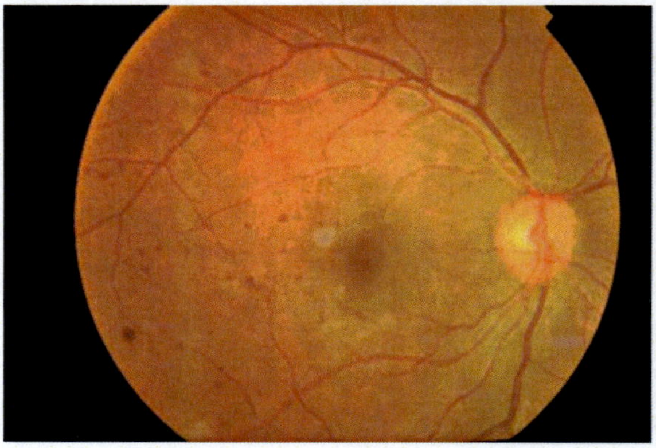

Fig. 1 Fundus eye sample figure 1

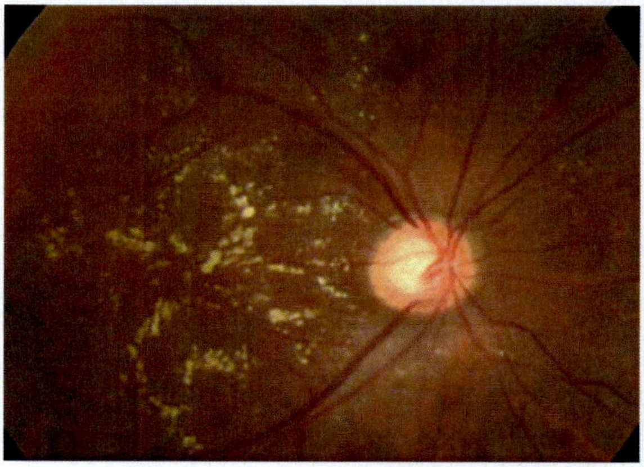

Fig. 2 Fundus eye sample figure 2

B. Data Preprocessing

The model takes eye image as input. These eye fundus images are divided into 5 classes: no DR (class 0), mild DR (class 1), moderate DR (class 2), severe DR (class 3), proliferative DR (class 4). Firstly, weights are assigned to each class. The images are the processed properly. It helps to extract important features from the images. There are several steps involved in image preprocessing. Image resizing is a critical preprocessing step as deep learning models train faster on smaller size images. All eye fundus images are cropped and resized into fixed sized images that is 512×512 pixels. Then, the images are transformed to tensors. A tensor is like a NumPy array. For accelerated computations, all the images are converted into tensors. Data is then normalized to a smaller range. It helps to improve the accuracy and integrity of the data and is generally preferred for classification algorithms. After normalization, tensors are given to the model for training and testing.

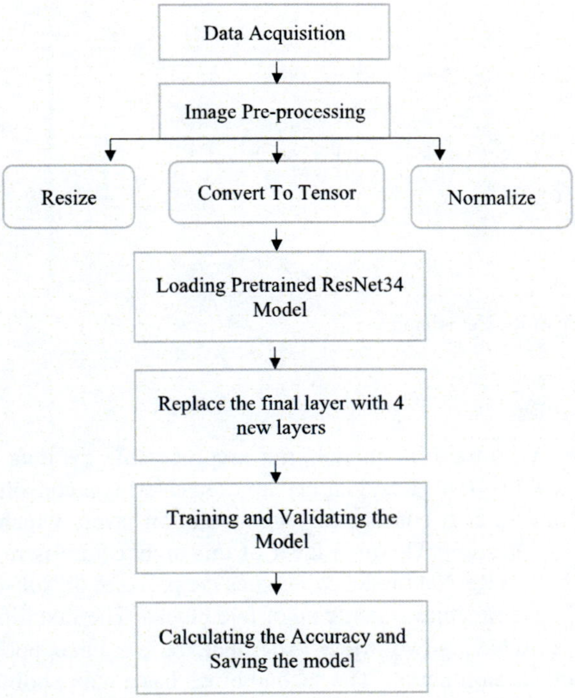

Fig. 3 Flowchart

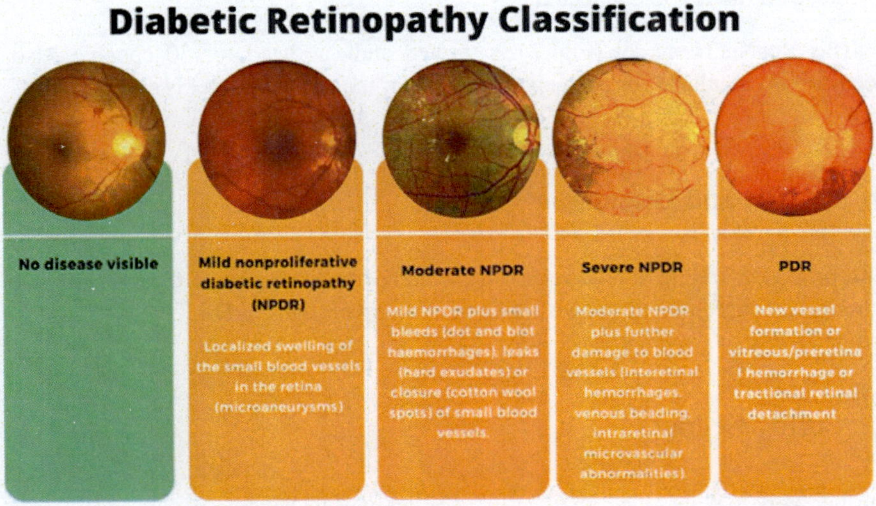

Fig. 4 Stages of diabetic retinopathy

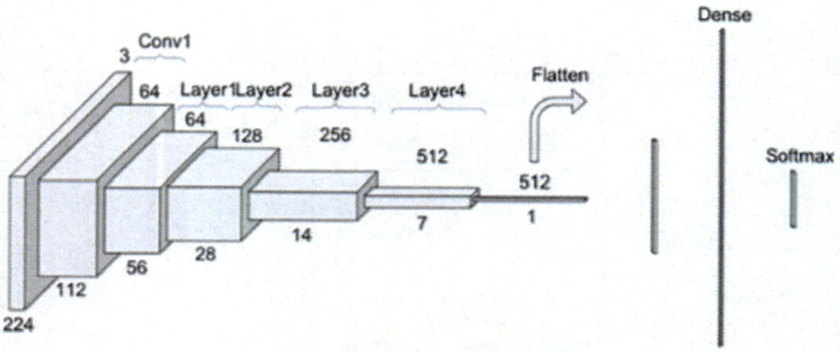

Fig. 5 Architecture of ResNet34 model

C. **Model Training**

The next step is to train the model after preprocessing the data. A pretrained ResNet34 model is used. Residual network (ResNet) is a convolutional neural network architecture. It consists of 34 convolution layers which can be used for image classification. The final layer of this architecture is replaced with 4 new layers. Use of ResNet model overcomes the problem of vanishing gradient. Every ResNet architecture is made up of five blocks. The first block has a total of 64 filters, each having two strides. After that, there is a max pooling layer and the ReLU activation function. The second block has a max pooling layer and a 3 * 3 kernel size. Third, fourth, and fifth block have kernel sizes of 3 * 3 with input channels 64,256 and 512, respectively. Linear activation function helps to keep all the layers connected.

This ResNet34 model (Fig. 5) is trained and validated for 30 epochs. Also, accuracy is calculated for each epoch, and then, the trained model is saved.

4 Results and Discussion

The proposed architecture successfully achieved an accuracy of 0.82. It helps to classify the unseen input images into one of the five stages of DR.

5 Future Scope

Further work may include utilizing more comprehensive behavioral data and altering the layers of the neural network. Specific models can also be trained to increase the overall accuracy.

6 Conclusion

In recent years, diabetes has become one of the fastest-growing illnesses. According to numerous studies, a diabetic patient has a 30% probability of developing diabetic retinopathy (DR). Also, manual detection of DR requires ophthalmologists and consumes a lot of time. So, with the knowledge of data mining and deep learning, we developed an architecture for automatic DR detection. The used ResNet34 architecture successfully achieved a accuracy of 0.82. It helps to classify the unseen input images into one of the five stages of DR. The findings demonstrate that the proposed ensemble model outperforms the previous state-of-the-art approaches and can detect all phases of DR. To improve the accuracy of early stages in future, we propose to train models for various phases and then ensemble the results.

References

1. Zengi X, Cheni H, Luo Y, Ye W (2016) Automated diabetic retinopathy detection based on binocular siamese-like convolutional neural network. IEEE Trans J 6
2. Shah S, Khan A, Shamshirband A, Ur Rehmani Z, Khan IA, Jadoon W (2019) A deep learning ensemble approach for diabetic retinopathy detection, October. IEEE Access
3. Mishra S, Hanchate S, Saquib Z (2020) Diabetic retinopathy detection using deep learning. Institute of Electrical and Electronic Engineers (IEEE)
4. Kauppi T, Kalesnykiene V, Kamarainen JK, Lensu L, Sorri I, Raninen A, Voutilainen R, Uusitalo H, Kälviäinen H, Pietilä J. DIARETDB1 diabetic retinopathy database and evaluation protocol
5. Kaur P, Chatterjee S, Singh D (2019) Neural network technique for diabetic retinopathy detection. Int J Eng Adv Technol (IJEAT)
6. Firke SN, Jain RB (2021) Convolutional neural network for diabetic retinopathy detection. In: Proceedings of the international conference on artificial intelligence and smart systems (ICAIS)
7. Jayakumari C, Lavanya V, Sumesh EP (2020) Automated diabetic retinopathy detection and classification using ImageNet convolutional neural network using fundus images. In: Proceedings of the international conference on smart electronics and communication (ICOSEC)
8. Palavalasa KK, Sambaturu B (2018) Automatic diabetic retinopathy detection using digital image processing. In: International conference on communication and signal processing
9. Chetoui M, Akhloufi MA, Kardouchi M (2018) Diabetic retinopathy detection using machine learning and texture features. In: IEEE Canadian conference on electrical & computer engineering (CCECE)
10. Oh K, Kang HM, Leem D, Lee H, Seo KY, Yoon S (2021) Early detection of diabetic retinopathy based on deep learning and ultra-wide-field fundus images

Comparative Analysis Using Data Mining Techniques to Predict the Air Quality and Their Impact on Environment

Rahul Deo Sah, Neelamadhab Padhy, Nagesh Salimath, Sibo Prasad Patro, Syed Jaffar Abbas, and Raja Ram Dutta

Abstract People are altogether dependent on air as a wellspring of energy. Without air, people would die. Inspecting and safeguarding air quality have become one of the most fundamental government exercises in numerous modern and metropolitan regions today. Individuals' well-being is affected by the release of these hurtful mixtures into the environment, particularly in metropolitan regions. As the total populace extends, deforestation is turning out to be progressively far and wide. The nature of the air will diminish. The climate is turning out to be more dirtied, and the temperature is climbing too. A wide scope of medical conditions has been distinguished. A few strategies have been created. In this examination, information mining approaches are utilized to gage the temperature focus in the climate. The skin and mucous membranes of the eyes, nose, throat, and lungs are affected by the disease.

Keywords Deep learning · Air quality · Temp · Prediction · Gasses

R. D. Sah (✉)
Department of CA & IT, Dr. SPM University, Ranchi, Jharkhand, India
e-mail: rahuldeosah@gmail.com

N. Padhy · S. P. Patro
Department of CSE, Giet University, Gunupur, Odisha, India
e-mail: sibofromgiet@giet.edu

N. Salimath
Department of CSE, Poojya Dodappa Appa College of Engineering, Kalaburagi, India
e-mail: nageshsalimath@pdaengg.com

S. J. Abbas
Department of CSE, Jharkhand Rai University, Ranchi, Jharkhand, India
e-mail: sjaranchi@gmail.com

R. R. Dutta
BIT Mesera, Ranchi, Jharkhand, India
e-mail: rajaramdutta@bitmesra.ac.in

K. A. Reddy et al. (eds.), *Proceedings of Fourth International Conference on Computer and Communication Technologies*, Lecture Notes in Networks and Systems 606,
https://doi.org/10.1007/978-981-19-8563-8_21

1 Introduction

Rapid population growth and monetary development in urban areas in agricultural countries like India have created environmental problems such as air pollution, water pollution, wave pollution, and many others. Air pollution has a direct impact on human well-being. In our country, it has risen to an open consciousness on this issue. Long-term air pollution impacts include unusual weather changes, heavy rainfall, and increased asthma. Projected air quality could also reduce the impacts of the most severe pollution on people and the climate. As a result, further development of air quality sensors has become one of the main concerns of the general public. Sulfur dioxide is a gas that can be found in the climate. It is one of the most commonly transported climate toxins. It has an unpleasant, intense odor and is dry. It reacts rapidly with various synthetic materials to produce hazardous mixtures such as sulfur corrosion and sulfur corrosion. When taken in, sulfur dioxide affects human well-being. Shortness of breath, wheezing, shortness of breath, or tightness in the chest are caused by disorders of the nose, throat, and airways. The level of sulfur dioxide in the environment can affect the suitability of the local flora to the natural environment as well as the life of living things. The proposed framework is equipped to predict the agglomeration of sulfur dioxide in the coming months/years.

2 Related Work

Understudy performed AI calculations to predict air quality (AQI) files for specific regions in this audit article to assess air quality in India. The air quality index is a regular measure for evaluating air quality. The office tracks gas focus such as SO_2, NO_2, CO_2, RSPM, and SPM. The analyst creates a model that predicts air quality inventory based on historical information from the previous year, and how gradient dip has helped with the multivariate relapse question for a given next year. They improved the model's performance by including cost estimates to evaluate the problem. This model guarantees this recorded information on foreign objects, which can actually audit the list of air quality lists in the entire province, state, or limited zone. [1] This study uses an artificial neural network to use an artificial neural network and uses an artificial neural network to provide a long model that uses an artificial neural network and an artificial neural network to expand the degree of air pollution in different places of Navi Mumbai and Navi Mumbai. The proposed model is performed with the indicators of the results with the participation of MATLAB for the ANN and R of Kriging [2–4]. The next day, this approach participated in the multilayer convolution (ANN) as well as reissue. This frame supports predictions of providing precision of dressing due to the main variable as well as the study of current pollution and the determination of future pollution. Time series analysis was further used to predict pollution levels and to recognize future informational elements [5]. The proposed framework serves two main goals such as (i) determine the PM2.5 level based on climate data and (ii) predict the PM2.5 level for a specific date. Strategic relapses are performed to determine whether information testing has been compromised. Given past PM2.5 estimates, autoregression is often used to predict future PM2.5 levels. Its main task is to predict air pollution levels in cities using a set of ground data [6]. An important goal of this article was to describe the vast research work and provide a useful overview of the latest technologies in materials for air quality assessment and prediction, vast information approaches, and AI procedures. The air quality guidelines were prepared and designed using information from Shenzhen, China.

ANN genetic algorithm model, arbitrary shield, selection tree, and deep convicted network have a calculation used, various advantages are presented, and the disadvantages of the model were presented [7]. In-progress tests are engaged in cutting measurable learning calculations to assess the prediction of air quality and pollution levels. The brain network was used [7–9] [Example: 10 micron (PM10)]. You must wait for the model until individual monopoly occurs, such as the particles of microns (PM10). To prepare these models, we used [7] reference vehicles (SVM) and fake brain tissue (ANN). Their best model ANN was almost 79% of personnel and 0.82% of fake positive rate, and the best SVM model has a characteristic of 80% and is also a positive tempo of 0.13%. For AQI class prediction [10], RAQ, an irregular backward method, is recommended. Since then, Leong et al. [3] have used deep brain tissue to predict design subcategories. To predict the AQI level, Frank et al. [11] used various settings that outperform K-nearest neighbors (KNNs), selection trees, and SVMs. Their ANN model outperformed all other evaluated calculations with an accuracy of 92.3%.

3 Dataset Observation

3.1 Dataset Collection

The CSV document design is utilized for the informational index. It is available to the general population on Kaggle. There are around 450,000 records in the assortment. In this distribution, the scientists center around PM10 (particulate matter), vaporous toxins like sulfur dioxide (SO_2), carbon monoxide (CO), nitrogen oxide (NO_2), ozone, and ecological temperature. Datasets were arranged into two segments: 75% for preparing and 25% for testing.

3.2 Preprocessing and Feature Selection

We only review and applied the algorithms on data. We have simplified the type attribute to contain only one of the six categories: PM10, CO, NO_2, SO_2, ozone, temp. So after preprocessing, our dataset contains 60,380 rows and 6 columns.

4 Technical Approach

In the Fig. 1 shown, data is processed in few steps:

- Structure data.
- Select the data subset.
- Preprocessed the data.
- Data transformation.
- Data mining techniques are used.
- Pattern extraction.
- Evaluation of different algorithms.
- Results.

The invention of a data science software platform received.

- Structure data for processing and descriptive analysis.
- Train the model utilizing k-fold cross-validation to reduce or
- estimate uncertainty in future predictions.
- To compare the models fixed, use classification performance measurements such as the area under curve, absolute number, root—mean—square error, classification error, and lambda statistic.

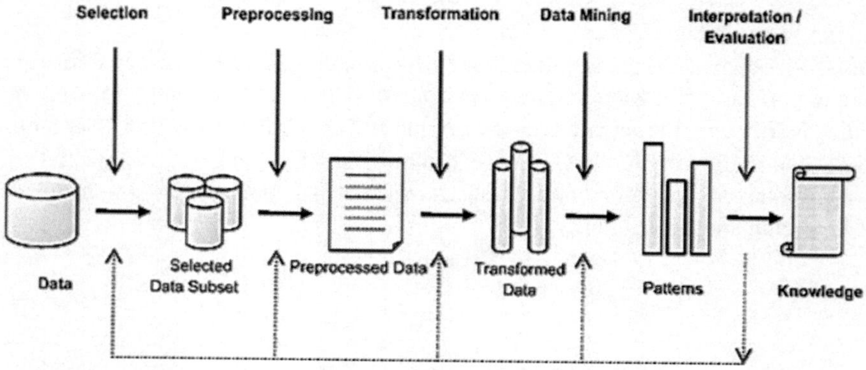

Fig. 1 Data processing

For equivalent dataset splitter paths, similar arbitrary seeds (887), and $k = 10$, 10x, the "mutual validation" strategy applies widely. To emphasize the validity of the various strategies of this characterization problem, it is used in relative studies using the selection tree method: Naive Bayes, generalized line method, logistic regression, fast large-scale method, deep learning, decision tree, gradient boosting trees, and support vector machines. Mutual recognition is a system for reducing variance when a conforming model is transferred to a dataset [12]. Finally, the findings collected are compared to this measurement. The area under the ROC bend is a graphical and measurable check of the feasibility of expectations. Tragically, the unique alignment method does not have the ability to evaluate and represent the area under precision and recall. This is very important for heterogeneous samples. The basic idea of profound learning is based on progressive learning methods, which are associated with calculated recurrences. Deep learning, decision tree, SVM classifier, naive Bayes, generalised line method, fast large-scale method, gradient strengthening tree, support vector machine of this situation, as well as perform the extensive learning calculations.

4.1 Bayesian Naive

It is a straightforward and fast philosophy, yet some way or another it expects that the indicator factors are regularly dispersed.

The choice tree order technique is one of the most widely recognized arrangement strategies since it reliably performs well [4].

4.2 Decision Trees

Decision trees are vulnerable to overfitting, and alternative methods are usually superior in quality of accuracy. In this situation, the decision tree method, also known as the bagging algorithm, is a powerful option for creating many trees to improve prediction accuracy and reduce the risk of overfitting [13].

4.3 Gradient Boosting Tree

For decreasing estimation bias, it is a wonderful alternative to bagging. Iteration is the process of building a new tree based on the previous trees' knowledge. It's a great alternative to bagging for getting rid of estimation bias. Iteration is the process of putting together a new tree based on what was learned from the previous trees. Aside from making or trying to make older models less inefficient [13, 14].

4.4 Logistic Regression

LR is a probabilistic strategy that responds well in binary problem classification and is based on the principle that the attributes and the dependent variable have an appropriate level. There is also no correlation between the variables in this approach.

4.5 Support Vector Machine

Support vector machine enables define category edges also using linear and non-linear metrics. This function is useful in classification tasks because it allows you to split observations into classes using a polynomial or wavelet transform function (non-linear). To classify relevant observations, it will use a linear function in SVM [15, 16].

4.6 Deep Learning with Algorithms

In the final part of the study, we used a hierarchical learning approach to implement the core idea of deep learning or neural network algorithms. The best performance algorithm for this dataset is based on a logistic regression algorithm and needs further development. This method applies the algorithms in the proper order, tracking the results of the previous iteration. Use the "deep learning" feature to use well-known deep learning algorithms. The "target" characteristics that indicate the results of the diagnostic test are analyzed [4].

5 Result Analysis with Different Data Mining Techniques

The invention of a data science software interface enables for a simple and quick approach to various models in the machine learning categorization discipline. Those processes compare the outcomes of various models' analysis techniques using conventional machine learning measures. The creation of a software interface for data science that makes it easy and quick to use different models in the field of machine learning categorization using standard machine learning measures, these steps compare the results of different models' analysis techniques. First, the process of analysing and preprocessing data so that descriptive statistics can be made.

In Fig. 2, we used different data mining techniques for comparing the receiver operating characteristics (ROC) and got deep learning method performance is better than other.

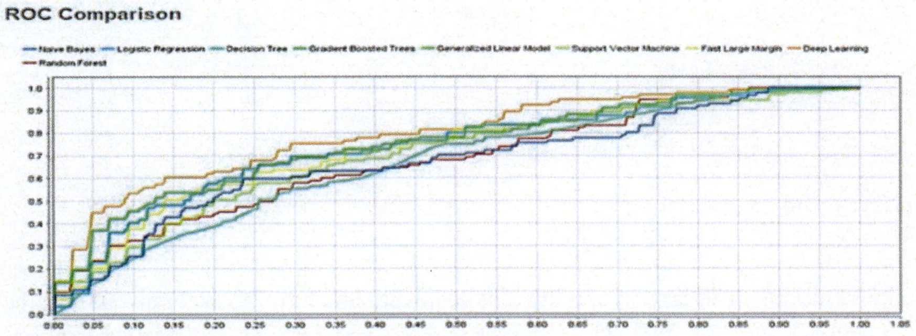

Fig. 2 ROC differentiate among data mining techniques

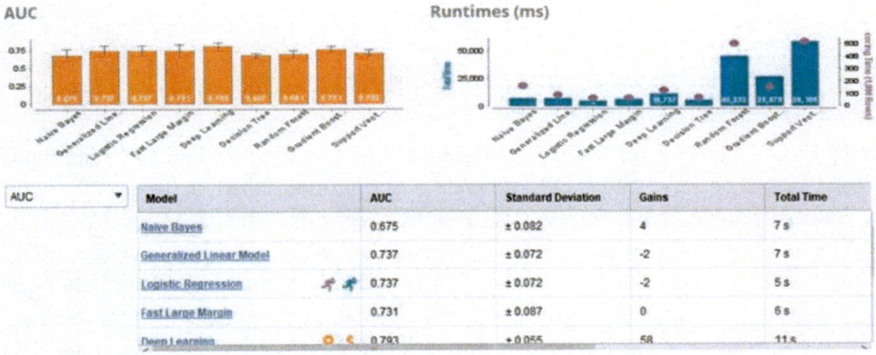

Fig. 3 Model comparison

5.1 Models Comparison

In Fig. 3, we used different models and got area under curve deep learning 0.783, standard deviation + 0.055 gains 58, and taking total time is 11 s. Scoring time (1000 rows) in 10.797 ms for deep learning compares to other.

5.2 Classification Errors Rate in Different Models

In Fig. 4, we have seen different models which have classification errors, and we found after building the models, classification errors rate 19.3% of deep learning is less than among others.

Lists model prediction accuracy and other performance criteria, depending on the type of classification problem. Performance is calculated using a holdout rate of 40%, which was not used to optimize the run model. This holdout set is then used as an input to a multi-hold initial validation that computes the powers of seven relatively

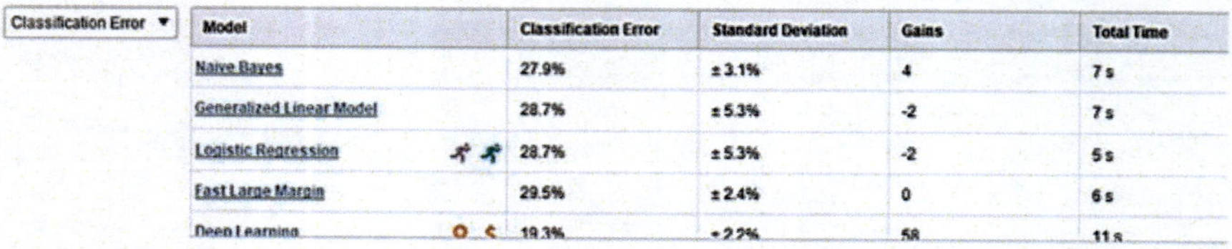
Classification Error ▾

Model	Classification Error	Standard Deviation	Gains	Total Time
Naive Bayes	27.9%	± 3.1%	4	7 s
Generalized Linear Model	28.7%	± 5.3%	-2	7 s
Logistic Regression	28.7%	± 5.3%	-2	5 s
Fast Large Margin	29.5%	± 2.4%	0	6 s
Deep Learning	19.3%	± 2.2%	58	11 s

Fig. 4 Classification error rates

disjoint subsets. The largest and highest achievements have been removed, and the average of the remaining 5 achievements is displayed here. This validation is not as thorough as full mutual validation, but this approach balances the quality of run-time and model validation.

5.3 *Hierarchical Learning and Deep Learning Algorithmic*

In the Fig. 6, statics values of temperature are shown range 1 and range 2, and there is total values of these statics. The detailed description is shown in Table 1.

Gains/lift Table (avg response rate: 29.03%, avg score: 30.12%) (Fig. 7).

The accuracy and execution time of the model are displayed in the overview. ROC comparison: It shows the ROC curves for all models in one graph. The better the model, the closer the curve is to the upper left corner. Only two classes of problems are displayed. After making changes for simulation, the dataset is stored electronically. This is the data that all modeling approaches and automated feature engineering can take as input. You can use only a subset of this data in your model or generate more columns. Text: Displayed only if feature extraction of text data is enabled. As a survey of tables and surveys, we show the words in the columns of text used in the

Table 1 Detailed description

Model metric type	Binomial
Description framework	Indicators reported for the complete training
Model ID	rmh2omodelmodel75
Frame ID	rmh2oframemodel75
MSW	0.11852029
RMSE	0.34426776
R^2	0.4247576
AUC	0.88920367
pr_auc	0.74335134
Log loss	0.37855956
mean_per_class_error	0.1754386
Default threshold	0.34164929389953613

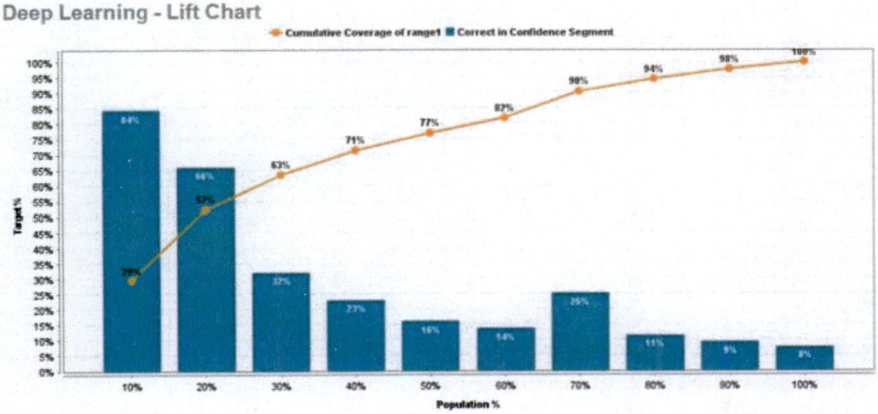

Fig. 5 Deep learning lift chart

Table 2 Range 1 and range 2 error rate values

	Range 2	Range 1	Error	Rate
Range 2	352	66	0.1579	66/418
Range 1	33	138	0.1930	33/171
Total	385	204	0.1681	99/589

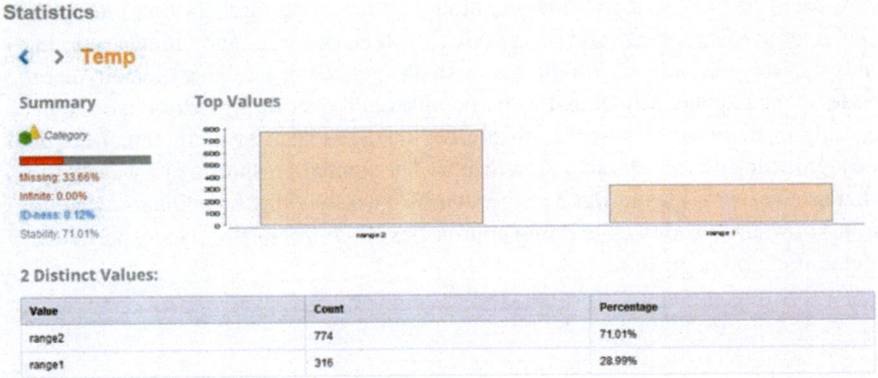

Fig. 6 Statics values of temp

analysis. Finally, if you allow emotional or language evaluation, you can examine the distribution of those values across columns of text. Correlated weights: The global relevance of each input data column to the value of the desired column, regardless of model-based algorithms or perhaps showcase techniques. For predictions, the weights are based on the correlation between the columns and the target columns. Model-specific weights, on the other hand, are the columns that have the greatest

Fig. 7 Variables' correlation, numeric values

Correlations

Attribut...	CO	NO	NO2	Ozone	PM10	SO2	Temp = ...
CO	1	0.184	0.006	0.099	-0.287	-0.221	0.075
NO	0.184	1	0.144	0.159	0.067	0.092	-0.130
NO2	0.006	0.144	1	0.110	0.022	0.112	0.001
Ozone	0.099	0.159	0.110	1	-0.011	-0.044	-0.094
PM10	-0.287	0.067	0.022	-0.011	1	0.434	-0.302
SO2	-0.221	0.092	0.112	-0.044	0.434	1	-0.071
Temp = r...	0.075	-0.130	0.001	-0.094	-0.302	-0.071	1

overall impact on each model's predictions. Correlation: A matrix showing how columns are related (Fig. 6).

This result demonstrates a link between the qualities: carbon monoxide, nitrogen oxide, ozone, PM10, SO_2, and temp.

6 Conclusion and Future Scope

There are basically two possibilities for this study to improve the results obtained with logistic algorithms. One is to improve the algorithm characteristics, and the other is to use the "deep rooted" approach in this algorithm.

Lambda parameters, also known as generalized linear factors, are an important component of logistic regression techniques and help researchers find the optimal combination of simplicity and complexity. In other words, a high lambda ratio indicates that the model is easy to fit, and a small number of lambdas indicate that the model is inadequate. Model that is too complex and over fitting. Moving on to deep learning, it demonstrates how sophisticated methods such as hierarchical learning can significantly boost classification results. The standard response rate was 28.95%, whereas the average score was 31.26%. It is also worth noting how this study demonstrates how advanced deep learning approaches can boost performance, particularly in classification algorithms.

References

1. Soundari AG, Jeslin JG, Akshaya AC (2019) Indian air quality prediction and analysis using machine learning. Int J Appl Eng Res 14(11). ISSN 0973-4562 (Special Issue)
2. Guttikunda K, Goel R, Pant P (2014) Nature of air pollution, emission sources, and management in the Indian cities. Atmos Environ 95:501–510
3. Leong WC, Kelan RO, Ahmad Z (2020) Prediction of air pollution index (API) using support vector machine (SVM). J Environ Chem Eng 8(3):103208
4. Kotsiantis SB, Kanellopoulos D, Pintelas PE (2006) Data preprocessing for supervised leaning. Int J Comput Sci 1(2):111–117

5. Han S, Qubo C, Meng H (2012) Parameter selection in SVM with RBF kernel function. In: Proceedings of world automation congress, Puerto Vallarta, Mexico, pp 1–4
6. Smola AJ, Scholkopf B (2004) A tutorial on support vector regression. Stat Comput 14(3):199–222
7. Arampongsanuwat S, Meesad P (2011) Prediction of pm10 using support vector regression. In: Proceedings of international conference on information and electronics engineering, Bangkok, Thailand, vol 6
8. Vong CM, Wong PK, Yang JY (2012) Short-term prediction of air pollution in Macau using support vector machines. J Control Sci Eng 2012
9. Sah RD, Sheetlani J (2017) Pattern extraction and analysis of health care data using rule-based classifier and neural network model. Int J Comp Technol Appl 8(4):551–556
10. Vapnik V et al (1997) Predicting time series with support vector machines. In: Proceedings of ICANN, Lausanne, Switzerland, pp 999–1004
11. Frank E, Hall MA, Pal CJ, Witten IH (2017) Data mining: practical machine learning tools and techniques, 4th edn. Elsevier/Morgan Kaufmann, Cambridge, Massachusetts, pp 147
12. Chicco D, Warrens MJ, Jurman G (2021) The coefficient of determination R-squared is more informative than SMAPE, MAE, MAPE, MSE and RMSE in regression analysis evaluation. PeerJ Comp Sci 7:623
13. Albon A (2018) Machine learning with Python cookbook: practical solutions from preprocessing to deep learning. O'Reilly, First edition. Kindle Edition, p. 91
14. Parbat D, Chakraborty M (2020) A python based support vector regression model or prediction of COVID19 cases in India. Chaos, Solitons Fractals 138:109942
15. Sah RD (2017) Review of medical disease symptoms prediction using data mining technique. IOSR J Comp Eng (IOSR-JCE) 19(3):59–70, Ver. I (May–June 2017). a-ISSN: 2278-0661, p-ISSN: 2278-8727
16. Weizhen H et al (2014) Using support vector regression to predict PM10 and PM2.5. Proc IOP Conf Ser: Earth Environ Sci 17:012268. Jakarta, Indonesia

Complexity Reduction by Signal Passing Technique in MIMO Decoders

Ramya Jothikumar and Nakkeeran Rangasamy

Abstract Breadth first tree search algorithms are intended to search its lattice points by using breadth first search method; guarantees optimal BER performance without the need for an estimate of SNR. However, one such breadth first signal decoder (BSIDE) algorithm, usually, searches more nodes in the tree and incurs a higher implementation complexity. A signal processing technique capable of minimizing the number of multipliers needed for realizing the processing unit of the breadth first signal decoder is proposed. The proposed signal passing technique reduces 86% of computational complexity for 2×2 and 99% for 4×4 multiple input multiple output (MIMO) systems with similar performance as that of BSIDE.

Keywords BSIDE · DFE · CMU · MIMO · ML · QAM

1 Introduction

Future generation communication networks apply multiple input multiple output (MIMO) systems as one among the important technology to attain increased capacity and high spectral efficiency [1, 2]. The main challenge in realizing the MIMO systems is handling the detection complexity at the receiver. Many decoding algorithms having variety in complexity-performance tradeoffs exist in the literature. The maximum likelihood (ML) [3–5] decoding is considered as an optimum scheme in respect of minimizing the bit error rate (BER). However, it is infeasible to implement due to its exponentially increasing computational complexity, particularly, in higher dimensional MIMO systems. Tree searching schemes are well known due to their capacity to realize efficient ML decoding with fewer complexity. Basically, tree search technique is divided into two categories by the way in which

R. Jothikumar (✉)
Department of Electronics and Communication Engineering, Sri Manakula Vinayagar Engineering College, Puducherry 605014, India
e-mail: mithulramya@gmail.com

N. Rangasamy
Department of Electronics Engineering, School of Engineering and Technology, Pondicherry University, Puducherry 605014, India

K. A. Reddy et al. (eds.), *Proceedings of Fourth International Conference on Computer and Communication Technologies*, Lecture Notes in Networks and Systems 606,
https://doi.org/10.1007/978-981-19-8563-8_22

search has been done, namely breadth and depth first. Efficient methods for instance sphere decoding (SD) [6, 7] and K-best [8–10] for depth and breadth first decoding method, respectively, have layer processing unit for symbol detection at the detector, where the constellation points of quadrature amplitude modulation (QAM) are realized using multipliers. These multipliers multiply a constant with the decomposed channel element [11] and are named as constant multiplication units (CMUs). These CMUs used at each layer of the tree are realized by parallel processing and increase the hardware requirement of symbol detection. Therefore, this paper proposes an alternate method that majorly contributes toward the reduction of hardware requirements needed for layer processing unit by employing a signal processing technique that involves serial computation. Thus, the proposed method shows improvement in complexity reduction with the similar logic delay.

2 System Model

The system model considered is a MIMO system comprising of N_{T} and N_{R} transmit and receive antennas, respectively. The signal being received at each time instant is written as

$$Y = HX + \tilde{n} \tag{1}$$

where H denotes channel of size $N_{\mathrm{R}} \times N_{\mathrm{T}}$. X is the signal transmitted, and $\tilde{n}$ is independent and identically distributed (i.i.d) complex additive white Gaussian noise (AWGN) vector. Assuming $N_{\mathrm{R}} = N_{\mathrm{T}}$ and applying QR decomposition for H, results in R an upper triangular matrix, and the Q an unitary matrix. The Eq. (1) can be written as

$$\begin{aligned} Y &= QRX + \overline{n} \\ Q^H Y &= RX + Q^H \overline{n} \\ \underline{\hat{Y}} &= RX + \hat{n} \end{aligned} \tag{2}$$

Considering H is known at the receiver, the ML signal vector X is determined and is given as

$$X_{\mathrm{ML}} = \underset{X \in \Omega^{N_{\mathrm{T}}}}{\arg\min} \| Y - HX \|^2 \tag{3}$$

where 'S' denotes the set of quadrature amplitude modulation (QAM) entries in the constellation. The received signal Y and transmitted signal X at the receiver part are transformed to real value representation with $N = 2N_{\mathrm{T}}$ and $M = 2N_{\mathrm{R}}$ elements, respectively, which in turn transforms H to an $M \times N$ matrix. Each X_i where $i = 1, \ldots N_{\mathrm{R}}$ may be one of the real number from the set Ω, for example, it can be + 1 or − 1 if 4-QAM modulation is considered and takes value from $\Omega = \{1, 3, -3, -1\}$for

16 QAM. Maximum likelihood detection method estimates the value $\left\| \hat{\underline{Y}} - RX \right\|^2$ and finalizes the minimum one to be the ML estimate. The ML distance is illustrated as

$$\hat{d} = \left\| \hat{\underline{Y}} - R\hat{X} \right\|^2$$
$$= \left(\hat{Y}_1 - \sum_{i=1}^{N} R_{1,i} \hat{\underline{X}}_i \right)^2 + \left(\hat{Y}_2 - \sum_{i=2}^{N} R_{2,i} \hat{\underline{X}}_i \right)^2 + \cdots + \left(\hat{Y}_N - R_{N,N} \hat{\underline{X}}_N \right)^2 \quad (4)$$

where an ML detector is capable of attaining optimal BER performance by obtaining the ML solution $\hat{X} = \left[\underline{X}_1, \underline{X}_2, \ldots, \underline{X}_N\right]^T$. Assuming $N_{\mathrm{T}} = N_{\mathrm{R}}$, H is QR decomposed so that Eq. (3) is written recursively in a transformed way as,

$$\| Y - HX \|^2 = \left| \sum_{i=1}^{N} \hat{Y}_i - \sum_{j=i}^{N} R_{ij} \underline{X}_j \right|^2 \quad (5)$$

where $\hat{Y}_i$ represents ith element of $\hat{Y} = Q^H Y$, R_{ij} denotes the upper triangular matrix, and $\underline{X}_j$ is the element of transmitted signal.

3 Existing Method

Consider a C-ary tree as shown in Fig. 1 with number of layer equal to N and C denotes the elements present in lattice. Let $\underline{X}^{(s)} = \left[\underline{X}_{1,l}^{(s)}, \underline{X}_{2,l}^{(s)} \ldots \underline{X}_{N-l+1,l}^{(s)}\right]^T$ be a vector corresponding to the sth node of the lth layer, where $1 \leq s \leq C^{N-l+1}$ and $1 \leq \mathrm{l} \leq N$. The BSIDE [12, 13] algorithm uses breadth first search strategy, in which search for minimum valued node is done at each layer. And the assessment for the received signal is taken at their appropriate corresponding level. The same procedure repeats until reach the first layer, to get the ML solution. The computational burden decreases when the parameters, $\{d_l\}_{l=N}^{2}$, are smaller. To realize it, BSIDE is used, which merges linear detection algorithm, namely decision feedback equalization (DFE) and nonlinear method ML with a result of, $\tilde{X} = \left[\tilde{X}_1, \tilde{X}_2, \ldots, \tilde{X}_N\right]^T$ and $\tilde{d} = \left\| \hat{Y} - R\tilde{X} \right\|^2$. The $\overline{X}$ and d_l give a solution of DFE and distance. The equation of d_l is given by

$$d_l = \min\left[d_{l+1}, \left\| \hat{Y} - Rq^{(l)} \right\|^2 \right] \quad (6)$$

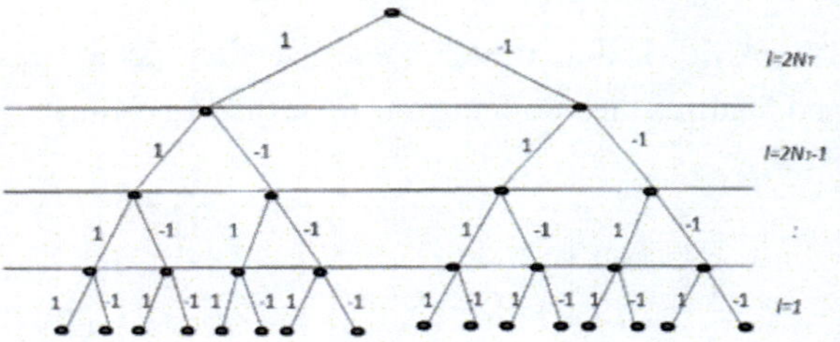

Fig. 1 Tree search representation with l layers and $\Omega = \{-1, 1\}$

With $d_{N+1} = \tilde{d}$ and $q^{(l)} = \left[\tilde{X}_1, \tilde{X}_2, \ldots, \tilde{X}_{l-1}, \underline{X}^{(1)}_{1,l}, \underline{X}^{(1)}_{2,l}, \ldots, \underline{X}^{(1)}_{N-l,l+1}\right]^T$. Let e_l be the chosen node without discarded at the lth layer with assuming condition $e_l + 1 = 1$. The distance of the node at the lth layer is mathematically given as

$$\Psi\left(\left[\underline{X}^{(j)}_{1,N}, \underline{X}^{(t)}_{1,l+1} \ldots \underline{X}^{(t)}_{N-l,l+1}\right]\right)^T = \left(\hat{Y}_l - R_{l,l}\underline{X}^{(j)}_{1,N} - \sum_{i=l+1}^{N} R_{l,i}\underline{X}^{(q)}_{i-l,l+1}\right)^2 + \cdots \sum_{j=l+1}^{N}\left(\hat{Y}_j - \sum_{i=l+1}^{N} R_{j,i}\underline{X}^{(t)}_{i-l,l+1}\right)^2 = \hat{Y}_l - R_{l,l}\underline{X}^{(j)}_{1,N} - \Theta\left(\underline{X}^{(t)}_{l+1}\right)^2 + \Psi\left(\underline{X}^{(t)}_{l+1}\right) \quad (7)$$

With $j = 1, 2, \ldots, C, t = 1, 2, \ldots, e_l + 1$. Where Ψ and Θ are the node distances and signal residue of an $(N - 1 + 1)$ dimensional vector. Since Eq. (3) can be illustrated as a tree with C-ary and 'l' layers, starting from a root; thus, the symbol detection procedure can be transformed in to BSIDE tree search detection problem. Figure 2 shows the conventional processing unit required for preceding the ith layer.

In accordance with Eq. (3), the hardware structure requires $N - i + 1$ multiplication units to compute $R_{i,j}\underline{X}_j$, $i \leq j \leq N$ in parallel for $\underline{X}_j \in \Omega$. The number of multiplication units will still increase as the constellation size increases. Therefore, in this paper, a signal passing technique that computes the value of $R_{i,j}\underline{X}_j$ in series for $\underline{X}_j \in \Omega$ that reduces the hardware requirement even the constellation size is increased was proposed.

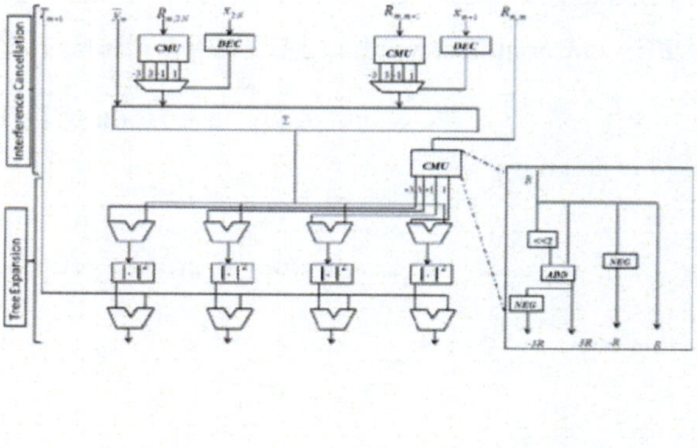

Fig. 2 Conventional processing unit for 16 QAM MIMO symbol detection, where DEC-decoder, ADD-adder and NEG-negator, and << 2—left shift operation [11]

4 Proposed Algorithm

A signal passing technique that exploits serial computation in CMU with reduced complexity in realizing the constellation was proposed. The requirement of CMU's in each layer of the processing unit is high, while reducing the complexity of a single CMU will have a significant impact over the total complexity of the processing unit. The proposed signal passing technique can be realized through cost metric defined below;

$$\left\|\hat{Y} - R\underline{X}\right\|^2 = \left\|\hat{Y} - RX'a_i\right\|^2 \tag{8}$$

where 'X' is written in terms of modulated signal, $S_i(t) = a_i\sqrt{E_0}\,\Psi(t) = a_i X'$, $\Psi(t)$ is the basis function, and $\sqrt{E_0}$ is the energy signal with lowest amplitude. Considering only a_i in the receiver and neglecting all other, we can reproduce the a_i of QAM to be

$$a_i = (2i - 1 - M) \quad i = 1, 2, \ldots M \tag{9}$$

where $M = 2^n$, n is number of bits in symbol.

The similarity in the constellation point makes it possible to divide the constellation set a_i as two of size smaller in nature, namely

$$\begin{aligned} \Omega_1 = a_i &= (2i - 1 - M) \quad i = 1, 2, \ldots \frac{M}{2} \\ \Omega_2 = a_i' &= (2i' - 1 - M) \quad i' = \frac{M}{2} + 1, \ldots M \end{aligned} \tag{10}$$

And it enforces to compute the cost metric with only one set (namely Ω_2). The cost metric of the other (Ω_1) can be realized through reflection. The proposed method passes the signal of first computed one to realize the next, so that the modified constellation is given as

$$\Omega_2 = a_i' = (2i' - 1 - M) \quad i = \frac{M}{2} + 1 \ldots M \tag{11}$$

Then, the set Ω_2 can be written as

$$\begin{aligned} \Omega_2 = \{a_i', \underbrace{a_i' + 2}_{\Downarrow\; a_i''}, \underbrace{a_i'' + 2}_{\Downarrow\; a_i'''}, \ldots\} \end{aligned} \tag{12}$$

where

$$\Omega_1 = -\Omega_2 \tag{13}$$

with the help of Eq. (13), the modified version of cost metric is illustrated as

$$\left\|\hat{Y} - RX'a_i\right\|^2 = \left\|\hat{Y} - RX'(2i - 1 - M)\right\|^2$$
$$L'_m = \hat{Y}_m - \sum_{k=m+1}^{2N} R_{m,k} \sum_{i=\frac{M}{2}+1}^{M} X'_k(2i' - 1 - M)$$
$$L'_m = \hat{Y}_m - \sum_{k=m+1}^{2N} R_{m,k}\left[X'_k\left(2\left(\frac{M}{2}+1\right) - 1 - M\right) + (X' + 2) + (X''' + 2) + \cdots\right]$$
$$\Downarrow \qquad \Downarrow \qquad \Downarrow$$
$$X'' \qquad X''' \qquad X'''' \tag{14}$$

Since 'R' remains identical for the respective layer of breadth first tree structure of MIMO, sign change technique can be applied. Thus, the constellation points are grouped into two sets, and the calculation of $RX'a_i$ can be made simple. Further to reduce the arithmetic computations required by the CMU, the proposed technique passes the present computed value of X' to compute the next, in serial manner. To understand this, a 64-QAM modulation scheme is considered, in which the set a_i is divided into $\Omega_1 = \{-1, -3, \ldots\}$ and $\Omega_2 = \{1, 3, \ldots\}$, where Ω_2 can be obtained by changing the sign of Ω_1. Let the processing unit computes the fourth layer of the symbol detection for 64-QAM system for which the input–output relation is given as

$$\overline{X}_4 = \underset{\overline{x}\in\Omega^N}{\arg\min} \left\|\hat{y}_4 - R_{44}\underline{X}_4\right\|^2 \tag{15}$$

$\underline{X}_4$ substitutes the entries from the set Ω_1 and Ω_2; the Eqs. (11–12) with respect to proposed model becomes as follows

$$\begin{aligned}
\overline{X}_{4,1} &= \hat{Y}_4 - R_{44}(-1) + \overline{n}_4 && \text{w.r.t. } \underline{X} = -1\\
\overline{X}_{4,2} &= \hat{Y}_4 - (R_{44}(-1) + R_{44}(-2)) + \overline{n}_4 && \text{w.r.t. } \underline{X} = -3\\
&\swarrow\\
\overline{X}_{4,3} &= \hat{Y}_4 - (R_{44}(-3) + R_{44}(-2)) + \overline{n}_4 && \text{w.r.t. } \underline{X} = -5\\
\overline{X}_{4,4} &= \hat{Y}_4 + R_{44}(-1) + \overline{n}_4 && \text{w.r.t. } \underline{X} = 1\\
\overline{X}_{4,5} &= \hat{Y}_4 + (R_{44}(-1) + R_{44}(-2)) + \overline{n}_4 && \text{w.r.t. } \underline{X} = 3\\
&\swarrow\\
\overline{X}_{4,6} &= \hat{Y}_4 + (R_{44}(-3) + R_{44}(-2)) + \overline{n}_4 && \text{w.r.t. } \underline{X} = 5
\end{aligned} \tag{16}$$

Thus, the computational complexity of the multiplier unit is cut down by the signal passing technique. The hardware configuration for the proposed multiplier unit is represented in Fig. 3. The feedback encountered in the proposed method introduces delay, which is trivial when compared to reduction in computational complexity.

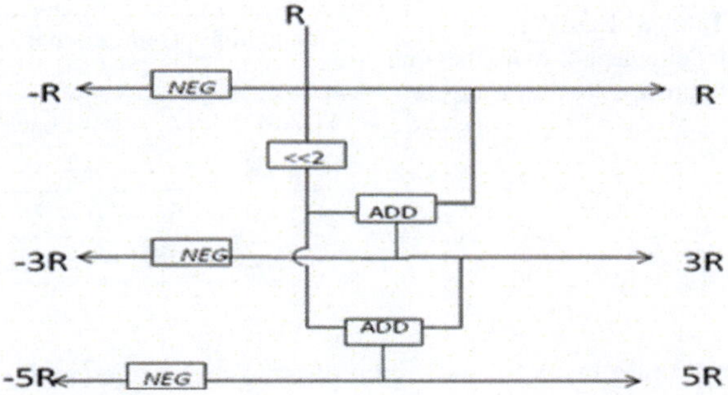

Fig. 3 Proposed multiplier unit for 64-QAM MIMO symbol detection

5 Evaluation

The modified cost metric is evaluated by employing signal passing technique in series; so that the computational complexity encountered in layer processing unit of MIMO detection can be reduced. The multipliers are implemented in MIMO systems for various constellation sizes through simulation set up. The quality measure for the hardware complexity is measured by the number arithmetic units required for implementation. Table 1 shows the tabulation on the number of computation units required for the existing and proposed multiplier circuits. The number of multiplier units needed at the ith layer among N layers is $N - i + 1$ [9]. The proposed technique reduces the computation units required to perform the multiplication that results in significant reduction in hardware resource required for each multiplier unit. This reduction in arithmetic unit increases as the constellation size increases.

A reduction of 53.3% is observed for 256 QAM. Table 2 shows the reduction in total hardware complexity in layer processing unit. From Table 2, it is observed that the hardware requirement follows a descending trend as the size of the constellation increases. Let U_l and D_l be the computational complexity in-spite of multiplications required to find $\{\Theta(\underline{X}_{l+1}^{(t)})\}_{t=1}^{e_l+1}$ and D_l, respectively. The count of the multiplication

Table 1 Hardware requirement for each multiplier unit

Order of modulation	Technique used	Number required		Logic delay	
		ADD	NEG	ADD	NEG
4-QAM	Existing	0	1	0	1
	Signal passing technique	0	1	0	1
16-QAM	Existing	1	2	1	1
	signal passing technique	1	2	1	1
64-QAM	Existing	3	4	1	1
	signal passing technique	2	3	1	1
256-QAM	Existing	6	7	2	1
	signal passing technique	3	4	2	1

Table 2 Hardware requirement of multiplier unit to process 'N' layers of tree

Modulation	Conventional	Proposed	Percentage of reduction (%)
4 QAM	$N - i + 1$	$N - i + 1$	–
16 QAM	$3N - 3i + 3$	$3N - 3\text{i} + 3$	–
64 QAM	$7N - 7i + 7$	$5N - 5i + 5$	28.5
256 QAM	$15N - 15i + 15$	$7N - 7i + 7$	53.3

at the lth layer for the signal passing technique can be given as

$$\sum_{l=1}^{N} \frac{C}{4} + \frac{C}{2} + e_{l+1} + U_l + D_l \tag{17}$$

The parameters e_l, U_l, and D_l are given as

$$\begin{aligned} e_l &= 2 \\ U_l &= N - l \\ 0 \leq D_l &\leq N + \sum_{i=1}^{N} i \end{aligned} \tag{18}$$

This proposed algorithm works well for increased constellation size with substantial reduction in complexity.

6 Simulation Results

Figure 4 illustrates the error performance characteristics of the proposed signal passing technique compared with BSIDE algorithm.

From the simulation results, it is worth to note that there is no degradation in the performance of the proposed method with respect to BSIDE MIMO systems. The outcome for 2 × 2 MIMO systems shows a reduction in complexity of 86% and for

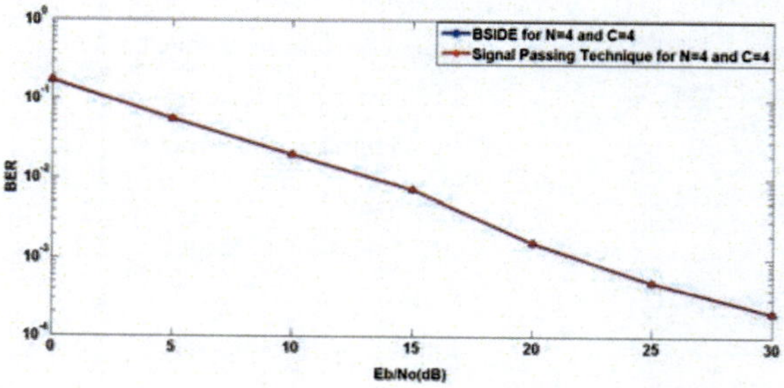

Fig. 4 Performance analysis of BSIDE with signal passing technique

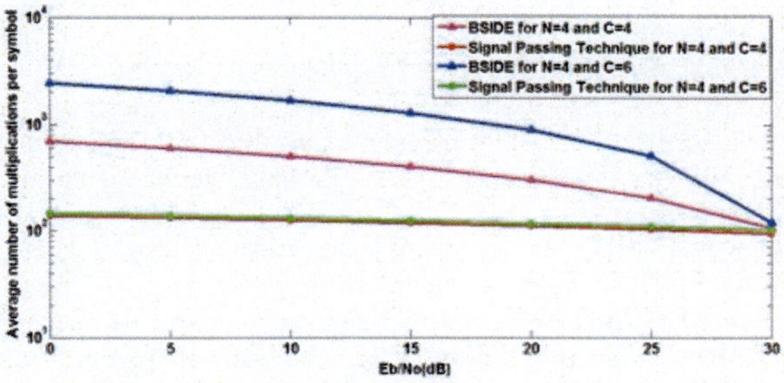

Fig. 5 Complexity analysis of 2 × 2 MIMO using 16, 64-QAM

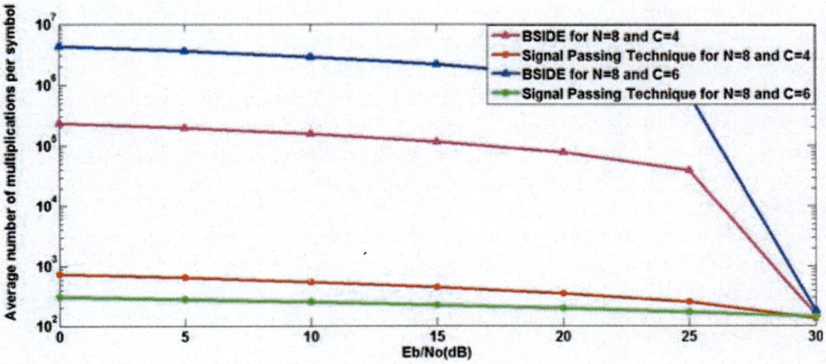

Fig. 6 Complexity analysis of 4 × 4 MIMO using 16, 64-QAM

a 4 × 4 is 99%. Figures 5 and 6 show comparison of the complexity curves for both 2 × 2 and 4 × 4.

7 Conclusion

A signal passing technique is proposed, which utilizes the previously estimated value to perform multiplication in the constant multiplication unit. A reduction in complexity is obtained, and also, it is observed that the proposed method works well for higher order constellation size.

References

1. Gesbert D, Shafi M, Shiu DS, Smith PJ, Naguib A (2003) From theory to practice: an overview of MIMO space-time coded wireless systems. IEEE J Sel Areas Commun 21(3):281–302
2. Yang S, Hanzo L (2015) Fifty years of MIMO detection: the road to large-scale MIMO. IEEE

Commun Surv Tutorials 17(4):1941–1988
3. Chang MX, Chang WY (2017) Maximum-likelihood detection for MIMO systems based on differential metrics. IEEE Trans Signal Process 65(14):3718–3732
4. Hijazi H, Haroun A, Saad M, Al Ghouwayel AC, Dhayni A (2021) Near-optimal performance with low-complexity ML-based detector for MIMO spatial multiplexing. IEEE Commun Lett 25(1):122–126
5. Chen RH, Chung WH (2012) Reduced complexity MIMO detection scheme using statistical search space reduction. IEEE Commun Lett 3(16):292–295
6. Mansour M, Alex SP, Jalloul LMA (2014) Reduced complexity soft output MIMO sphere detectors-part I: algorithmic optimizations. IEEE Trans Signal Process 21(62):5505–5520
7. Shen CA, Eltawil AM, Salama KN, Mondal S (2012) A best-first soft/hard decision tree searching MIMO decoder for a 4 x 4 64-QAM system. IEEE Trans Very Large Scale Integr (VLSI) Syst 8(20):1537–1554
8. Cher Q, Wu J, Zheng YR, Wang Z (2013) Two stage list sphere decoding for under determined multiple input multiple output systems. IEEE Trans Wirel Commun 12(12):6476–6487
9. Han S, Cui T, Tellambura C (2012) Improved K-best sphere detection for uncoded and coded MIMO systems. IEEE Wirel Commun Lett 5(1):472–475
10. Kim TH, Park IC (2010) High-throughput and area-efficient MIMO symbol detection based on modified Dijkstra's search. IEEE Trans Circ Syst I: Reg Papers 57(7):1756–1766
11. Roger S, Ramiro C, Gonzalez A, Almenar V, Vidal AM (2011) Practical aspects of preprocessing techniques for K-Best tree search MIMO detectors. Elsevier J Comp Electr Eng 4(37):451–460
12. Kim TH (2015) Low-complexity constant multiplication for layer processing in MIMO symbol detection. IET Electron Lett 51(13):989–991
13. Kang HG, Song I, Oh J, Lee J, Yoon S (2008) Breadth-first signal decoder: a novel maximum likelihood scheme for multi input multi output systems. IEEE Trans Veh Technol 3(57):1576–1583

A New Approach to Improve Reliability in UART Using Checksum Algorithm

Kshitiz Rathore, Mamta Khosla, and Ashish Raman

Abstract Embedded systems often rely on universal asynchronous receiver and transmitter (UART) transmission protocol to govern their data flow. UART is one of the most widely used interfaces because its characteristics, such as its capability to transmit the data over long distances, and its low cost makes it a desirable interface. The standard UART protocol contains one start bit (logic 0), 8 bits representing its input data, one bit for parity (even or odd parity bit as per design requirement), and one-stop bit (logic 1). So, the total data frame is of 11 bits. The primary disadvantage of parity bit is that it may fail to catch errors. Mismatch of the baud rates, transmission of data over long distances and electromagnetic radiation can alter the bits. So, if two data bits are corrupted; for instance, the parity will not detect the error. This limitation is removed in this paper by introducing checksum bits into a standard UART protocol. Based on Verilog HDL, we present the design of a 12-bit UART module with checksum bits. The proposed UART module can detect the bit flips of data frames automatically which is not possible in standard UART protocol. This paper showcases a novel UART interface design approach that aims to overcome the bottlenecks present in the existing model. Furthermore, the paper explains the functions of each individual sub-module and how the design works on the Xilinx ISE simulator.

Keywords Serial communication protocol · Baud rate generator · Negative edge detector · Checksum · Universal asynchronous receiver–transmitter

K. Rathore (✉) · M. Khosla · A. Raman
Dr. B.R. Ambedkar National Institute of Technology, Jalandhar, Punjab, India
e-mail: kshitiz19may@gmail.com

M. Khosla
e-mail: khoslam@nitj.ac.in

A. Raman
e-mail: ramana@nitj.ac.in

K. A. Reddy et al. (eds.), *Proceedings of Fourth International Conference on Computer and Communication Technologies*, Lecture Notes in Networks and Systems 606,
https://doi.org/10.1007/978-981-19-8563-8_23

1 Introduction

A micro-controller or computer system will usually have many serial data ports that are used for communicating with input–output devices such as computer serial communication ports that are compatible with UART, serial printers, key-boards, Bluetooth-UART devices, and so on. As its names implies, a UART is a serial communication protocol that receives and transmits the data serially [1, 2]. UART is an acronym of universal asynchronous receiver and transmitter. It works as a data transmission protocol that facilitates serial communication among devices. The UART protocol supports full-duplex, half-duplex, and simplex transmissions between any transmitter–receiver pair. In the simplex mode of communication systems, the data bits are transmitted only from the source end only. In the half-duplex mode of communication systems, the data transmission is possible from both directions, but at a time, only one of the two users can perform data transmission. If the receiver receives the data, then the transmitter is in an idle state and vise-versa. In the full-duplex mode of communication systems, both users can actively participate and exchange data at the same time. UART consists of two modules, namely transmitter and receiver. The transmitter module converts the bytes into serial bits and transmits the data serially. The receiver performs serial-to-parallel conversion on the asynchronous data frame received from the serial data input [3]. UARTs are asynchronous in nature since the transmitter and receiver modules transfer the data without support from an external clock signal. To synchronize the received data frame, the clock is not required. Instead, UART's transmitter and receiver module operates at equal baud rates. A baud rate is a rate at which unit data is transmitted through a communication channel, usually in bits-per-second (bps). Standard baud rates for UART are 1200, 2400, 4800, 9600, 19200, 38400, 57600, and 115200 bps. In order to transmit data effectively over UART, both the transmitter and the receiver must use the same baud rate [4–6]. The standard UART data frame consists of 1 start bit, 8 transmits data bits, 1 parity bit, and 1 stop bit. The parity bit is optional which depends upon the designer's requirement whether they want to consider even parity or odd parity. In order to produce a parity bit in the UART protocol, all 8 bits of the data byte are added up, and the evenness or the oddness of the sum decides whether the bit is set or not. So, this low-level error checking mechanism makes the whole system less reliable because if two data bits are corrupted; for instance, the parity will not detect the error. Thus to remove this limitation, we are introducing checksum bits in the standard UART protocol [7]. Checksums are counts of the bits that are also transmitted with the payload. This helps the receiver to ensure that the number of bits received are equal to the number of bits transferred by the sender. If both counts are equal, then the transmission is judged as a success, else error detection mechanism is initiated [8, 9]. In this paper, we proposed the architecture of a UART transmitter block and receiver block that consists of a checksum generator and a checksum checker, respectively, and these blocks have been synthesized and simulated using Verilog hardware descriptive language [10].

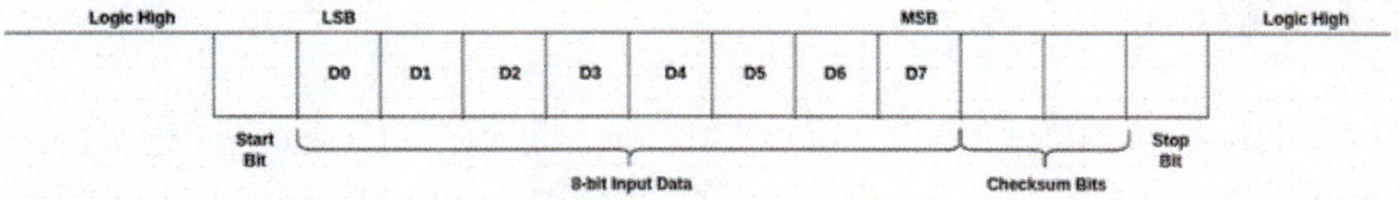

Fig. 1 Modified UART data frame

2 Modified UART Data Frame

Whenever the data transmission is initiated using the UART module, it always generates a data frame. Now, to manage the serial transmission of this data, the transmitter adds certain bits, namely start bit (one), stop bit (one), and checksum (two bits) initially. So, total 12 bits are present in the data frame at the time of its creation out of which only bits represent the actual data. During the reset condition, the data transmission remains high, i.e., logic 1. At the time of transmission, first start bit is sent which has logic 0; after that 8 bits of data are transmitted followed by 2 bits of checksum, and at last, stop bit is sent which has a logic 1 (Fig. 1).

3 Proposed UART Transmitter Architecture

The proposed UART transmitter architecture is collection of various modules including a baud rate generator module, checksum generator module, transmitter finite state machine and parallel in serial out register module, which collaborate together to ensure fast and accurate data transmission (Fig. 2).

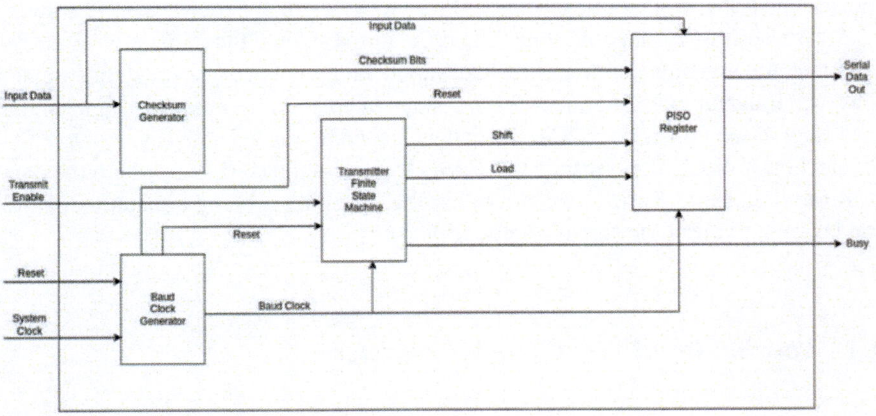

Fig. 2 Modified UART transmitter architecture

3.1 Checksum Generator

The checksum is added to this protocol to eliminate the problem of corrupted data bits, which is not possible while using even/odd parity. Using this method, the receiver can check if the output is correct as it is more reliable. The checksum generator first divides the transmitted 8-bit input data into 4 chunks of 2 bits each; then, an add operation is implemented on these 4, 2-bit chunks. After that, take 1's complement of the result. Afterward, the checksum bit will be attached to the 8-bit input data as the final result. The checksum generator operates on a mechanism formed by full adders and NOT gates.

3.2 Baud Rate Generator

A baud rate generator (BRG) actually operates as a frequency divider circuit. This BRG module has an active-low reset and a system clock, which act as inputs, and the baud clock acts as an output. In this design, the BRG produces a frequency clock that is 8 times of the baud rate clock. In this way, the asynchronous serial data at the receiver is sampled precisely.

3.3 Transmitter Finite State Machine

The transmitter FSM changes the transmitter's state. This module consists of three inputs and three outputs. The transmit enable signal, the active-low reset, and the baud clock act as input signals, and the load, shift, and busy act as outputs of a transmitter FSM. In this transmitter FSM module, there are four states: idle state, load state, shift state, and hold state. Idle is the initial state of the transmitter FSM. In this state, the transmit enable signal, load signal, busy signal, and shift signal remain low. The transmitter FSM moves to load state when transmit enable is high. In the load state of the transmitter FSM, the data is loaded before a frame is generated. On the next baud clock, the transmitter's FSM changes to the shift state where the data is transmitted serially one clock at a time till all the data is being transmitted. Hold state is used to clean the signal's value (Fig. 3).

3.4 Parallel Input Serial Output Register

To transmit serial data, PISO registers are used. The baud clock, the load signal, the active-low reset, the shift signal, the checksum bits, and the input 8-bit transmitted data act as inputs of the PISO register. Serial out acts as an output of the PISO

Fig. 3 Transmitter FSM

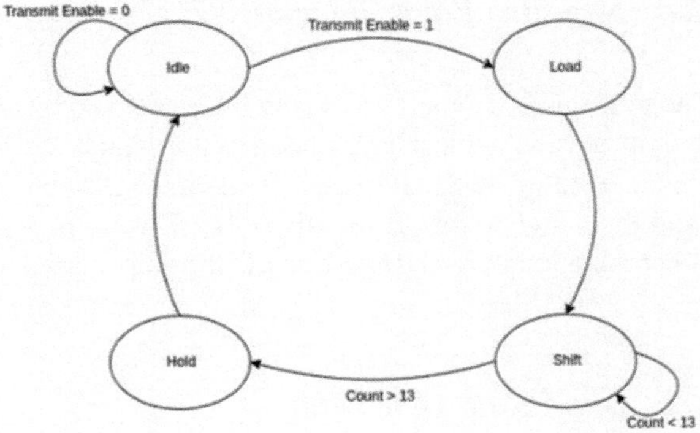

register. The data frame is created when the load signal is high, meaning that the required additional bits (start, parity bit, and stop bits) are appended to the data bits. Furthermore, as the signal reaches a high value, serial data transmission starts.

4 Proposed UART Receiver Architecture

The proposed UART receiver architecture is an integration of a baud rate generator module, negative edge detector module, receiver finite state machine, checksum checker module, and serial in parallel out register module (Fig. 4).

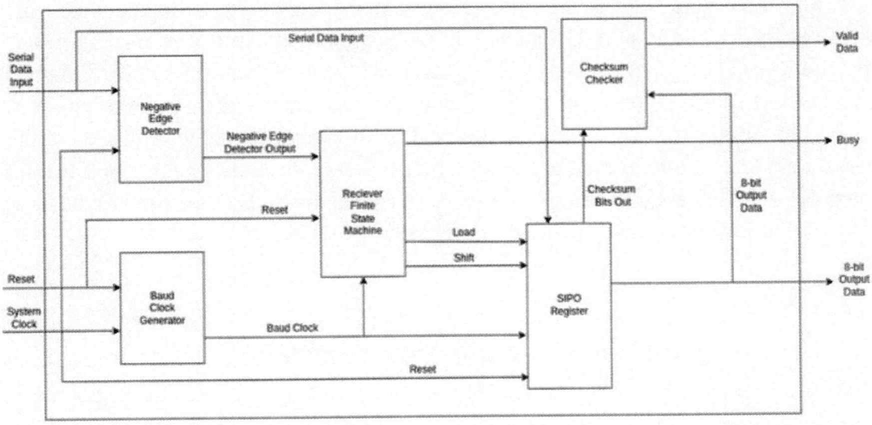

Fig. 4 Modified UART receiver architecture

4.1 Negative Edge Detector

The transmission data frame's start bit is detected by a negative edge detector. Prior to transmission, logical high is used as the default transmitted data signal. Serial bits are received by the UART receiver when the start bit appears, then the signal shifts from logic high to low. A negative edge detector is useful for detecting the start bit. A combination of a AND gate and D flip-flop is used to design such edge detector.

4.2 Baud Rate Generator

A baud rate generator actually operates as a frequency divider circuit. This baud rate generator module has an active-low reset and a system clock, which act as inputs, and the baud clock acts as an output. Both the transmitter and receiver operate at the equal rates (baud).

4.3 Receiver Finite State Machine

The receiver FSM changes the receiver's state. This module consists of three inputs and three outputs. The negative edge detector signal, the active-low reset, and the baud clock act as input signals, and the load, shift, and busy act as outputs of a receiver FSM. In this receiver FSM module, there are four states: idle state, shift state, load state, and hold state. Idle is the initial state of the receiver FSM. In this state, the negative edge detector signal, load signal, busy signal, and shift signal remain low. The start bit is being detected by the negative edge detector module, which signals the receiver to start. Once the receiver reaches the shift state, shifting operations start until all bits have been received. On the next baud clock, the receiver moves to the load state. Here, 8 bits of data are loaded by removing start bit, checksum bits, and a stop bit. On the next baud clock, the receiver moves from the load state to the hold state, and hold state is used to clean the signal's value (Fig. 5).

4.4 Serial Input Parallel Output Shift Register

To receive the serial data, SIPO registers are used. The baud clock, the load signal, the active-low reset, the shift signal, and the received 8-bit serial data act as inputs of the SIPO register. Parallel data out acts as an output of the SIPO register. One bit of data is shifted on the positive edge of the baud clock when the shift signal is set to 1. After removing the extra bits, 8 bits carrying the actual data are sent to the receiver's output during high load signal.

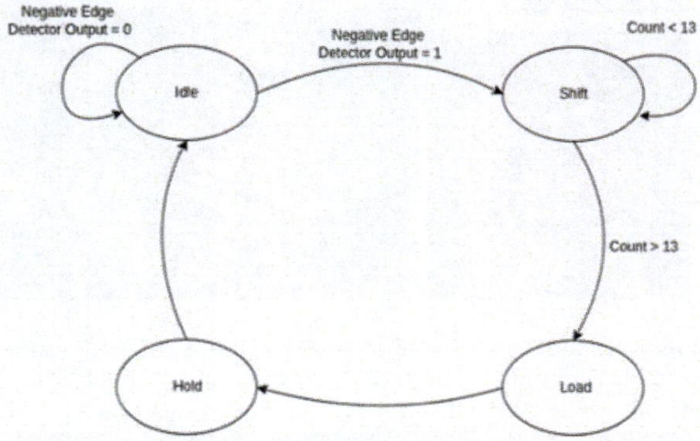

Fig. 5 Receiver FSM

4.5 Checksum Checker

The checksum checker module validates the correctness of received data. This module consists of two inputs and one output. The 8-bit input data signal and checksum bits signal from the SIPO register act as inputs, and the data valid signal acts as an output for the checksum checker. If the value of the data valid signal is 00, then the received data is correct, otherwise not.

5 Results and Discussions

We have simulated our UART architecture design on the Xilinx ISE. Figure 6 represents the waveform simulation of the UART transmitter module. In this paper, 8-bit input data, i.e., 10101100 is transferred serially using the UART module which utilizes the baud clock's positive edge and is shown using blue color in Fig. 6. Moreover, the data out signal is present in violet color. By default, the serial data out remains high (logic 1). At the time of high load signal, the PISO register creates a new data frame using one start bit (logic 0), 8 input bits (10101100), two checksum bits (10), and at last one-stop bit (logic 1). When the signal is high, shifting operations is initiated, and the start bit shifts (logic zero) on the next positive edge of the baud clock. Similarly, least significant bit of the 8 data bits shifts, and it is followed by the shifting of checksum bits and stop bit. When transmitter in the shift state, the counter counts the bits until all bits are not transmitted serially. Finally, serial output shows that 010101100101 (12 bits of the data or one modified UART frame) transmitted successfully. If the FSM will not detect the start bit, it will remain in its idle state. Once it detects a valid start bit (in this case logic 0), the FSM will move toward shift state. Moreover, the shift signal will hold the high value until all the serial bits including data and extra bits are saved in the temporary register. When the load signal is high, stored data (10101100) is the load to the receiver output. The 8-bit receiver output (10101100) and checksum bits signal (10) from the SIPO

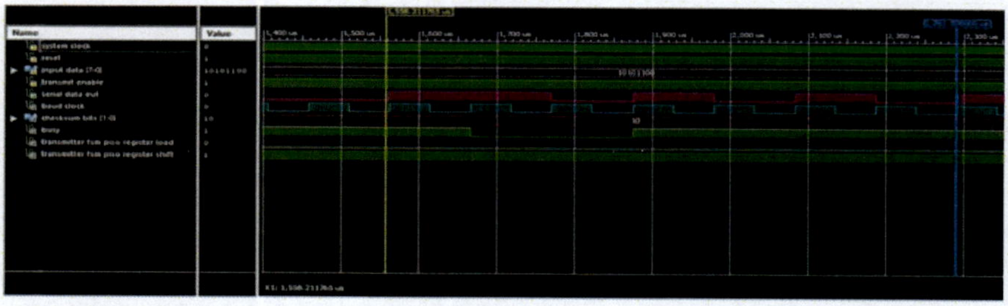

Fig. 6 Output waveform of proposed modified UART transmitter

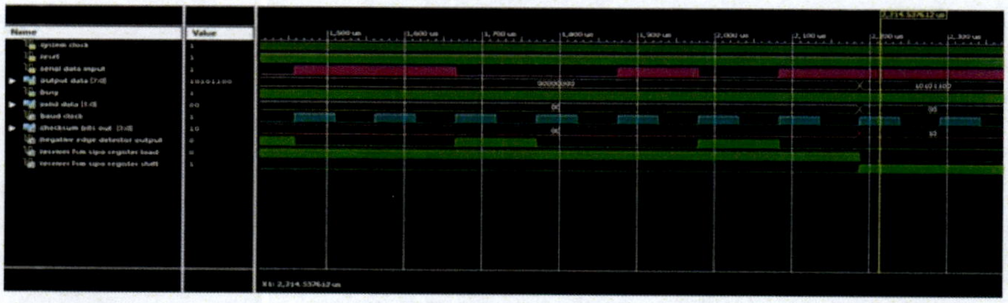

Fig. 7 Output waveform of proposed modified UART receiver

register act as inputs for the checksum checker. The checksum checker will divide the 10-bit data (including 8 bits of input data and 2 bits of checksum) into 5, 2 bits of chunks. Then, an add operation is implemented on these 5, 2-bit chunks. After that, take the 1's compliment of the result. Now, the value of the data valid signal is 00, which shows yellow color in Fig. 7 that means the received data is correct.

Now, let us consider another example. In this, we will intentionally corrupt some bits of the input data signal (i.e., the output of the UART transmitter), and this signal will act as an input for the UART receiver. On the transmitter side, the user provides 8 bits of input data (11001000) to the UART transmitter; it is shown using the orange color in Fig. 8. After applying the respective module's logic, the transmitter can transmit the 12 bits of data serially, which includes one start bit (logic 0), 8 bits of data (11001000), two checksum bits (01), and one-stop bit (logic 1). During transmission of data, 2 bits of the input data signal get corrupted, specifically the 2nd (D2) and 6th (D6) bit. Now, the UART receiver receives 11101010 as an input instead of 11001000, which includes one start bit (logic 0), checksum bits (01), and stop bit (logic 1). Therefore, 1 data frame concatenates to (011101010011), which acts as serial input data for the UART receiver. The data valid signal gives 1 0 as an output, and it is shown using the yellow color in Fig. 9, which leads to the contradiction checksum algorithm (data valid = 00). So, the received signal is found to be corrupted.

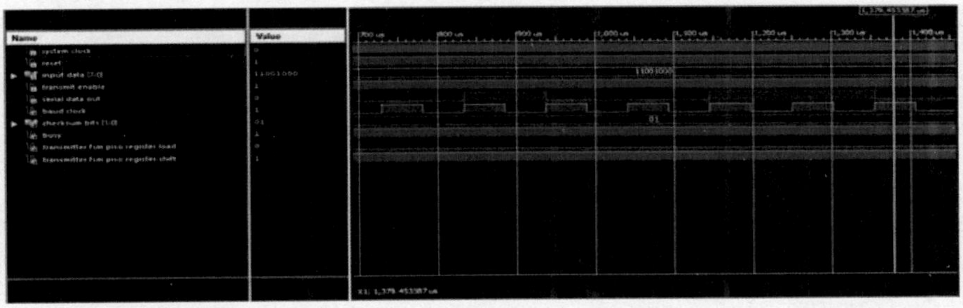

Fig. 8 Output waveform of proposed modified UART transmitter

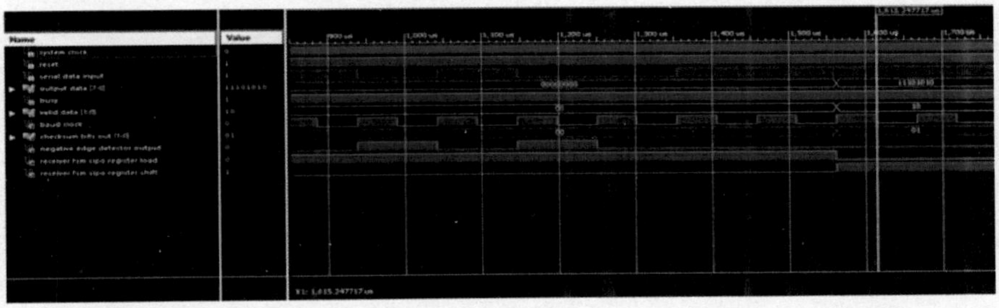

Fig. 9 Output waveform of proposed modified UART receiver

6 Conclusion

All most all the UART protocols use even/odd parity as an error detection technique. This low-level error checking mechanism makes the whole system less reliable. To fix this limitation, an enhanced version of UART has been presented with the introduction of a checksum. This modified UART protocol performs verification by simulating the transmitter and receiver waveforms on Xilinx ISE. Using this modified UART protocol could significantly enhance the efficiency of the serial data transmission protocol, and it also adds reliability, stability, and flexibility to the standard UART design that is often used in embedded systems and digital circuit applications.

References

1. Fang Y-Y, Chen X-J (2011) Design and simulation of uart serial communication module based on vhdl. In: 2011 3rd International workshop on ıntelligent systems and applications. IEEE, pp 1–4
2. Nanda U, Pattnaik SK (2016) Universal asynchronous receiver and transmitter (uart). In: 2016 3rd İnternational conference on advanced computing and communication systems (ICACCS), vol 1. IEEE, pp 1–5

3. Daraban M, Corches C, Taut A, Chindris G (2021) Protocol over uart for real-time applications. In: 2021 IEEE 27th ınternational symposium for design and technology in electronic packaging (SIITME). IEEE, pp 85–88
4. Wang Y, Song K (2011) A new approach to realize uart. In: Proceedings of 2011 international conference on electronic & mechanical engineer-ing and information technology, vol 5. IEEE, pp 2749–2752
5. Anjum F, Thakre MP. Vhdl based serial communication interface inspired by 9-bit uart
6. Mahure B, Tanwar R (2012) Uart with automatic baud rate generator and frequency divider. J Inf Syst Commun 3(1):265
7. Fletcher J (1982) An arithmetic checksum for serial transmissions. IEEE Trans Commun 30(1):247–252
8. Tong XR, Sheng ZB (2012) Design of uart with crc check based on fpga. In: Advanced materials research, vol 490. Trans Tech Publication, pp 1241–1245
9. Wakhle GB, Aggarwal I, Gaba S (2012) Synthesis and implementation of uart using vhdl codes. In: 2012 International symposium on computer, consumer and control. IEEE, pp 1–3
10. Priyanka B, Gokul M, Nigitha A, Poomica J (2021) Design of uart using verilog and verifying using uvm. In: 2021 7th International conference on advanced computing and communication systems (ICACCS), vol 1. IEEE, pp 1270–1273

Modified VHDL Implementation of 128-Bit Rijndael AES Algorithm by Asymmetric Keys

Soham Das, Nitesh Kashyap, and Ashish Raman

Abstract Using electronic means to transfer data exposes the data to risk of attack. The increasing usage of electronic media has pushed security into the spotlight. Cryptography's relevance has risen dramatically in recent years as a result of the rise of electronic data transfers. This paper gives us a scenario about the commonly used and well reliable advanced encryption standard (AES) algorithm. It also throws light on the information about functional cipher operation. Since digital data is being exchanged at such a rapid rate, the security of information in data storage and transmission becomes significantly more important. The security of information transmitted over wireless networks is of the utmost importance. Security of the data is ensured in wireless communication by encryption and decryption of the data. Security is provided through encryption algorithms used in the transmission channels. Developed as a Federal Information Processing Standard (FIPS) of the United States, AES is an algorithm that can protect electronic data by encrypting it. The AES algorithm for cryptography is a block cipher that encrypts and decrypts information by means of asymmetric keys.

Keywords AES algorithm · Encryption · Decryption

1 Introduction

With the expansion of data communications, security systems and devices that safeguard personal information transmitted over transmission channels have become more necessary. A cryptosystem is much more appropriate for protecting large amounts of data. Cryptography is already becoming increasingly important in

S. Das (✉) · N. Kashyap · A. Raman
Department of ECE, Dr. B.R. Ambedkar National Institute of Technology, Jalandhar, India
e-mail: sohamrules101@gmail.com

N. Kashyap
e-mail: kashyapn@nitj.ac.in

A. Raman
e-mail: ramana@nitj.ac.in

K. A. Reddy et al. (eds.), *Proceedings of Fourth International Conference on Computer and Communication Technologies*, Lecture Notes in Networks and Systems 606,
https://doi.org/10.1007/978-981-19-8563-8_24

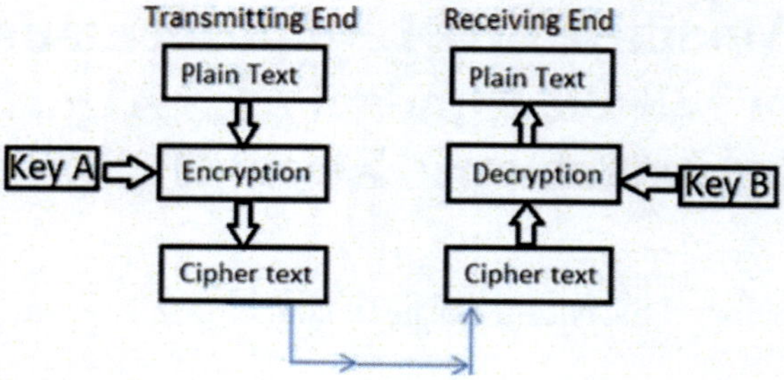

Fig. 1 Block diagram of AES cryptography

embedded systems innovation due to the rapid increase in devices and apps sending and receiving data; data transfer rates are increasing. Any organization or academic organization should analyze the cipher strength as part of their security risk assessment [1]. The NIST of the USA has approved the AES algorithm to succeed DES (FIPS-197, 2001). Here, for the encryption and decryption purpose, we use separate keys (Key A and Key B). Both the keys (A and B) are given prior to their respective inputs. Key A is used for the encrypting the plaintext, and Key B is used for decrypting the cipher text (Fig. 1).

The block size for this encryption algorithm is 128 bits, while the key size is 128, 192, or 256 bits (Table 1).

Because of its great soundness and is dependability in both software and hardware, AES is extensively used [2]. Despite the availability of several technology solutions, they are too sluggard for fast-paced operations such as for wireless communication networks. For wide range of applications, a number of AES optimized designs and modifications have been presented. AES analysts state that, out of 10 rounds, about 8 can be brute force successfully on today's modern hardware systems. In addition, the remaining 2 rounds cannot be broken within enough time that would allow the attacker to make the attack on the system impactful [3, 4].

Table 1 AES key lengths and its parameters

	Block size N_b words	Key length N_k words	Number of rounds N_r
AES—128 bits key	4	4	10
AES—192 bits key	4	6	12
AES—256 bits key	4	8	14

1.1 Literature Review

Numerous studies have been conducted for adjusting cryptography and maneuver the big data using cloud servers. An article describes the use of unexpected confidentiality as a security solution with AES-based storage needs and less storage [5]. AES was created due to its lower storage requirements and faster execution time as compared to previous approaches. The study presented a secure method basis of two private keys, with the secondary key (extra) being used for both encrypting and decrypting. According to the conclusions, this enhances the security while maintaining the performance index close to the original AES [6]. To protect the cloud computing paradigm, a similar study effort with a special emphasis on cloud computing and the field of cloud computing is examined in [7], where a reconfiguration of AES is outlined that offers protection over data stored in the cloud by leveraging a new key generation procedure as well as the involvement of a transpose matrix to construct ciphertexts that are buried deep from the eyes of third parties, providing security to access crucial data for over cloud. Reena et al. [8] offered a study that focused on key expansion and shift row transformation to maintain a high degree of security. The purpose was to prevent and safeguard the information against cyber-attacks. Their experiment also cuts the time it took to encrypt photos and produced a better outcome than AES. They also helped to improve bandwidth efficiency.

1.2 Methodology

Inputs and outputs: The AES algorithm uses a single 128-bit sequence as both outputs and inputs. An AES cipher key is 128 bits in length. A byte is used in the AES algorithm as its basic unit of computation, so the input bits are converted into byte sequences prior to processing. After that, a two-dimensional array of bytes (known as the State) is created. A state array is organized into four rows of bytes. Each row contains N_b bytes, where N_b is the block size divided by 32. The State array will go through all of the processes core process (Cipher and inverse cipher), after which its final result will be transmitted to the output [9].

Key Schedule: To construct a key schedule, the AES system accepts the cipher key as an input and runs it through a key expansion procedure. N_b $(N_r + 1)$ is generated via the key expansion. There are broadly two types of keys used during the algorithm process.

Symmetric Key: Secret-key or shared-key cryptography is other name for symmetric key cryptography. In this sort of system, the transmitter and receiver utilize the same key for both decryption and encryption. The framework is dependent on self-certification, which implies that the key is self-certified. This type of cryptographic technology is necessary since it allows for speedier service without consuming a lot of resources [10].

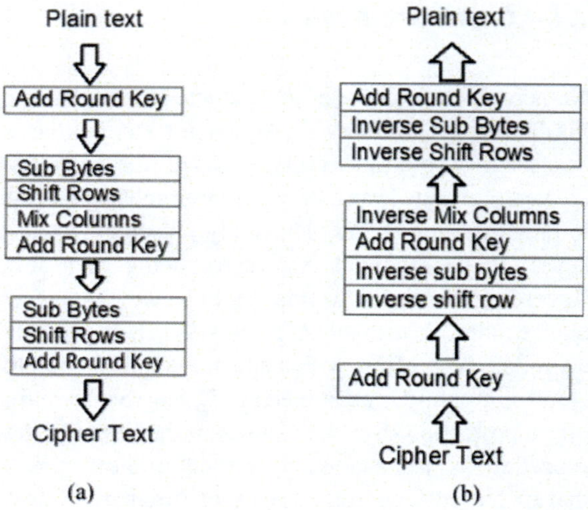

Fig. 2 **a** Block diagram of AES encryption (cipher) **b** block diagram of AES decryption (inverse cipher)

Asymmetric Key: Asymmetric-key cryptography denotes public-key cryptography. The sender uses the recipient's public key to encrypt the message, and the receiver uses his private key to decrypt it. Instead of employing self-certification, the keys are authenticated using digital signatures. This technique is far more suitable and allows for improved authentication while yet ensuring privacy [11].

Key expansion: As a result of the round keys to be utilized in every round originating completely from the secret input key, the spawning maneuver is a key expansion process. It is the original round key that is used in encryption. The original key is always the last generated key by key expansion during decryption. Before repetitious encryption or decryption can begin, the secret round key would be inserted as a second input to the plaintext input. During the generation of 128 bits key size, ten round keys of 24 bytes will be generated (Fig. 2).

Cipher (Encryption): The cipher turns input data (which is initially transferred to the State array) to an unreadable form termed ciphertext using the round function, which is made up of four separate byte-oriented transformations. State is first added to a round-key array, and then, before repetitious encryption or decryption can begin, the secret round key would be inserted as a second input to the plaintext input. The round function employs a one-dimensional array of four-byte words as a key schedule (the Round Key) derived from a key expansion routine. In the final round, the mix columns transformation does not apply transformations, and to the State array, apply an add round Key transformation.

Sub-Bytes: It executes nonlinear substitution on each byte of a State through the use of a substitution table. Implementing AES hardware with S-box design is crucial to optimizing its performance [12].

Shift Rows: Deals in shifting rows cycles through different numbers of bytes (offsets) for the last three rows in the State column. There is no shift on the first row,

$r = 0$, while the second shifts by one byte, third by two bytes, and finally fourth rows move cyclically left by three bytes.

Mix Columns: A Galois field multiplication is used to achieve this transformation. Each byte in a column is given a new value depending on a concatenation of all four bytes in the column.

Add Round Key: When compared to the previous encryption and decryption rounds, the add round key procedure is done once more. As a hardware implementation, it uses a simple exclusive-or operation with the 128-bit data and key.

Inverse Cipher: This is accomplished by first copying the input (ciphertext) into the State array, and then performing three inverse transformations, and adding an add round key transformation to the State array. After adding the first round key to the State array, a round function is constructed, with the final round again differing slightly from the first round likewise encryption. A key expansion routine is used to derive a one-dimensional array of four-byte words (Round key) that are used to parameterize the round function. Except for the last round, which does not involve the transformation of inverse mix columns, there are no distinctions between the rounds of N_r [13].

Add Round Key: Because of its inverse nature, add round key is exactly its inverse. To select the round keys, reverse the order in which they are entered.

Inverse Shift Rows: Inverse shift rows have similar properties to that of shift rows. The first row is not changed, but the second is shifted to the right by one bit, third by two, and fourth by three bits.

Inverse Sub Bytes: It automates the substitution process by making use of a previously calculated substitution table called inverse S-box. 256 numbers (from 0 to 255) and their values are stored in the inverse S-box table.

Inverse Mix Columns: It operates similarly to mix columns in the encryption part; however, it has a different matrix. Polynomial transformations of degree less than 4 over GF (2 * 8) are used for the inverse mix columns transformation. The coefficients of these polynomials are the columns of the state multiplied by the mix columns matrix.

2 Results and Discussion

The VHDL programming language was used to code the suggested design, and the ISE Design Suite software was used to analyze the results. Here, we have taken the hexadecimal value of nitjalandhargood as an input, and then, we have carried out the simulation results. Figure 3a shows 3c4fcf098815f7aba6d2ae2816157e2b is the private key, and 10 rounds of encryption have been performed to the plaintext of 6e69746a616c616e64686172676f6f64. Figure 3b shows the message bit has been encrypted at the end of 10th round for the given input of 6e69746a616c616e64686172676f6f64, and at the end of encryption algorithm, we have successfully encrypted the input message of 6e69746a616c616e64686172676f6f64 in the form of cipher text e666a864aadfdbed1466867b4c97ea77 (Table 2).

The timing waveforms for the decryption are shown in Figure 3c and 3d. Figure 3c shows inclusion of private key a60c63b6c80c3fe18925eec9a8f914d0 and 10 rounds of decryption to the ciphertext input of e666a864aadfdbed1466867b4c97ea77. Figure 3d shows the message bit has been decrypted at the end of 10th round and for the given input of e666a864aadfdbed1466867b4c97ea77, and at the completion of decryption algorithm, we have successfully decrypted the message in the form of plaintext—6e69746a616c616e64686172676f6f64 (Table 3).

As a result, we were able to effectively encrypt and decrypt a random text input using AES 128-bit encryption and decryption. The technology has already been put through its paces on a dedicated test bench, with the system being triggered for both decryption and encryption for a certain pulsating inputs. Each program is tested using some of NIST's sample vectors. With the Xilinx Spartan Family Device XC3S500, the throughput for both encryption and decryption processes exceeds 352 Mbit/sec.

2.1 Conclusion

Software can easily implement the AES algorithm. Software implementations are the cheapest, but they have the least physical security and are the slowest. Apart from the increasing amounts of high, increased secure communications mixed with physical security, cryptography is now being implemented efficiently. Cryptography is now becoming particularly crucial in today's society. As a result, the frequency is by far the most important aspect in order to minimize the time duration. We have addressed the basics of the AES algorithm as well as the implementations of its modules in VHDL in this study.

2.2 Future Scope

More research is needed into optimization methodologies for implementations that enable varied key lengths and modes of operation. We can make necessary amendments to the inverse substitution box and substitution box, which are already existing and accessible. Inverse mix columns and mix columns lookup tables can also be revised. We need to focus on reducing more latency and implementing the algorithm in many applications.

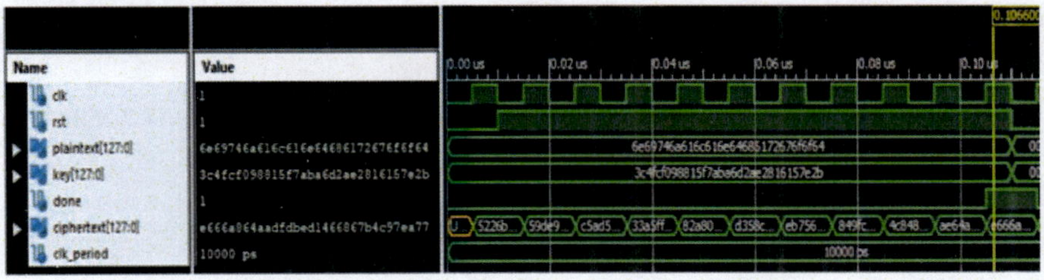

(a)

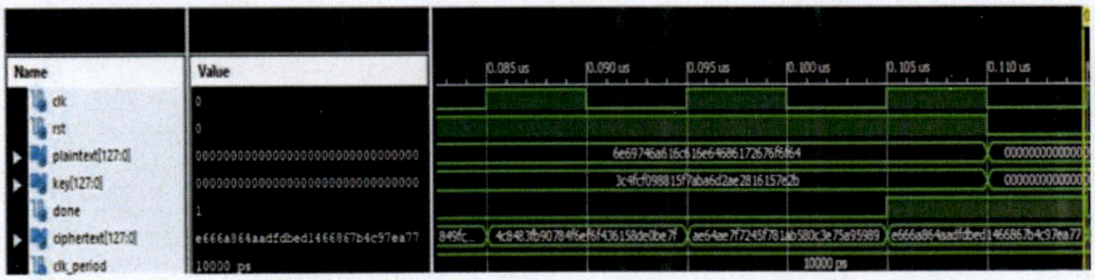

(b)

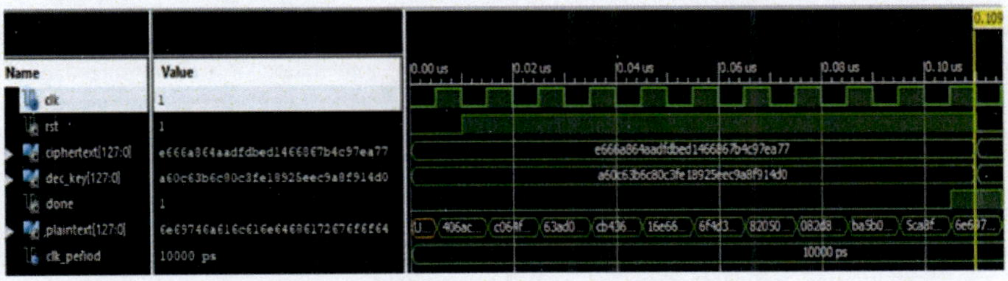

(c)

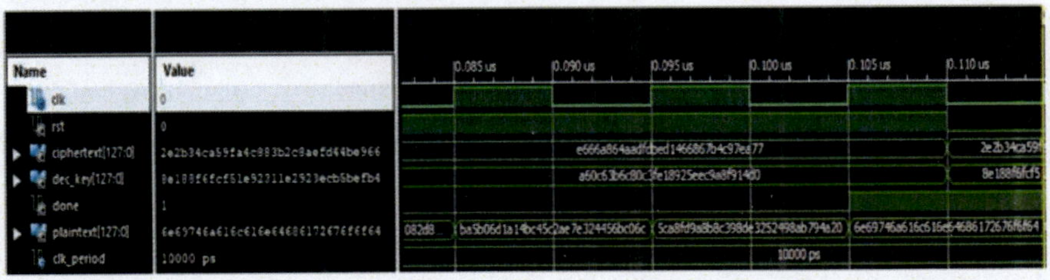

(d)

Fig. 3 **a** Simulation results of AES encryption algorithm; **b** simulation results of AES encryption algorithm; **c** simulation results of AES decryption algorithm; **d** simulation results of AES decryption algorithm

Table 2 Test vector for AES encryption algorithm

Text to be encrypted	nitjalandhargood
Plaintext (hexadecimal value of text to be encrypted)	6e69746a616c616e64686172676f6f64
Private key for encryption	3c4fcf098815f7aba6d2ae2816157e2b
Ciphertext (encrypted message)	e666a864aadfdbed1466867b4c97ea77

Table 3 Test vector for AES decryption algorithm

Ciphertext (Message to be decrypted)	e666a864aadfdbed1466867b4c97ea77
Private key for decryption	a60c63b6c80c3fe18925eec9a8f914d0
Plaintext (Decrypted message)	6e69746a616c616e64686172676f6f64
Original text	nitjalandhargood

References

1. Sharma N (2017) A review of information security using cryptography technique. Int J Adv Res Comp Sci 8(4)
2. Luo AW, Yi QM, Shi M (2011, May) Design and implementation of area-optimized AES based on FPGA. In: 2011 International conference on business management and electronic information, vol 1. IEEE, pp 743–746
3. Jun Y, Jun D, Na L, Yixiong G (2010, March) FPGA-based design and implementation of reduced AES algorithm. In: 2010 International conference on challenges in environmental science and computer engineering, vol 2. IEEE, pp 67–70
4. Deshpande AM, Deshpande MS, Kayatanavar DN (2009, June) FPGA implementation of AES encryption and decryption. In: 2009 International conference on control, automation, communication and energy conservation. IEEE, pp 1–6
5. Roy S, Das AK, Chatterjee S, Kumar N, Chattopadhyay S, Rodrigues JJ (2018) Provably secure fine-grained data access control over multiple cloud servers in mobile cloud computing based healthcare applications. IEEE Trans Industr Inf 15(1):457–468
6. Fadul IMA, Ahmed TMH (2013) Enhanced security of Rijndael algorithm using two secret keys. Int J Secur Appl 7(4):127–134
7. Pancholi VR, Patel BP (2016) Enhancement of cloud computing security with secure data storage using AES. Int J Inno Res Sci Technol 2(9):18–21
8. Mehla R, Kaur H (2014) Different reviews and variants of advance encryption standard. Int J Sci Res (IJSR), ISSN (Online), pp 2319–7064
9. Daemen J, Knudsen L, Rijmen V (1997, Jan) The block cipher Square. In: International workshop on fast software encryption. Springer, Berlin, Heidelberg, pp 149–165
10. Terec R, Vaida MF, Alboaie L, Chiorean L (2011) DNA security using symmetric and asymmetric cryptography. In: The society of digital information and wireless communications (vol 1, No 1, pp 34–51). IEEE, Piscataway, NJ, USA
11. Wang CH, Chuang CL, Wu CW (2009) An efficient multimode multiplier supporting AES and fundamental operations of public-key cryptosystems. IEEE Trans Very Large Scale Integration (VLSI) Syst 18(4):553–563
12. Cheng H, Ding Q (2012, Dec) Overview of the block cipher. In: 2012 Second international conference on instrumentation, measurement, computer, communication and control. IEEE, pp 1628–1631

13. Jing MH, Chen YH, Chang YT, Hsu CH (2001, Nov) The design of a fast inverse module in AES. In: 2001 International conferences on info-tech and info-net. Proceedings (Cat. No. 01EX479), vol 3. IEEE, pp 298–303

A Computationally Inexpensive Method Based on Transfer Learning for Mobile Malware Detection

Saket Acharya, Umashankar Rawat, and Roheet Bhatnagar

Abstract With the broad usage of Android smartphones, malware growth has been rising exponentially. The high prominence of Android applications has roused attackers to target them. In the past few years, most scientists and researchers have researched detecting Android malware through machine learning and deep learning techniques. Though these traditional techniques provide good detection accuracy, they need high configuration machines such as GPUs to train complex datasets. To resolve this problem, the transfer learning approach is presented in this paper to efficiently detect Android malware with low computational power requirements. By transferring the necessary features and information from a pre-trained source model to a target model, transfer learning lowers the computational cost. In this paper, we initially performed Android malware detection using traditional models such as convolutional neural networks and then we applied the transfer learning technique to reduce the computational cost. Additionally, we evaluated how well the suggested strategy performed against other cutting-edge malware detection methods. The proposed method achieved an accuracy of 97.5 with 2.2% false positive rate. In addition, the overfitting problem and high computational power requirements are also reduced.

Keywords Android malware · Transfer learning · Deep learning · Machine learning · Computational cost

S. Acharya · U. Rawat (✉) · R. Bhatnagar
Department of Computer Science and Engineering, Manipal University Jaipur, Dahmi Kalan, India
e-mail: umashankar.rawat@jaipur.manipal.edu

R. Bhatnagar
e-mail: roheet.bhatnagar@jaipur.manipal.edu

K. A. Reddy et al. (eds.), *Proceedings of Fourth International Conference on Computer and Communication Technologies*, Lecture Notes in Networks and Systems 606,
https://doi.org/10.1007/978-981-19-8563-8_25

1 Introduction

Mobile malware has been increasing drastically from the last few years. The rapid growth of this malware has become the motivation for attackers to target smartphones, especially Android. According to Zimperium[1] mobile threat report, in 2021, more than 10 million mobile phones were impacted by various threats in more than 214 countries. During 2019–21, more than 50 million phishing websites were examined and 250% of the websites were mobile-specific. There is a huge increase in the percentage of phishing web pages based on HTTPS from 2019 to 2021. This makes it tough to differentiate between legitimate and malicious sites. Mobile threats along with network attacks have dominated the malware ground.

Conventional mobile malware detection methods were limited by pattern matching, and hence, it becomes difficult to identify novel variants. The detection methods are based on artificial intelligence algorithms to provide more accurate and robust results in the recent times. Moreover, the probability of getting false positive results with these algorithms is also less compared to traditional detection methods. Malware detection methods based on AI algorithms have two common phases: preprocessing and classification. The first phase deals with feature extraction and the second phase utilized the extracted features to train the machine learning or deep learning model. The feature extraction methods are further classified into static and dynamic feature extraction [1]. The features are extracted without executing the mobile application [2]. Static features include dex files, XML files, bytecode, API calls, application permissions, etc. The major objective of static feature extraction is to disassemble the application to get the source code. Ahma d Firdaus et al. [3] proposed a technique based on static approach to extract static features. To choose the features among 106 strings, the authors further employed genetic search (GS), which is a query for genetic algorithm. In contrast to static feature extraction, the dynamic feature extraction methods rely on executing the applications in a virtual emulator. Hence, the obtained features provide more accurate information.

To classify malware, machine learning and deep learning techniques are commonly utilized. The machine learning models require detailed knowledge of feature selection. Some of the machine learning algorithms used for malware classification are support vector machine (SVM), random forest, k-nearest neighbor (KNN), and so on. Deep learning models provide more accurate results during the classification stage as compared to classical machine learning techniques. However, these models require heavily configured machines for training and testing. The most common deep learning techniques used to detect mobile malware are convolutional neural networks (CNNs) and recurrent neural networks (RNNs).

In our study, we have statically obtained the bytecode features of Android applications. Further, grayscale images are recreated using an autoencoder. In the end, the overall malware features are generated with the help of an autoencoder. The experiments have been conducted using CCCS-CIC-AndMal-2020, Drebin, AAGM, and

[1] https://www.zimperium.com/global-mobile-threat-report/.

hybrid datasets, respectively. The experimental outcomes provide good accuracy and outperform various machine learning and deep learning models for detecting Android malware.

The rest of the article is structured as follows: Sect. 2 discusses the related literature study. Section 3 demonstrates the proposed method. Section 4 presents the experimental results. In the end, Sect. 5 concludes the paper.

2 Related Work

The scholarly world has done a lot of research in malware detection. Most of the studies utilized machine learning algorithms to detect and classify Android malware. The results obtained with machine learning algorithms are promising but the major drawback of using machine learning algorithms is that they require domain-level knowledge for feature selection and extraction. Hence, deep learning algorithms were introduced to automatically extract critical features and classify mobile malware more efficiently as compared to machine learning approaches. The Maldozer framework, proposed by Karbab et al. [4], can automatically detect Android malware and offer familial categorization. To discover malicious applications, the authors used deep learning algorithms. They derived numerous features from the dataset's API code sequences. Over 30 k hazardous samples out of 70 k samples were used in the dataset used to evaluate the framework. Low false positive rates were attained by the authors, however, the framework needed complex calculations to operate more effectively.

The authors in [5] have suggested DL-Droid, a deep learning model that uses dynamic analysis and stateful input generation to detect malware in Android platforms. A study found that 94% accuracy (dynamic features only) and 95% accuracy (dynamic+static features) may be attained. The method employs an automated framework for running Android apps and extracting their functionality. DL-Droid uses these features as inputs for categorization. The DynaLog dynamic analysis framework was used to test numerous apps.

The authors in [6] proposed a system that can be used with mobile phones. It saves money by utilizing flexible computing resources. They use a convolutional neural network to plot an API call graph to determine whether or not an application is malicious (CNN). Using a simple classifier, it distinguishes between API call networks used for malicious actions and API call graphs used by apps. They were successful in achieving a high degree of accuracy. The technique uses API call graphs from both harmful and helpful applications to train datasets. The next step is to use Grad-CAM to discover high-weight API call graphs that are used by rogue apps.

Feng et al. proposed MobiTive [7], a real-time and responsive malware detection system for mobile phones. It protects by utilizing specialized deep neural networks. This environment should be pre-installed and ready to use on mobile phones. There are two parts to the functionality, i.e., model preparation, Dl training model, model migration, and model quantization mobile phone deployment using migrated/quantized models. Instead of decompiling APKs, API calls are retrieved directly from binary code. Performance-based feature selection mechanisms are combined with behavior-based feature update approaches.

Abdelmonim Naway and Yuancheng LI [8] provided a comprehensive review on detecting Android malware using deep learning techniques. The authors also performed a comparative analysis of the techniques based on accuracy, false positives, and error rates.

Dongfang Li et al. [9] suggested a deep learning-based automatic Android malware detection engine. The authors combined the techniques and automated them into a tool which can be used to detect static and dynamic malicious applications. The authors used fine-grained features during classification phase.

The authors in [10] have tried constructing a malware detection system that can help determine and examine the impact of giving applications unnecessary access. Least square support vector machine (LSSVM) is used to create the model, which is coupled to three different kernel functions (linear, radial, and polynomial). The detection rate of LSSVM with radial kernel was around 98.8%.

Most of the above-mentioned studies focused on using deep learning techniques and achieved great accuracy, however, these approaches needed a high amount of computation resources to train and test the detection model. To resolve this issue, the transfer learning approach has been proposed in this paper. The proposed method is described in the next section.

3 Proposed Method

To detect and classify Android malware, a method based on transfer learning approach is proposed in this paper. The proposed model is illustrated in Fig. 1. Initially, the Android applications are converted into grayscale images by disassembling the APK files. The binary data is extracted from application methods and they are converted into bytes. These bytes are filled with pixels to generate the images. Further, the grayscale images are fed to the CNN model to classify benign and malicious applications. CNN models are very good for classifying images because they cover the entire geographical area of an image to output important features [11]. However, these models require a lot of computational power and resources to provide good accuracy. The transfer learning approach removes this drawback and it is capable of providing good detection accuracy without having the requirement of high computational resources. The feature sets are transformed to feature vectors, which are then sorted in order of relevance. For instance, the importance of the parameter "version" is less important than the parameter "Permissions". Similar to that, each feature parameter is ranked according to relevance. Filtering can be done by deleting the irrelevant features and sorting the parameters. After being filtered, the important feature sets are translated into the binary images.

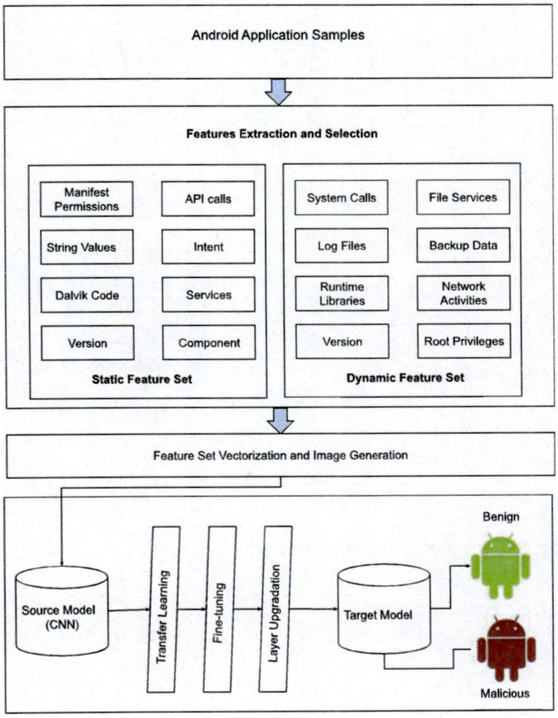

Fig. 1 Proposed methodology

3.1 Image Generation

The APK files are visualized best by utilizing static features. To extract the binary images from files, the files are converted into binary vector pixels. The entire APK file data is treated as a byte stream and is stored in a matrix called binary vector matrix. The APK files are extracted to produce 8-bit binary data files which are further transformed into grayscale images. This transformation is depicted in Fig. 2. Every byte in the binary vector matrix has been transformed into a pixel value because it might have a value between 0 and 255.

The steps to generate the images are given below:

- The files AndroidManifest.XML, Resources.arsc, Classes.dex, jar files, and others are retrieved from datasets containing APK archives.
- Disassembling the obtained files yields 8-bit binary files. The binary vector streams are formed once the data in the files is interpreted as binary data.
- An 8-bit array vector matrix is created from the binary vector streams.
- The array vector matrix is used to create grayscale images, which are then saved in an image dataset.

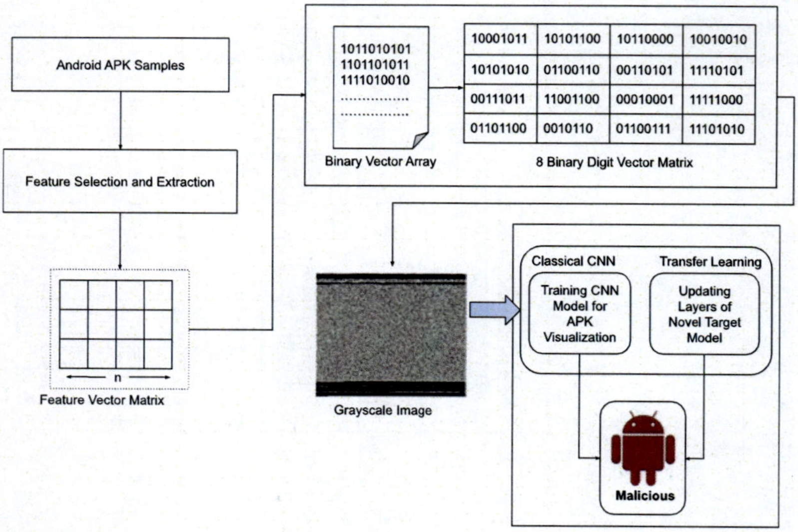

Fig. 2 Image generation process

Fig. 3 Transfer learning

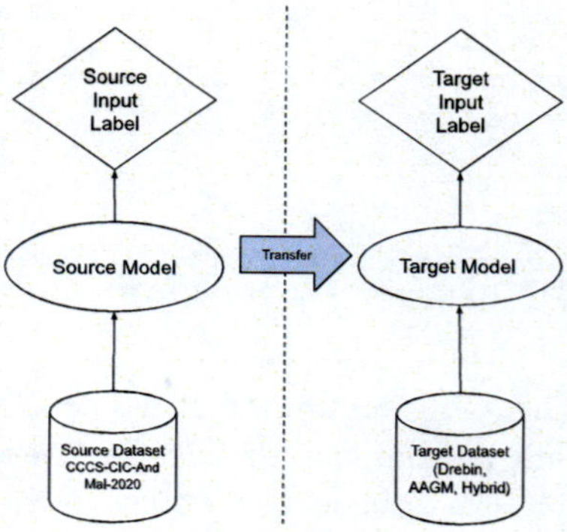

3.2 *Malicious Android Application Detection Based on Transfer Learning*

The transfer learning approach generates a model by inheriting the features of pre-trained neural networks. In this paper, we initially trained the dataset CCCS-CIC-AndMal-2020 on the CNN model. After that, we applied the transfer learning approach by transferring the last few layers of the CNN model to the target model. The basic aim of not transferring the entire layers is that we need to reduce the high computational cost requirement. By using the transfer learning method, a target model can be trained on smaller and hybrid size datasets. This reduces the computational complexity and needs for several GPUs.

The transfer learning method is depicted in Fig. 3.

To resolve the issue of overfitting, some of the layers of the pre-trained source model are fine-tuned. Fine-tuning helps to restrict the generalized features to be trained again and again. These general features can be APK version, history, software information, temporary files, etc. To achieve a better fine-tuning mechanism, we freeze the initial few layers of the source model.

4 Experimental Setup, Datasets, and Evaluation

In our study, we initially trained CNN model to classify benign and malicious apps. For training, we used the CCCS-CIC-AndMal-2020 dataset containing 400K samples. Further, we transferred some of the features of the trained model to the target transfer learning model and performed the classification with Drebin, AAGM, and hybrid datasets. To select the transferrable features, we used the feature selection method which is described below.

Feature Selection Method: Many of the features produced by the feature extraction are irrelevant. We use attribute selection to pick the most important features from the ones that were extracted. During attribute selection, we calculate the information gain of each feature to determine its value. The entropy decrease caused by classification is depicted as information gain, which captures feature efficacy in relation to the class. Formally, let F_s be a set of features to be classified into C classes and F_n denotes the nth subclass. Then, the entropy of F_s will be:

$$E(F_s) = -\sum_{n \in C} \frac{|F_n|}{|F|} \times log_2 \frac{|F_n|}{|F|} \tag{1}$$

Let F_x denotes the sample subset with feature value x for A for a feature F with x as the set of its potential values. The information gain can be calculated as:

$$\text{Info}_{\text{gain}}(F_s, f) = E(F_s) - \sum_{x \in x(f))} \frac{|F_X|}{F} \times E(F_x)) \tag{2}$$

4.1 Dataset Description

In our study, we utilized four different datasets, namely CCCS-CIC-AndMal-2020, Drebin, AAGM, and hybrid datasets. The hybrid dataset is constructed by taking the benign samples from Google Play and malicious samples from Drebin and AAGM datasets, respectively. The datasets are described in Table 1. The CCCS-CIC-AndMal-2020 is one of the recent datasets for Android malware detection having 400,000 benign and malicious samples, respectively. The malicious samples are taken from a total of 191 malware families. The Drebin dataset is having a relatively

Table 1 Datasets description

Dataset	Genuine apps		Malicious apps		
	Source	#Samples	Source	#Samples	#Families
CCCS-CIC-AndMal-2020	CCCS-CIC-AndMal-2020	200 K	CCCS-CIC-AndMal-2020	200 K	191
Drebin	Drebin public dataset	9476	Drebin public dataset	5560	179
AAGM	CIC-AAGM2017	1500	CIC-AAGM2017	500	42
Hybrid	Google Play	15,000	Malicious samples from Drebin and AAGM datasets	10,000	191

smaller size than the AndMal-2020 dataset. It consists of a total of 9476 benign samples and 5560 malicious samples, respectively. The malicious samples belong to 179 different families. The AAGM dataset consists of 1500 benign apps and 500 malware apps from 42 different malware families. Further, we constructed a hybrid dataset using 15,000 benign applications from Google Play with the help of crawler tools, and 10,000 malicious apps from the Drebin and AAGM datasets respectively.

4.2 Evaluation

In the first stage, the classification test is done for the CCCS-CIC-AndMal-2020 dataset containing more than 399 k samples. We used three well-known parameters to evaluate the process: "Recall rate (Rec)", "Precision (Prec)", and "F-Score". The following formulas are used to define these parameters:

$$\text{Prec}(bi) = \frac{N_{bi}}{N_i} \tag{3}$$

$$\text{Rec}(b, i) = \frac{N_{bi}}{N_b} \tag{4}$$

$$F_{\text{Score}}(b, i) = 2 * \left[\frac{\text{Rec}(b, i) * \text{Pre}(b, i)}{\text{Rec}(b, i) + \text{Pre}(b, i)}\right] \tag{5}$$

The experimental results provide an efficiency of 94.2% with a false positive rate of 5.7%.

In the next stage, transfer learning is applied by fine-tuning the feature sets of the CNN layers. The classification test is done for Drebin, AAGM, and Hybrid

Table 2 Classification results using classical CNN approach for CCCS-CIC-AndMal-2020 dataset

Sample	Precision	F-score	Rec	Support
Genuine apps	0.933	0.915	0.90	1200
Malicious apps	0.918	0.917	0.92	1563

Table 3 Classification results using transfer learning approach for hybrid dataset (Drebin/AAGM/Google Play)

Type	Precision	F-score	Rec	Support
Genuine apps	0.963	0.935	0.93	1100
Malicious apps	0.968	0.957	0.95	1423

Table 4 Classification results using transfer learning approach (Drebin)

Type	Precision	F-score	Rec	Support
Genuine apps	0.971	0.956	0.94	1331
Malicious apps	0.970	0.967	0.94	1399

Table 5 Categorization outcomes using transfer learning (AAGM dataset)

Sample	Precision	F-score	Rec	Support
Genuine apps	0.980	0.969	0.96	1352
Malicious apps	0.970	0.970	0.97	1563

datasets, respectively. In comparison with the conventional CNN model, we found that the transfer learning approach offers a greater detection accuracy of 97.5% with a false positive rate of 2.2%. Moreover, the issues related to overfitting and high computational power requirement were also resolved as only a few layers need to be trained, and the complete model does not need to be trained from the beginning. The evaluation results are given in Table 6.

We used sevenfold cross-validation to disperse the dataset evenly. Genuine and malicious APK images can both enter the CNN network in a single shot since the grayscale images are supplied into the network in a randomized manner. We used the "recall" score mentioned in Eq. (4) to reduce the false positives. We used 50 epochs for the Drebin dataset and 10 epochs for the hybrid dataset to get the best blend of hyper-parameters. Table 3 gives the classification results for the hybrid dataset. Tables 4 and 5 give the classification results for the Drebin and AAGM datasets, respectively. When compared to CNN classification precision values, the precision values are better (Table 2).

Finally, We compared the overall average accuracy and performance of the CNN and transfer learning model. The dataset evaluation result yields a score of 94.2%, with a false positive rate of 5.7%. This cross-validated score is calculated every fold

Table 6 Comparative analysis of the CNN and transfer learning approach

Technique	Accuracy (%)	False positive ratio (%)	Computational power requirements	Overall complexity
Simplest CNN	94.2	5.7	More	More
Transfer learning	97.5	2.2	Less	Less

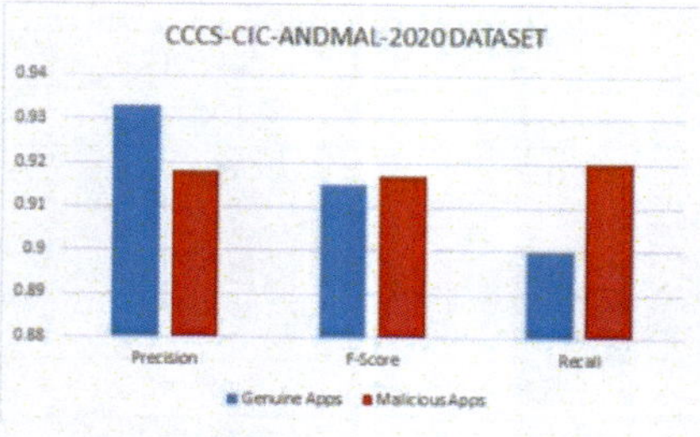

(a) CCCS-CIC-AndMal-2020 Dataset

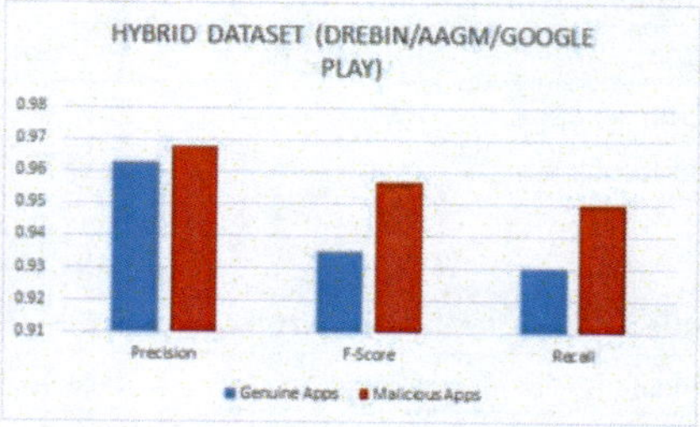

(b) Hybrid Dataset

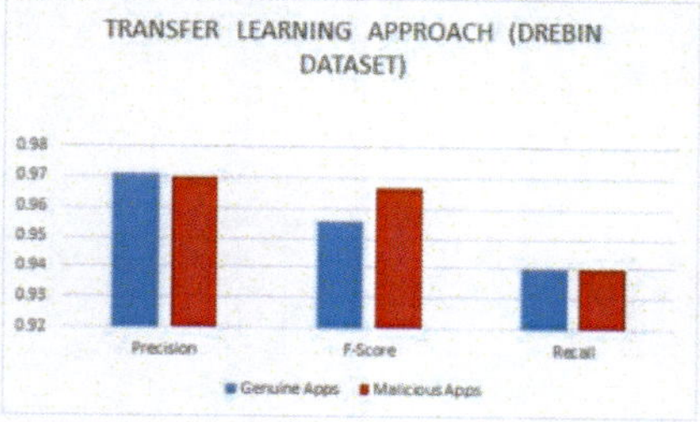

(c) Drebin Dataset

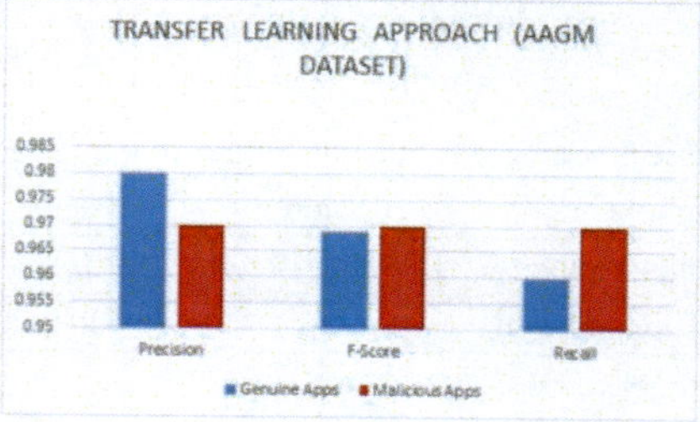

(d) AAGM Dataset

Fig. 4 Classification results for various datasets

and distributed equally throughout the dataset. We then used the transfer learning strategy, which resulted in a cross-validated score of 97.5%. We changed the configuration file and fine-tuned the hyper-parameters of the CNN layer and dense layer while using transfer learning. In comparison with the classic CNN model, the transfer learning approach achieves superior performance and fewer false positives. Table 6 gives the results of the performance evaluation. It can be observed that the transfer learning strategy outperforms the other two in terms of efficiency, computational requirements, and FPR as well as no overfitting concerns. The transfer model's convergence rate is also quick because the entire model re-training is not required (Figs. 4 and 5).

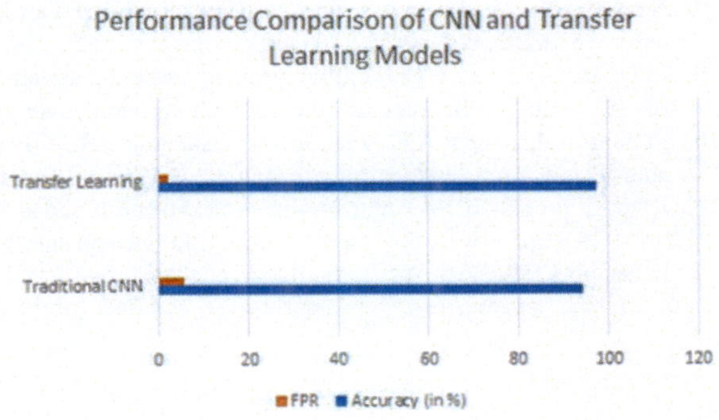

Fig. 5 Performance comparison of CNN and transfer learning models

5 Conclusion

Malware has been a part of smartphones since their inception. Malware applications continue to succeed in eluding security models as the popularity of Android grows. We explored how to detect and categorize Android malware using classic CNN and transfer learning approaches in this article. The application of CNN on malware images has become essential due to the widespread use of CNN in image processing. A two-stage method for converting Android APKs into binary grayscale images was suggested. The standard CNN model is fed these images as input. We used the transfer learning strategy to the trained model, freezing the first layers of the pre-trained model, to avoid the difficulties of overfitting, complexity, and computing expense. The results of the evaluation demonstrate that the transfer learning strategy has a higher accuracy of 97.5%.

References

1. Kambar MEZN, Esmaeilzadeh A, Kim Y, Taghva K (2022) A survey on mobile malware detection methods using machine learning. In: 2022 IEEE 12th annual computing and communication workshop and conference (CCWC). IEEE, pp 0215–0221
2. Pan Y, Ge X, Fang C, Fan Y (2020) A systematic literature review of android malware detection using static analysis. IEEE Access 8:116363–116379
3. Firdaus A, Anuar NB, Karim A, Razak MFA (2018) Discovering optimal features using static analysis and a genetic search based method for android malware detection. Front Inform Technol Electron Eng 19(6):712–736
4. Karbab EB, Debbabi M, Derhab A, Mouheb D (2018) Maldozer: automatic framework for android malware detection using deep learning. Digital Invest 24:S48–S59
5. Alzaylaee MK, Yerima SY, Sezer S (2020) Dl-droid: deep learning based android malware detection using real devices. Comput Sec 89:101663
6. Kim J, Ban Y, Ko E, Cho H, Yi JH (2022) Mapas: a practical deep learning-based android malware detection system. Int J Inform Secur 1–14
7. Feng R, Chen S, Xie X, Meng G, Lin S-W, Liu Y (2020) A performance-sensitive malware detection system using deep learning on mobile devices. IEEE Trans Inf Foren Secur 16:1563–1578

8. Naway A, Li Y (2018) A review on the use of deep learning in android malware detection. arXiv:1812.10360
9. Li D, Wang Z, Xue Y (2018) Fine-grained android malware detection based on deep learning. In: 2018 IEEE conference on communications and network security (CNS). IEEE, pp 1–2
10. Mahindru A, Sangal A (2021) Fsdroid—a feature selection technique to detect malware from android using machine learning techniques. Multimed Tools Appl 80(9):13271–13323
11. Xiao X, Yang S (2019) An image-inspired and cnn-based android malware detection approach. In: 2019 34th IEEE/ACM international conference on automated software engineering (ASE). IEEE, pp 1259–1261

A Statistical Approach for Extractive Hindi Text Summarization Using Machine Translation

Pooja Gupta, Swati Nigam, and Rajiv Singh

Abstract Automatic text summarization (ATS) is a challenging problem for Indian languages. The outcomes of ATS research are the development of summarization systems that consistently generate high-accuracy summaries and have extensive coverage to handle a wide range of languages. Many ATS systems and techniques are available now in a variety of rich-resource languages. However, for Indian languages such as Hindi, less attention has been received due to resource limitations. To address this issue, in this work, we propose an extractive summarization technique for Hindi using different machine translation engines. For this purpose, we have used translated corpus in Hindi for benchmark datasets for BBC News stories, CNN News stories, and the DUC 2004 datasets and used a statistical approach for summary generation using sentence ranking. We have computed the *F*-score and ROUGE-3 metrics to evaluate the proposed system which has been found comparable for different machine translation engines.

Keywords Extractive summarization · Language modeling · Statistical machine translation · ROUGE

1 Introduction

This study is based on the results of ATS systems for various MT engine outputs that are not affected by human involvement. ATS systems are becoming increasingly popular and widely used [1, 2]. Numerous languages are available on the Internet with great variations. But, except a few languages like English, other languages are of low resources in terms of datasets, modeling techniques, and hence, accurate methods for summary generation are less. To overcome and address this problem, we propose a solution for automatic extractive text summarization using different machine translation engines. We have used machine translation engines to translate benchmark BBC News [3], CNN News [4], and DUC 2004 [5] datasets into Hindi

P. Gupta · S. Nigam · R. Singh (✉)
Department of Computer Science, Banasthali Vidyapith, Banasthali, India
e-mail: jkrajivsingh@gmail.com

K. A. Reddy et al. (eds.), *Proceedings of Fourth International Conference on Computer and Communication Technologies*, Lecture Notes in Networks and Systems 606,
https://doi.org/10.1007/978-981-19-8563-8_26

language to overcome the under-resource issue. These datasets are very popular and easily accessible on Internet. We have focused on Google [6], Microsoft Bing [7] and Systran translators [8] for English–Hindi translation and corpus generation for the proposed method.

The proposed framework consists of preprocessing, summary extraction, summary generation, and postprocessing steps. This work uses sentence ranking using maximum likelihood estimation (MLE) and generated summaries using ranking scores. Also, this method can check the similarity score or closeness of the output summary. We have evaluated our system by calculating the ROUGE-3 and F-score.

Rest of the paper is organized as follows: a brief overview of related work is given in Sect. 2. Section 3 describes the proposed work. Results and evaluation are shown in Sect. 4. Finally, the conclusion of this work is discussed in Sect. 5.

2 Related Work

Extractive text summarization is a method to summarize single or multiple text documents to obtain relevant contents from the source files [1, 2]. Summarization of the text documents is based on extraction of important ideas, contents, and relevant texts. This extraction of text may be manual or automatic. As manual text summarization is very difficult to perform due to large amount of data and requires great efforts and time, therefore, ATS has been considered for single and multiple documents [9].

There are a number of ATS systems are present for different rich-resource languages like English [10], Arabic [11], Chinese [12], French [13], and German [14]. However, for resource-limited languages like Hindi [15], a very little attention was received from the researchers for text summarization. ATS can be classified into several approaches such as statistical [1, 16], machine learning [17, 18], deep learning [19], and graph-based techniques [20]. All of these approaches are used to perform summarization for English language.

A statistical extractive text summarization approach has proposed in 2007 for multilingual text documents for newspaper articles [21]. A hybrid text summarization approach has developed for Hindi and Punjabi text documents. They extracted the features for effective sentence generation [22]. Several machine learning methods in biomedical domain were discussed for extractive text summarization [23]. An approach for hybrid text summarization has proposed for utilizing various features such as domain-specific, statistical and semantic similarity for Arabic documents [24].

SMT system stands for statistical machine translation that makes the use of probabilistic approach. An N-gram language model can be used to calculate the frequency of the output by using the probabilistic approach [25]. Language model is an essential part of SMT. N-gram language models are used to calculate the occurrences of the words in the sentences or documents [26].

Further, the evaluation has been performed for summaries using different metrics such as precision, recall for computing the F-score and ROUGE-3 score [27]. ROUGE is a set of metrics, rather than one metrics. ROUGE measures by the different levels of N-grams, where $N = 1, 2, 3$ denotes unigrams, bigrams, and trigrams, respectively [28].

3 Proposed Work

This research work basically divided into three phases—preprocessing, automatic text summarization, and summary generation, explained in Fig. 1.

3.1 Datasets and Translated Hindi Corpus Generation

In this paper, we have used BBC News [3], CNN News [4], and DUC 2004 [5] datasets in English language. We have collected English sentences from among three datasets, and then, we have translated these sentences into Hindi language using three machine translators, given in Table 1. We have extracted the unigrams, bigrams, and trigrams of given datasets for translated Hindi text documents.

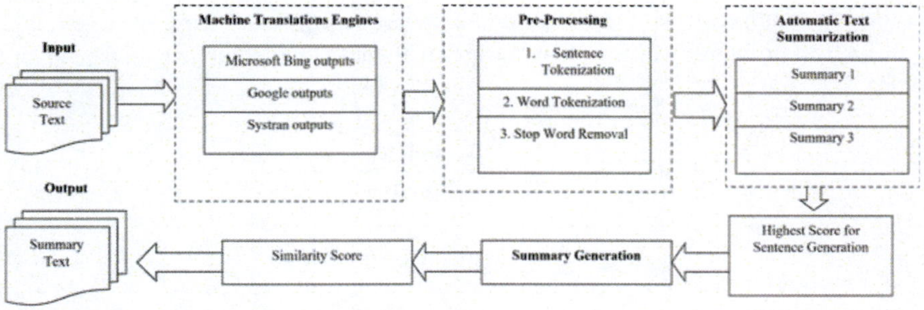

Fig. 1 Proposed summarization methodology

Table 1 MT systems

Engine No.	Description
Engine 1	Microsoft Bing machine translator [6]
Engine 2	Google machine translator [7]
Engine 3	Systran machine translator [8]

3.2 Preprocessing

Initially, we get the source or input documents $D = (D_1, D_2, \ldots, D_n)$ from the translated Hindi datasets obtained from BBC News, CNN News, and DUC 2004 datasets.

Sentence tokenization. In this step, we have identified the beginning and ending of the sentence from the given document. An input document can be tokenized into sentences using language modeling.

Word tokenization. After sentence tokenization, each sentence is tokenized into separate words, i.e., called word tokenization. It is required for the calculating the score of separate words.

Stop word removal. In this step, we have eliminated all the unwanted or unimportant stop words from the input text documents that have no relevant meaning. For this, we have used NLTK stop word list [29] for both the languages.

3.3 Maximum Likelihood Estimation

The score for output text has been calculated using MLE. We have used trigram language model for calculating the probability for each trigram of Hindi text using the Markov chain approach for computation of occurrences score and coherence factor. For example, if we want to compute the probability of a string $W = (w_1, w_2, \ldots, w_n)$, then probability estimated of a trigram on these given sentences can be given by Eq. 1.

$$P(w_{n-2}w_{n-1}w_n) = \frac{\text{Count}(w_{n-2}w_{n-1}w_n)}{\text{Count}(w_{n-2}w_{n-1})} \tag{1}$$

3.4 Automatic Text Summarization

ATS defines the overall working of generating the score of translated text documents. The probabilities of each translated sentence are computed using MLE. Furthermore, we have applied ranking algorithm to find the score of each sentence. These scores will be computed for each sentence of given datasets.

3.5 Summary Generation

In this step, summary generation has been done with help of the generated scores of the sentences. The sentences with highest scores have been considered as the output summary of the text documents.

4 Results and Discussions

We conduct our experiments on three datasets BBC News, CNN News articles, and DUC 2004 documents which have been translated into Hindi for experiments. The details of these datasets are shown in Table 2. The summary length chosen for this work is 3, 10, and 15 for the BBC News dataset, CNN News dataset, and DUC 2004 datasets, respectively.

To obtain the accuracy for our training model, we use ROUGE-3 [33] for evaluation of the proposed method. This includes the comparison of summary generated from our approach and existing summaries. ROUGE-3 measures the overlapping trigrams from predicted and reference summaries. By selecting the top ranked sentences from the documents, we obtain the output summary.

To evaluate the performance of overall approach, we have tested our system on 25 documents from given datasets. These documents are preprocessed by the proposed method. We have found 14 and 11 correct summaries from 25 documents for BBC and CNN datasets, respectively. For DUC dataset, 7 correct summaries from 15 documents have been retrieved. These observations are summarized in Fig. 2. The proposed approach achieves an accuracy of 56% for BBC News dataset, 44% for CNN News dataset, and 46% for DUC 2004 dataset.

Table 3 shows the *F*-score of the 10 documents for BBC News and CNN News datasets. *F*-score is measured by ROUGE-3 metrics extracted from machine-translated Hindi text documents. The obtained results for BBC and CNN News datasets have been shown in Figs. 3 and 4, respectively.

Table 2 Details of the size of datasets and extracted summaries

Datasets	Source sentences	Avg. length in sentences	Avg. length in words	Testing documents	Summary sentences avg. length
BBC News	500	6	2500	25	3
CNN News	500	20	5000	25	10
DUC 2004	500	30	3500	15	15

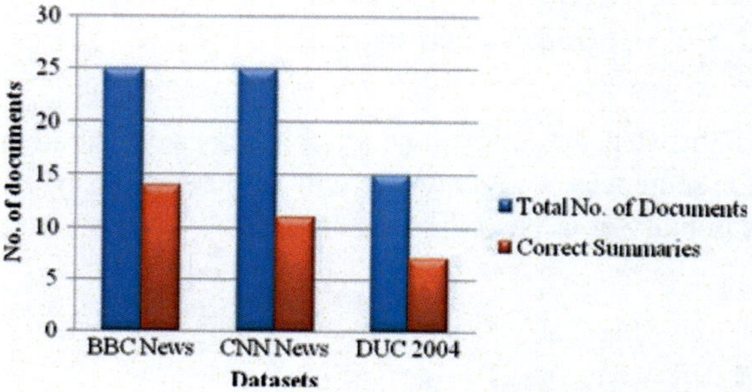

Fig. 2 Evaluation of generated summaries

Table 3 *F*-score for BBC and CNN News datasets for translated summaries in Hindi

Documents	BBC News			CNN News		
	Bing	Google	Systran	Bing	Google	Systran
D1	0.77	0.75	0.79	0.63	0.62	0.63
D2	0.31	0.31	0.24	0.91	0.93	0.91
D3	0.68	0.71	0.42	0.67	0.74	0.72
D4	0.59	0.57	0.65	0.92	0.81	0.93
D5	0.17	0.81	0.39	0.82	0.82	0.88
D6	0.53	0.89	0.82	0.58	0.42	0.54
D7	0.27	0.6	0.01	0.59	0.58	0.57
D8	0.97	0.48	0.89	0.92	0.88	0.89
D9	0.94	0.83	0.83	0.34	0.32	0.38
D10	0.65	0.59	0.46	0.16	0.14	0.15

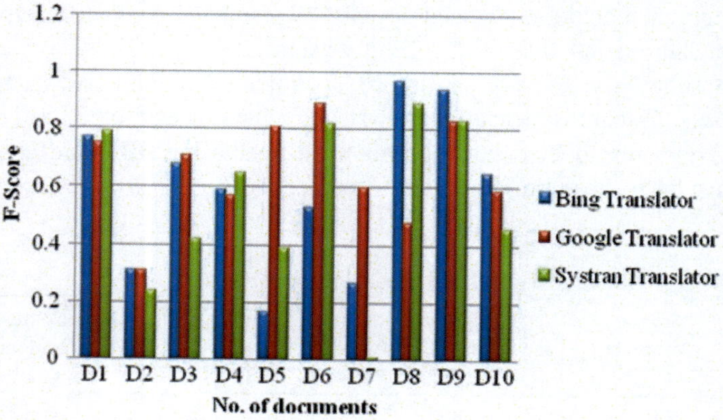

Fig. 3 Comparison of MT engines for Hindi summary generation for BBC News dataset

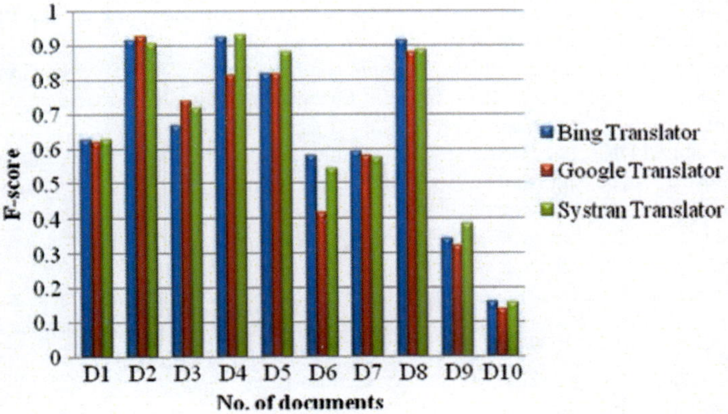

Fig. 4 Comparison of MT engines for Hindi summary generation for CNN News dataset

5 Conclusion

In this study, we have proposed an extractive text summarization approach for multiple translated Hindi text documents using Microsoft Bing, Google, and Systran translators. The proposed model is implemented by using *N*-gram language model approach which computes the ROUGE-3 scores for summary generation. We have performed experiments on BBC News, CNN News, and DUC 2004 datasets. This experimental study gave the results summaries with accuracy of 56%, 44%, and 46% for BBC, CNN, and DUC datasets, respectively. The proposed method can be extended to use multi-language datasets for translated summary generation and implement a uniform language model for accurate summaries.

References

1. Luhn HP (1958) The automatic creation of literature abstracts. IBM J Res Dev 2(2):159–165
2. Maybury M (1999) Advances in automatic text summarization. MIT Press
3. https://www.kaggle.com/pariza/bbc-news-summary
4. https://www.tensorflow.org/datasets/catalog/cnn_dailymail
5. https://www.kaggle.com/datasets/usmanniazi/duc-2004-dataset
6. https://www.microsofttranslator.com
7. https://translate.goolge.com
8. https://www.systran.net/en/translate/
9. Aggarwal CC (2018) Machine learning for text, vol 848. Springer, Cham
10. Radev DR, Allison T, Blair-Goldensohn S, Blitzer J, Celebi A, Dimitrov S, Zhang Z (2004) MEAD-a platform for multidocument multilingual text summarization
11. Abdulateef S, Khan NA, Chen B, Shang X (2020) Multidocument Arabic text summarization based on clustering and Word2Vec to reduce redundancy. Information 11(2):59
12. Oufaida H, Blache P, Nouali O (2015) Using distributed word representations and mRMR discriminant analysis for multilingual text summarization. In: International conference on applications of natural language to information systems, pp 51–63

13. Kaljahi R, Foster J, Roturier J (2014) Semantic role labelling with minimal resources: experiments with French. In: SEM@ COLING, pp 87–92
14. Kabadjov M, Atkinson M, Steinberger J, Steinberger R, Goot EVD (2010) NewsGist: a multilingual statistical news summarizer. In: Joint European conference on machine learning and knowledge discovery in databases, pp 591–594
15. Rani R, Lobiyal DK (2022) Document vector embedding based extractive text summarization system for Hindi and English text. Appl Intell:1–20
16. Edmundson HP (1969) New methods in automatic extracting. J ACM (JACM) 16(2):264–285
17. Srivastava R, Singh P, Rana KPS, Kumar V (2022) A topic modeled unsupervised approach to single document extractive text summarization. Knowl-Based Syst 246:108636
18. Yang K, He H, Al Sabahi K, Zhang Z (2019) EcForest: extractive document summarization through enhanced sentence embedding and cascade forest. Concurr Comput: Pract Exp 31(17):e5206
19. Yousefi-Azar M, Hamey L (2017) Text summarization using unsupervised deep learning. Expert Syst Appl 68:93–105
20. El-Kassas WS, Salama CR, Rafea AA, Mohamed HK (2020) EdgeSumm: graph-based framework for automatic text summarization. Inf Process Manage 57(6):102264
21. Patel A, Siddiqui T, Tiwary US (2007) A language independent approach to multilingual text summarization. Large scale semantic access to content (text, image, video, and sound), pp 123–132
22. Gupta V (2013) Hybrid algorithm for multilingual summarization of Hindi and Punjabi documents. In: Mining intelligence and knowledge exploration, pp 717–727
23. Mishra R, Bian J, Fiszman M, Weir CR, Jonnalagadda S, Mostafa J, Del Fiol G (2014) Text summarization in the biomedical domain: a systematic review of recent research. J Biomed Inform 52:457–467
24. Al-Radaideh QA, Bataineh DQ (2018) A hybrid approach for Arabic text summarization using domain knowledge and genetic algorithms. Cogn Comput 10(4):651–669
25. Koehn P (2010) Statistical machine translation. Cambridge University Press
26. Barzilay R, Lee L (2004) Catching the drift: probabilistic content models, with applications to generation and summarization. arXiv preprint cs/0405039
27. Lin CY (2004) Rouge: a package for automatic evaluation of summaries. In: Text summarization branches out, pp 74–81
28. Jing H, Barzilay R, McKeown K, Elhadad M (1998) Summarization evaluation methods: experiments and analysis. In: AAAI symposium on intelligent summarization, pp 51–59
29. https://www.nltk.org/nltk_data/

Semantic Parser Using a Sequence-to-Sequence RNN Model to Generate Logical Forms

Sanyam Jain and Yash Bhardwaj

Abstract Neural networks have been shown to replicate neural processing and in some cases intrinsically show features of semantic insight. It all starts with a word; a semantic parser converts words into meaning. Accurate parsing requires lexicons and grammar, two kinds of intelligence that machines are just starting to gain. As the neural networks get better and better, there will be more demand for machines to parse words into meaning through a system like this. The goal of this paper is to introduce the reader to a new method of semantic parsing with the use of vanilla or ordinary recurrent neural networks. This paper briefly discusses how mathematical formulation for recurrent neural networks (RNNs) could be utilized for tackling sparse matrices. Understanding how neural networks work is key to handling some of the most common errors that might come up with semantic parsers. This is because decisions are generated based on data from text inputs. At first, we present a copying method to speed up semantic parsing and then support it with data augmentation.

Keywords Semantic parsers · Neural networks · Vanilla RNNs · Machine learning

1 Introduction

A key area where natural language processing (NLP) is transforming the era of artificial intelligence is semantic parsing. The goal of a prevailing extensible semantic parsing system is to have parsers and phrase-structure grammars in less of a classical AI fashion, but rather a more neural network-Esque approach where the parser is provided much more inputs before coming up with an answer. Furthermore, recurrent neural networks are predominantly used for prediction functions in speech recognition, handwriting recognition, and language understanding. The basic architecture

S. Jain
Maharaja Agrasen Institute of Technology, Delhi, India
e-mail: sanyam.04514804919@eee.mait.ac.in

Y. Bhardwaj (✉)
Jodhpur Institute of Engineering and Technology, Mogra, India
e-mail: yash.17jiec062@jietjodhpur.ac.in

K. A. Reddy et al. (eds.), *Proceedings of Fourth International Conference on Computer and Communication Technologies*, Lecture Notes in Networks and Systems 606,
https://doi.org/10.1007/978-981-19-8563-8_27

of a recurrent neural network is a loop; although it does contain loops, it does not constitute an infinite loop. In fact, the input to this network is based on its previous values. The outputs of a recurrent neural neuron constitute the input to the next time step. So, the natural question is, can we use RNNs to build an accurate semantic parser?

Two major challenges stand in the way. First, semantic parsers must be able to generalize to a large set of entities that may not appear during training.

Second, semantic parsers must understand compositionality: they must be able to recognize hard alignments between fragments of utterances and logical forms and know about the predictable ways in which these fragments can be combined. RNNs do not intrinsically have a concept of compositionality and can only learn about these crisp structural regularities by observing data.

In this paper, we present the first semantic parser that uses a sequence-to-sequence RNN model to generate logical forms. Our contributions are twofold. First, we introduce an attention-based copying mechanism that allows our RNN model to generalize to unseen entities. Second, to teach the model about compositionality, we introduce compositional data augmentation, which induces a high-precision grammar from the training data and augments the training data with new examples sampled from this grammar.

2 Literature Review

Recent literature acknowledges that current models for natural language processing (NLP), though delivering progress, still have lots of room for improvement. And we specifically find in recent work on deep recurrent neural networks (RNNs) with predictive models like conditional random fields can help us to overcome these kinds of limitations due to their robustness and performance [1].

Semantic parsers or RNNs are computer systems capable of understanding the meaning of a human statement, returning a human-readable explanation of that meaning—and then outputting a response [2].

They perform well in "sentiment analysis," which is defined as estimating if these sentences are to be interpreted as happy or sad [3]. The servers read in the sentences and predict the embedding vectors, which are found digitally in relation with probability estimation process [4]. Using clustered vector goal projection for modeling expressions to help spot words and phrases that could have multiple meanings due to inserted space by prepositions, connectors, or transposed letters [5].

In natural language processing, semantic parser refers to the mechanism for understanding bi-tagged sentences which can offer a broader context and its representation in the reading framework [6]. "Review" is guided by a sense inventory, which is done separately as a strongly annotated corpus [7]. This review's potential goals mainly include analysis of recent research and state-of-the-art results on how to use neural network architectures such as bi-directional recurrent neural network (BDRNN), convolutional neural network (CNN), and recursive artificial neural network (RANN) [8].

Semantic parsers have reached a threshold of competent human translation; now, anybody can get a truly human interface without even speaking a word [9]. The self-feeding of language corpora (or the iteration of data in machine learning models) is the bread and butter of generative algorithms that are outside the strict purview of neural networks. However, recent advances in recurrent neural network architectures point to some new and fascinating applications [10].

Classically, recurrent neural networks (RNNs) have been mostly confined to modeling numbers, making them ill-suited for research and industry alike when the processing of natural languages based on more complex soft phenomena is desired. Our study is focused on blending two different mechanisms through RNNs. First, we alter the model so that it can more easily handle a particular type of crisp regularity: words that can be copied from input to output. Second, we generate synthesized training examples to teach the model about the rules that govern how smaller fragments of language can be composed to form larger units. Understanding the trade-offs between these two paradigms—designing new models and generating new data—is an important open challenge.

3 Task

In this paper, we take semantic parsing and consider it a generic sequence-to-sequence task. We define a simple encoding for the input sequence of natural language words and show that it is possible to train an end-to-end system to map from sequences of words to sequences of vector representations without any explicit mention of semantics or meaning.

3.1 Datasets

One of the things we looked for when testing machine intelligence was how it would score on 3 standard datasets:

- **GeoQuery** (GeoQuery) comes with 600 questions about US geography and is paired with database queries. It has a split of training testing examples, meaning the app includes 600 examples, including all the questions you will see during training.
- **Regular Expressions** (Regular Expressions) contain natural language descriptions of regular expressions paired with associated regular expressions. We evaluate on a test set of 164 examples selected randomly from the dataset.
- **ATIS** (ATIS) Here, the query is translated into SQL, and the database is queried with the corresponding result set.

The scope of this research is limited to extracting knowledge from logical forms. We, therefore, do not use any semantic parsing datasets that only include denotations, such as WebQuestions.

4 RNN Model

We are using a standard recurrent neural network model that is backed by our generic sequence-to-sequence framework. It combines existing neural machine translation models with our novel copying mechanism.

At a high level, our system consists of two main modules:

1. **Encoder Module**. It transforms a string of words $x_1, \ldots, x_m$ into context-sensitive representations b_1, ..., b_m, where each b_i is a real-valued fixed-dimensional vector.
2. **Decoder Module**. This module takes in the input sequence and context-sensitive embeddings and generates a probability distribution over sequences $y = y_1$, ..., y_n, where each $y_j \in$. It writes the output tokens one at a time, maintaining a hidden state s_j at each time j.

This can be further decomposed into four modules:

1. **Initialization Module**: Takes in the context-sensitive embeddings b_1, ..., b_m, and outputs the initial decoder hidden state s_0.
2. **Attention Module**: Takes in b_1, ..., b_m and the current state s_j, and outputs an attention score vector e_j of length m.
3. **Output Module**: Takes in b_1, ..., b_m, s_j, e_j, and x, and outputs a probability distribution for y_{j+1}, the next word to write.
4. **Update Module**: Takes in b_1, ..., b_m, s_j, e_j, and y_{j+1}, and outputs the new state s_{j+1}.

Figure 1 illustrates how these modules are connected to form the overall RNN model. In the next sections, we describe these modules in greater detail.

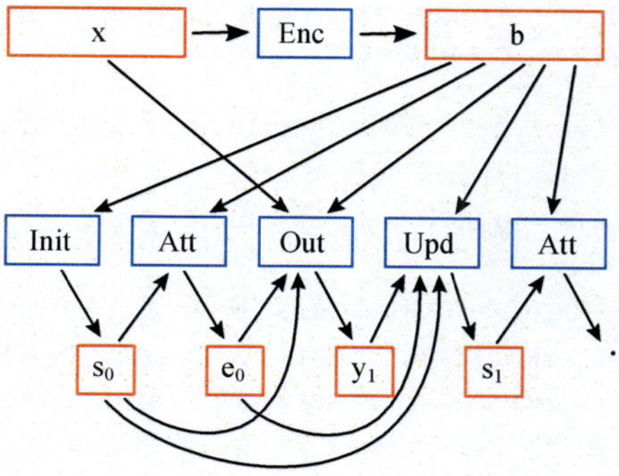

Fig. 1 Overview of our RNN model. Enc, Encoder; Init, Decoder initialization; Att, Decoder attention; Out, Decoder output; Upd, Decoder update

4.1 Encoder Module

The first step is to replicate each word into an individual vector.

There are a couple of RNNs involved. The input comes in as a vector and is fed to both the forward and backward RNN, with the output from each being used as the input to the other.

The backward RNN similarly generates hidden states h_m^B, …, h_1^B by processing the input sequence in reverse order.

4.2 Decoder Initialization Module

Let h be the concatenation of h_m^F and h_1^B. The decoder's initial state s_0 is

$$s_0 = \tanh\left(W^{(i)} h\right), \tag{1}$$

where $W^{(i)}$ is a parameter matrix.

4.3 Decoder Attention Module

At each time step j, and for each word x_i in the input, we compute an attention score e_{ji}. We use the general content-based scoring function:

$$e_{ji} = s_j^T W^{(a)} b_i, \tag{2}$$

where $W^{(a)}$ is a parameter matrix.

4.4 Decoder Update Module

AT each time step j, the scores e_j from the attention module are converted to a probability distribution over $\{1, \ldots, m\}$ with a softmax:

$$\alpha_{ji} = \frac{\exp\left(e_{ji}\right)}{\sum_{i=1}^{m} \exp\left(e_{ji}\right)}, \tag{3}$$

α_{ji} is known as the attention weight, and can be interpreted as the amount of attention paid to the i-th input word at time step j. Then, a context vector c_j is computed as a weighted average of the b_i's:

$$c_j = \sum_{i=1}^{m} \alpha_{ji} b_i \tag{4}$$

The current input vector v_{j+1} is computed as the concatenation of (y_{j+1}) and c_j, where is another word embedding function. Finally, the state is updated according to the recurrence

$$s_{j+1} = \text{LSTM}\big(v_{j+1}, s_j\big). \tag{5}$$

4.5 Decoder Output Module

Finally, we describe two decoder output modules: a baseline module, and a more sophisticated module that performs attention-based copying.

Baseline The baseline output module uses a simple softmax overall output vocabulary words. At each time step j, it first computes the context vector c_j, as in the update module.

Attention-based Copying One of the main contributions of this paper is a novel attention-based copying mechanism, which improves upon the baseline output module. This new mechanism is motivated by the need to generalize well to a large set of entity names, including ones that were not seen at training time. Oftentimes, these entity names can be copied directly from the input to as well as quoted strings in regular expressions.

Our proposed attention-based copying mechanism enables the network to copy a word directly from input to output, with probability determined by the amount of attention paid to that input word. Formally, we have

$$P(y_j + 1 = w|x, y_{1:j}) \propto \tag{6}$$

$$\exp\big(M_w s_j + U_w c_j\big) + \sum_{i=1}^{m} \mathbb{I}[x_i = w] \exp\big(e_{ji}\big) \tag{7}$$

4.6 All Models

We define a total of three models (one main model and two baselines):

- **Attention-Based Copying** This is our full model with attention-based copying.
- **Attention** This is the same as attention-based copying, except with the baseline output module.
- **Encoder-Decoder** This baseline is an encoder-decoder model that uses the baseline decoder output module. This can be thought of as a variant of the attention model where the decoder initialization module just returns $s_0 = h_m^F$, and the context vector c_j is artificially set to always be 0.

4.7 Learning

We train the model using stochastic gradient descent. Gradients are computed automatically using Theano.

5 Compositional Data Augmentation

The strength of deep learning models lies in their flexibility. However, this flexibility also presents a challenge: because neural models make fewer assumptions about the task, they can be at a disadvantage compared to specialized systems that have domain knowledge baked in.

Our solution to this problem is to augment our training datasets with new examples generated from the original training examples. This approach allows us to inject prior knowledge into our system, as the new examples can be generated in a way that leverages domain knowledge.

For semantic parsing, one important phenomenon to model is compositionality. There are often hard alignments between fragments of the input and output, and these units can be composed with each other in predictable ways. We, therefore, propose a *compositional data augmentation* scheme that uses an induced grammar to generate new, highly structured examples. We focus primarily on applying this to the GeoQuery domain. More details are shown in Fig. 2.

This procedure begins by identifying *high-precision alignments* between pieces of an utterance and associated logical form. First, for each (x, y) pair, there is a trivial alignment that matches the entire utterance with the entire logical form (e.g., what states border Illinois? aligns to an entire logical form). We write some manual rules to convert questions into noun phrases by stripping things like question marks and "wh" words (e.g., to create states border Illinois). Finally, we match the entity mentioned in the input and output based on simple string matching (e.g., Illinois).

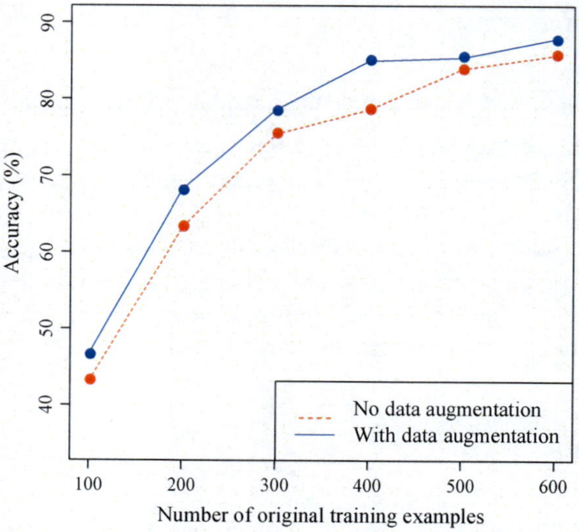

Fig. 2 Accuracy on GeoQuery as a function of number of training examples. Data augmentation gives a consistent accuracy boost, regardless of original dataset size

These high-precision alignments allow us to induce simple grammar over pairs of utterances and logical forms. This grammar can replace individual entity mentions with other entities of the same type, or with entire phrases that evaluate a set of entities of the same type. We then generate new examples from this grammar and add these to our original training dataset.

Our augmentation scheme is notable in that it transforms both the inputs and outputs simultaneously. In contrast, other data augmentation techniques used in computer vision only transform inputs without changing any output labels.

Our procedure generates examples that are on average longer than the examples in the training and test sets. Later, we explore in greater detail the ramifications of augmenting the training dataset with examples drawn from a distribution that is very different from the test distribution.

5.1 Augmentation on Regular Expressions

Regular expressions and ATIS have less nesting structure, making them less suited for the compositional data augmentation scheme described above. However, we can still use high-precision alignment rules to perform a simpler form of data augmentation. We do this on regular expressions by looking for quoted strings and integers. We generate new examples by swapping quoted strings and integers in one example for other quoted strings or integers.

Note that unlike our synthesized examples for GeoQuery, these synthesized examples are more like additional (non-independent) samples from the probability distribution that generated the training data.

6 Experiments

We evaluate our system based on the following grounds. Denotation match assesses the relevance of content in relation to a specific keyword. Match accuracy has to find the similarity in their surrounding context. A regular expression or regular expressions are an elegantly easy way of matching a single text string with an ordered sequence. Unlike denotation match, this solution can evaluate the user's response based on character-level similarities.

6.1 Specifics of Execution

We tokenize (punctuation) logical forms in a domain-specific manner, according to the syntax of the formal language being used. We ensure that entity names can be easily copied from input to output. At the same time, we perform name mangling on predicate names, so that the model cannot cheat by copying these as well.

We run all experiments with a hidden size of 400 units. Word vector sizes were chosen individually for each domain: we used 50 for regular expressions, 100 for GeoQuery, and 200 for ATIS. We randomly initialized all the parameters, but we tried to make sure they were between -0.1 and 0.1. We set the learning rate at 0.1, 0.05, and 0.025 for three separate sessions, about 100 iterations per session. We replace words that only occur once in the training set with a universal word vector.

Another important hyperparameter is the number of new examples to generate when performing data augmentation. For GeoQuery, we train on the original dataset, plus 300 randomly sampled new examples. For regular expressions, we train on the original dataset, plus 200 examples generated by swapping integers and 200 examples generated by swapping quoted strings.

At test time, we use beam search with beam size $K = 10$. In tests, our team consists of 10 lexicons. We use beam search with a size of 10. We automatically balance the right parentheses. We will then pick the most accurate logical form. Whenever the regex parsing fails, we use the highest scoring parsing, which is successful. Alternatively, when converting a regex to a DFA, the conversion will not cause any error. Beams have predictions on them. When we go to grab the next prediction, we always just pick it up from the front of the beam. We compare the results that our system obtained to what is usually achieved. The old system was more accurate, but it needed a "seed lexicon" (a list of predicates) to work. This is not necessarily a problem, as we do not deliberately include prior knowledge in the system.

First, we evaluate our system trained on the original dataset alone, with no data augmentation.

Note that we are roughly competitive with the state-of-the-art on regular expressions although the numbers are not directly comparable as the other work evaluates different data. However, we lag behind on GeoQuery and ATIS.

We see that our compositional data augmentation improves our accuracy on GeoQuery by more than two percentage points. In contrast, we do not see accuracy gains on regular expressions, where we performed a less compositional form of data augmentation.

6.2 Effect of Data Augmentation

To further measure the effects of compositional data augmentation, we trained our system both with and without data augmentation on various random subsets of the GeoQuery dataset. In Fig. 2, we plot test accuracy as a function of the number of real training examples. As a heuristic, when doing data augmentation on n real examples, we always generate $\frac{n}{2}$ new examples.

From this plot, we see that data augmentation consistently boosts accuracy across all dataset sizes. In some cases, the synthesized examples prove to be almost as helpful as real examples: for example, training on 400 real and 200 synthesized examples gets nearly the same accuracy as training on the full set of 600 examples.

6.3 Out-of-Domain Augmentation

Compositional data augmentation on GeoQuery helped accuracy even though the generated examples were not guaranteed to be in the support of the test distribution; meanwhile, data augmentation on regular expressions proved ineffective even though the generated examples were similar to the test examples. One possible explanation is that by generating longer examples on GeoQuery, we biased the model to focus on examples that are similar to the harder examples in the test set. However, an interesting alternative explanation is that data augmentation can in fact be most beneficial when the examples generated do not match the test distribution.

More specifically, we wish to test two hypotheses. First, we hypothesize that i.i.d. examples drawn from the test distribution may not help the model as much as i.i.d. examples that are longer than those in the test distribution. Second, we hypothesize that a similar claim holds when performing data augmentation, which generates examples that are correlated with the initial training set.

Intuitively, longer examples could be helpful simply because they carry more information content per example, or because their length forces the network to learn better alignments between input and output tokens.

It is impossible to construct any artificial reality without specifying these two bases, which are the entities and relationships between them.

Meta-data is another important component of data collection. For example, it provides insights on deeply learning models that one has used like the amount of nodes in such model. However, its importance varies over different levels of the research process in machine learning.

We ran the experiments where we added different types of binary relations between our entities and found out that there are many paths to adding relations. It turns out that it is more complicated than an exhaustive enumeration search, e.g., taking every possible relation between any two entities and adding it.

AI could be used to take the results of such experiments and extract patterns from them later on in order to provide insights on how these would work in machine learning systems and what constructs can be used for data construction.

We use four different types of examples in our training set:

- **Same-domain**: Arbitrarily picked depth-2 examples.
- **Out-of-domain, independent**: Randomly chosen depth-4 examples.
- **Same-domain, correlated**: Take the given training examples and swap out entity mentions for different entity mentions. This is similar to our augmentation strategy for regular expressions. Note that each new example is in fact a sample from the test distribution though these samples are correlated with the training examples.
- **Out-of-domain, correlated**: Take the given training examples and swap out one entity mentioned in one example for another complete example. On top of this, swap out the entity in the second example for a new entity. The result is a new depth-4 example. This is similar to our augmentation strategy for GeoQuery.

Figure 3 results confirm both of our hypotheses. Independent, long, out-of-domain examples are in fact more efficient at getting the model to achieve perfect accuracy than independent same-domain examples. Additionally, while both data augmentation strategies proved helpful, the out-of-domain strategy was the more successful of the two.

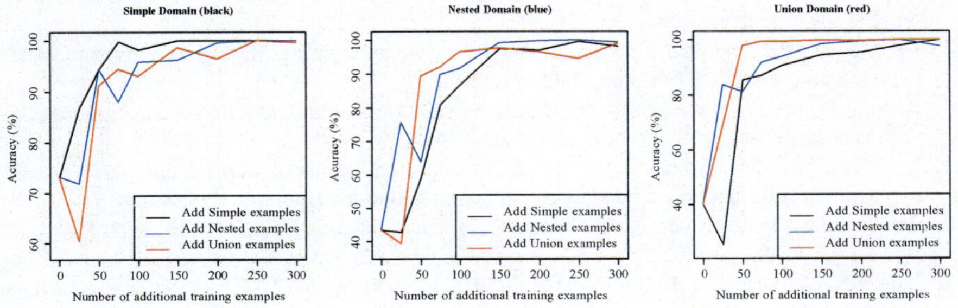

Fig. 3 Comparing augmentation methods on artificial data

7 Conclusion

Our research introduces the first sequence-to-sequence RNN model for semantic parsing. Our model is easy to train and gives good accuracy on several semantic parsing datasets when trained with logical form supervision. Furthermore, we propose a compositional data augmentation scheme to inject prior knowledge about compositionality into our model.

One limitation of our current approach is that it uses annotated logical forms as supervision.

An alternative direction would be to incorporate the execution step itself into the network. Our model includes a novel attention-based copying mechanism to deal with unseen words such as entity names. Our attention-based copying can be used for both rare and common words, so our model can learn when it is best to perform copying.

We used a small set of high-precision manual rules to perform data augmentation. It is possible that an automatic grammar induction approach could expand the recall of our grammar while keeping precision high.

Our experiments on artificial data show that compositional data augmentation can help the model learn even when the new examples look different than the examples seen at test time.

Tree-structured recursive neural networks leverage the structure of a syntactic parse tree to compositionally build representations of sentences. Their focus on soft representations contrasts with our goal of modeling hard relationships between fragments of sentences and logical forms.

References

1. Li Z, Wu Y, Peng B, Chen X, Sun Z, Liu Y, Paul D (2022) Setransformer: a transformer-based code semantic parser for code comment generation. IEEE Trans Reliab
2. Sales N, Efson J (2022) An explainable semantic parser for end-user development. Ph.D. dissertation, Universität Passau
3. Demlew G (2022) Amharic semantic parser using deep learning. In: Proceeding of the 2nd deep learning Indaba-X Ethiopia conference 2021
4. Rongali S, Arkoudas K, Rubino M, Hamza W (2022) Training naturalized semantic parsers with very little data. arXiv preprint arXiv:2204.14243
5. Li B, Fan Y, Sataer Y, Gao Z, Gui Y (2022) Improving semantic dependency parsing with higher-order information encoded by graph neural networks. Appl Sci 12(8):4089
6. Arakelyan S, Hakhverdyan A, Allamanis M, Hauser C, Garcia L, Ren X (2022) Ns3: neuro-symbolic semantic code search. arXiv preprint arXiv:2205.10674
7. Lam MS, Campagna G, Moradshahi M, Semnani SJ, Xu S (2022) Thingtalk: an extensible, executable representation language for task-oriented dialogues. arXiv preprint arXiv:2203.12751

8. Lukovnikov D (2022) Deep learning methods for semantic parsing and question answering over knowledge graphs. Ph.D. dissertation, Universitäts und Landesbibliothek Bonn, 2022
9. Marton G, Bilotti MW, Tellex S. Why names and numbers need semantics
10. Yang L, Liu Z, Zhou T, Song Q (2022) Part decomposition and refinement network for human parsing. IEEE/CAA J Automatica Sinica 9(6):1111–1114

NFF: A Novel Nested Feature Fusion Method for Efficient and Early Detection of Colorectal Carcinoma

Amitesh Kumar Dwivedi, Gaurav Srivastava, and Nitesh Pradhan

Abstract Colorectal cancer is one of the most common cancer types and causes of death due to cancer in the world. Wireless curated endoscopy is used to diagnose and classify colorectal carcinoma. However, the major drawback of wireless curated endoscopy is that it presents many images to be analyzed by the medical practitioner. Therefore, many studies have been performed to automate the detection and classification of colorectal carcinoma using machine learning and deep learning models. Studies vary from traditional image classification techniques to image processing algorithms combined with data augmentation combined with pre-trained neural networks for early detection and type classification of colorectal carcinoma. In this manuscript, we proposed a novel nested feature fusion method to fuse the deep features extracted by the pre-trained EfficientNet family to devise an approach for early detection and classification of colorectal carcinoma. We have used the WCE curated colon disease dataset, which consists of 4 classes: normal, ulcerative colitis, polyps, and esophagitis. Our proposed method and experimental results outperformed compared to the state of the art with the fused model showing an accuracy of 94.11%. Medical centers can use the proposed method to detect colorectal cancer efficiently in real life.

Keywords Colorectal carcinoma · Deep learning · Feature extraction · EfficientNet · Nested feature fusion

1 Introduction

Colorectal carcinoma (CRC) is ubiquitous and is the underlying cause of death due to cancer worldwide [1, 2]. Unfortunately, colorectal carcinoma is mainly discovered in very late stages in patients for its effective treatment [3]. Mainly, colonoscopy is used to detect the various types of CRCs. However, such methods also impose risks to the patient, such as bleeding, negative consequences of sedation, colonic perforation, and

A. K. Dwivedi · G. Srivastava · N. Pradhan (✉)
Department of Computer Science and Engineering, Manipal University Jaipur, Rajasthan, India
e-mail: nitesh.pradhan@jaipur.manipal.edu

K. A. Reddy et al. (eds.), *Proceedings of Fourth International Conference on Computer and Communication Technologies*, Lecture Notes in Networks and Systems 606,
https://doi.org/10.1007/978-981-19-8563-8_28

other clinical risks [4, 5]. Furthermore, due to wide-ranging variation in data from one patient to another, traditional learning methods of diagnosis are not extremely reliable [6].

Biomedical image processing is the mainstay of scientific research and an essential part of medical care, which is being highly sought after in the field of deep learning [7]. Although clinical detection of diseases based on traditional medical imaging methods has provided factual accuracy, developments in machine learning have pushed deep learning research developments in biomedical medical imaging [6].

To augment the process of colorectal carcinoma detection, a tremendous amount of research is focused on detecting CRCs through medical image processing and computer-aided diagnosis.

Machine learning methods have provided accurate classification and prediction abilities and have been deployed to be used for the diagnosis and prognosis of various medical ailments and health conditions due to their data-backed method of analysis, which unifies diverse risk factors into a classification/prediction algorithm [8–10]. However, deep learning methods are more effective than conventional machine learning methods due to their ability to process a high number of available samples during the training stage [11], their ability to execute feature engineering on its own, and their need for less human intervention while training which is highly suitable for datasets with a large number of samples. Furthermore, deep neural network models and frameworks can be retrained using a custom dataset compared to traditional computer vision algorithms, which are highly domain-specific. This provides much flexibility in deep learning compared to traditional machine learning algorithms [12].

With deep learning, an image dataset with object classes annotated to each image is presented to the machine to facilitate end-to-end learning [13], which is, in comparison with traditional computer vision techniques where parameters have to be fine-tuned by the CV engineer, is much easier.

The remaining contents of the proposed experimentation can be summarized as follows: Sect. 2 briefs about the previous academic works of various scholars in detecting colorectal carcinoma. Section 3 explores EfficientNet models, other deep learning strategies, and the materials and methods used. Section 4 describes the deep feature extraction and model training. Finally, Sect. 5 presents the experimentation and results from the mentioned experimentations.

2 Related Works

A variety of research has been performed on the automated detection and classification of colorectal cancer using machine learning and computer vision algorithms. Recently, deep learning has become the state-of-the-art approach for performing the classification of colorectal cancer due to its current popularity in biomedical image classification experimentations.

The study presented by Jesmar et al. proposed a model that integrates EfficientNet, MobileNetV2, and ResNetV2 into a single feature extraction pipeline called multi-fused residual convolutional neural network (MFuRe-CNN) with Auxiliary Fusing Layers (AuxFL) and a Fusion Residual Block (FuRB). The fusion model along with the Alpha Dropouts diagnosed a diverse set of endoscopic images of gastrointestinal ailments and handled conditions such as ulcerative colitis, esophagitis, polyps, and a healthy colon. The datasets used in the experimentation consisted of KBASIR and ETIS-Laris PolyDB. The fusion model showed an accuracy of 97.25% with only 4.8 million parameters. Furthermore, the FuRB and Alpha Dropouts substantially contributed to reducing overfitting and performance saturation [14]. However, due to lack of testing on other datasets, the proposed model does not immediately guarantee similar results for other medical images. Additionally, FuRB and Alpha Dropouts tend to slow down interference. In another study conducted by Khan et al. [15], an automated system is used to distinguish gastrointestinal infections based on WCE imaging. Automated functions within the research experiment included preprocessing ulcer frames with a dark channel, decorrelation, optimization of saliency-based segmentation to improve ulcer visibility, feature extraction using deep learning, selection of best frames, and classification of the selected features. A multi-class cubic SVM was used to classify the selected features, which attained an accuracy of 98.40%. However, in this study, if the segmentation of ulcers is incorrect, then the deep learning model can be mistrained.

Furthermore, in a study by Poudel et al., a neural network for endoscopic image classification is provided using an adequate dilation in convolutional neural networks (CNNs). To deal with overfitting and extraneous noise and miscellaneous features, DropBlock, a regularization technique has been used. The proposed study compares its proposed model's efficiency with that of other state-of-the-art models such as VGG16, InceptionResnetV2, Xception, ResNet, DenseNet, and NasNet. Using the proposed model, 95.7% accuracy is achieved, and an F1 score of 0.93 is achieved with the colorectal dataset, and an F1 score of 0.88 is obtained with the KVASIR dataset. The achieved accuracy gives better results than the traditional methods. However, the model has not been tested on other medical image datasets for classification purposes [16]. In the study presented by Silva et al. [17] likely polyps within image samples were withdrawn using geometric shape features. Further, the regions containing polyps were boosted using textural features. Evaluation of this method was conducted on datasets that contained 300 images of polyps and 1,200 images without polyps. According to the proposed method, 91.2% sensitivity, 95.2% specificity, and a deceit detection rate of 4.82% were achieved, which are comparable to the analysis systems developed for online video colonoscopy images. In the study presented by Fan et al., AlexNet convolutional neural network was used and trained to a database containing more than ten thousand images of wireless capsule endoscopy images to detect ulcers and erosion. Based on the proposed model, the accuracy was 95.16% and 95.34%, the sensitivity was 96.80% and 93.67%, and the specificity was 94.79% and 95.98%, correspondingly. Despite the fact that the method used in this experiment had great results in detecting ulcers and erosions, and it was unable to identify some ulcers, erosions, and other WCE images. After the experimentation, approximately 5% of images were not incorrectly [18].

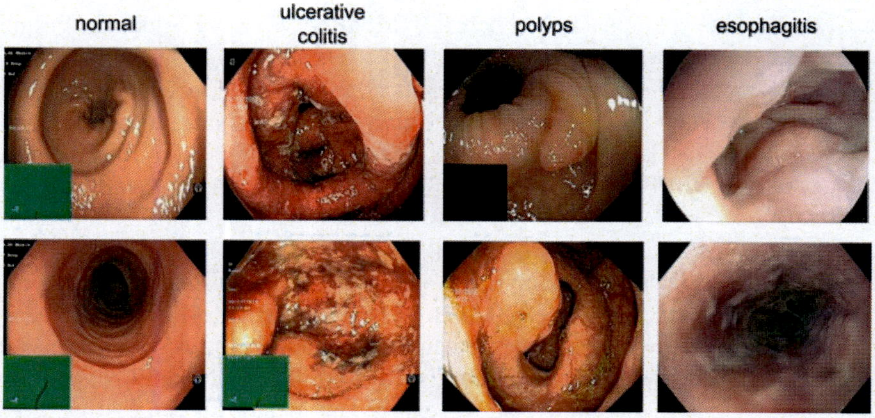

Fig. 1 Dataset samples of normal, ulcerative colitis, polyps, and esophagitis

We observed that most experiments focused on metrics such as accuracy, sensitivity, and F1 score and observed negligence in the area of efficiency. Thus, we decided to propose a novel and efficient neural network to tackle the early detection of colorectal carcinoma (Fig. 1).

3 Materials and Methods

3.1 Data Collection

WCE curated colon disease dataset deep learning is an image dataset for gastrointestinal tract or simply, a colon disease image dataset [19, 20]. These are images of the gastrointestinal tract captured during the procedure of wireless capsule endoscopy, which in the scope of current experimentation, will be used to devise a deep learning model for the early detection of colorectal carcinoma. The dataset contains 6000 colored images and the dataset contains four classes: Normal, ulcerative colitis, polyps, and esophagitis as given in Table 1.

Table 1 Dataset description

Classes	Normal	Ulcerative colitis	Polyps	Esophagitis
No. of samples	1500	1500	1500	1500

3.2 Data Preprocessing

Data preprocessing is an essential step for deep learning model training. It outlines the processes required to alter or encode data so the model can parse it effectively. In neural networks, the model expects the input image to be the same size. However, the images gathered are not the same size or form. The images in our dataset originally ranged in size from 400×300 to 936×768 pixels. We converted all the images into a common size of 128×128 pixels as a preprocessing step before training because the dataset's images were not homogeneous and came in varied sizes. After applying RGB reordering to all images, the model's final input was delivered as a $128 \times 128 \times 3$ matrix.

While downscaling the images, we can sometimes lose some vital information, so this has to be done carefully by observing the dataset. For example, suppose we have a dataset of MRI scans for brain tumor classification. In that case, if we downscale the images to a minimal size, the tumor will almost disappear from MRI scans, which can impact training accuracy. Also, resizing the image to a very large size like 512×512 can exceed the GPU memory. Therefore, to make it both memory efficient and not lose any critical information from the image, we have to choose the best image size based on the experiments.

We scaled and ran our trials on all 128×128, 196×196, and 256×256 image sizes in this study, and we found that the accuracy is similar for all three image sizes. However, training time is considerably shorter on 128×128, saving significant computational efforts.

3.3 Dataset Division

A deep learning model may obtain a 99% accuracy rate, but it fails when evaluated on real-world images. In order to prevent model selection bias and overfitting, it is ethical to divide the dataset into training, validation, and testing sets. Furthermore, our parameter estimations are more variable when we have a scant amount of data. Similarly, our performance measure will be more variable if we have fewer testing data. As a result, we should split the data so that no variances are excessive.

Adding more data to the final testing set ensures the method's resilience and minimizes the chance of failure in real-world tests. As a result, as given in Table 2, we partitioned the entire dataset into three sections: 70% training, 10% validation, and 20% testing.

Table 2 Dataset division

Classes	Normal	Ulcerative colitis	Polyps	Esophagitis
Training set	1050	1050	1050	1050
Validation set	150	150	150	150
Testing set	300	300	300	300

3.4 Transfer Learning

Transfer learning was initially discussed in the NeurIPS (Conference on Neural Information Processing System), which talked about using previously learned knowledge to augment further future learning. Deep transfer learning (DTL) combines deep learning architecture with transfer learning. Deep neural networks (DNNs) provide a powerful way to learn features, making them useful in feature-based transfer learning. Methods based on latent feature spaces utilize DNNs to discover a common latent feature space where both source and target data can exhibit the same probability properties. Consequently, the source data can be used as a training set for target data in the latent feature space, which improves the model's performance with target data [21].

3.5 EfficientNet

EfficientNet is a simple convolutional neural network that is known for its profound effective compound scaling method that helps researchers to scale up a convolutional neural network to any target resource constraints in a highly fundamental way, quickly. Unlike other architectures, EfficientNet uniformly scales network resolution, depth, and width. EfficientNets are also highly used in transfer learning which is why they are being used in the scope of this experiment [22].

3.6 Proposed Nested Feature Fusion Method

In order to construct a CNN, you need to extract features and classify them. The model's first layers may be considered as descriptors of image features, whereas the latter layers are associated with specific categories. In feature extraction, many convolution layers are utilized, followed by max-pooling and an activation function. A fully connected layer and a softmax activation function are standard components of a classifier. Since the number of classes in a dataset is directly proportional to the number of features in a model to learn, to learn complex features, the feature extraction component of the convoluted neural network should be more complex and deeper.

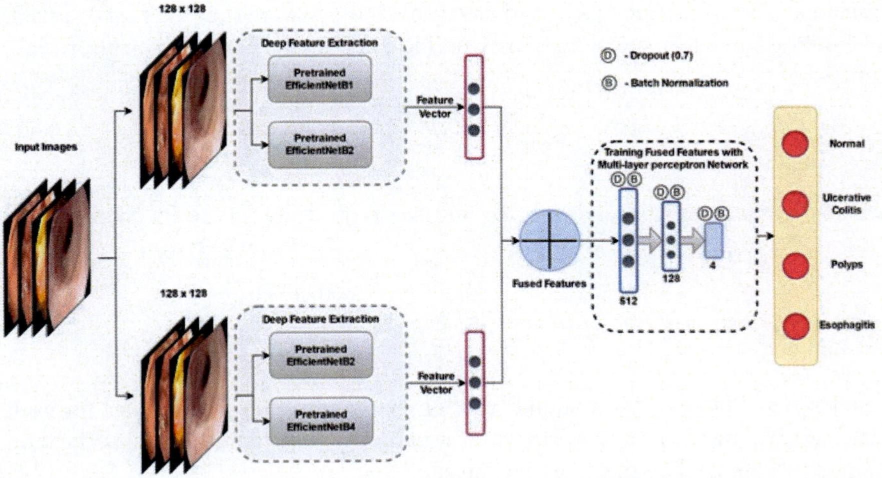

Fig. 2 Graphical abstract of the proposed nested feature fusion method

A feature in an image is a component or pattern of an object that helps with identification. In computer vision and image processing, it is a piece of information regarding the content of an image, usually pertaining to whether a particular section of the image contains specified properties. Different structures in an image, such as points, edges, or objects, are examples of features. Each CNN produces a feature vector with a distinct set of features extracted. They can overlap, but they are not always the same, which is why the accuracy varies from time to time. The key idea behind this proposed nested fusion model is that each CNN will produce a feature vector. By integrating those feature vectors, we will not miss any features the model ignores, resulting in a significant set of features being omitted.

So, in the proposed method, we first fused EfficientNetB1 and EfficientNetB2 and EfficientNetB2 and EfficientNetB4 individually. Both provide two output feature vectors, which we fuse further to create our final model. After the feature extraction, we use a multi-layer perceptron network with a softmax activation function to classify the input image into their respective categories. The proposed methodology is depicted in Fig. 2.

4 Deep Feature Extraction and Model Training

4.1 Loss Function: Categorical Cross-Entropy

The loss function is used to measure the deviation of the estimated value from the true value. It is a computational procedure to assess how the algorithm used models the data. In this experiment, cross-entropy loss function is used because of its ability to increase in magnitude when predicted probability skews from the actual results. The following mathematical Eq. 1 explains the computation of the cross-entropy loss function:

$$L_{\text{CE}} = -\sum_{i=1}^{n} t_i \log (p_i)\,, \text{ for n classes,} \tag{1}$$

where t_i is the truth label and p_i is the Softmax probability for the ith class.

4.2 *Classifier: Softmax*

The softmax classifier is an output function that outputs the probabilities for each class label in the form of a vector. It is usually used for multi-class classification purposes. Softmax function is defined in Eq. 2.

$$\sigma(\mathbf{z})_i = \frac{e^{z_i}}{\sum_{j=1}^{K} e^{z_j}} \tag{2}$$

where σ = softmax, $\mathbf{z}$ = input vector, e^{z_i} = standard exponential function for input, K = number of classes in the multi-class, and e^{z_j} = standard exponential function for output.

4.3 *Learning Rate Decay*

Learning rate decay is an actual practical technique that is used to instruct modern neural networks. It initializes with an enormous learning rate and then declines multiple times: Decomposition of learning rate—decay. It is used to enhance optimization and generalization in the experimentation process. Learning rate decay can be time-based, step-based, and exponential.

5 Experimental Results and Discussion

5.1 *Experimental Setup*

All the models mentioned in the proposed research were implemented with TensorFlow in Python. Further, Kaggle was used to train the models mentioned, with the following specs - GPU Tesla P100-PCIE-16GB compute capability: 6.0 and 16 GB GPU RAM.

Table 3 Experimental results of efficientnet family

Model	Training accuracy	Validation accuracy	Testing accuracy
EfficientNetB0	95.79	91.71	91.25
EfficientNetB1	95.95	91.71	92.42
EfficientNetB2	95.43	92.45	92.93
EfficientNetB3	96.10	91.87	92.09
EfficientNetB4	94.55	92.04	93.09
EfficientNetB5	94.90	92.87	90.40
EfficientNetB6	93.79	91.04	91.92
EfficientNetB7	95.76	93.20	91.08

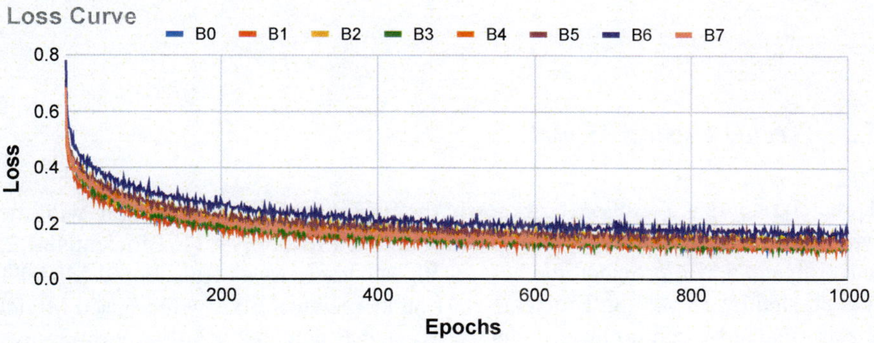

Fig. 3 Loss curve

5.2 *Classifier Performance*

The first and most crucial step in constructing a deep learning model is to define the network architecture. We prefer to use pre-trained networks to extract deep features as they have been initially trained on a large-scale ImageNet dataset. Therefore, we save a lot of computational power when adjusting weights to match our WCE dataset. In this study, we have used pre-trained networks of the EfficientNet family for feature extraction. The extracted deep features were then trained with a multi-layer perceptron network with a softmax activation function. The accuracy achieved on each of the networks is reported in Table 3. The loss and accuracy curve of the training of EfficientNet family are shown in Figs. 3 and 4, respectively.

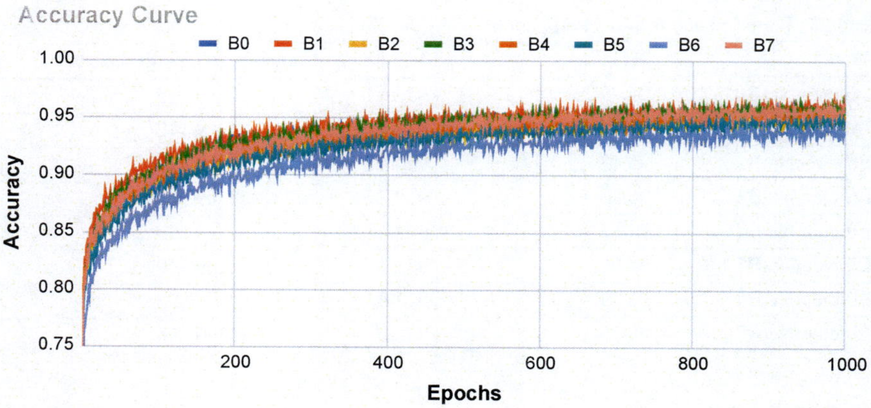

Fig. 4 Accuracy curve

5.3 *Nested Fusion Model*

Three classifiers are required to generate the fusion model. After working with the whole EfficientNet family, it was discovered that EfficientNetB1, EfficientNetB2, and EfficientNetB4 provided the best testing accuracy. As a result, Fused Model 1 was created by combining EfficientNetB1 and EfficientNetB2, while Fused Model 2 was created by combining EfficientNetB2 and EfficientNetB4. Furthermore, we have fused models 1 and 2 together to generate our final nested fusion model.

On the test dataset, combining the EfficientNetB1 and EfficientNetB2 generated an accuracy of 93.43%, while combining the EfficientNetB2 and EfficientNetB4 gave an accuracy of 93.63%. Finally, when the previous two fused models were combined, an accuracy of 94.11% was achieved on the test dataset as given in Table 4. The loss and accuracy curve of the training of fusion models are shown in Fig. 5. The confusion matrix and AUC-ROC plots of each fusion model are shown in Fig. 6.

Table 4 Experimental results of nested feature fusion model

Fused model	Training accuracy	Validation accuracy	Testing accuracy
EfficientNetB1 and EfficientNetB2	98.71	93.28	93.43
EfficientNetB2 and EfficientNetB4	98.76	93.45	93.63
Final fused model	99.50	93.95	94.11

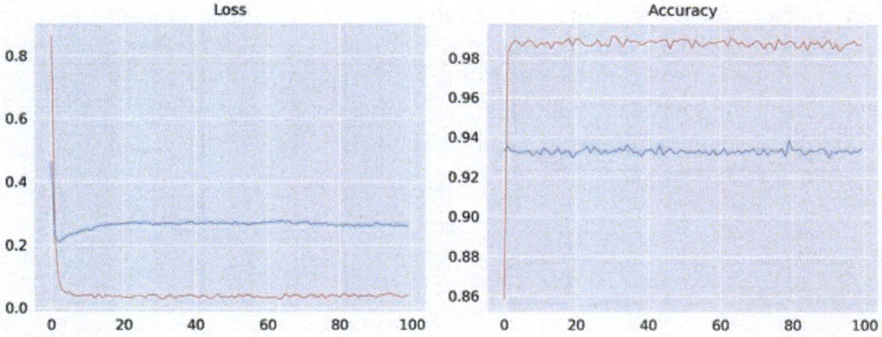

Fig. 5 Loss and accuracy curve of the training of final fused model

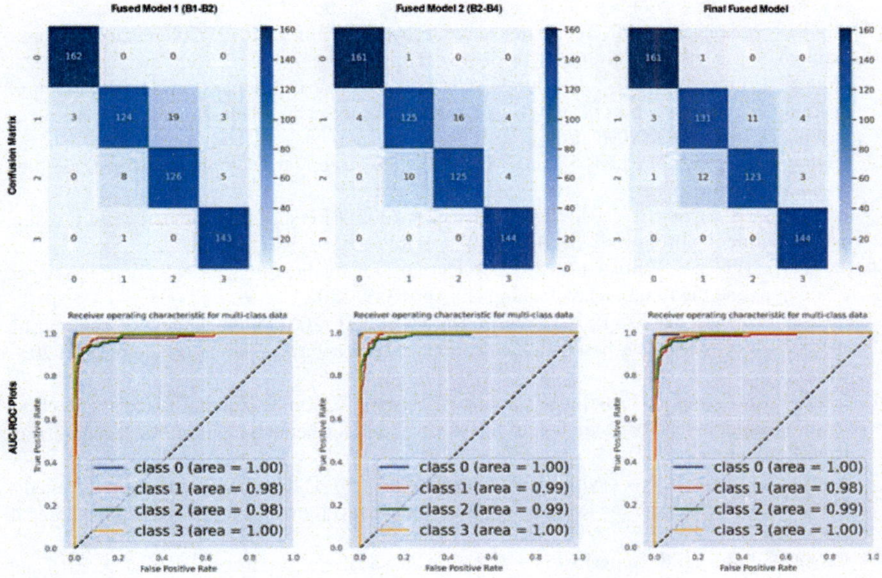

Fig. 6 Confusion matrix and AUC-ROC plots of each fused model

6 Conclusion and Future Directions

Early stage detection of colorectal carcinoma is essential for correctly diagnosing and curing the disease. Our research experimentations establish that a nested fused model can be used to predict colorectal carcinoma in its early stages accurately and can also perform classification upon the type of colorectal carcinoma in its early stage. First, we use pre-trained networks of the EfficientNet family for feature extraction. Later, the deep features are trained in a multi-layer perceptron network with a softmax activation function. We experimented with pre-trained networks of the EfficientNet family. Afterward, we fused EfficientNetB1 and EfficientNetB2 and

EfficientNetB2 and EfficientNetB4 and developed a model and a novel approach for early detection and classification of colorectal carcinoma. Our proposed model gives a testing accuracy of 94.11%. This is a novel approach to early stage detection and classification of colorectal carcinoma. Furthermore, this method can also be used in other biomedical classification tasks for fast and automated detection and classification of diseases.

References

1. Ponzio F, Macii E, Ficarra E, Cataldo SD (2018) Colorectal cancer classification using deep convolutional networks. In: Proceedings of the 11th international joint conference on biomedical engineering systems and technologies, vol 2, pp 58–66
2. Matthew F, Sreelakshmi R, Tatishchev Sergei F, Wang Hanlin L (2012) Colorectal carcinoma: pathologic aspects. J Gastrointest Oncol 3(3):153
3. Wan N, Weinberg D, Liu T-Y, Niehaus K, Ariazi EA, Delubac D, Kannan A et al (2019) Machine learning enables detection of early-stage colorectal cancer by whole-genome sequencing of plasma cell-free DNA. BMC Cancer 19(1):1–10
4. Young Patrick E, Womeldorph Craig M (2013) Colonoscopy for colorectal cancer screening. J Cancer 4(3):217
5. Su H, Lin B, Huang X, Li J, Jiang K, Duan X (2021) FFNet: multi-branch feature fusion network for colonoscopy. Front Bioeng Biotechnol 515
6. Razzak MI, Naz S, Zaib A (2018) Deep learning for medical image processing: overview, challenges and the future. Classification BioApps 323–350
7. Isensee F, Jaeger PF, Kohl SAA, Petersen J, Maier KH (2021) nnU-Net: a self-configuring method for deep learning-based biomedical image segmentation. Nature Methods 18(2):203–211
8. Liyan P, Guangjian L, Fangqin L, Shuling Z, Huimin X, Xin S, Huiying L (2017) Machine learning applications for prediction of relapse in childhood acute lymphoblastic leukemia. Sci Rep 7(1):1–9
9. Konstantina K, Exarchos Themis P, Exarchos Konstantinos P, Karamouzis Michalis V, Fotiadis Dimitrios I (2015) Machine learning applications in cancer prognosis and prediction. Comput Struct Biotechnol J 13:8–17
10. Passos IC, Mwangi B, Kapczinski F (2016) Big data analytics and machine learning: 2015 and beyond. Lancet Psychiatry 3(1):13–15
11. Dinggang S, Guorong W, Heung-Il S (2017) Deep learning in medical image analysis. Annual Rev Biomed Eng 19:221
12. O'Mahony N, Campbell S, Carvalho A, Harapanahalli S, Hernandez GV, Krpalkova L, Riordan D, Walsh J (2019) Deep learning vs. traditional computer vision. In: Science and information conference. Springer, Cham, pp 128–144
13. Montalbo Francis Jesmar P (2022) Diagnosing gastrointestinal diseases from endoscopy images through a multi-fused CNN with auxiliary layers, alpha dropouts, and a fusion residual block. Biomed Signal Process Control 76:103683
14. Poudel S, Kim YJ, Vo DM, Lee S-W (2020) Colorectal disease classification using efficiently scaled dilation in convolutional neural network. IEEE Access 8:99227–99238
15. Khan MA, Kadry S, Alhaisoni M, Nam Y, Zhang Y, Rajinikanth V, Sarfraz MZ Computer-aided gastrointestinal diseases analysis from wireless capsule endoscopy: a framework of best features selection. IEEE Access 8:132850–132859
16. Juan S, Aymeric H, Olivier R, Xavier D, Bertrand G (2014) Toward embedded detection of polyps in wce images for early diagnosis of colorectal cancer. Int J Comput Radiol Surgery 9(2):283–293

17. Fan S, Lanmeng X, Fan Y, Wei K, Li L (2018) Computer-aided detection of small intestinal ulcer and erosion in wireless capsule endoscopy images. Phys Med Biol 63(16):165001
18. Chenjing C, Shiwei W, Youjun X, Weilin Z, Ke T, Qi O, Luhua L, Jianfeng P (2020) Transfer learning for drug discovery. J Med Chem 63(16):8683–8694
19. Pogorelov K, Randel KR, Griwodz C, Eskeland SL, de Lange T, Johansen D, Spampinato C et al (2017) Kvasir: a multi-class image dataset for computer aided gastrointestinal disease detection. In: Proceedings of the 8th ACM on multimedia systems conference, pp 164–169
20. Juan S, Aymeric H, Olivier R, Xavier D, Bertrand G (2014) Toward embedded detection of polyps in wce images for early diagnosis of colorectal cancer. Int J Comput Radiol Surgery 9(2):283–293
21. Pan SJ, Yang Q (2009) A survey on transfer learning. IEEE Trans Knowl Data Eng 22(10):1345–1359
22. Tan M, Le Q (2019) Efficientnet: rethinking model scaling for convolutional neural networks. In: International conference on machine learning. PMLR, pp 6105–6114

Arrhythmia Classification Using BiLSTM with DTCWT and MFCC Features

Shaik Munawar, A. Geetha, and K. Srinivas

Abstract Heart disease is the number one causes of mortality all over the world. Electrocardiogram (ECG) is a valuable powerful tool in the diagnosis of cardiac disorders and detection of arrhythmia. In this study, a new feature set proposed by combining MFCC and DTCWT-based feature for accurate identification and classification of arrhythmia. First, various filters and wavelet transform are used to remove noise from the ECG signals. Latter, R-peak locations are then detected, and ECG segments are generated. From these ECG segments, MFCC and DTCWT-based features were extracted and provided to BiLSTM to implement the classification. The arrhythmia classification carried out according to the Association for the Advancement of Medical Instrumentation (AAMI) criteria. Our model attained an average sensitivity of 94.59%, precision of 94.97%, and overall accuracy of 99.12% on class-oriented arrhythmia classification scheme.

Keywords Arrhythmia · ECG · BiLSTM · MFCC · DTCWT

1 Introduction

According to the WHO, cardiovascular disease (CVD) is the leading cause of death worldwide [1]. Heart disease is very difficult to cure in the later stages. Therefore, it is important to diagnose and treat cardiovascular disease in advance.

One type of heart disease is arrhythmia. It is a disorder of the frequency or rhythm of heartbeats [2]. During arrhythmia, the heart may not be able to pump enough blood to the body. Due to this circulatory failure, the brain, heart, and other organs may be damaged and can lead to death. Types of arrhythmia are broadly classified into two categories. The first category includes life-threatening arrhythmias such as tachycardia and ventricular fibrillation. These arrhythmias need prompt defibrillator

S. Munawar (✉)
Annamalai University, Chidambaram, India
e-mail: shaikmunawar215@gmail.com

A. Geetha · K. Srinivas
CSE, CMR Technical Campus, Hyderabad, India

K. A. Reddy et al. (eds.), *Proceedings of Fourth International Conference on Computer and Communication Technologies*, Lecture Notes in Networks and Systems 606,
https://doi.org/10.1007/978-981-19-8563-8_29

therapy. Even though the other group contains arrhythmia that may not be immediately life-threatening, they require apt treatment or therapy to avoid additional complications in future [3].

ECG is an important modern medical tool that can record the process of cardiac activity. A careful examination of ECG can help to diagnose a cardiac function issue [4]. The occurrence of abnormal beats in ECG may not be regular. So, ECG signals need to be monitored for long duration. Monitoring such a big volume of data manually is not practicable [5]. As a result, automated approaches for ECG signal processing and analysis are essential.

Arrhythmia classification from ECG typically consists of three stages: preprocessing the ECG signal, extracting features from the preprocessed signal, and classifying arrhythmia beats using machine learning techniques [6]. The preprocessing step is primarily concerned with detecting and attenuating unwanted frequencies from the ECG signal. Latter, features are extracted from preprocessed ECG signals. The extracted features can be frequency based, statistical based, ECG morphology based, or auto-extracted. The collected features are then supplied as input into machine learning-based classification algorithms. Deep learning (DL) is a high-performance and effective machine learning algorithm that is gaining popularity. DL is frequently employed in image processing, signal processing, voice and natural language processing operations. Actually, DL is a neural network topology that uses additional hidden layers to handle deeper feature levels to improve classification performance [7].

The aim of this paper is to classify the arrhythmia beats according to AAMI standards. Initially, the ECG signal is denoised and then features extracted using dual-tree complex wavelet transform (DTCWT) and Mel-Frequency Cepstral Coefficient (MFCC). These features are feed to BiLSTM to classify the beat type.

The rest of this paper is organized as follows. The ECG database used in proposed work is introduced in Sect. 1. Section 2 covers noise removal from the ECG signal and obtaining ECG segments, while Sect. 3 describes the feature extraction process. Section 4 outlines the proposed model, as well as its training process and parameters. Section 5 presents results and discussion, and finally, conclusion of the article is presented in Sect. 6.

2 Database Used

The MIT-BIH Arrhythmia Database [8] is the widely used and openly available ECG database for heartbeat classification. This database is used to assess. The database has 48 ECG recordings such that each with duration of half-hour with a sampling rate of 360 Hz. According to the AAMI standards, the fifteen approved arrhythmia classes from the MIT-BIH Arrhythmia Database are divided into five super-classes [9]. These are: N (normal), V (ventricular), S (superventricular), F (fusion), and Q (unclassified) beats. The performance of the proposed ECG classification model is assessed using these five AAMI class beats.

3 Preprocessing

3.1 ECG Signal Denoising

ECG signals commonly contain technological and physiological noises such as power-line interference, baseline wander, and high-frequency noise such as electromyography signals [10]. The arrhythmia classification is influenced by all these types of noises. So, various denoising techniques are applied to remove these noises from ECG signal.

The baseline wander is a low-frequency noise (often less than 1 Hz) generated by body movement, respiration, and poor electrode contact [11]. It corrupts the ST segment and other low-frequency ECG components. Wavelet decomposition (WD) is a powerful method for removing baseline drift. The WD with "db8" and a soft threshold is employed to decompose the noisy ECG signal. Then, baseline drift noise-free signal is reconstructed using the threshold wavelet detail coefficients.

The electromyography frequency noise caused due to muscles contraction other than heart muscles. Due to significant overlapping of EMG frequency with ECG signals, local ECG waves are altered [11]. These frequency distortions are removed using a second-order Butterworth low-pass filter with a cut-off frequency of 30 Hz. The power-line interface is a high-frequency noise (usually around 48 to 50 Hz) induced by the electromagnetic field created while recording the ECG signal. It modifies the morphological aspects of the ECG, such as the duration, amplitude, and P-wave shape [11]. The power-line noise is filtered using an adaptive band stop filter with a stop band corner frequency $W_s = 50$ Hz. The stage-wise denoised sample signal is shown in Fig. 1.

3.2 ECG Segmentation

Detecting R-peak location and forming ECG segments are crucial for arrhythmia beat classification performance. However, detecting R-peaks position is beyond the scope of this work. The already indexed R-peak locations in each ECG record from MIT-BIH arrhythmia database is used in this work. An ECG segment having 359 samples to the left and 360 samples to the right is created from each indexed R-peak. In another way, each ECG segment has 720 samples or data of two seconds duration. Our method largely mimics the way doctors scans the ECG. On the other hand, compared to previous ECG segmentation strategies, each segment obtained in this work always has more ECG data than a single heartbeat cycle. However, this segmentation strategy requires additional processing time to train the proposed model, while capturing hidden ECG features to improve classification performance.

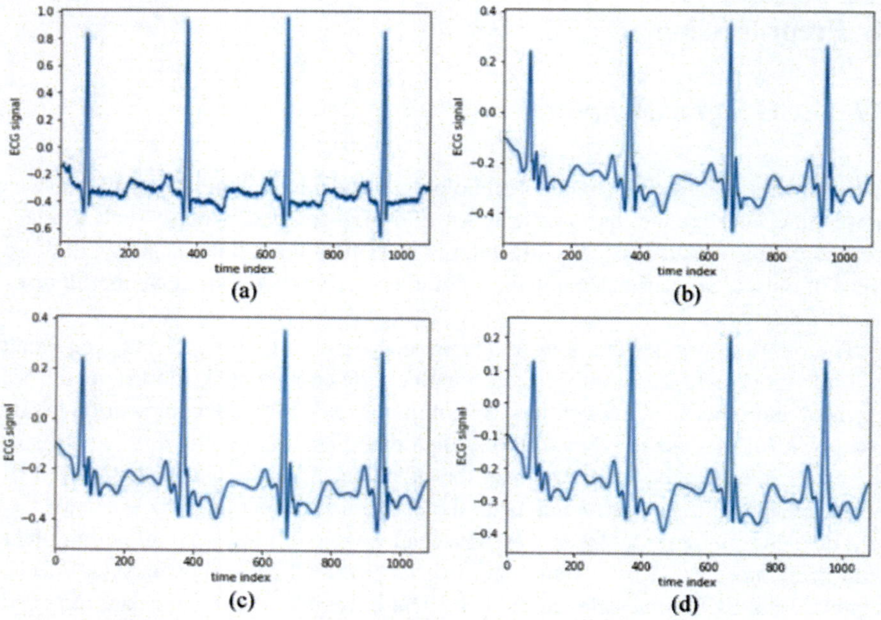

Fig. 1 **a** Sample raw ECG signal. **b** After removal of baseline wander. **c** After removal of high-frequency noise. **d** After removal of power-line interference

4 Feature Extraction

4.1 DTCWT-Based Features

The dual-tree complex wavelet transform (DTCWT) concept is depending up on the use of two parallel trees. The first tree provides real part or the odd samples, and the second tree provides imaginary part or even samples produced at the first level. These trees yield the signal delays needed for each level, therefore eliminating aliasing problems and achieving shift invariance [12]. The vertical and horizontal sub-bands are further subdivided into six separate sub-bands: ±15, ±45, ±75. The benefits of DTCWT, including as directional sensitivity and shift invariance, produce better fusion outcomes than the discrete wavelet transform (DWT). The steps of calculating the MFCC features are illustrated.

(i). Apply the 1D DTCWT transform to the ECG segment, decomposing it up to six scales.

(ii). Select the sixth and fifth scale detail coefficients as features. The sixth and fifth scale detail coefficients of real part and imaginary part are obtained from upper tree and the lower tree, respectively.

(iii). Obtain the absolute coefficients values for the above selected real and imaginary coefficients.

(iv). Apply 1D FFT to the absolute coefficients values and compute the logarithm of the Fourier spectrum.

4.2 MFCC-Based Features

The Mel-Frequency Cepstral Coefficient (MFCC) is a sophisticated algorithm that is frequently utilized, especially in the signal processing field, particularly in the applications of voice recognition [13]. It was a linear expression of a short-duration cosine transform of the logarithmic power spectrum of a voice signal over a nonlinear Mel-frequency scale. The core concept of the MFCC is depending on Mels criterion, the nature of speech intelligibility and human hearing perceptions. The steps of calculating the MFCC features are illustrated.

(i). ECG segment of N samples represented in time domain are converted into frequency domain using fast Fourier transform.

(ii). Apply Mel filter, which is a bank of band-pass filters that overlaps with each other to filter the power coefficients. The mathematical form of Mel scaling is.

$$\text{Mel} = 2595 * \log_{10}\left(1 + \frac{1}{f}\right) \tag{1}$$

where and f is the filter-bank input and Mel is the output. 700 and 2595 are the predefined values that have been used by many researchers.

(iii). Calculate the N features with discrete cosine transform (DCT) to generate the MFCC.

5 Proposed Methodology

In the classification phase, we have used BiLSTM for classifying the arrhythmia types. The best architecture of the BiLSTM is usually obtained using a trial-and-error process. Therefore, after running many simulations, the architecture of BiLSTM classifier fixed with two layers of BiLSTM, each containing 50 hidden units followed by a flatten layer. Then follows two dense layers such that the first one contains 128 neurons with "RELU" as activation function, and the second one contains 5 neurons with "softmax" activation function. The second dense layer gives classification output. The proposed model focuses on solving the objective function in terms of maximizing the accuracy, sensitivity, and precision of the arrhythmia classification. The aim of the developed model is indicated in Eq. (2).

$$F_2 = \underset{\{\text{HN}_b^{\text{blstm}}, \text{ep}_c^{\text{blstm}}\}}{\operatorname{argmin}} \left(\frac{1}{\text{acr} + \text{sen} + \text{prc}}\right) \tag{2}$$

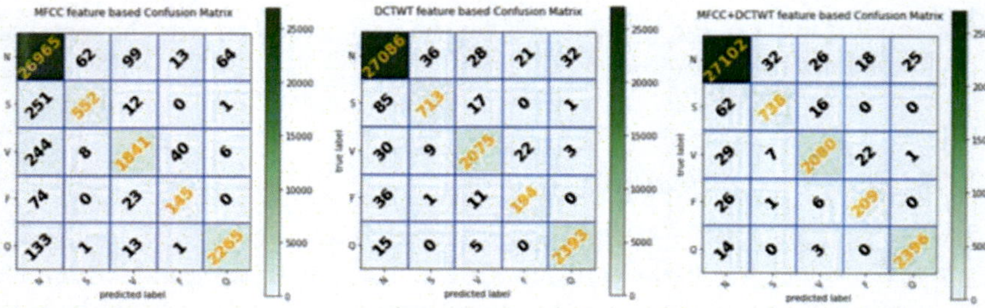

(a) (b) (c)

Fig. 2 Confusion matrix of BiLSTM with **a** MFCC features, **b** DTCWT features, **c** MFCC + DTFCT features

Here, the term $\text{ep}_c^{\text{blstm}}$ and $\text{HN}_b^{\text{blstm}}$ denotes the number of epoch and hidden neurons of classifiers. The classification accuracy, sensitivity, and precision are expressed as acc, sen, and prc, respectively. We conducted experiments in accordance with the guidelines of the AAMI intra-patient scheme to demonstrate the performance of the proposed work. The deep network in this work was created and tested in Google Colab with Keras and Tensorflow as the backend.

The proposed BiLSTM model was evaluated and compared using three distinct schemes in total. The first scheme uses just MFCC-based features; the second employs only DTCWT-based features, and the third uses both MFCC and DTCWT-based features. The AAMI class beats were randomly divided into training and testing sets using the 70:30 stratified splitting approach in each scheme. Figure 2 depicts the confusion matrix for each scheme on the test dataset.

The confusion matrices of all the three schemes are compared and analyzed. The values in confusion matrix presented in Fig. 2 indicate the classification results ratio; the color depth represents the classification results proportion; the darker the color, the bigger the proportion. The comparison of the three schemes performed based on four basic evaluation metrics. They are computed from the confusion matrix such as, classification accuracy (ACC), specificity (SPE), sensitivity (SEN), and positive predictive value or precision (PPV).

Table 1 provides the performance of three schemes compared in this study from four evaluation indicators; each of them is calculated by the mean of the five category beat classes. The results demonstrate that the average accuracy of the MFCC feature model, DTCWT feature model, and MFCC + DTCWT feature model are 98.73%, 99.57%, and 99.65%, respectively. For the MFCC + DTCWT feature scheme, all the four evaluation metrics are higher as compared to other two schemes. It can be observed that for proposed model with fusion of MFCC and DTCWT-based features, classes N, V, and Q have higher classification sensitivity, which are greater than 97%. But, class S has a sensitivity of roughly 90%, whereas class F has a sensitivity of less than 90%. When the characteristics of the AAMI dataset are compared, it is clear that the numbers of class S and class F are very less in the overall data, which will have an influence on the classification effect to some extent. The graphical presentation of sensitivity and precision of all three schemes is given in Fig. 3.

Table 1 Comparison of AAMI arrhythmia classification using three schemes

Model	Metric (%)	N	S	V	F	Q	AVG
Weighted BiLSTM with MFCC features	ACC	97.14	98.98	98.64	99.54	99.33	98.73
	SEN	99.13	67.65	86.07	59.92	93.87	81.32
	SPE	87.49	99.78	99.52	99.83	99.77	97.28
	PPV	97.46	88.6	92.61	72.86	96.96	89.7
Weighted BiLSTM with DTCWT features	ACC	99.14	99.55	99.62	99.72	99.83	99.57
	SEN	99.57	87.38	97.01	80.17	99.17	92.66
	SPE	97.04	99.86	99.8	99.87	99.88	99.29
	PPV	99.39	93.94	97.14	81.86	98.52	94.17
Weighted BiLSTM with MFCC + DTCWT features	ACC	99.29	99.64	99.66	99.78	99.87	99.65
	SEN	99.63	90.44	97.24	86.36	99.3	94.59
	SPE	97.66	99.87	99.83	99.88	99.91	99.43
	PPV	99.52	94.86	97.61	83.94	98.93	94.97

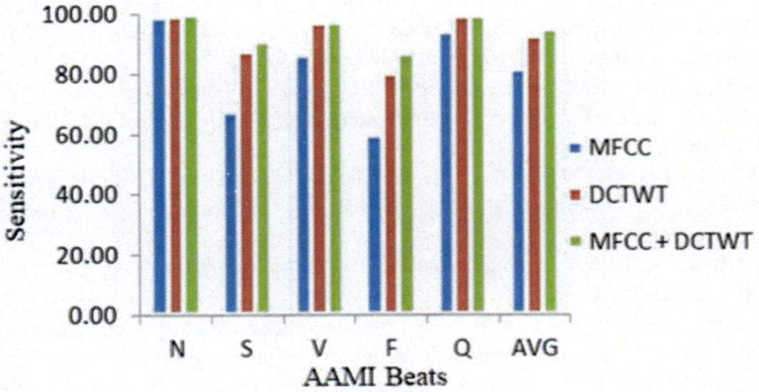

(a)

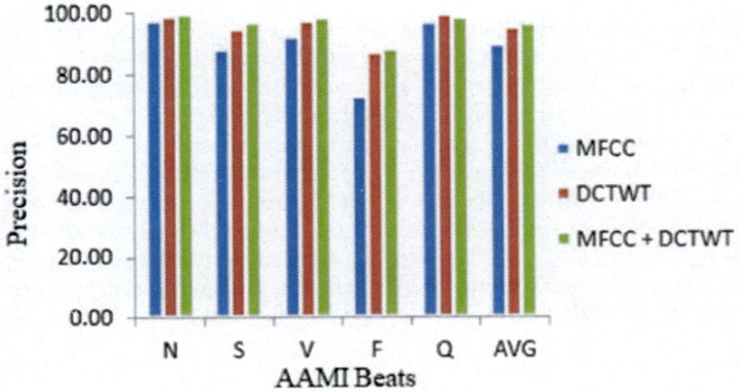

(b)

Fig. 3 Comparison of AAMI classes **a** sensitivity, **b** precision

6 Conclusion

This study investigated the use of a BiLSTM classifier to classify ECG beats accurately. A robust approach is proposed for cardiac arrhythmia identification and classification using MFCC and DTCWT time–frequency-based features. The classification scheme started with denoising of ECG signals and extracting important morphological features using MFCC and DTCWT. The combined features are provided as input to BiLSTM classifiers to perform classification of the arrhythmia according to AAMI standard. The results show that the BiLSTM classifier has the best detection accuracy of 99.12%, indicating its superiority in detecting cardiac arrhythmia. As a result, the presented automated approach can be used to detect cardiac arrhythmias effectively.

References

1. https://www.who.int/news-room/fact-sheets/detail/cardiovascular-diseases-(cvds)
2. Essa E, Xie X (2021) An ensemble of deep learning-based multi-model for ECG heartbeats arrhythmia classification. IEEE Access 9:103452–103464
3. Karraz G, Magenes G (2006) Automatic classification of heartbeats using neural network classifier based on a Bayesian framework. In: 2006 international conference of the IEEE engineering in medicine and biology society. IEEE
4. Pandey SK, Janghel RR (2019) ECG arrhythmia classification using artificial neural networks. In: Proceedings of 2nd international conference on communication, computing and networking. Springer, Singapore
5. Acharya UR et al (2017) A deep convolutional neural network model to classify heartbeats. Comput Biol Med 89:389–396
6. Ebrahimzadeh A, Khazaee A (2009) An efficient technique for classification of electrocardiogram signals. Advances in Electrical and Computer Engineering 9(3):89–93
7. Cai J et al (2021) Real-time arrhythmia classification algorithm using time-domain ECG feature based on FFNN and CNN. Mathematical Problems in Engineering 2021
8. Moody GB, Mark RG (2001) The impact of the MIT-BIH arrhythmia database. IEEE Eng Med Biol Mag 20(3):45–50
9. Yang H, Wei Z (2020) Arrhythmia recognition and classification using combined parametric and visual pattern features of ECG morphology. IEEE Access 8:47103–47117
10. Jagtap SK, Uplane MD (2012) A real time approach: ECG noise reduction in chebyshev type ii digital filter. International Journal of Computer Applications 49(9)
11. Mogili R, Narsimha G (2021) A study on ECG signals for early detection of heart diseases using machine learning techniques. J Theor Appl Inf Technol 99(18):4412–4424
12. Yang Y et al (2014) Dual-tree complex wavelet transform and image block residual-based multi-focus image fusion in visual sensor networks. Sensors 14(12):22408–22430
13. Yusuf SAA, Hidayat R (2019) MFCC feature extraction and KNN classification in ECG signals. In: 2019 6th international conference on information technology, computer and electrical engineering (ICITACEE). IEEE

Anomaly-Based Hierarchical Intrusion Detection for Black Hole Attack Detection and Prevention in WSN

Voruganti Naresh Kumar, Vootla Srisuma, Suraya Mubeen, Arfa Mahwish, Najeema Afrin, D. B. V. Jagannadham, and Jonnadula Narasimharao

Abstract The wireless sensor network (WSN) is the network of sensors which might be deployed in the surroundings for sensing any kind of physical phenomenon. Further sensed data is transmitted to base station (BS) in order to processes the data. During data processing, routed data security is most vital and is very challenging in WSN. The black hole is a most malicious attack, and it targets the routing protocols of sensors. This type of attack may have devastating impacts over hierarchical routing protocols. In this paper, anomaly-based hierarchical intrusion detection for black hole attack detection and prevention in WSN is presented. The black hole attack may happen if the intruder catches and reprogrammed a node set in a network for blocking the packets as opposed to transmitting them to the BS in WSN. Here, the active trust routing model concept is utilized for defining the black hole attacks in data packets routing. The results can demonstrate that, this presented system enhances the security with the prolonged life time of network and less energy utilization and high-efficiency throughput and packet delivery ratio (PDR) the life time of network.

Keywords Black hole attack · Wireless sensor network (WSN) · Energy utilization · Attack prevention

V. N. Kumar (✉) · N. Afrin · J. Narasimharao
Department of CSE, CMRTechnical Campus, Hyderabad, Telangana, India
e-mail: nareshkumar99890@gmail.com

V. Srisuma · A. Mahwish
Department of IT, CMR Technical Campus, Hyderabad, Telangana, India

S. Mubeen
Department of ECE, CMR Technical Campus, Hyderabad, Telangana, India

D. B. V. Jagannadham
Department of ECE, Gayatri Vidya Parishad College of Engineering, Madhurawada, Visakhapatnam, India

K. A. Reddy et al. (eds.), *Proceedings of Fourth International Conference on Computer and Communication Technologies*, Lecture Notes in Networks and Systems 606,
https://doi.org/10.1007/978-981-19-8563-8_30

1 Introduction

WSN is a low-cost network which contains smaller sensing devices, namely sensors. The sensor nodes have a unique identity with the capabilities of sensing, processing and sharing the information to other devices. The WSNs might be limited to small sensing device components, e.g., temperature sensing device to the most critical and complex jet-engine parts. The WSN is a self-organized and least-cost devices network. Such devices utilize actuators and sensors which can minimize the interactions of human. Smart home appliances such as air conditioners (AC) adjust the temperature of room by sensing the temperature of room. Motion detection devices can alert the user about suspicious activities. The nodes of WSN are low cost and simpler for deployment through the wireless medium to communicate. However, the sensor nodes can be limited in terms of battery power, computations and processing. These devices are not protected by traditional cryptographic algorithms. This resource-constrained behavior and wireless medium make them as vulnerable to various attacks.

The sensor nodes would be deployed in an unattended and hostile field in which nodes might always be prone to various security attacks. The WSN can be most susceptible to the security breaches because of its inherent nature, limited resources, unattended hostile environment and open environment. Security is one of the most vital threats among all other aspects in a network. Earlier security methods are not much effective because of their limitations such as energy, memory and nodes accessing after deployment. However, the security aspect is one of the most challenging issues which will deserve much attention in WSNs.

The data packets routing from source to the sink via network have gained more attention from the researcher's in WSN fields. One of the major impacts is energy sources limitation since energy is a vital fundamental element in routing protocols designing. In addition, to lessen the same data superfluous transmissions, the sensed data coalition will be required to be considered in the routing protocols of WSN [1]. Many of the present routing protocols may endeavor at the parameters such as responsiveness, energy preservation, robustness and reliability. But, the feasible security obstacles non-forbearance in routing is perilous since in most of the application fields where WSNs can be utilizes the sensors node will be deployed in unfavorable and inimical environments, provide the opportunities to antagonist for launching some attacks against the sensor nodes.

The security solutions such as key management or cryptography and authentication would improve the protection in WSNs; however, such solutions alone won't forest all achievable attacks. A greater variety of attacks can be introduced with the compromised nodes in WSN nodes which can seem as legitimate inside the network but do not operating for third party; hence, a defense system such as intrusion detection system (IDS) is needed.

The security attacks in WSNs can be categorized as passive and active. In passive attacks, generally, the attacker can be disguised (cover up) and may tap the related connection for gathering the information or devastates the system working elements.

The active attacks can be classified as jamming, Sybil types, denial-of-service (DoS), flooding and hole attacks (sinkhole, wormhole) [2].

In the black hole attack, a malicious node attacks the entire traffic. While advertising that it possess shorter direction in a network. Thus it generates a symbolic black hollow with a malicious node or an adversary with inside the middle. The black hollow dropped complete packets which might be acquired from different nodes. During this attack, a compromised node is trying for pulling the entire traffic from the encompassing nodes. This compromised node creates the false direction data to the neighborhood nodes. This should occupy the entire traffic to the malicious node. A malicious node publicized that it contains high remaining energy. While publicizing this, the malevolent node obtained as cluster head (CH) in every round. All nodes transmit the packets to malicious nodes because it goes as CH. The malicious nodes collect the entire packets and do not send to BS.

Further, the paper is arranged as: Sect. 2 demonstrates the literature corresponds to presented work, and Sect. 3 discussed about the Black hole Attack Detection and Prevention in WSN. Section 4 discussed about presented security solutions performance, and finally, paper is concluded in Sect. 5.

2 Literature Survey

Liu et al. [3] presented a new security and trust routing system relying on active detection. This system achieved higher scalability, anticipation and successful routing security. This active trust system might be able for sensing a nodal trust and even able for stopping doubtful nodes. In addition, design system is greatly energy effective; it utilizes silt energy to pro-create the multiple detecting routs. The author carried out a test run for results verification. Das et al. [4] presented an algorithm and made the dynamic formation of cluster and CHs based on the nodes distance from the cluster node using a genetic algorithm and sensor nodes trust. The cluster information is passed to each node, and after that real-time routing may appear. Motamedi and Yazdani [5] proposed unmanned aerial vehicle (UAV) to find black hollow assaults in WSN. In a black hollow attack, a terrible node shows that the direction to the vacation spot is brief and can be possible, that could appeal to a huge quantity of visitors and drop the whole packet. The device makes use of the UAV to validate the node and makes use of the sequential chance ratio take a look at version as a dynamic threshold device to avoid malicious nodes.

Geethu and Mohammed [6] designed a novel multipath transmission system. This built-up method is used as protection approach in opposition to selective forwarding attacks. In this system, during routing time, when a node senses the packet then that packet would be passed through an alternate node. Due to the resending method, the routing mechanism reliability can be maximized. Satyajayant et al. [7] presented multiple BSs to improve the data in black hole attacks presence. But, these multiple BSs produce additional overhead and increased the memory cost and communication.

In addition, the black holes strategic position not taken, the region of black hole which is closed to base station captures all the packets with higher probability.

Tan et al. [8] present a new model for achieving the confidentiality in multi-hop code destination. In multi-hop protocol, the authors integrated confidentiality as well as DoS-attack resistance. Depending on Deluge, state-of-art code dissemination protocol and an open source is presented for WSNs. They also provided the evaluation of performance in this approach with real Deluge and current secured Deluge.

3 Black Hole Attack Detection and Prevention in WSN

The flow diagram of anomaly-based hierarchical intrusion detection for black hole attack detection and prevention in WSN is represented in Fig. 1. This system contains two major protocols, namely data routing protocol and active detection routing protocol.

Different types of attacks like data type attack, selective forwarding attack and black hole attacks will be detected and prevented using this system. First, the network is implemented while entering the number of nodes. Next, the user selects source and destination. All the possible multiple paths would be computed from source to destination after the generation of network. The detection packet (DP) is transmitted via each path. The DP consists of certain data which is as follows:

$$\text{DP} = \{\text{Source ID, Destination ID, Path length}\}$$

where the path length defines the number of hopes to the destination.

If destination receives the DP, then every node in the path transmits feedback path (FP) to the source node.

$$\text{FP} = \{\text{Source ID},\ \text{Destination ID},\ \text{Path length}\}$$

Here, the threshold must be calculated for each path, and the path with the lowest threshold is considered the safest path for routing data. To achieve this, each node contains its own trust value and is calculated as follows:

$$\text{NodeTrust} = \{C_A^B = \sum_{i=1}^{w} \left\{ \frac{\Delta_A^B(t_i)}{\Delta_A^B(t_i).\left(\frac{h_i}{w}\right)} \right\}, \ldots w \neq 0 \tag{1}$$

$$\text{NodeTrust} = 0, \ldots w = 0 \tag{2}$$

For every node, the distance between the node and destination would be calculated. The threshold value is computed for every node by the equation which is as follows

$$X = \text{Threshold}_{\text{Node}} = \frac{\text{Trust}}{\text{Distance}}. \tag{3}$$

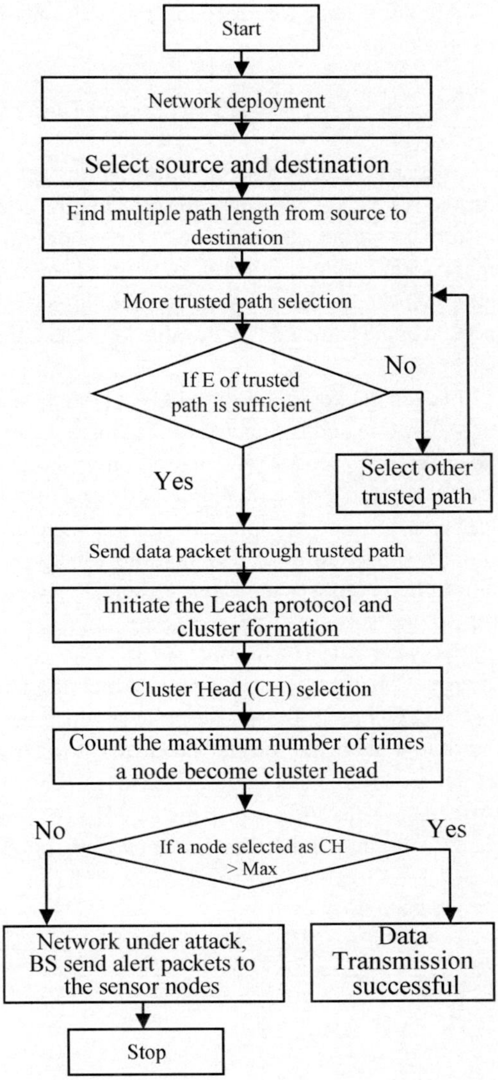

Fig. 1 Flowchart of detection method

Using this equation, the threshold will be calculated for every path.

$$\text{Threshold}_{\text{Path}} = \sum_{\text{from node0ton}} X \tag{4}$$

The above formula is the sum of the thresholds for all the nodes in the path. Finally, the path with the lowest threshold is adopted as the safest and most reliable path for routing data.

After the selection of route, the path data is transmitted via that path, and it is expressed as:

$$PI = \{\text{Type of Data}, \ \text{Size of Data}\}$$

Whenever a node gets the information packet, then it verifies the information with the data of packet. If in case, the information doesn't healthy, then it is far decided that information kind assault is took place in pervious node, in such cases the node drops that packet and transmits rest of the packets to further nodes. In selective forwarding attack, if the data size doesn't match, then the node will recover the data from past node which is attacked by the attacker. Thus, the packet loss ratio is less in presented system.

In each cluster of sensor nodes, a node is selected as a local base station for fixed time duration and is selected as the cluster head for that particular cluster. The sensor node sends its sensor information only to related CH. If all the members of cluster are communicating with single node which is their CH, thus the CH requires more transmission and computation than sensor node members. The LEACH protocol utilizes the rotation of CH randomly between the sensor nodes for avoiding the cluster head rapid dying. Thus all sensor node energy is equally consumed, and alive time of networks is increased. Using local data fusion technique at each cluster head, compressed data is transmitted to BS by each CH. The CH selection is based on energy probability distribution in which the CH nodes would broadcast their status as being CH to all the sensor node in sensing networks, so that every member node can know about the cluster head that it belongs. The formation of cluster is done as per the strength of signal. The leach protocol is used to observe that how many times a specific node is became as CH in overall duration. If a CH is repeated more than the highest limit, then the network is in black hole attack. The BS transmits an alert packet to all the sensor nodes. Or else transmission of data is done successfully across the network.

4 Result Analysis

The presented system is implemented on windows platform using Java framework (version jdk 6). Here, NetBeans (version 6.9) is utilized as development tool. The Jung tool is utilized for network creation. This system doesn't need any particular hardware for running, any standard machine which is capable to run the applications.

The consumption of energy, packet delivery ratio (PDR) and throughput are the parameters used for performance evaluation. These parameters are calculated by using below equations:

Energy Consumption: As whole sensor nodes are operated with battery and they lose their energy very quickly, the consumption of energy is occurred because of the computation, and it must be monitored periodically. Sensor nodes need to be energy efficient because the power consumption of sensor devices should be kept to a minimum, and limited power resources have reached the end of their life. Energy consumption for sending a k-bit message at distance d:

$$E_{Tx}(k, d) = E_{\text{elec}} * k + \in_{\text{amp}} + k * d^2. \tag{5}$$

where E_{Tx} is the total energy consumed during the transmission of data, $\in_{\text{amp}}$ is the energy of amplifier, d indicates the distance, k is a constant and E_{elec} is transmitter energy.

Throughput: The network throughput is referred as the rate of successfully delivered messages on the communication channel. The throughput is measured as bit/second. The packets are successfully delivered during flooding and selective forward attacks because the earlier one floods only the undesired packets, so desirable packets would be successfully delivered. In the next case, only certain packets are dropped so that the throughput is not much affected compared black hole attack where all the packets are dropped.

$$\text{Throughput} = \frac{\text{Total no. of packets delivered at destination}}{\text{Total simulation Time}} * 100 \tag{6}$$

Packet Delivery Ratio (PDR): The PDR depends on created and received packets which are recorded in trace file. Usually, the PDR can be defined as the ratio of total number of received packets by the destination to the total number of generated packets at the source. The PDR is expressed as

$$\text{PDR in \%} = \frac{\text{Total number of packets at destination}}{\text{Total no. of packets generated at Source}} \times 100 \tag{7}$$

The comparative analysis performed in between presenting method as hierarchical intrusion detection for black hole attack detection and prevention in WSN (HID-based black hole attack detection model) and 'black hole attack in WSN with UAV' (Detection of black hole attack using UAV).

Figure 2 represents the consumption of energy of both the methods. The consumption of energy is less in HID-based black hole attack detection method, since the black hole attack is identified before routing the original data, and data is recovered after dropping the packet. Throughputs of two models are represented in Fig. 3.

Throughput is very high in HID-based black hole attack detection method, because black hole attacks are detected and prevented; thus the number of packets delivered to the destination side is increased. The PDR of described two models is represented in Fig. 4.

Packet delivery ratio (PDR) is high in HID-based black hole attack detection model than the black hole attack detection using unmanned aerial vehicle (UAV). One trusted path is chosen from multiple trusted paths. Data is transmitted via that chosen path. However, during the transmission of data from the base system, whenever any node have insufficient energy, then the entire data is not transmitted to destination; this results load imbalanced case. If this case is occurred in presented system, then

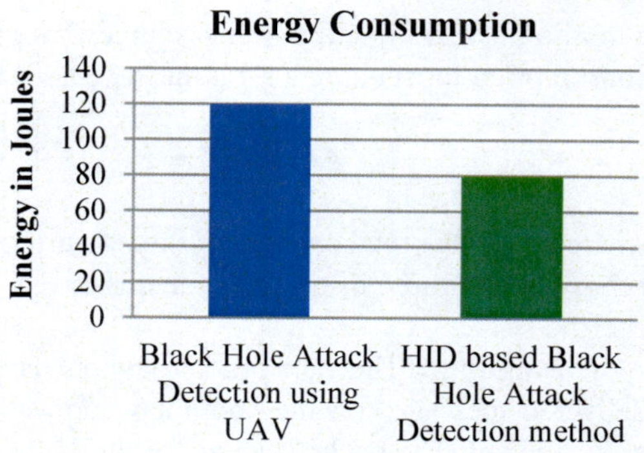

Fig. 2 Represents the comparative analysis of energy consumption (Energy consumption graph)

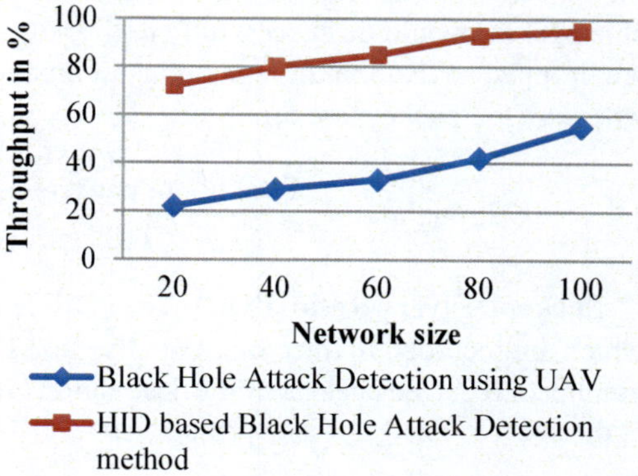

Fig. 3 Throughput comparison graph

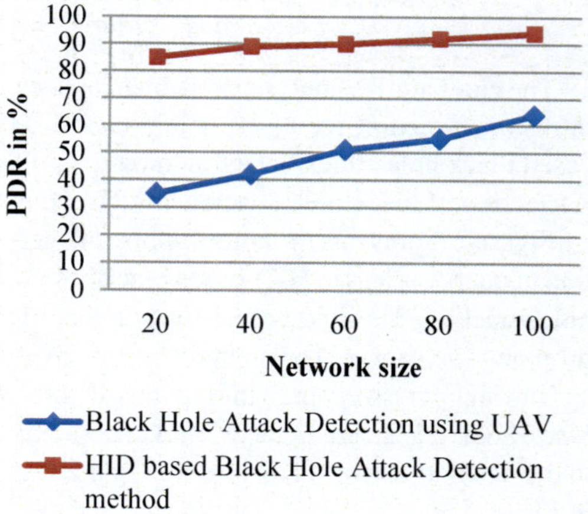

Fig. 4 'PDR' comparison graph

the data is transmitted via another trusted path; hence, in this manner, entire data reaches the destination. Therefore, from results, it is clear that described model is detected and prevented the black hole attacks efficiently than previous models.

5 Conclusion

In this paper, anomaly-based hierarchical intrusion detection for black hole attack detection and prevention in WSN is described. One of the most challenging issues in WSN is security. For detecting and preventing the black hole attacks, this system utilized updated active trust model and data routing method with the data type checking method at the time of routing. Modified low-energy adaptive clustering hierarchy (*LEACH*) protocol is used for black hole attack simulation on WSN. The black hole attacks impact is analyzed by the parameters PDR, throughput and consumption of energy. Comparative analysis in between the 'HID-based Black Hole Attack Detection method' and 'Black Hole Attack Detection' using UAV resulted that, minimum energy consumption, high throughput and high PDR which indicates that, great efficiency of HID-based black hole attack detection model.

References

1. Abdul-Wahab Y, Alhassan A-B, Salifu A-M (2020) Extending the lifespan of wireless sensor networks: a survey of LEACH and non-LEACH routing protocols. International Journal of Computer Applications 975:8887
2. Sikora M, Fujdiak R, Kuchar K, Holasova E, Misurec J (2021) Generator of slow denial-of-service cyber attacks. Sensors 21(16):5473
3. Liu Y, Dong M, Ota K, Liu A (2016) ActiveTrust: secure and trustable routing in wireless sensor networks. IEEE Trans Inf Forensics Secur 11(9):2013–2027
4. Das S, Barani S, Wagh S, Sonavane SS (2016) Energy efficient and trustable routing protocol for wireless sensor networks based on genetic algorithm (E2TRP). In: 2016 international conference on automatic control and dynamic optimization techniques (ICACDOT), Pune, pp 154–159
5. Motamedi M, Yazdani N (2015) Detection of black hole attack in wireless sensor network using UAV. In: 2015 7th conference on information and knowledge technology (IKT), Urmia, pp 1–5
6. Geethu PC, Mohammed AR (2013) Defense mechanism against selective forwarding attack in wireless sensor networks. In: 2013 fourth international conference on computing, communications and networking technologies (ICCCNT), Tiruchengode, pp 1–4
7. Satyajayant M, Kabi B, Guoliang X (2011) BAMBi: blackhole attacks mitigation with multiple base stations in wireless sensor networks. In: IEEE ICC proceedings
8. Tan H, D Ostry H, Zic J, Jha S (2009) A confidential and DoS-resistant multi-hop code dissemination protocol for wireless sensor networks. ACM WiSec'09, Zurich, Switzerland, 16–18 March 2009

A Reliable Novel Approach of Bio-Image Processing—Age and Gender Prediction

A. Swathi, Aarti, V. Swathi, Y. Sirisha, M. Rishitha, S. Tejaswi, L. Shashank Reddy, and M. Sujith Reddy

Abstract Image processing in its field has many applications. With the advancement in the deep learning, many researchers experimented on various traits recognition of face. One of the best applications is age prediction. Using various location points on the face, the age is predicted from the face. Similarly gender also. Age and gender prediction is which allows us to predict age and gender from a texture image or real time video. Important application of age and gender prediction is in Biometrics which is used for security purposes. This paper presents the results on gender prediction and age estimation system based on convolutions neutral networks by extracting features from given input image and reorganization by taking large data set and dividing it into training data (80%) and testing data (20%). The proposed system can get accurate results by taking large sets of training data. The proposed method used ResNet architecture using facial points identification, to classify the age group and gender of the input subject. The experimentation achieved the accuracy of 84% in predicting the age and 71% predicting the gender.

Keywords Age prediction · Gender prediction · Facial images · ResNet

1 Introduction

Biometric is used to analyze the characteristics of each individual for their identity age and gender prediction is mainly used in Biometrics for security purposes from which gender prediction and estimation will be done from a facial image or from a real time video. Face recognition has been one of the most interesting and important tasks in prediction age and gender from face images. Many techniques were applied for gender prediction from face images from the last few years convolution neural network in deep learning were used which has powerful ability to estimate and extract

A. Swathi (✉) · V. Swathi · Y. Sirisha · M. Rishitha · S. Tejaswi · L. S. Reddy · M. S. Reddy
Sreyas Institute of Engineering and Technology, Hyderabad, India
e-mail: swathigowroju@sreyas.ac.in

Aarti
Lovely Professional University, Phagwara, India

K. A. Reddy et al. (eds.), *Proceedings of Fourth International Conference on Computer and Communication Technologies*, Lecture Notes in Networks and Systems 606,
https://doi.org/10.1007/978-981-19-8563-8_31

feature from the given input image or real time video and get accurate results. The main aim is to develop intelligent system which are able to learn efficiently and recognizing objects.

The proposed system will split the data into training data and test data and we will apply the model sequential and test the predictions. In this machine learning project we are training convolutional neural networks to predict age and gender. After the increase in social network and social media are being concerned with automatic age classification in social interaction the most fundament facial qualities are age and gender. Tensor flow and open source library used for math, data flow and specific machine learning application. Coevolution neural network is a deep learning algorithm which takes input images on different aspects and can differentiate from one image [1–6]. Convolution networks takes less processing fire compared to other algorithms. Prediction algorithm that is implemented will work in a way so the model will able to predict age and gender.

2 Literature Survey

In the survey analyzed by Zafeiriou et al., the introduction of feature extraction methods such as SIFT, HoG's, local Binary Pattern (LBP) and SURF, that combines the features with integral images are used to describe face appearance [7–11]. The combination of pre-trained DResNet models with DPMs are resulting in best performance [12]. Through the dilation of pupil, Gowroju et al. [13–15] experimentally analyzed in predicting the age of the person. The UNet architecture build using optimization technique performed the segmentation of pupil effectively hence the age prediction became easier. As age grows the size of pupil increases to certain and after 45 years of age the size of pupil is shown to decrease. Due to this fact, the pupil is also used as the biometric trait to determine age of the person.

2.1 Proposed Method

The proposed scheme is to fill the gap between automatic age and gender prediction. We will first introduce the basic structure of CNN then we will describe the ResNet model for training data to classify gender and age then the result will be obtained by these data using a trained model. Primary method of the proposed system is to recognize the gender and age from the human face images. By using the set of facial features in the real time application extraction of features from face images is an important part in this method. Figure 1 explains the classification of age and gender using ResNet model. The binary classification is used for gender prediction as we need to classify in to two groups. Multi-class classification and regression models are used for the age classification techniques.

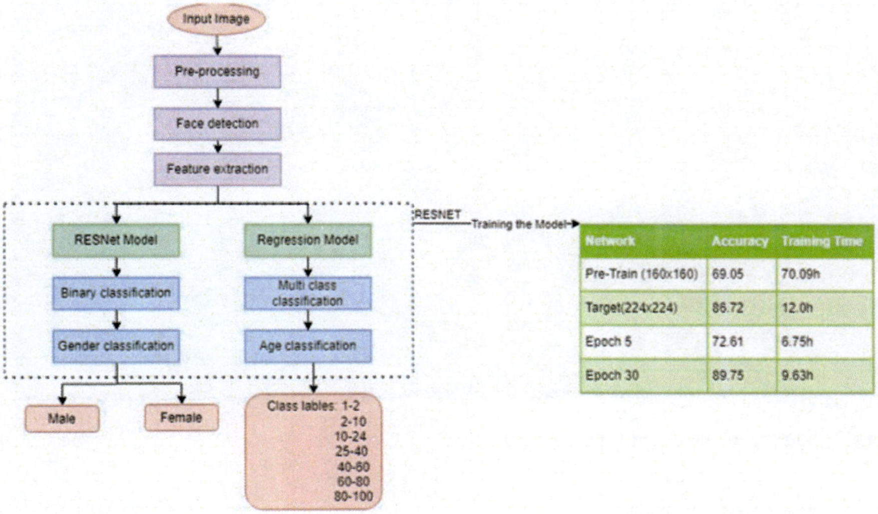

Fig. 1 Proposed classification flowchart

The proposed architecture is shown in Fig. 1. The model is trained using the ResNet architecture. These layers results in training deep neural network with 50 layers on FGNet dataset. These connections skip the training from few layers and connects directly to the output. Hence, if $H(x)$ is initial mapping of the network fit,

$$F(x) := H(x) - x \tag{1}$$

Equation (1), which gives $H(x)\colon = F(x) + x$ in the skip connection. The steps to proceed in the proposed method are explained in the following subsections.

The training and testing accuracies for the two networks in comparison to the number of training examples and training hours. Expressing accurateness of epochs enables one to assess how quickly the network learns as data is submitted to it, whereas representation in terms of days allows one to evaluate how quickly the network learns as data is taught to it. In terms of training time, the factor to be optimized is reflected. Figure 1 shows that the increase in test accuracy ceases a few phases after the previous modification in learning rate. As a result, we consider both systems to be given training at the conclusion of 30 epochs, resulting in top test accuracies of 69.09% (epoch 5) for the targeted system and 89.75% for the reference network (epoch 30).

Fig. 2 Sample images from UTK face dataset

2.2 Face Detection

The model is built using the CNN, mobilenet_1.0 model. The model is depicted in Fig. 2. It is used to extract the face area from the background because this background can be confusing and fail to recognize the expressions. It involves segmentation and extraction of facial features from the uncontrolled backgrounds.

Face extraction plays an important role in gender object detection. It includes shapes, color, texture, movements of the facial image. It also reduces information of image, which requires less storage. The geometric separation of two reference points. Following the identification of the eye centers, 11 correlated points are obtained from the provided face input picture. The crucial points are identified three locations from eye, as well as the lateral endpoints, are put in the face; as well as the nose's vertical midway, the lip's midpoint and two points on the lateral ends of the lips. This procedure works with the face is upfront, color pictures and consistent lighting are used with the sample image is either indifferent or smiling.

2.3 Features Classification

Classification is also known as feature selection stage. Classification is a complicated process because it can be playing role in many areas. It deals with exchange of essential information and connects them in certain parameters. The following points explain the algorithmic approach for the classification. Model is trained with age and gender dataset which is then divided into training data and testing data. Training data is used to train the ResNet model and testing data is for testing the model after training. Train dataset and validation dataset will be given to ResNet model training, train the ResNet model and given input image to trained model then performance evaluation will done to get output as age and gender.

3 Results

The proposed model is built using ResNet architecture with new kernel obtained from Eq. (2), defined as

$$f(x) := f(x) - x \text{ which gives } f(x) := f(x) + x \quad (2)$$

filter size, pooling layers and the proposed system use convolution layers to evaluate the impact of the CNN depth and size of filter on gender prediction. The layer dataset is given as input and the dataset given is divided into train data and test data where train data is of 80% and test data is of 20% if given with more train data then there are better chances of good accuracy. Hence the proposed system used the own dataset obtained from university and the existing dataset-UTK to compare. First the system will be trained using train data and then given input image from which age and gender should be predicted then the given input image will be processed that is image processing will be one and then it will be given to facial model and then it will be given to final model and then results will be obtained the data which it trained will be given to final model in training of data also image processing will be done and then CNN algorithm will be applied where features will be extracted and classification will be done and it is given to final model.

Preferably, the proposed method would be able to completely train a small network, upsize its kernels and instantaneously attain the target network's test accuracy. However, there was a drop in learning rate and the elimination of weight decay lead to a rise in overfitting, imposing certain limits on this basic technique.

3.1 Training Parameters

The proposed system is executed on GPU-based system with tensor flow package of python. In order to map to a logistic regression model, the proposed system defined a classification threshold up to 0.5. For each step of threshold, the accuracy, F1-score, recall, precision, false prediction rate and true prediction rates were analyzed. The values are tabularized in Table 1. At a threshold of 0.5 the system is showing the better accuracy compared to the other threshold values. The accuracy, precision and recall were calculated using true positive, true negative, false positive and false negative values obtained from the confusion matrix while training the dataset. The corresponding accuracy, precision and recall were noted in Table 2.

It is a noticeable fact that the large number of datasets are available publicly. Among them the MORPH-II and FG_Net are greatly used datasets. The proposed

Table 1 Hyper parameters of proposed system

Threshold	Recall	Precision	F1	tpr	fpr
0.0	1	0.5	0.66	1	1
0.1	0.95	0.50	0.66	0.96	0.94
0.1	0.91	0.53	0.68	0.91	0.80
0.3	0.81	0.57	0.68	0.85	0.66
0.4	0.79	0.64	0.71	0.80	0.50
0.5	0.74	0.76	0.73	0.75	0.32

Table 2 Training parameters of proposed method

	Accuracy	Precision	Recall
Resnet50	88.9	76.5	74.8

method experimented on FG-Net and obtained the results with respect to FACES. The proposed system evaluated the gender and age for the loss value of 0.59 in the model. The proposed system is checked using python-tensor flow library on CPU-based system.

Resultant images are shown in Fig. 3. Figure 3a, b, c, d are predicting the age and gender. Using multi-class classification the results are figured in Fig. 3e, f, g, h. The proposed system is executed using input video and also on the images of the test set. The proposed system has shown the accuracy of 88.9% at the epoch of 24. Accuracy for the training dataset is little less compared to the trained set as 69%. Although the difference is too tiny to be called an improvement (0.11%), it does demonstrate that the upper-bound is reachable with the suggested strategy while eliminating 30.7 h while training is a great improvement compared to the existing state of art systems by 11.41% (Fig. 4).

4 Conclusion

CNN can be used to provide improved results of age and gender classification even by considering the much smaller size unconstrained image sets labeled for age and gender. The simplicity of the model implies that more elaborated system using more or large training data may be capable of improving results and gender accurate results. Using regression model for age and gender prediction instead of classification if enough data is available. Mainly that can be drawn is that age and gender from face reorganization are very popular among research which can be used in social network and advertising to implement an intelligent system that can achieve good and robust results in the accuracy of recognition deep learning algorithm, convolution neural network to propose a study contain various ResNet models in gender classification, and trained in well-known datasets then to apply an efficient model for age estimation.

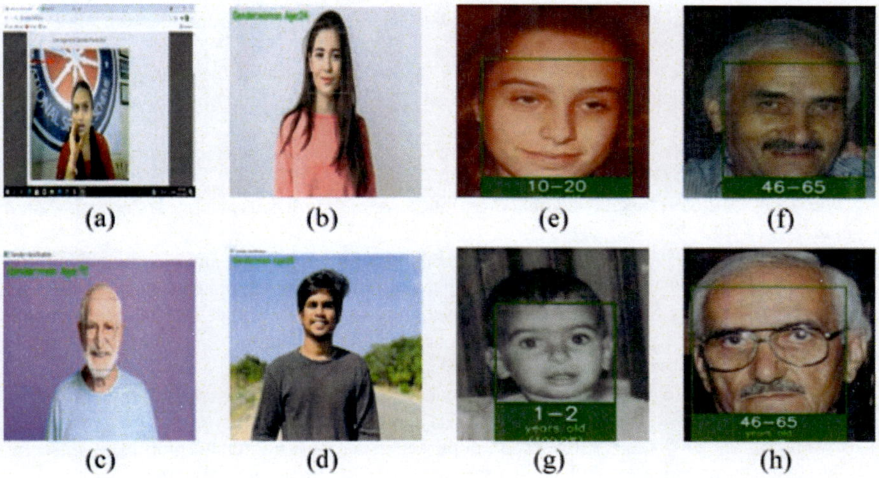

Fig. 3 Resultant images after predicting the age

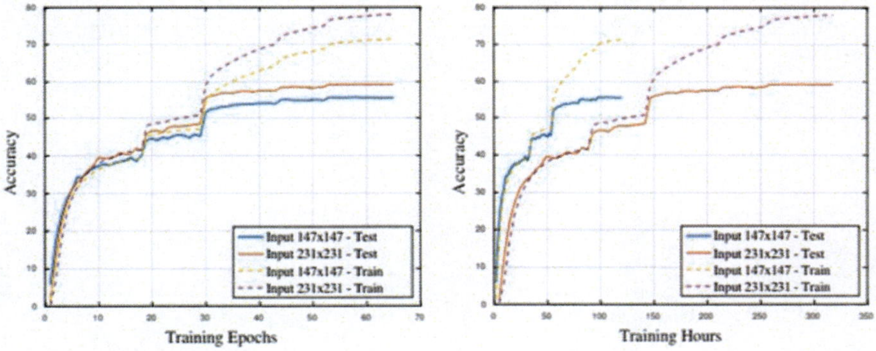

Fig. 4 Accuracy plot obtained for both original image with 231 × 231 resolution with its pre-trained image of 147 × 147 resolution

References

1. Fu Y, Guo G, Huang TS (2010) Age synthesis and estimation via faces: a survey. IEEE Trans Pattern Anal Mach Intell 32(11):1955–1976
2. Dhimar T, Mistree K (2016) Feature extraction for facial age estimation: a survey. In: 2016 international conference on wireless communications, signal processing and networking (WiSPNET). IEEE, pp 2243–2248
3. Dantcheva A, Elia P, Ross A (2015) What else does your biometric data reveal? A survey on soft biometrics. IEEE Trans Inf Forensics Secur 11(3):441–467
4. Fu S, He H, Hou Z-G (2014) Learning race from face: a survey. IEEE Trans Pattern Anal Mach Intell 36(12):2483–2509
5. Zafeiriou S, Zhang C, Zhang Z (2015) A survey on face detection in the wild: past, present and future. Comput Vis Image Underst 138:1–24

6. Ng C-B, Tay Y-H, Goi B-M (2015) A review of facial gender recognition. Pattern Anal Appl 18(4):739–755
7. Sariyanidi E, Gunes H, Cavallaro A (2014) Automatic analysis of facial affect: a survey of registration, representation, and recognition. IEEE Trans Pattern Anal Mach Intell 37(6):1113–1133
8. Ding C, Tao D (2016) A comprehensive survey on pose-invariant face recognition. ACM Trans Intell Syst Technol (TIST) 7(3):1–42
9. Wu Y, Ji Q (2019) Facial landmark detection: a literature survey. Int J Comput Vision 127(2):115–142
10. Savchenko AV (2019) Efficient facial representations for age, gender and identity recognition in organizing photo albums using multi-output ConvNet. Peer J Computer Science 5:e197
11. Gowroju S, Kumar S (2020) Robust deep learning technique: U-Net architecture for pupil segmentation. In: 2020 11th IEEE annual information technology, electronics and mobile communication conference (IEMCON). IEEE, pp 0609–0613
12. Swathi A, Kumar S (2021) A smart application to detect pupil for small dataset with low illumination. Innovations Syst Softw Eng 17(1):29–43
13. Swathi A, Kumar S (2021) Review on pupil segmentation using cnn-region of interest. In: Intelligent communication and automation systems. CRC Press, pp 157–168
14. Gowroju S, Kumar S (2021) Robust pupil segmentation using UNET and morphological ımage processing. In: 2021 international mobile, intelligent, and ubiquitous computing conference (MIUCC). IEEE, pp 105–109
15. Gowroju S, Aarti KS (2022) Review on secure traditional and machine learning algorithms for age prediction using IRIS image. Multimed Tools Appl. https://doi.org/10.1007/s11042-022-13355-4

Restoration and Deblurring the Images by Using Blind Convolution Method

Jonnadula Narasimharao, Bagam Laxmaiah, Radhika Arumalla, Raheem Unnisa, Tabeen Fatima, and Sanjana S. Nazare

Abstract Because of the camera shaking or motion, blurring the image is introduced in the digital photography. Another reason for blurriness of the image is less shutter speed or background light intensity. So because of this, the image important information is significantly degraded. In order to deblur these affected images, there are different techniques. One of the techniques is blind image deblurring and works in different cases as less information or unavailability of point spread function (PSF). The blurred image deblurred process of deconvolving is simple with the help of any deblurring filter when it estimated the PSF. The proposed deblurring process is to be used even in the case of no idea about blur type information. With the help of estimated PSF re-blurred, the deblurred image is takes place. Then, the quality of deblurred image in between the re-blurred image and original blurred image is calculated by the peak signal-to-noise ratio (PSNR). The deblurred images are containing the noise, which is produced by the deblurring filters. Every iteration of this method is uses the algorithm of Richardson-Lucy along with blurred image computation and by using PSF image is restored.

J. Narasimharao (✉) · B. Laxmaiah · R. Unnisa · T. Fatima · S. S. Nazare
Department of Computer Science and Engineering, CMR Technical Campus, Hyderabad, Telangana, India
e-mail: jonnadula.narasimharao@gmail.com

B. Laxmaiah
e-mail: laxmaiah.cse@cmrtc.ac.in

R. Unnisa
e-mail: raheemunnisa.hb@gmail.com

T. Fatima
e-mail: ftabeen@gmail.com

S. S. Nazare
e-mail: sanjananazare9907@gmail.com

R. Arumalla
Department of Information Technology, B.V. Raju Institute of Technology, Narsapur, Hyderabad, Telangana, India
e-mail: radhika.a@bvrit.ac.in

K. A. Reddy et al. (eds.), *Proceedings of Fourth International Conference on Computer and Communication Technologies*, Lecture Notes in Networks and Systems 606,
https://doi.org/10.1007/978-981-19-8563-8_32

Keywords Re-blurred image · Point spread function (PSF) · Blind image deblurring · Peak signal-to-noise ratio (PSNR)

1 Introduction

With the fast growth in modern digital technology, using digital image as digital information carrier has been the people's attention. The digital images are used in various areas, such as medical, military and transportation, microscopy imaging and photography deblurring [1]. The recorded image consisting a noise and blur version of original picture [2].

Different techniques are used in identification of colors and shades in several pictures, and these are not really recognized by human eye. A huge amount of information is conveyed through the single images than speaking more words. The main aim of the image capturing is, captured image is cannot differentiated with the original or real image. Even though sometimes, images are attacked by the interference or disturbance in the form of blurriness. Then, the actual information in the image is disturbed. Outside interference or camera physical properties can results to occurrence of disturbance in the original image.

The camera or object movements are the main reasons of image blurring in the capturing time, using wide-angle lens and long exposure times, etc., the process of getting real image from the corresponding blurred image is called image restoration [3, 4]. Image processing technology uses this technology widely [5]. So from the distorted image, the original image is retrieved by this image restoration process. In many situations, the process of eliminating the blur from the image is difficult and causes a great damage to the original image.

In general terms, image sharpness and contrast deviating situations are called as blurring. Image restoration is the best solution of this type of problems in the images. From the distorted images, the blur can be eliminated by using the process of image deblurring. In this process, sharpness has been given to the degraded image with clear appearance. The noise function and degradation function are expresses the blurred image in the degradation model of the image [6].

The process of image recovering is disturbed by the linear degradation is called as image deblurring commonly known as inverse problem [7]. The first challenge in the blurred images is blurred kernel [8] and point spread function (PSF) approximation because blurred kernels estimation is too difficult in the blurred image. If the dynamic scene or camera rotation in the image, then noise estimation is too hard because it is spatially invariant. Second challenge is noise elimination from the blurred image in order to get a noise free image. From the scene, noise has done the attenuation high frequency information and neighboring pixels averaging. The sharpness of image is estimated by the blind motion deblurring from blurred image [9]. The image blurriness is defined in this model as

$$B = K * S + n \tag{1}$$

where blurry image is denoted by *B*, blur kernel as *K*, noise as '*n*' and latent sharp image as *S*. In case of blind motion deblurring, blur kernel value is undefined. Therefore, the blur kernel value and latent sharp image is to be calculated for the given image *B*.

2 Literature Survey

Optical aberration, atmosphere scattering, sensor spatial and temporal integration and lens defocus are the different sources of getting blurred images. The mechanisms are partially understood by the humans, while visual systems are easily recognized. Therefore, the blur image estimation is too difficult. Inaccurate focusing of camera and movement in camera results the blur in the image. An aperture can causes the shallow field depth which results the blur in image subsequently non-sharp the image. Blind deconvolution algorithm gives the effective results even in the situation of no information as regards to the noise or blur of the image.

In the form of space invariant or space variant, the blurring degradation is present. Two types of image deblurring methods are present as blind and non-blind. In the blind type, unknown factor is blurring operator. The blurring operator is known factor in non-blind type. Image bandwidth reduction is called as blurring which is because the process of imperfect image formation. Between the original image and camera, relative motion can results the imperfect image formation. Recovering image in the blind image deconvolution is too difficult because degrading PSF knowledge is little in this process. Therefore, the blind deconvolution algorithm can performs the point spread function restoration simultaneously. In each iteration, Richardson-Lucy method is used. The improvement in the quality of image restoration is achieved with additional optical system characteristics, for example, input parameters of camera. The PSF constraints can be passed in the user specified function.

2.1 *Iterative Phase Retrieval*

The phase recovery also called phase retrieval. $\hat{k}(\omega)$ phase component estimation is required by the power spectrum $\left|\hat{k}\right|^2$ kernel k recovering. However, this procedure only obtains the spectrum information; the phase information is still unknown because it iteratively switches between Fourier and real-space domains. Unique solution is not guaranteed by the spatial constraints and input $\left|\hat{k}\right|^2$. A hybrid input–output method is used to estimate the blur kernel in iterative phase retrieval procedure under the appropriate frequency/spatial domain constraints. Therefore, based on iterative phase retrieval algorithm, the blur kernel can be recovered. Thus, the blurry image can be deblurred through a deconvolution.

2.2 Measure of Blur Kernel Quality

As known above, it can obtain n blur kernels after iterating n times. Each of NSM values for deconvolution by using the corresponding kernel can be calculated. It is obvious that the symmetric relationship exists among blur kernels and the estimated blur kernel for each iteration is also different. Hence, the measure of blur kernel quality will test the symmetry of blur kernel and will make a score. In order to estimate the symmetric characteristic of blur kernel, it has to calculate the NSM score twice. For example, if there are thirty kernels, it will calculate the NSM for sixty times. After computing NSM value, the smaller NSM score, the better the reconstructed image.

2.3 Image Quality Estimation

Natural image signals are highly structural information, such as pixels exhibiting strong dependencies and containing important information about the objects structure in the visual scene. In order to estimate the reconstructed image structural performance after deconvolution, we adopt the structural-similarity-based image quality measure (SSIM) instead of the mean squared error (MSE). The SSIM mainly computes the structural similarity between the reference and the distorted signals. However, one usually requires the overall image quality measure, a mean SSIM (MSSIM) derived from SSIM is used to achieve this measure, which is showing the good visual appearance with the best consistency.

3 Restoration of Blur Images

Proposed blind convolution deblurring method is represented in Fig. 1. The details of the procedures are described in the following.

3.1 Methodology

The blurred image is obtained with various reasons of camera properties and movements. In this process, blurred image is formed with the intermixing of real PSF (h) and true image (f). The blurred image (g) is deblurred when it is passed through the restoration filter. This deblurred image is estimates the true image with candidate PSF (h') which is extracted from the PSF's list. So the blur in (g') which is same as restoration filter is produced by the real PSF (h) which as or similar to candidate PSF so less noise is produced by the restoration filter. The produced blurring image is

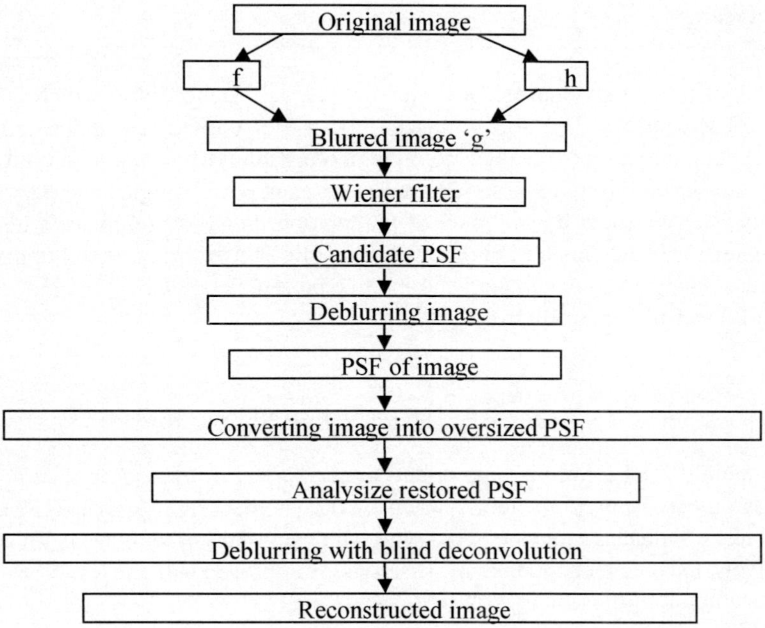

Fig. 1 Proposed block diagram

similar to real blur image and not measured by the peak signal-to-noise ratio (PSNR) value in between the re-blurred and blurred image. The image point spread function (PSF) is derived in the next step of this process. Then, four times pixels shorter undersized PSF is modified from the colored image. In the next step, this under sized color image is over sized with four times higher pixels to the initial image. At last initial, PSF is converted from the colored image and is having the same pixel size. Then, analysis is done for the PSF images and stored.

Every iteration of this method is uses the algorithm of Richardson-Lucy along with blurred image computation. Input parameters of camera are used as additional optical system characteristics, which are improve the quality of image restoration. The PSF constraints are passed in specified function of user. The definition of blind deblurring method is represented through the below equation as

$$n(x, y) = \text{PSF} * f(x, y) + \eta(x, y) \tag{2}$$

where observed function is denoted by $n(x, y)$

Constructed image as $f(x, y)$.

Additive noise term as $\eta(x, y)$ and

Point spread function as PSF.

3.2 Wiener Filter

The blur in images is removed or eliminated by the Wiener filter, which is most important techniques. The blur is formed because of unfocussed optics or linear motion. Poor sampling is resulted from the photograph linear motion, and blurring is also introduced through this signal processing standpoint. Single stationary point intensity is represented by the pixel of photograph digital representation in front of the camera. If the camera is in motion and slow shutter speed, then the pixel is intensities amalgam from camera's motion line presented points.

The Wiener filter is given by

$$G(u, v) = \frac{H^*(m, n)}{[H(m, n)]^2 + \text{NSR}} \tag{3}$$

where signal-to-noise ratio is represented by the NSR. For getting optimal results, NSR parameter is used to adjust when unknown the original signal. Noise is completely eliminated when the NSR value is high and deblurred image is smoothed extensively. On the other hand, image sharpness is improved with the less NSR value and less amount of noise is present in this case.

3.3 Point Spread Function (PSF)

Point spread function is defined as the amount of degree of blur or spread point of light in any optimal system. Point spread function (PSF) Fourier transform is denoted by the optical transfer function (OTF), and it is a frequency domain function. The impulse response of invariant and linear system is defined by the OTF. Conversely, OTF inverse Fourier transform is denoted as PSF. Point spread function (PSF) is given by the light emission pattern, which is diffracted from the point source. One of the image fundamental units is PSF. Convolution integrals can give the blurring and represented as

$$(r) = \int h(r, s) f(s) \mathrm{d}s \tag{4}$$

$h(r, s)$ is denoted for the image position of r, and object brightness distribution of point spread function is $f(s)$. The equation can be simplified through the same coordinates of r, s.

Centered PSF is given by

$$(q, s) \equiv h(q + s, s) \tag{5}$$

Modified equation is as follows:

$$(r) = \int k(r - s, s) f(s) \mathrm{d}s \tag{6}$$

Point object or point source imaging system response is explained by the point spread function (PSF), and it is systems impulse response representation.

3.4 The Richardson-Lucy (R-L) Algorithm

The blurred images restoring process can be done by the method of Richardson-Lucy (R-L) algorithm and widely used method. The deblurring images and its restoration are very compatible in this R-L method because it is more and more characteristics.

The Poisson statistics are adopted this R-L method for best probability solutions for its data. The images are restored as non-negative at the local and global iteration, and flux is conserved by the R-L method. Strong characteristics are gained in the restored images point spread function (PSF). Certain calculations are required for storing the image in R-L algorithm. The derivation of R-L algorithm iterations is very comfortable for the Poisson statistics equation.

4 Results

4.1 Deblurring Results

The results are obtained from deblurring process is same as real images. This process includes the images as motion blur case and atmospheric turbulence. Figure 2a shows a blurred frame and deblurred image using atmospheric turbulence.

The camera can captures the observed scene, which is affected by the atmospheric turbulence as considered blur because of the refractive index of medium fluctuations.

(a) (b)

Fig. 2 **a** Blurred video frame, **b** Deblurred using estimated atmospheric turbulence psf

Atmospheric turbulence blur of OTF for long exposures under some conditions are given as

$$H(u, v) = \exp^{-\lambda(u^2+v^2)5/6} \tag{7}$$

Gaussian function can estimates the atmospheric turbulence blur for long exposures as

$$d(i, j; \sigma_G) = C\exp\left(-\frac{x^2 + y^2}{2\sigma_G^2}\right) \tag{8}$$

where the blur variance is denoted as σ_G^2. Uniform blur is represented in Fig. 2a so in this case PSF is utilized. Fig. 2b explains the deblurred image with high sharpness. 0.79 is the blur variance in this method (Fig. 3).

This proposed method is used even in case of distortion type is unknown and efficiently deblur the colored images. PSF can analysis the image, then after blind deconvolution process is done. Every iteration of this method is uses the damped algorithm of Richardson-Lucy. Colored images are also deblurred by using this method, and image is taken by the point spread function. Under sized color image is over sized with four times higher pixels to the initial image. At last initial, PSF is converted from the colored image and is having the same pixel size. The PSF constraints are passed in specified function of user. Figure 4 shows the appropriate results.

The blurred image because of motion is depicted with unreadable book name in Fig. 5a. In the figure, a man is holding the book and moving across the camera. In this process, image is captured with an ordinary camera. The deblurred image with estimation of PSF is depicted in Fig. 5b. Minus 4° angle with 71 pixel long is considered in this proposed PSF estimation approach. In the kurtosis-based scheme,

Fig. 3 Blurred image

Fig. 4 Deblurring with oversized (4X pixel) 'PSF'

1degree angle and 76 pixel long is considered with PSF estimation. The book title is readily readable.

The performance evaluation factors used for the comparison are peak signal-to-noise ratio (PSNR) and mean square error (MSE) with respect to the ground truth. The values PSNR and MSE for reconstructed image according to threshold values are given in Table 1.

In Table 1, PSNR and MSE values are calculated. The result shows that PSNR value is maximum at threshold 0.2 and corresponding MSE value is minimum at this threshold. The graphical representation of Table 1 is in Fig. 6.

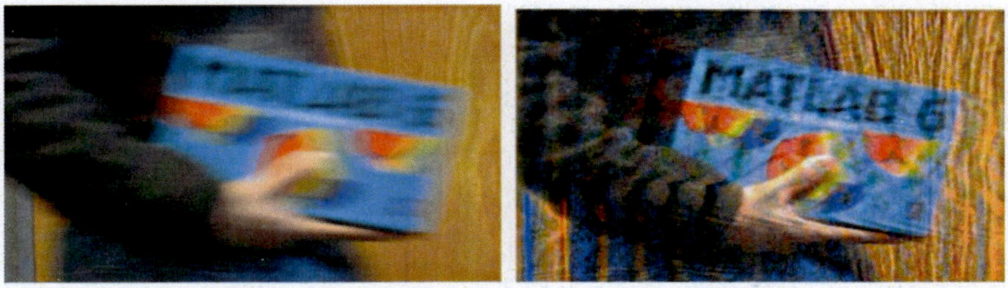

(a) (b)

Fig. 5 **a** Image under motion blur. **b** Deblurred with PSF of length 71 pixels and angle minus-4

Table 1 Performance factors of reconstructed image

Threshold value	PSNR	MSE
0	55	94
0.1	62	49
0.2	97	27
0.3	82	35
0.4	86	39
0.5	82	36

Fig. 6 Comparison plot

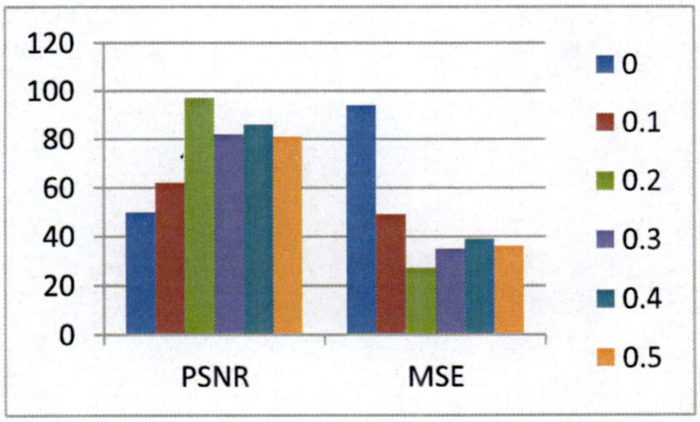

5 Conclusion

A novel method for deblur the image based on PSF is discussed in this paper even in the case of distortion type is unknown and efficiently deblur the colored images. The implementation of this process is very easy and efficiently works. Different types of blurring situations are applicable for this method as motion blur or atmospheric turbulence. Deblurred image ringing effect and noise are eliminated by using various restoration filters and significantly effects the characteristics of image. Through this blind convolution method, the blurred images are deblurred efficiently with less computational time. Moving objects images are effectively restored through this algorithm presented in this paper by the low quality surveillance cameras with a good visual quality. It could enhance the image quality shown in many kinds of the view devices.

References

1. Li B, Cheng Y (2021) Image segmentation technology and its application in digital image processing. In: 2021 IEEE Asia-Pacific conference on image processing, electronics and computers (IPEC)
2. Lu X, Gu C, Zhang C, He Y (2021) Blur removal via blurred-noisy image pair. IEEE Trans Image Process 30
3. Tao S, Dong W, Chen Y, Xu G (2021) Blind deconvolution for poissonian blurred image with total variation and L0-norm gradient regularizations. IEEE Trans Image Process 30
4. Rajagopalan AN, Purohit K, Suin M (2021) Degradation aware approach to image restoration using knowledge distillation. IEEE Journal of Selected Topics in Signal Process 15(2)
5. Tang M (2020) Image segmentation technology and its application in digital image processing. In: 2020 international conference on advance in ambient computing and intelligence (ICAACI)
6. Chen J, Wu G, Wang W, Zeng L, Cai W (2020) Robust prior-based single image super resolution under multiple Gaussian degradations. IEEE Access 8
7. Li H, LuoW, Zhang K, Ma L, Zhong Y, Liu W, Stenger B (2020) Deblurring by realistic blurring. In: 2020 IEEE/CVF conference on computer vision and pattern recognition (CVPR)
8. Li Y, Zhang H, Zhang Z, Wu Y (2020) BID: an effective blind image deblurring scheme to estimate the blur kernel for various scenarios. IEEE Access 8
9. Lee D, Seo D, Kim H, Cha D, Jung J (2019) Blind motion deblurring for satellite image using convolutional neural network. In: 2019 digital image computing: Technique and Apps (DICTA)

Interpretation of Brain Tumour Using Deep Learning Model

J. Avanija, Banothu Ramji, A. Prabhu, K. Maheswari, R. Hitesh Sai Vittal, D. B. V. Jagannadham, and Voruganti Naresh Kumar

Abstract Brain tumour analysis without human involvement is a crucial field of study. Convolutional neural networks, on the other hand, excelled at solving computer vision and other challenges such as visual object recognition, detection, and segmentation. It aids in the diagnosis of brain tumours by improving brain pictures utilising segmentation algorithms that are extremely resistant to noise and cluster size sensitivity issues, as well as automated area of interest (ROI) detection. One of the key arguments for using CNNs is that they have a high level of accuracy and do not require human feature extraction. Detecting a brain tumour and correctly identifying its kind is a difficult undertaking. Because of its widespread use in image recognition, CCN performs better than others. Providing assistance to diagnose brain tumours becomes difficult if performed manually. Furthermore, it becomes difficult process when there

J. Avanija
Department of CSE, Sree Vidyanikethan Engineering College, Tirupati, Andhra Pradesh, India
e-mail: avans75@yahoo.co.in

B. Ramji
Department of CSE (DS), CMR Technical Campus, Hyderabad, Telangana, India
e-mail: vasramji@gmail.com

A. Prabhu (✉) · K. Maheswari · V. N. Kumar
Department of CSE, CMR Technical Campus, Hyderabad, Telangana, India
e-mail: prabhuak@gmail.com

K. Maheswari
e-mail: mahi.kirubakaran@gmail.com

V. N. Kumar
e-mail: nareshkumar99890@gmail.com

R. H. S. Vittal
Hyundai Mobis, Hyderabad, Telangana, India
e-mail: hiteshvittal@gmail.com

D. B. V. Jagannadham
Department of ECE, Gayatri Vidya Parishad College of Engineering, Madhurawada, Visakhapatnam, India
e-mail: dbvjagan@gmail.com

K. A. Reddy et al. (eds.), *Proceedings of Fourth International Conference on Computer and Communication Technologies*, Lecture Notes in Networks and Systems 606,
https://doi.org/10.1007/978-981-19-8563-8_33

is a huge amount of data to assist. Extracting tumour from the images becomes difficult. To overcome this drawback, the proposed method uses convolutional neural network-based model using MobileNet for detection of brain tumours given MRI images.

Keywords Convolutional neural network · Brain tumour · MobileNet · Backpropagation

1 Introduction

Today, we live in an era where illnesses are on the rise, necessitating the advancement of treatment quality. Tumours are an irregular bulge that can appear anywhere on the body and are one of the most hazardous illnesses. The most dangerous of all cancers is the brain tumour, which can develop in any area of the brain. It is primarily described as aberrant cell proliferation in the brain. These aberrant cells can cause damage to healthy brain cells, resulting in brain dysfunction. There are several distinct forms of brain tumours. These tumours can be either malignant (cancerous) or benign (not cancerous). Detecting a brain tumour and correctly identifying its kind is not an easy process. CNN [1] outperforms the competition due to its widespread application in image recognition. It is essentially a collection of neurons with weights that may be learned. They are also noted for their exceptional precision and performance. Because of the noise and abnormalities in the picture, human's observation in predicting the tumour might be misleading. This drives our efforts to develop a tumour prediction algorithm. This contains methods for identifying tumours and categorising them as benign, malignant, or normal. A new horizon for radiology has opened up with the emergence of technologies to quantitatively analyse gliomas using computational methodologies. It is critical for radiologists to stay up to date on machine learning developments. The college of radiologists in New Zealand has recently updated its curriculum to include machine learning in the part I applied imaging technology examinations.

Quantitative analytic methods will complement the traditional visual study of pictures. This will enable statistical examination of characteristics that are not visible to the naked eye. Radiomics is rapidly evolving as a way of forecasting survival durations using imaging parameters such as the shape of a region of interest. With the advancement of these approaches, the necessity for automatic segmentation has grown. Inconsistencies in the first and second authors' blinded hand segmentation of brain tumours are considered. The Srensen–Dice coefficient, which was determined using the StudierFenster calculator, is a measure of picture segmentation consistency (available).The result obtained from the first and second authors' segmentation was 0.91, demonstrating the disparity in hand segmentation. Convolutional neural networks (CNNs) [2, 3] work on the principle of human brain and it is a machine learning method. Machine learning is rapidly evolving, with increasing representation at major conferences. Radiologists require an educated viewpoint. This research strikes a unique mix between teaching and a comprehensive review of convolutional neural networks in glioblastoma.

2 Related Work

This section focuses on the background analysis that was done in this domain. As there are variety and complexity of tumours, detecting MRI brain tumour pictures is a tough process. This study introduces two detection techniques: the first is edge detection and segmentation, and the second is artificial neural network proficiency. The proposed strategy for brain tumour identification and segmentation is more accurate and successful in this study [4]. First, while all interscale correlations were statistically significant, they were modest, indicating that the scales were measuring different aspects of the quality of life concept [5]. Due to the variety and complexity of tumours, detecting MRI brain tumour pictures is a tough process. This study introduces two detection techniques: the first is edge detection and segmentation, and the second is artificial neural network proficiency [6, 7]. The data set collected of ePROs through the cancer clinics that gave the monitoring of patent care a survey of the validated symptoms with 78 questions [8].

Patients who are diagnosed with cancer frequently experience uncertainty and a lack of control over their circumstances, which has a poor impact on their health outcomes. Patients' quality of life is further harmed by cancer therapy. Patients frequently rely on their doctors for social/interpersonal, informational, and decisional support during their cancer experience. An increasing amount of evidence suggests that doctors' communication style has a favourable influence on patient health outcomes. As a result, the patient–physician contact is extremely important in the delivery of cancer care. It is great to see that cancer researchers are paying attention to research in this field, which is generally dominated by primary care studies. A review of significant data tying physician conduct to cancer patient health outcomes follows a discussion of several techniques to evaluate physician behaviour [9, 10]. Finally, the shortcomings of the existing work possible shortcomings are mentioned, as well as opportunities for future research.

Alternative approaches have been used to diagnose brain tumours, including pre-trained models, different designs of convolutional neural networks, and ensemble models that combine many models. The existing methods had issues with noises such as light fluctuations, blurring, and occlusion, and some of the existing systems failed to identify real time due to limited data sets.

3 Proposed Method

The proposed system uses convolutional neural networks to diagnose brain tumour which can handle scalability of images through the architecture including input layer, convolution layer, rectified linear unit (ReLU), pooling layer, and fully linked layer.

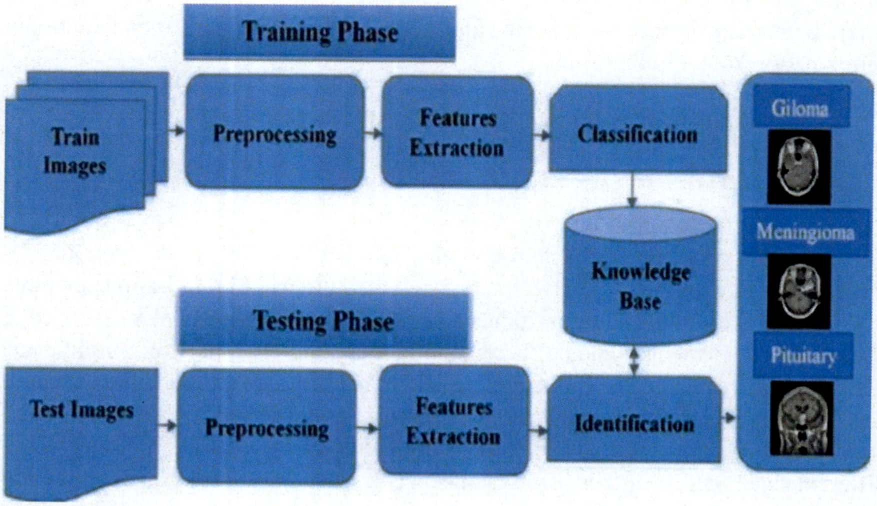

Fig. 1 Architecture of the proposed system

The architecture of the proposed approach is specified in Fig. 1. During the training phase, the images from the data set will be pre-processed to remove noise and outliers. Next step is to extract the features from pre-processed images and then perform classification of images using convolutional neural network.

Convolutional neural network is a deep learning neural network which is mainly used for image processing and classification. CNNs are feedforward networks in that information flow takes place in one direction only, from their inputs to their outputs. It is an algorithm which takes an image and is able to differentiate one from another with minimal pre-processing compared to other classification algorithms. Automatic detection of features without any human supervision is the main advantage of CNN compared to others. CNN architecture [11, 12] is built by using three types of layers: convolutional layer, pooling layer, and fully connected layer. A convolutional layer can be followed by additional convolutional and pooling layers, and the final layer is a fully connected layer. These layers are stacked together to form a deep model. The convolution layer divides the supplied input image into smaller parts. The ReLU [13] layer activates each element individually. The pooling layer is optional. The network architecture consists of fully connected layer to compute the scores for class, label value, which is based on a probability values ranging from 0 to 1.

Convolutional layer acts as a feature extractor to extract the features from the input image. It contains learnable filters called kernels, which is a matrix of integers (trainable weights). The filter shifts by a stride throughout the image and performs a dot product with that portion of the image on which the filter is hovering in order to produce a feature map. Various categories of feature maps in the same layer of convolutional network contain different weights, and at each location, several features are extracted. In order to reduce the dimensionality of the feature maps by selecting the best features, a pooling layer is used. In the pooling layer, the pooling

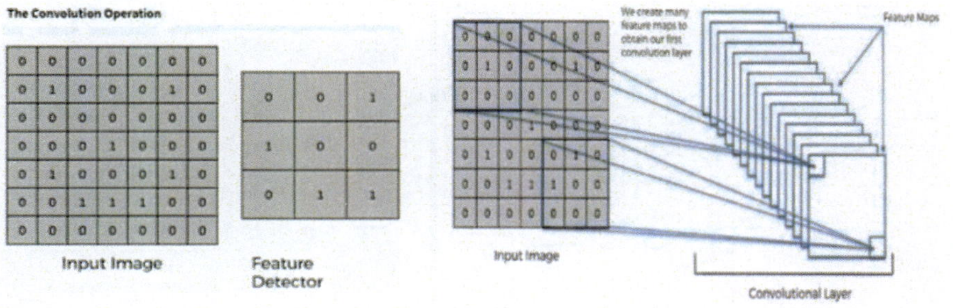

Fig. 2 Convolution matrix representation of input image

operation will sweep the filter throughout the entire input but it does not contain any weights like convolution layer. The filter applies an aggregation function to the values in respective fields and produces an output array. Fully connected layer [14] with softmax, sigmoid activation function is used for image classifications. Softmax activation function uses probability distributions to classify the images (Fig. 2).

The training of convolution neural network specified in Fig. 3 is divided into two stages such as forward propagation and backpropagation. During forward propagation, the sample x and its label y are extracted where x will be the input given to the network and the dimension of y will be 7 and specified as vector. The output of the previous layer will be the input to the current one. Activation function is applied to calculate the output which will be passed to the layers at lower level. At last, the model finds the output of softmax layer. After completing the forward propagation, the error between the output y and the softmax layer will be calculated and propagated back. Based on the error value, weight adjustment takes place. MobileNet model is used which works same as the convolution network to apply the image filters but the depth of the convolution varies from the normal representation. Rectified linear unit (ReLU) function is used which has a derivative function and allows for backpropagation while simultaneously making it computationally efficient. The neurons will only be deactivated if the output of the linear transformation is less than 0.

4 Experiments

4.1 Data set

The data set used in the proposed system is brain tumour MRI scan images an open-source data set publicly shared in Kaggle. The data set consists of 3264 images, out of which, 2860 are used for training and the rest for testing. The images of both training and testing are demonstrated into four classes. Class-1 is glioma_tumour, class-2 is meningioma_tumour, class-3 is pituitary_tumour, and class-4 is no tumour.

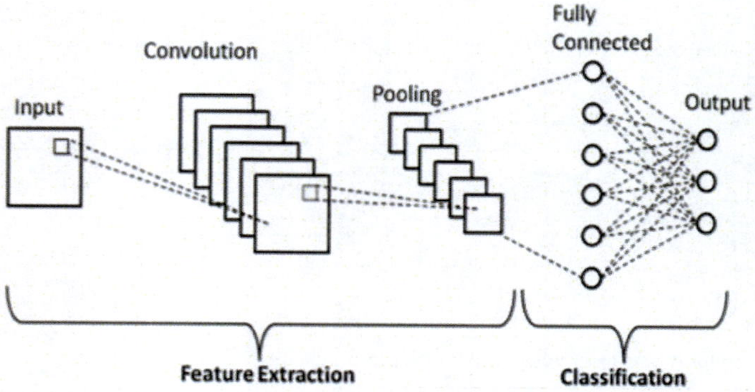

Fig. 3 Convolutional neural network architecture

Table describes the number of images used for training and testing. Training and test data description are specified in Table 1.

Collection of a data set consisting of images. (In this case, the data set is brain tumour MRI scan images an open-source data set of 3264 images, greyscale images of brain each labelled with one of 4 classes such as glioma_tumour, meningioma_tumour, pituitary_tumour, and no tumour) [15]. Experimentation was carried out using python libraries in GoogleColab. The image data set is pre-processed using function in ImageDataGenerator() and classification is performed using the CNN model with 3 layers such as convolution, pooling, and dense. Model fitting is performed by calling mode.fit_generator() with parameters train data set and setting epochs to 35. This model is validated on test data set during training. During training, the phases such as forward and backward propagation will be performed on the pixel values. After the model is trained, evaluation of model on test data is performed. The trained model predicts the classes for the test data. A test run of the system is performed to remove defects before implementing the new system activity or capability. Figure 4 shows the sample input and output images. Table 2 gives the evaluation metrics considered to measure the performance of the model. Comparison of various existing models along with proposed model is given in Table 3. The proposed CNN-based MobileNet model shown accuracy of 96.6% which is better when compared to other models as specified in Table 3.

Table 1 Training and testing data set images

Training data set		Test data set	
Tumour type	No. of images	Tumour type	No. of images
Glioma	826	Glioma	100
Meningioma	822	Meningioma	115
Pituitary	827	Pituitary	74
No tumour	395	No tumour	105

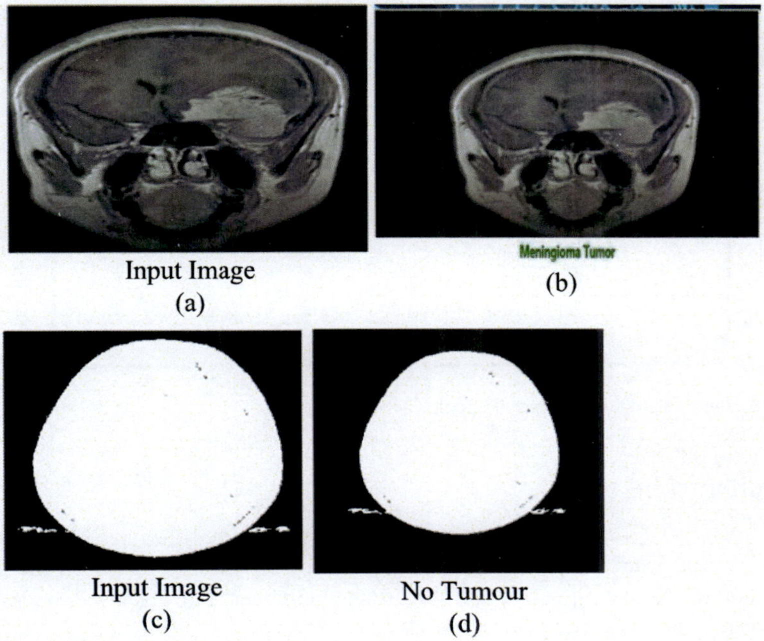

Fig. 4 Input and classified images

Table2 Evaluation metrics

Precision	TP/(TP + FP)
Recall	TP/(TP + FN)
F-measure	(2*precision*recall)/(precision + recall)
Accuracy	(TP + TN)/(TP + TN + FP + FN)

Table 3 Comparison of existing methods

Features	Model	Accuracy (%)
Model based	CapsNet	86.56
Model based	CNN	84.19
CNN	NN	91.90
CNN	MobileNet	96.6

The accuracy of the proposed approach based on number of epochs is given in Fig. 5.

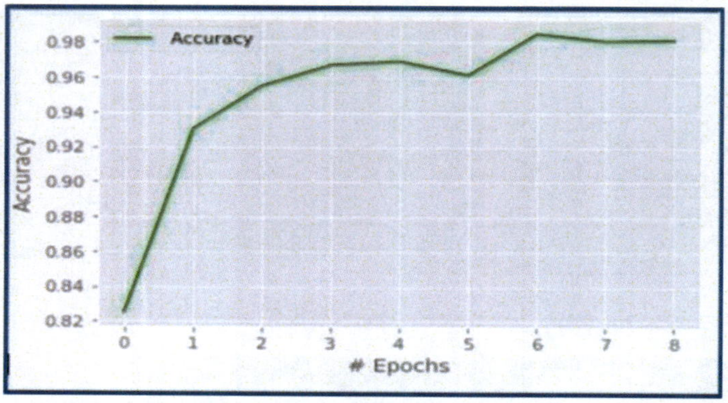

Fig. 5 Accuracy based on number of epochs

5 Conclusion

The main aim of the proposed work is to detect brain tumour based on the given image data set consisting of patients MRI scanned images. The proposed model uses convolutional neural network with MobileNet to classify the images. This model even performs well in detecting glioma tumour, meningioma tumour, pituitary tumour, and non-tumour scans among the opted images (MRI scans). To categorise the tumours, image enhancing methods, a CNN model, and a softmax classifier were used to achieve a 96.6% accuracy which is better compared to existing methods. Future work is to identify and use optimum deep learning network architecture and also to expand the model to detect other types of tumours.

References

1. Vijayakumar T (2019) Neural network analysis for tumor investigation and cancer prediction. Journal of Electronics 1(02): 89–98. https://doi.org/10.36548/jes.2019.2.004
2. Hassan M, DeRosa MC (2020) Recent advances in cancer early detection and diagnosis: role of nucleic acid based aptasensors. TrAC, Trends Anal Chem 124:115806. https://doi.org/10.1016/j.trac.2020.115806
3. Pandian P (2019) Identification and classification of cancer cells using capsule network with pathological images. Journal of Artificial Intelligence and Capsule Networks 01(01): 37–44. https://doi.org/10.36548/jaicn.2019.1.005
4. Siegel RL, Miller KD, Jemal A (2017) Cancer statistics, 2017. CA: A Cancer Journal for Clinicians 67(1): 7–30. https://doi.org/10.3322/caac.21387
5. Razzak MI, Imran M, Xu G (2019) Efficient brain tumor segmentation with multiscale two-pathway-group conventional neural networks. IEEE J Biomed Health Inform 23(5):1911–1919. https://doi.org/10.1109/jbhi.2018.2874033
6. Khan HA, Jue W, Mushtaq M, Mushtaq MU (2020) Brain tumor classification in MRI image using convolutional neural network. Math Biosci Eng 17(5):6203–6216

7. Alzubaidi L, Zhang J, Humaidi AJ et al (2021) Review of deep learning: concepts, CNN architectures, challenges, applications, future directions. J Big Data 8:53. https://doi.org/10.1186/s40537-021-00444-8
8. El Hamdaoui H, Benfares A, Boujraf S, Chaoui NEH, Alami B, Maaroufi M, Qjidaa H (2021) High precision brain tumor classification model based on deep transfer learning and stacking concepts. Indonesian Journal of Electrical Engineering and Computer Science 24(1):167–177
9. Gunasekara SR, Kaldera HNTK, Dissanayake MB (2020) A feasibility study for deep learning based automated brain tumor segmentation using magnetic resonance images. arXiv preprint arXiv:2012.11952
10. Yu H, Yang LT, Zhang Q, Armstrong D, Deen MJ (2021) Convolutional neural networks for medical image analysis: state-of-the-art, comparisons, improvement and perspectives. Neurocomputing 444:92–110
11. Amin J, Sharif M, Haldorai A, Yasmin M, Nayak RS (2021) Brain tumor detection and classification using machine learning: a comprehensive survey. Complex & Intelligent Systems. https://doi.org/10.1007/s40747-021-00563-y
12. A. Sinha, Annesh RP, Nazneen S (2021) Brain tumour detection using deep learning. In: 2021 seventh international conference on bio signals, images, and instrumentation (ICBSII), pp 1–5. https://doi.org/10.1109/ICBSII51839.2021.9445185
13. Gajja M (2020) Brain tumor detection using mask R-CNN. Journal of Advanced Research in Dynamical and Control Systems 12(SP8):101–108. https://doi.org/10.5373/jardcs/v12sp8/20202506
14. Shivdikar A, Shirke M, Vodnala I, Upadhaya J (2022) Brain tumor detection using deep learning. International Journal for Research in Applied Science and Engineering Technology 10(3):621–627. https://doi.org/10.22214/ijraset.2022.40710
15. Kuraparthi S et al (2021) Brain tumor classification of MRI images using deep convolutional neural network. Traitement du Signal 38(4):1171–1179. https://doi.org/10.18280/ts.380428.J

An Improved Blind Deconvolution for Restoration of Blurred Images Using Ringing Removal Processing

U. M. Fernandes Dimlo, Jonnadula Narasimharao, Bagam Laxmaiah, E. Srinath, D. Sandhya Rani, Sandhyarani, and Voruganti Naresh Kumar

Abstract One of the most difficult challenges in image processing is restoring a defocused image by reducing blur and noise. Blurring characterizes image deterioration, and recovery is accomplished by point spread function estimation and ideal image estimation processing was repeated. Ringing, or wavelike artifacts that arise along strong edges, is a difficult challenge in latent image restoration. Therefore, this paper will introduce an improved blind deconvolution for restoring blurry images using ringing removal process. This paper provides an improved deconvolution technique that uses blur kernel prediction based on dark channels before achieving clear image recovery. To hold picture information and beautify the rims of the ringing impact created at some point of the authentic clean picture recuperation process, an easy bilateral clear out is used. The ringing removal method L_0 regularization is used with the restoration process, which can estimate a sharper image. By removing the difference map from the final deconvolution result, it is possible to get a clearer

U. M. F. Dimlo
Department of CSE, Sreyas Institute of Engineering and Technology, Hyderabad, Telangana, India
e-mail: umfernandesdimlo@gmail.com

J. Narasimharao (✉) · B. Laxmaiah · D. S. Rani · V. N. Kumar
Department of CSE, CMR Technical Campus, Hyderabad, Telangana, India
e-mail: jonnadula.narasimharao@gmail.com

B. Laxmaiah
e-mail: laxmaiah.cse@cmrtc.ac.in

D. S. Rani
e-mail: davu.sandhya@gmail.com

V. N. Kumar
e-mail: nareshkumar99890@gmail.com

E. Srinath
Department of CSE, Keshav Memorial Institute of Technology, UGC Autonomous, Hyderabad, Telangana, India
e-mail: eslavath.sri77@gmail.com

Sandhyarani
Department of CSE (Data Science), CMR Technical Campus, Hyderabad, Telangana, India
e-mail: apsandhya21@gmail.com

K. A. Reddy et al. (eds.), *Proceedings of Fourth International Conference on Computer and Communication Technologies*, Lecture Notes in Networks and Systems 606,
https://doi.org/10.1007/978-981-19-8563-8_34

picture without ringing. Finally, the results are presented in terms of performance parameters such as signal-to-noise ratio (SNR), mean squared error (MSE), and peak signal-to-noise ratio (PSNR).The results show that the performance parameters of the improved blind deconvolution model are superior compared to existing image blur removal algorithms.

Keywords Blur image · Blind deconvolution · Ringing effect · Bilateral filter · Kernel estimation

1 Introduction

Digital photographs are utilized in a lot of fields, together with medical, military, transportation, microscopy imaging, and images deblurring, among others. The recorded picture is a noise and blurred model of the unique picture. Images are affected by blurring and noise in various areas of applied science.

Blurring is a problem created by an imaging system (caused, for example, by diffraction, aberrations, etc.), whereas noise is a part of the detecting process. As a result, picture deconvolution is essentially a post-processing of the recognized image with the target of decreasing blur and noise.

Convolution, which is frequently associated with the band-limited nature of acquisition technology, and contamination by additive Gaussian noise, which may be attributable to the electronics of the recording and transmission processes, is well-known sources of signal/image degradation in many practical situations. For example, the blur effect in remote sensing images is caused by the limited aperture of satellite cameras, optical system, and mechanical vibrations. A blurred image is created by convolution a sharp image with a blur kernel or point spread function (PSF). To get the crisp image, first extract the blur kernel from the sharp image. On the other hand, the problem is estimation of the blur kernel. Deconvolution is the estimation of an unknown blur kernel.

These concepts are used by the majority of deblurring techniques. A data restoration process is frequently required before any further processing to remove these artifacts. Many papers have been written on the deconvolution of noisy signals [1]. Inverse problems related to practical interest are often badly encountered so it is difficult to devise appropriate deconvolution methods. Deconvolution techniques are a computationally intensive image processing technique that is widely used to improve digital image contrast and resolution [2]. The basis of deconvolution is mainly designed using a set of methods to remove blurring of an image. Therefore, the deconvolution method is often recommended as a good choice to reduce the effects of visual blurring of captured images. In addition, image processing using a resolution technique offers an advantage in cases where images are captured using a pin hole aperture [3].

2 Literature Survey

Various works have been done on deblurring algorithms on different platforms and with different assumptions. Some related works are discussed below. Satoshi Motohashi [4] was responsible for the gradient reliability map (Rmap). This paper describes the implementation of a new algorithm based on a two-step blind deployment process. The latent image recovery phase employs an overall variation sort and a bump filter to reduce texture components and noise while improving edges. After that, the gradient reliability map is used to reduce the margins, which has a significant impact on the PSF rating. In this paper, the author [5] uses the frequency domain for image analysis because of the convenience of transforming the blur model between the spatial domain and the transmission domain. They first applied a Butterworth bandpass filter to the image to avoid signal noise. They use layer-specific data as an image before getting better results. To generate the latent image, this algorithm selects the true values of the frequency components from the training image. The latent image is built by combining the bandpass components of the training image. As a result, this technique provides high image accuracy while also dealing with blurred images lacking high-frequency detail. Although is algorithm has a limitation that it can only blur images containing only single objects, in practice that is assumed for a week.

In this paper, author [6] uses patch-based image priorities learned from a set of clean images in a particular class. Use a weak blur precursor to restore the various filters. Use a denoiser based on a Gaussian Mixed model (GMM). The proposed method is also incorporated into the alternating direction method of multipliers (ADMM) optimization algorithm for estimating both images and blur filters. This algorithm has several advantages which can process noisy text images and can be used for various blur filters. This algorithm also has limitations on the internal ADMM algorithm and the setting of regularization parameters and stop criteria for external iterations.

The author [7] proposes a new image based on elastic net regularization of singular values computed from similar image sections in this paper. They altered the algorithm to account for non-uniform blurring. The deblurring model is built on the missed approach point (MAP) framework. To choose the leading edge, the proposed method does not require a sophisticated filtering strategy. After finding the blur kernel, a set of non-blind deconvolution methods based on iteratively reweighted least squares (IRLS) can be used to estimate the latent picture. If the image has rich textures and are located in most regions of this rich texture, this method will fail. Yang et al. [8] proposed a novel blind deblurring algorithm for predicting the blur kernel. In a clear image, the color distribution of the edges is more distinct than in a blurred image. Filters are recommended to clear the edges of blurred image which is used as a reference image.

Marapareddy [9] worked on Wiener filtering to restore blurred images that were degraded due to complex environments. First, determine the pattern of atmospheric turbulence degradation. After applying the inverted filtering and the minimum average square error, i.e., the wiener filtering, the blurry images are restored. Some authors have considered ringing issues and sought to minimize deconvolution scheme artifacts. Liu et al. [10] invented the ring detection method design and build the pyramid at various scales of the restored image and compute the gradient difference between each level of the pyramid. Such artificial ringing detectors can only be used to assign the quality of blurry images. It is not directly involved in efforts to produce blurry images free of artifacts. The original ringing artifacts are eliminated by applying a residual multi-scale deconvolution approach to the edge-preserving bilateral filter and the traditional reinforcement learning (RL) algorithm.

3 Improved Blind Deconvolution Using Ringing Removal Processing

This paper gives a stepped forward blind blur elimination set of rules primarily based totally on darkish channels, in addition to a bilateral clear out shared with the unique set of rules, to eliminate ringing and generate a blur removal image. For example, ringing is effective in suppressing deterioration for the gradient's prior probability. As a result, only gradient information is used to estimate ideal images. This method compares the previously estimated ideal image with the image obtained using gradient information and a two-sided filter. The following Fig. 1 shows steps used for improved blind deconvolution using ringing removal process.

3.1 Image Blind Deblurring

The pixels of the remote sensing pictures are uniformly blurred due to the jitter and blur of the distant sensor. This is known as the mathematically clear picture and the blur kernel noise convolution. It can be expressed as:

$$b = k * x + n \tag{1}$$

where b denotes a blurred image, x denotes a sharpened image, k denotes a blur kernel, and n denotes noise. The blind image cloning algorithm works in two steps. The first step is to evaluate the blur kernel, followed by image decoding to produce a clearer image.

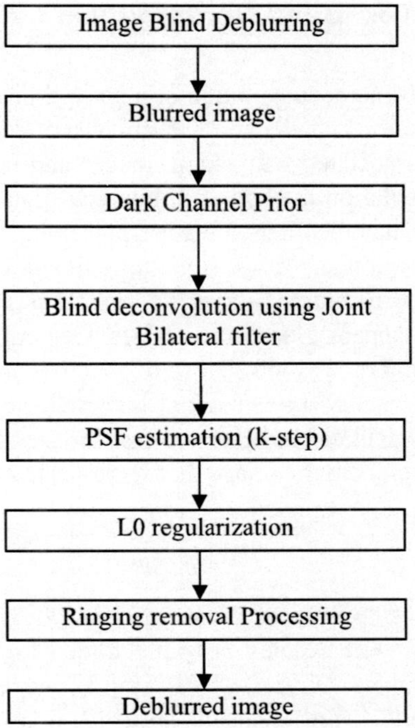

Fig. 1 Steps used for improved blind deconvolution using ringing removal process

3.2 *Dark Channel Prior*

The dark channels of a fog-free outdoor image have almost zero pixels, and the dark channels were previously applied to the image defog problem. Intuitively, the blurring process replaces the pixel values of very dark pixels with the weighted average of other bright pixels nearby, increasing the size of the very dark pixels. As a result, dark channel preferences can be used to enhance the dark channels potential for sharper images. The dark channel in image is defined as:

$$D(I)(X) = \min_{y \in N(x)} \left\{ \min_{c \in N(x)} I_C(y) \right. \tag{2}$$

If x and y are pixel positions, then $N(x)$ is the x-centered image point and I_C is the cth color channel. The dark channel solution will compute the smallest red, green, and blue (RGB) component per pixel, save it in the same grayscale image as the original image, and then scale it using Eq. (2).The size of the pop-up area determines the filter radius in the value filter.

3.3 Blind Deconvolution Using Joint Bilateral Filter

Blind deconvolution of a blurred image means deconvolution of the image without prior knowledge of the blur kernel and white Gaussian noise. The cost functions associated with sharp images and blur kernels must be determined by the deconvolution method. The bilateral filter is an advanced version of the Gaussian filter. Bilateral filters are edge-preserving filters with weights determined by the spatial region and range smoothing function. However, this method has its drawbacks due to the blurred edges. The weight of this bilateral clear out is unstable, and inversion happens close to the edges. General bilateral filtering, which is based on bilateral filters, was introduced to preserve image details while improving image edge functionality. In contrast to bilateral filtering, the values of a typical bilateral filter are calculated based on the input image rather than the improved image. The guide image refers to the image that serves as the filter's guide information.

$$\text{JBF}[I, D]_m = \frac{1}{k_p} \sum_{n=\Omega(x)} I_n f(\|m - n\|) g(\|D_m - D_n\|) \tag{3}$$

F represents the spatial filter, g represents the distance filter, I represents the input image, D represents the guide image, Ω represents the spatial support of kernel f, and k_p represents the normalization coefficient.

3.4 PSF Estimation

The problem of blind image recovery can be divided into two parts: Calculating the PSF from the degraded **image** (k-step) and **the best image** from the PSF (x-step). A different method is used to repair the damaged image. This blind image reconstruction process employs a deconvolution based fast reconstruction method. By regularizing overall variation, X-step reduces the effects of noise and improves edges with shock filters and ideal images. Also, Rmap only the strong edge component remains in the k-step. The process is then repeated in a series of PSF calculations using the estimated ideal image's derivative thresholding and the conjugate gradient method. The PSF obtained from the iteration is used for the final deconvolution. For error detection and prevention, the power value is calculated (4). As this value increases, the estimated PSF threshold changes. When the objective function converges and the energy value decreases, the reconstruction is successful.

$$e = \frac{|b - x \times k|^2}{\omega \times h} \tag{4}$$

where $x \times k$ is the estimated blurred image and w and h are the horizontal and vertical pixel counts of an image.

3.5 *Ringing Removal Processing*

The regularized probability L_0 of the image is used in this method to remove blur. Based on the prior probabilities of the pixel values and the prior probability of the regular pixel value, L_0 calculates the image's pre-probability $P(x)$ (5).

$$P(x) = \sigma P_t(x) + P_t(\nabla x) \tag{5}$$

$P_t(x)$ represents the previous probability of the pixel value and $P_t(\nabla x)$ represents the gradient's previous probability. Prior probabilities for gradients can help you control deterioration like ringing. As a result, while reducing ringing, set to $\sigma = 0$ and estimate the ideal image represented by using only the gradient information (6).

$$x = \mathcal{F}^{-1}\left(\frac{\overline{\mathcal{F}(k)}\mathcal{F}(b) + \beta\mathcal{F}(u) + \mu\mathcal{F}_G}{\overline{\mathcal{F}(k)} + \beta + \mu\overline{\mathcal{F}(\nabla)}\mathcal{F}(\nabla)}\right) \tag{6}$$

where u, β, and μ denote auxiliary variables, and $\mathcal{F}(\bullet)$ and $\mathcal{F}^{-1}(\bullet)$ are fast Fourier transform (FFT) and inverse FFT, respectively. $\overline{\mathcal{F}(\bullet)}$ denotes a complex conjugate operator which is expressed by FG (7).

$$\mathcal{F}_G = \overline{\mathcal{F}(\nabla_h)}\mathcal{F}(g_h) + \overline{\mathcal{F}(\nabla_u)}\mathcal{F}(g_u) \tag{7}$$

In (5), ∇_h and ∇_u are horizontal and vertical component operators, and g is a gradient auxiliary variable.

4 Results

The images were evaluated objectively for signal-to-noise ratio (SNR), mean squared error (MSE), and peak signal-to-noise ratio (PSNR). SNR is a simple metric used to assess the effectiveness of noise reduction techniques. Higher signal-to-noise ratios are regarded as a sign of effective noise reduction. The SNR is given as

$$\text{SNR(dB)} = 20\log\frac{\text{RMS signal}}{\text{RMS Noise}} \tag{8}$$

MSE is a metric used to assess denoising accuracy. Lower MSE values indicate that the noise reduction signal is more similar to the original signal. This is thought to result in better noise reduction. The MSE is given as

$$\text{MSE(dB)} = \frac{1}{mn}\sum_{i=0}^{m-1}\sum_{j=0}^{n-1}[1(i-j) - k(i-j)]^2 \tag{9}$$

PSNR is a metric similar to SNR, with higher values indicating more accurate noise reduction. The PSNR is given as,

$$\mathrm{PSNR(dB)} = 10\log_{10}\frac{\max j^2}{\mathrm{MSE}} \tag{10}$$

The results of objective evaluation are presented in Figs. 2 and 3. It shows the SNR and PSNR analysis of image at different steps of improved blind decovolution in Fig. 2. If it can be seen that the SNR has PSNR which has high values at after blind deconvolution compared to the stages at known PSF and in out blur image.

Figure 3 shows the MSE analysis of image at different steps of improved blind decovolution. It can be seen that the MSE has lower values at after blind deconvolution compared to the stages at known PSF and in out blur image.

Here, the performance of the image ringing removal scheme is evaluated. A series of blurry images was used for this purpose. Blurred images and their point spread function (PSF) pairs are used. An image containing motion blur (handshake) was captured by the camera. The PSF of the image was estimated using the blind deconvolution approach. Blurred images were deblurred by applying an improved blind deconvolution using a ringing removal process. The deblurring results are shown in

Fig. 2 Analysis of SNR and PSNR of image at different steps

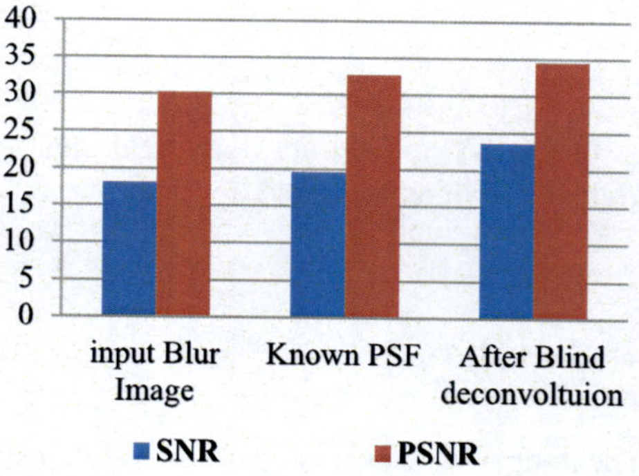

Fig. 3 MSE analysis of image at different steps

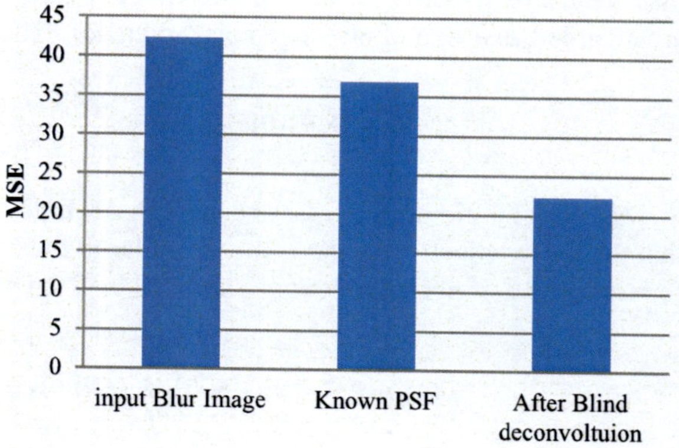

Fig. 4a–c. Input parameters for the deblurring algorithm such as rule weights and smoothing factors are chosen in such a way that they do not produce overly sensitive and cartoonish results. The deconvolution scheme, as seen in these figures, produces ringing artifacts, as shown in Fig. 4a.

To identify the artifacts generated during the deconvolution stage, the blurred image was subjected to a ringing removal process. It is used a filter to remove the ringing artifacts. The Gaussian parameters were determined during the ringing detection step. Because of the symmetry of the PSF Fourier transform, half of the detected minimum points are ignored. This cuts down on the number of filters needed during detection. Figure 4b illustrates the ringing artifact detection results. For example, the marked ring mask is superimposed on a yellow-colored blurred image. The algorithm identifies almost all ringing areas in the blurred image. Blind deconvolution was used to estimate the PSF used in the image deblurring process.

(a) DEBLURRING IMAGE AFTER DECONVOLUTION

(b) DETECTED RINGING REGIONS

(c) IMAGE AFTER RINGING REMOVAL

Fig. 4 Image deconvolution with ringing removal effect

5 Conclusion

This paper implemented an improved dark channel before image blur method with blind deconvolution and restored the image with a ringing removal process that targets the ringing effect of the image blur. Use common bilateral filtering during restoration to reduce ringing of the restored image, save edges more effectively, and enhance the image restoration effect. The signal-to-noise ratio (SNR), peak signal-to-noise ratio (PSNR), and mean squared error (MSE) of the performance parameters are calculated and graphically displayed. According to simulation results, when compared to non-blind deconvolution, the maximum signal-to-noise ratio and noise-to-signal ratio are higher, indicating that the signal information is higher, and the mean squared error is lower, indicating a lower amount of error. According to experimental results, this algorithm effectively eliminates motion blur in an image. The results obtained with this method show that the blind deconvolution method is superior. The simulation results show that the blind deconvolution technique performs better when reconstructing an image from an out-of-focus image.

References

1. Cheng L, Wei H (2020) An image deblurring method based on improved dark channel prior. J Phys: Conf Ser 1627(1):012017
2. Xu X, Zheng H, Zhang F, Li H, Zhang M (2020) Poisson image restoration via transformed network. Journal of Shanghai Jiaotong University (Science) 1–12
3. Kanwal N, Pérez-Bueno F, Schmidt A, Molina R, Engan K (2022) The devil is in the details: whole slide image acquisition and processing for artifacts detection, color variation, and data augmentation. A review. IEEE Access
4. Shamshad F, Ahmed A (2020) Class-specific blind deconvolutional phase retrieval under a generative prior. arXiv preprint arXiv:2002.12578
5. Barani S, Poornapushpakala S, Subramoniam M, Vijayashree T, Sudheera K (2022) Analysis on image restoration of ancient paintings. In: 2022 international conference on advances in computing, communication and applied informatics (ACCAI). IEEE, pp 1–8
6. Sarbas CHS, Rahiman VA (2019) Deblurring of low light images using light-streak and dark channel. In: 2019 4th international conference on electrical, electronics, communication, computer technologies and optimization techniques (ICEECCOT). IEEE, pp 111–117
7. Wang H, Pan J, Su Z, Lianga S (2017) Blind image deblurring using elastic-net based rank priors. In: Computer vision and image understanding, Elsevier, pp 157–171
8. Yang F-W, Lin HJ, H Chuang HJ (2017) Image deblurring, IEEE smart world, ubiquitous intelligence and computing, advanced and trusted computed, scalable computing and communications, cloud and big data computing, internet of people and smart city innovation
9. Marapareddy R (2017) Restoration of blurred images using wiener filtering. International Journal of Electrical, Electronics and Data Communication
10. Liu Y, Wang J, Cho S, Finkelstein A, Rusinkiewicz S (2013) A no-reference metric for evaluating the quality of motion deblurring. ACM Transactions on Graphics (SIGGRAPH Asia)

A Review on Deep Learning Approaches for Histopathology Breast Cancer Classification

Rathlavath Kalavathi and M. Swamy Das

Abstract Deep learning (DL) is the most rapidly expanding in the current scenario. For image analysis and categorization, deep neural networks (DNNs) are presently the most extensively utilized technology. DNN designs include GoogleNet, residual networks, and AlexNet, among others. Breast cancer is seen as a major problem that endangers the lives and health of women. Ultrasonography or MRI scanning methods are used to diagnose breast cancer disease. Imaging methods used for diagnosis include digital mammography, ultrasonography, magnetic resonance imaging, and infrared thermography. The primary objective is to investigate different deep learning algorithms for recognizing breast cancer-affected imageries. The best models provide accuracy for the 2, 4, and classifications on cancer datasets. No previous research is carried out for the current model investigation. Early detection and screening are critical for effective therapy. The following is a synopsis of recent progress in mammograms and identification, as well as a discussion of technological advancements. An effective test result should meet the following requirements: performance, sensitivity, specificity, precision, recall, and low cost. The experimental settings for every study on breast cancer histopathology images are thoroughly reviewed and deliberated in this article.

Keywords Medical images · Deep learning · Optimization · Breast cancer · Classification

R. Kalavathi (✉)
Research Scholar, Department of Computer Science and Engineering, Osmania University, Hyderabad, India
e-mail: rathodkalavathi005@gmail.com

M. Swamy Das
Department of Computer Science and Engineering, Chaitanya Bharati Institute of Technology, Hyderabad, India
e-mail: msdas_cse@cbit.ac.in

K. A. Reddy et al. (eds.), *Proceedings of Fourth International Conference on Computer and Communication Technologies*, Lecture Notes in Networks and Systems 606,
https://doi.org/10.1007/978-981-19-8563-8_35

1 Introduction

As per the note of National Cancer Institute (NCI), women are facing breast cancer problems [1]. It is envisaged that all the advanced cases of breast cancer would be recognized and cured timely. It plays a vital part in the indicative process; it is also hoary for differentiating between malignant and benign tissues, separating them as in situ and invasive carcinoma [2].

Hematoxylin and Eosin were used to stain tissue samples (H&E). Pathologists will then evaluate the samples using light microscopy. However, due to the complexity of the visible structures, and photographic estimation of tissue microstructure, the overall arrangement of centers in histological pictures takes time and is too specific. As a result, computer-assisted diagnostic methods that work automatically are critical for minimizing expert labor by enhancing diagnostic efficiency and minimizing subjectivity in illness categorization [3]. Many approaches for object detection in medical diagnostics have been developed. Deep learning-based approaches have recently been proven to outperform traditional machine learning methods in several image analysis tasks. Most of the image processing using DL methods has shown promise in the detection of breast cancer [4–7].

The volume and size of medical datasets are continually rising, yet the majority of these data are not evaluated for important and hidden knowledge. Extracted data and correlations can be discovered using powerful data mining algorithms [8].

Simulations derived from these approaches can help healthcare professionals make sound judgments. Breast cancer is viewed as a severe danger to the health and lives of women. Breast cancer is one of the most frequent kinds of cancer in women all over the world [9]. Mammography produces high-quality pictures of the breast's interior architecture. Breast cancer can be detected with mammograms, and out of one with architectural deformities and the other with macrocalcifications.

It is very critical for detecting primary tumors, although architectural aberrations are less relevant specifically related to masses and MCs [10]. Various authors have recently developed ML algorithms for diagnosing breast abnormalities in mammography imageries. Singh and Gutte [11] developed a group categorization based on common polling. On the Wisconsin breast cancer dataset (WBCD), many ML algorithms were proposed to identify and classify the cancer data and were evaluated with an accurateness of 99.42%. The [12] employed photograph handing out to eliminate the pectoral muscle from the digital mammogram database for the mammographic image analysis society (MIAS) [13] and the digital mammogram dream challenge dataset [14]. The characteristics were retrieved and identified by the researchers using conventional and multiple classifications based on statistical measures. The maximum achievable accuracy was 99.7% [12, 13]. To categorize the MIAS dataset samples, [15] employed Fourier analysis, PCA, and SVM. The achieved accuracy was 92.16%. Furthermore, certain articles, such as Refs. [16–21], acknowledged conventional CAD systems that utilized ML approaches.

2 Breast Cancer Approaches

2.1 *Self and Clinical Examinations and Mammography*

Screening for breast cancer is done using breast self-examination (BSE) and clinical breast examination (CBE). The sensitivity of CBE is 57.14%, and the specificity is 97.11% [22]. Although it cannot be used to identify cancer with certainty, it can be used to detect worrisome breast lesions. Reference [23] discovered no difference in breast cancer death tolls between those who were tested with BSE and CBE and those who were not, despite the fact that persons who were screened had twice quite so many biopsies. Other studies indicate that many professors and healthcare professionals, i.e., people who influence young women, are either uneducated or unable to perform BSE properly [24]. In one study, 99% of nurses felt able to conduct a BSE, but 26% performed BSE every month [25]. The methods BSE and CBE are very much useful in screening cancer. The sensitivity and specificity of the cancer are influenced by the parameters such as age, HRTs, BMI, menstrual phase, and genetics [26, 27]. The research results from women who utilized HRTs provided a mammographic specificity of around 91.7% [27]. Mammography is less sensitive in women who have thick radiographic breasts. Sensitivity ranges from 62.9% in highly dense-breasted women to 87% in extremely fatty-breasted women, while specificity ranges from 89.1 to 96.9% [27].

Using the leukemia dataset, Ref. [28] employed the Bayesian model for feature selection and then used ANN, KNN, and SVM classifiers. In 2004, the researchers [29] employed uncorrelated linear discriminant analysis (ULDA) for feature selection and found that it outperformed previous approaches in terms of classifier accuracy. The authors [30] used SVM-RFE to choose features and a kernel-based fuzzy technique to classify them. Reference [31] employed the subset information gain strategy for feature selection in 2012, repeatedly gaining an informative gene subset with the subset merge and split procedure. Reference [32] employed a discriminant kernel-based classifier with ANOVA, a statistical technique for feature selection. Reference [33] used slide photographs to diagnose metastatic breast cancer using a deep learning technique. Reference [34] used deep belief networks to construct a breast cancer classification model with 99.68% accuracy. Skin infections are a fairly prevalent type of infection; yet, they are difficult to identify and forecast their targets. To categorize skin illnesses, [35] presented a deep learning technique. Reference [36] suggested a method for identifying and diagnosing cancer kinds based on unsupervised feature learning. They employed deep learning to extract characteristics automatically by merging different forms of cancer gene expression data. The majority of the offered techniques see feature selection as a pre-classification activity. Reference [37] suggested a hybrid method for feature selection that combines correlation and optimization approaches. They tested their method on multi-class benchmark gene expression cancer datasets including MLL, Lymphoma, and SRBCT. Reference [38] developed an architecture for detecting and visualizing basal cell carcinoma. To achieve balanced accuracy, they applied fivefold cross-validation procedures on the BCC dataset.

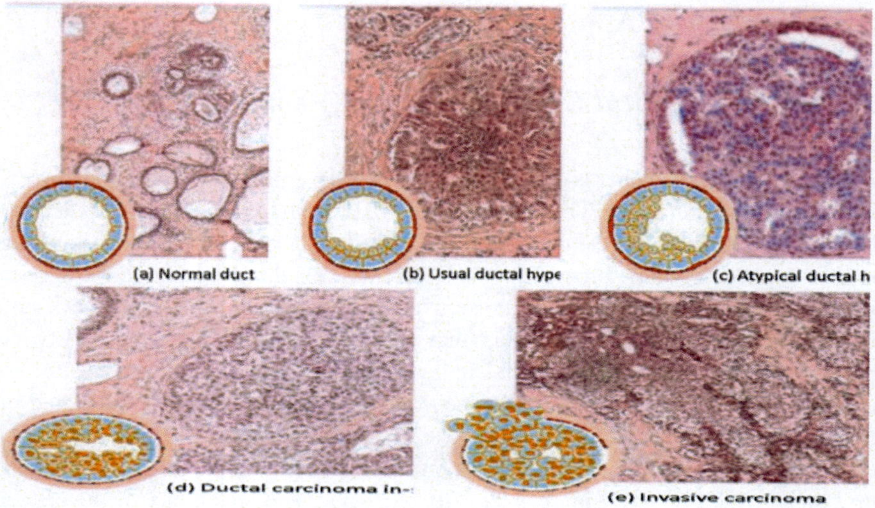

Fig. 1 Number of stages of breast cancer

The following is how the paper is organized: The third section presents a breast cancer outline, then datasets, augmentation, preprocessing, and a few approaches have been described in Sect. 4.

3 Breast Cancer and Types

Although there are around 20 primary kinds of breast cancer, the bulk of them may be divided into two histological classes: Invasive Ductal Carcinoma (IDC) and Invasive Lobular Carcinoma (ILC) [38, 39]. Researchers are focusing more on IDC than on the other two kinds of breast cancer. The various stages of breast cancer are seen in the following Fig. 1 such as (a) normal duct, (b) usual ductal hyperplasia, (c) atypical hyperplasia, (d) ductal carcinoma in situ, and (e) invasive cancer.

4 Datasets and Processing Methods

4.1 Databases

- Natural databases
  - ImageNet
  - Object-centric database

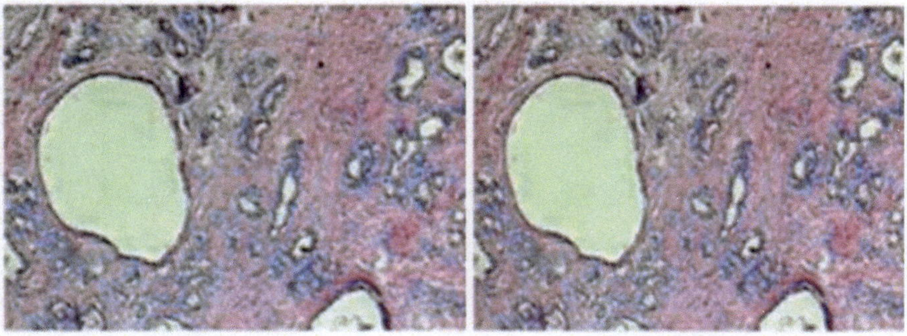

Original Image Normalized image

Fig. 2 Histological image stain normalization

- Pathology datasets
  - Cancer Metastases in Lymph Nodes (Camelyon)
  - Breast Cancer Histopathological Image Classification (BreakHis)
  - Bio-Image Semantic Query User Environment (BISQUE)
  - Tissue Microarray (TMA) from Stanford
  - Breast cancer histopathology (BACH)

4.2 Data Augmentation

Cropping, rotation, color change, flipping, translation, and intensity are data augmentation procedures used in breast cancer histopathology.

4.3 Preprocessing

The preprocessing methods such as resizing, rebalancing the classes, normalization, image contrast enhancement, segmentation with the resolution, stain normalization, and color transformations among images have been used on the images of histopathology to report issues such as low perseverance and strident images (Fig. 2).

5 Comparative Analysis

This section compares previously published material on deep learning models for histopathology pictures, as indicated in Tables 1 and 2.

Table 1 Classification study of BreakHis database

Model	Dataset BreakHis	Data augmentation	Preprocessing	Extraction optimization	Accuracy (%)
Inception ResNet-V2	Even/total: 27,262	Clockwise rotation	Normalization	Feature extraction Adam	99.42
Inception ResNet-V2	Uneven/total: 7909	–	Normalization	Feature extraction Adam	97.12
Inception-V3	Uneven/total: 7909	–	Normalization	Feature extraction Adam	96.20
VGG-16	Even/total: 4960	Rotation, flip, shift, zoom	Undersampling SVMs	Last 3 layer, Adam	97.00
AlexNet	Uneven/total: 7909	–	–	Fine-tuned last 3 layers, SGD, and L2 regularization	90.50
VGG-M	Uneven/total: 7909	–	–	Feature extraction, SGD, and dropout 0.5	86.40
AlexNet, VGG-16, SVM	Uneven/total: 7909	–	–	Feature extraction, SGD, and L2 regularization	87.3
NASNet-A-Large	Uneven/total: 7909	Rotation, flip	Normalization	Fine-tuned last 3 layers, Adam, dropout 0.5	99.24
SENet-154	Uneven/total: 7909	Rotation, flip	Normalization	Fine-tuned last 3 layers, Adam, dropout 0.5	99.97
DualPathNet-131	Uneven/total: 7909	Rotation, flip	Normalization	Fine-tuned last 3 layers, Adam, dropout 0.5	99.74
Inception ResNet-V2	Uneven/total: 7909	Rotation, flip	Normalization	Fine-tuned last 3 layers, Adam, dropout 0.5	99.74

Table 2 Classification study of BACH database

Model	Dataset BACH	Data augmentation	Preprocessing	Extraction optimization	Accuracy (%)
Inception ResNet-V2	Even/total: 400	Random, rotation, flip, crop	–	Fine-tuned last 2 layers, SGD, dropout	87.00
Inception-V3	Even/total: 400	Horizontal and vertical flipping rotation	Normalization	Fine-tuned last 2 layers, SGD	97.08
NASNet-A-Large	Even/total: 400	Rotation, flip	Resizing, normalization	Feature extraction, Adam	97.5
SENet-154	Even/total: 400	Rotation, flip	Resizing, normalization	Feature extraction, Adam	97.5
DualPathNet-131	Even/total: 400	Rotation, flip	Resizing, normalization	Feature extraction, Adam	97.5
Inception ResNet-V2	Even/total: 400	Rotation, flip	Resizing, normalization	Feature extraction, Adam	97.5
ResNeXt-101	Even/total: 400	Rotation, flip	Resizing, normalization	Feature extraction, Adam	97.5

6 Conclusion

In conclusion, this research shows that when models are analyzed at different resolutions, different results are obtained. DL models are prone to low perseverance and high noise, according to this distinction. As a result, dealing with high-resolution and high-quality breast cancer histopathology images is crucial. One of the challenges is gathering high-resolution photographs through cutting-edge scanners and data storage. As a consequence, the researchers should study and investigate the performance of the DL models after super-resolution techniques. Future research should focus on evaluating the performance of deep learning models after using models to analyze pathology images.

References

1. Eastland TY (2017) Prostate cancer screening in the African American community: the female impact
2. Tasnim Z, Shamrat FMJM, Islam MS, Rahman MT, Aronya BS, Muna JN, Billah MM (2021) Classification of breast cancer cell images using multiple convolution neural network architectures. International Journal of Advanced Computer Science and Applications 12(9)

3. Vesal S, Ravikumar N, Davari AA, Ellmann S, Maier A (2018)Classification of breast cancer histology images using transfer learning, image analysis and recognition. Springer, Cham
4. Robertson S, Azizpour H, Smith K, Hartman J (2017) Digital image analysis in breast pathology—from image processing techniques to artificial intelligence. Translational Research 1931–5244
5. Javed S, Mahmood A, Ullah I, Bouwmans T, Khonji M, Dias JMM, Werghi N (2022) A novel algorithm based on a common subspace fusion for visual object tracking. IEEE Access 10:24690–24703
6. Hwang Y, Cho E, Park N (2022) Development of teaching-learning contents for AI core principles at the elementary school level: with a focus on convolutional neural network. Webology 19(1)
7. Jurisica I (2022) Integrative computational biology, AI, and radiomics: building explainable models by integration of imaging, omics, and clinical data. In: Artificial intelligence/machine learning in nuclear medicine and hybrid imaging. Springer, Cham, pp 171–189
8. Vibert F, Martel C, Ionescu RA, Mathelin C, Ame S (2022) A new modality for breast cancer diagnosis during the COVID-19 pandemic: a case report. European Journal of Breast Health 18(1):91
9. Abed GA, Wahab SDA, Elamrosy SH, Hamied MMA (2020) Effect of breast cancer on psychological status among breast cancer patients. International Journal of Novel Research in Healthcare and Nursing 7(2):393–402
10. Dabass J (2020) Pectoral muscle and breast density segmentation using modified region growing and K-means clustering algorithm. In: Data communication and networks. Springer, Singapore, pp 331–339
11. Singh A, Gutte V (2022) Classification of breast tumor using ensemble learning. In: Mobile computing and sustainable informatics. Springer, Singapore, pp 491–507
12. Alfi IA, Rahman MM, Shorfuzzaman M, Nazir A (2022) A non-invasive interpretable diagnosis of melanoma skin cancer using deep learning and ensemble stacking of machine learning models. Diagnostics 12(3):726
13. Masek M, Christopher JS, Attikiouzel Y (2003) Automatic breast orientation in mediolateral oblique view mammograms. In: Digital mammography: IWDM 2002—6th international workshop on digital mammography. Springer, p 207
14. Digital mammography dream challenge dataset, n.d. https://www.synapse.org/#!Synapse:syn4224222/wiki/401743. Accessed 3 Nov 2019
15. Jia W, Jiang Y (2017) Comparison of detection methods based on computer vision and machine learning. In: 2017 international conference on mechanical, electronic, control and automation engineering (MECAE 2017). Atlantis Press, pp 386–390
16. Boryczko K, Kurdziel M, Yuenb DA (2007) Detecting clusters of microcalcifications in high-resolution mammograms using support vector machines. Poland: Institute of Computer Science, USA: Minnesota Supercomputing Institute
17. Walia H, Kaur P (2021) A quantitative analysis for breast cancer prediction using artificial neural network and support vector machine. In: International conference on soft computing and signal processing. Springer, Singapore, pp 59–82
18. Bacha S, Abdellafou KB, Aljuhani A, Taouali O, Liouane N (2022) Early detection of digital mammogram using kernel extreme learning machine. Concurrency and Computation: Practice and Experience, e6971
19. Toprak A (2018) Extreme learning machine (elm)-based classification of benign and malignant cells in breast cancer. Medical Science Monitor: International Medical Journal of Experimental and Clinical Research 24:6537
20. Marinovich ML, Wylie E, Lotter W, Pearce A, Carter SM, Lund H, Waddell A et al (2022) Artificial intelligence (AI) to enhance breast cancer screening: protocol for population-based cohort study of cancer detection. BMJ Open 12(1):e054005
21. Javed R, Rahim MSM, Saba T, Sahar G, Awan MJ (2022) An accurate skin lesion classification using fused pigmented deep feature extraction method. In: Prognostic models in healthcare: AI and statistical approaches. Springer, Singapore, pp 47–78

22. Ratanachaikanont T (2005) Clinical breast examination and its relevance to the diagnosis of a palpable breast lesion. J Med Assoc Thai 88(4):505–507
23. Kosters JP, Gotzsche PC (2003) Regular self-examination or clinical examination for early detection of breast cancer. Cochrane Database of Systematic Reviews 2, Article ID CD003373
24. Amoah C, Somhlaba NZ, Addo F-M, Amoah VMK, Ansah EOA, Adjaottor ES, Amankwah GB, Amoah B (2021) A preliminary psychometric assessment of the attitude of health trainee undergraduate students towards breast-self examination in Ghana
25. Madubogwu CI, Madubogwu NU, Azuike EC (2021) Practice of breast self-examination among female students of Chukwuemeka Odumegwu Ojukwu University, Awka. Journal of Health Science Research 10–18
26. Hanis TM, Islam MA, Musa KI (2022) Diagnostic accuracy of machine learning models on mammography in breast cancer classification: a meta-analysis. Diagnostics 12(7):1643
27. Sadovsky R (2003) Factors affecting the accuracy of mammography screening. Am Fam Physician 68(6):1198
28. Dai X, Fu G, Reese R, Zhao S, Shang Z (2021) An approach of Bayesian variable selection for ultrahigh dimensional multivariate regression. Stat e476
29. Wang Z, Sun X, Sun L, Qian X (2013) Tissue classification using efficient local fisher discriminant analysis. Przegląd Elektrotechniczny 89(3b):113–115
30. Hernandez JCH, Duval B, Hao J-K, A counting technique based on SVM-RFE for selection and classification of microarray data. Advances in Computer Science and Engineering 99
31. Koul N, Manvi SS (2020) Ensemble feature selection from cancer gene expression data using mutual information and recursive feature elimination. In: 2020 third international conference on advances in electronics, computers and communications (ICAECC). IEEE, pp 1–6
32. Syafiandini AF, Wasito I, Mufidah R, Veritawati I, Budi I (2018) Prediction of breast cancer recurrence using modified kernel based data integration model. Journal of Theoretical and Applied Information Technology 96(16):5489–5498
33. Broadwater DR, Smith NE (2018) A fine-tuned inception v3 constitutional neural network (CNN) architecture accurately distinguishes between benign and malignant breast histology. 59 MDW San Antonio United States
34. Dandil E, Selvi AO, Çevik KK, Yildirim MS, Süleyman UZUN (2021) A hybrid method based on feature fusion for breast cancer classification using histopathological images. Avrupa Bilim ve Teknoloji Dergisi 29:129–137
35. Liao H (2016) A deep learning approach to universal skin disease classification, CSC 400-Graduate Problem Seminar-Project Report
36. Oh J (2020) Potential of disease prediction using deep learning algorithms. Science 5(4):283–286
37. Namwongse P, Limpiyakorn Y (2012) Learning Bayesian network to explore connectivity of risk factors in enterprise risk management. International Journal of Computer Science Issues (IJCSI) 9(2):61
38. Zavareh PH, Safayari A, Bolhasani H (2021) BCNet: a deep convolutional neural network for breast cancer grading. arXiv preprint arXiv:2107.05037
39. de Boo LW, Jóźwiak K, Joensuu H, Lindman H, Lauttia S, Opdam M, van Steenis C et al (2022) Adjuvant capecitabine-containing chemotherapy benefit and homologous recombination deficiency in early-stage triple-negative breast cancer patients. British Journal of Cancer 126(10):1401–1409

IoT-Based Smart Agricultural Monitoring System

Rama Devi Boddu, Prashanth Ragam, Sathwik Preetham Pendhota, Maina Goni, Sumanth Indrala, and Usha Rani Badavath

Abstract Agriculture is critical to the Indian economy and people's survival. The intention of this work is to build an embedded-based soil surveillance system and to assist farmers in identifying appropriate crops to plant on the soil. The pH value of the soil, temperature, and humidity level in the air all have an impact on crop output. Using the Node MCU ESP8266 and the ThingSpeak server, this architecture enables to decrease physical field monitoring and to receive information through mobile or laptop. The technique is designed to assist farmers in increasing their agricultural output. The soil is evaluated using a pH sensor, as well as the humidity content and temperature values are collected using a DH11 sensor. Depending on the values sensed, these parameters are fed into a machine learning technique called decision tree regression, which aids in accurately determining the crop that best suits the soil. Farmers can plant the optimum crop for the soil type.

Keywords ThingSpeak cloud server · Machine learning · Soil monitoring · Node MCU ESP8266 · Raspberry Pi · Decision tree regression

1 Introduction

Farming has been performed for centuries in every country. Agriculture is the science and skill of growing plants. Agriculture was a pivotal event in the evolution of sedentary human society. Agriculture was always done by hand. As the world moves toward new technologies and applications, agriculture must keep up. The Internet of Things (IoT) is essential in smart agriculture [1–8]. Sensors in the Internet of Things can collect data on agricultural lands. We proposed a solution for automated IoT and smart agriculture. Adequate soil moisture is required for proper plant structure and

R. D. Boddu (✉) · S. P. Pendhota · M. Goni · S. Indrala · U. R. Badavath
Department of ECE, Kakatiya Institute of Technology and Science, Warangal, Telangana 506015, India
e-mail: brd.ece@kitsw.ac.in

P. Ragam
School of Computer Science and Engineering, VIT-AP University, Vijayawada, Andhra Pradesh 522237, India

K. A. Reddy et al. (eds.), *Proceedings of Fourth International Conference on Computer and Communication Technologies*, Lecture Notes in Networks and Systems 606,
https://doi.org/10.1007/978-981-19-8563-8_36

high crop yields. Water acts not only as a moisture repellent, but also as a temperature regulator in the plant. During the process of thermo-regulation, the plant evaporates up to 99% of its total water content while using only 0.2–0.5% to build vegetable weight. As a result, it is effortless to see how a plant's humidity requirements vary depending on the climate and growth stage. Whenever the IoT-based farm monitoring system is activated, it runs a set of tests. A smart farm monitoring project based on the IoT, Raspberry Pi, and Node MCU is presented to enhance the efficiency of crop production and effectiveness. Agriculture provides a significant source of income for India's largest population and contributes significantly to the Indian economy. Crop improvement has been minimal in the agricultural industry over the last decade. Food prices have risen steadily as crop yields have declined. A variety of factors, including water, contributed to this. The fundamental purpose of the Internet of Things is to ensure that the appropriate information is sent to the appropriate persons at the proper time. Hence, IoT integrated with agriculture gives an excellent solution and with addition machine learning algorithm decision tree regression will end this problem. The cultivation of the suitable crop for the soil is becoming more difficult to the humans due to the either atmospheric conditions or the instability of standard pH value of the soil, but by using the decision tree regression algorithm, it is quite easy for the farmers to grow the exact crop that matches the soil.

2 Types of Sensors

2.1 DHT11 Sensor

The "DHT11" is a temperature and humidity sensor. This sensor is widely used in many of the applications due to its accuracy and simple architecture. DHT11 sensor senses the humidity content in the air and temperature. The sensor has a specialized NTC for temperature measurement as well as a 8-bit microcontroller for serial output data of temperature and humidity values. After calibration, the sensor is ready to connect to other microcontrollers (Fig. 1).

Fig. 1 DHT11 sensor

Fig. 2 pH sensor

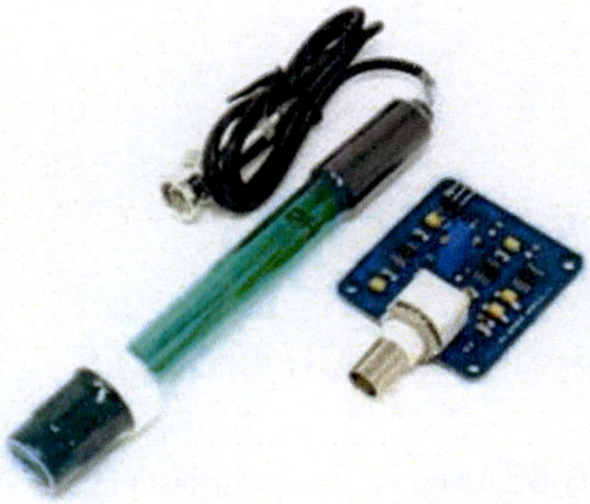

2.2 *pH Sensor*

The concentration of hydrogen ions in the soil is measured by a pH sensor. It is used to determine if a soil is acidic or alkaline. Water quality and other characteristics must be measured. Here is an analog pH meter with built-in connections and features that are easy, convenient, and useful. It comes with a power indicator LED, a BNC connector, and a "PH2.0" sensor interface. The pH sensor board requires an additional DC power supply of 9 V to sense the pH value of soil accurately. The pH sensor used in this project is an analog pH sensor which senses the values with great accuracy (Fig. 2).

3 Raspberry Pi 4, Model B

Raspberry Pi is a small computer that can run a variety of apps, when connected to regular monitors and peripherals. Traditional desktop operations like file creation, storage, and Internet streaming are available on Raspberry Pi models, which are barely larger than a credit card and include hardware components. The Raspberry Pi base contributes to the Linux kernel and other open source developments, as well as providing open source software for its own products (Fig. 3).

Fig. 3 Raspberry Pi 4 Model B

Fig. 4 Node MCU ESP8266

4 Node MCU ESP8266

The data can be transmitted via the Wi-Fi protocol utilizing the ESP8266-based Node MCU platform. Wi-Fi communication module: The ESP8266 is a low-cost Wi-Fi module that may be used with a UART serial connection to add Wi-Fi functionality. Among the features are the 802.11 b/g/n protocol, as well as an integrated TCP/IP protocol stack. Module for Wi-Fi Node MCU is a low-cost open source IoT platform. It came with firmware that operated on Espressif Systems' ESP8266 Wi-Fi SoC, as well as hardware based on the ESP-12 module (Fig. 4).

5 Decision Tree Regression Algorithm

A decision tree generates regression or classification models in the form of a tree structure. It decomposes a data set into smaller and smaller chunks while also creating a decision tree to go with it. The end result is a tree containing decision nodes and leaf nodes. A decision node can have two or more branches, each of which represents a value for the property under consideration. A leaf node represents a numerical target choice. The best predictor is the root node, which is the topmost decision node in a tree. Decision trees can handle both category and numerical data (Fig. 5).

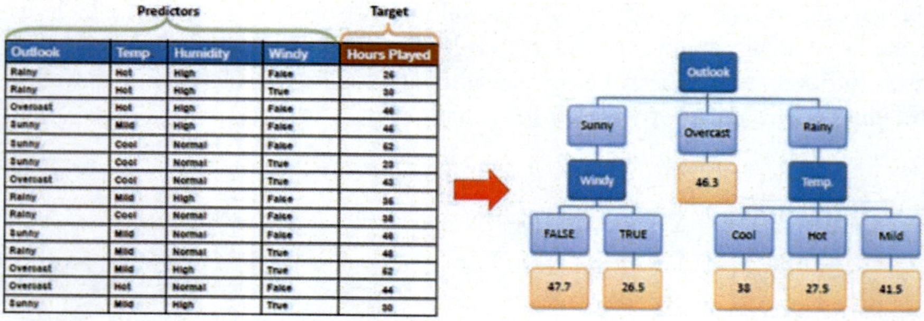

Fig. 5 Decision tree regression flowchart

Fig. 6 ThingSpeak visualization

6 ThingSpeak Server

ThingSpeak is an IoT open data platform and API that lets you gather, store, evaluate, monitor, and act on sensor data. ThingSpeak is a cloud-based platform that allows to combine, display, and study the data streams. ThingSpeak contains a range of useful capabilities, providing the ability to setup devices to submit data to it using standard IoT protocols. Real-time evaluation of sensor data (Fig. 6).

7 Flowchart

See Fig. 7.

8 Connection of Node MCU with pH Sensor and DHT11

pH sensor board requires an external voltage of 9 V DC power supply for accurate measurement of pH values and the output pins of pH sensor board connected to analog pin of Node MCU.DHT11 sensor consist of three pins namely 3 V, ground, DATA pins. These are connected to 3 V, GND, digital pin of the "Node MCU", respectively (Fig. 8).

9 Connection of ESP8266 with Raspberry Pi

Raspberry Pi requires power supply to run the device and Node MCU is activated by serial communication to the Raspberry Pi (Fig. 9).

Fig. 7 Flowchart

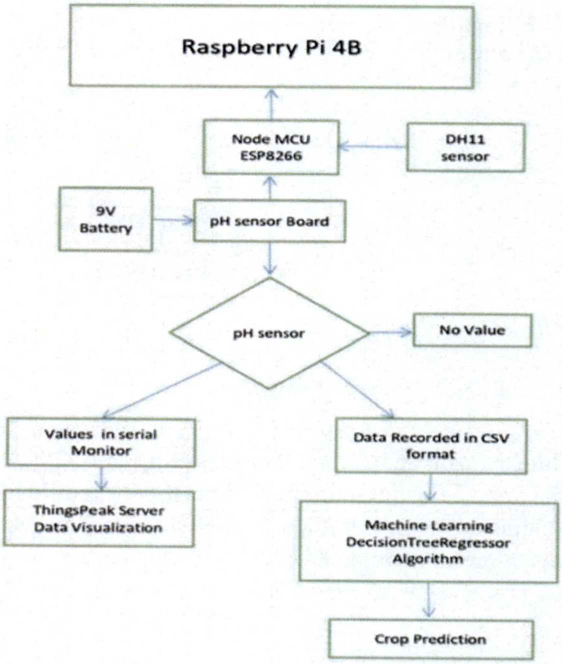

Fig. 8 Connection of Node MCU with pH and DHT11 sensors

10 Methodology

Firstly, to configure the Raspberry, we need to install the Raspberry Pi OS into a memory card, then insert it into Raspberry Pi and connect the Raspberry Pi to Internet. Then, by using IP address of Raspberry Pi, we can connect to VNC viewer and access Raspberry Pi. After the connections are given, installing the Arduino IDE software into Raspberry Pi and downloading the required libraries such as "ESP8266

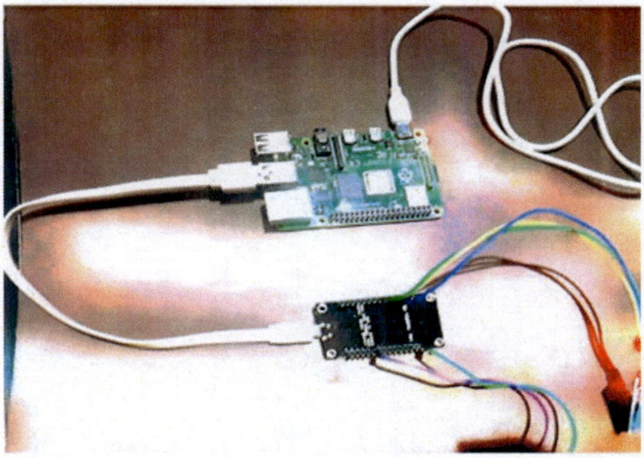

Fig. 9 Connection of ESP8266 with Raspberry Pi

Node MCU and DHT11" into it. Now by executing the code in Arduino IDE software, we can see the results in serial monitor.

We can use Raspberry Pi as a storage device to store the sensed values from the sensors. Then, these values undergo decision tree regression machine learning algorithm to find the exact crop that need to be grown on that soil with high accuracy.

11 Results

The pH, humidity, and temperature values of various fields are sensed. The resulted values were send to cloud with help of Node MCU. Data send to cloud useful to analyze the values in order to give best suitable crop to the soil. Then, the data set of pH, temperature, and humidity is exported in .CSV (comma separated values) format and trained with "decision tree regression" Machine learning algorithm to obtain the accuracy and ideal crop that is to be grown (Figs. 10, 11, 12).

12 Conclusion

We used a Raspberry Pi, a Node MCU ESP8266 (a Wi-Fi module), pH, and DH11 sensor in this IoT-based smart agriculture monitoring system. We will know the soil pH value, as well as the temperatures and humidity in a specific region, using this system, so that the irrigation system and fertilizer usage can be monitored and controlled. IoT is not restricted to a single application but may develop and explore new trends, and it is utilized in a variety of agricultural sectors to improve time efficiency, pest control, soil production management, and varied ways. This project reduces human effort while increasing crop yield. Farmers can benefit from this smart farming, which has a high level of precision.

	A	B	C
1	Temperature	Humidity	pH values
2	31.2	65	6.47
3	31.2	65	6.47
4	31.2	65	6.47
5	31.2	65	6.47
6	31.2	65	6.47
7	31.2	65	6.46
8	31.2	65	6.46
9	31.2	65	6.46
10	31.2	65	6.46
11	31.2	65	6.45
12	31.2	65	6.45
13	31.2	65	6.45
14	31.2	65	6.45
15	31.2	66	6.45
16	31.2	66	6.45
17	31.2	66	6.45
18	31.2	66	6.45
19	31.2	66	6.44
20	31.2	66	6.44
21	31.2	66	6.44
22	31.2	66	6.44
23	31.2	66	6.44
24	31.2	67	6.44
25	31.2	67	6.44
26	31.2	67	6.43
27	31.2	67	6.43
28	31.2	67	6.43
29	31.2	67	6.43
30	31.2	67	6.43

data_set1

Fig. 10 Data set of values from sensors

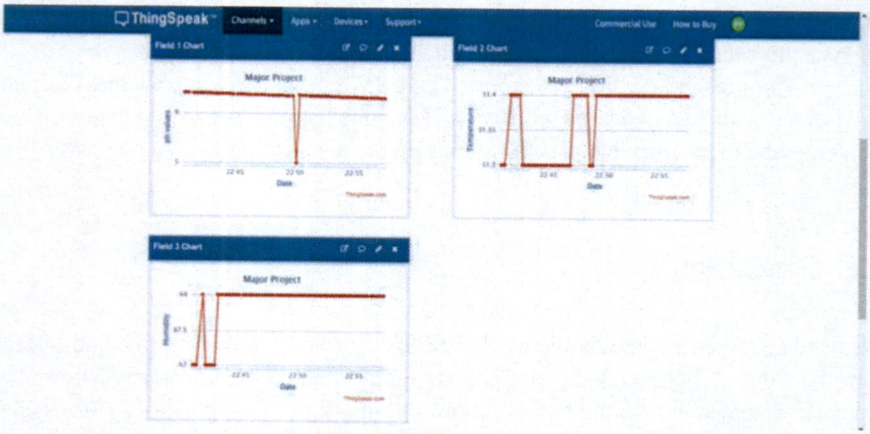

Fig. 11 Data visualization from ThingSpeak cloud server

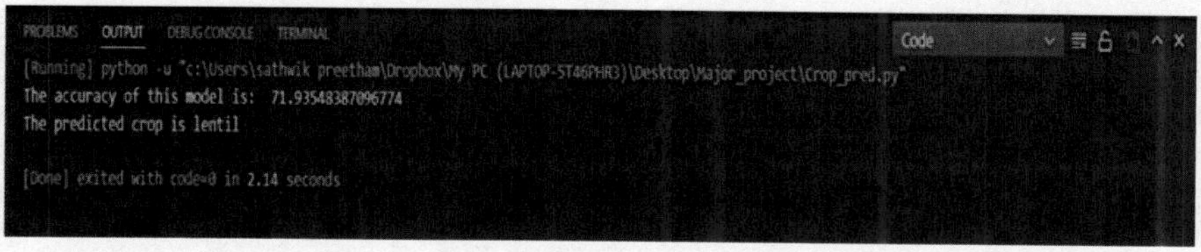

Fig. 12 Output after sensed values are subjected to decision tree regression algorithm

References

1. Sakthipriya N (2014) An effective method for crop monitoring using wireless sensor network. Middle-East J Sci Res 20(9):1127–1132
2. Hade AH, Sengupta DM (2014) Automatic control of drip irrigation system & monitoring of soil by wireless. IOSR Journal of Agriculture and Veterinary Science (IOSR-JAVS). e-ISSN, 2319–2380
3. Kuenzer C, Knauer K (2013) Remote sensing of rice crop areas. Int J Remote Sens 34(6):2101–2139
4. Sanjukumar RK (2013) Advance technique for soil moisture content based automatic motor pumping for agriculture land purpose. International Journal of VLSI and Embedded Systems 4:599–603
5. Giri M, Kulkarni P, Doshi A, Yendhe K, Raskar S (2014) Agricultural environmental sensing application using wireless sensor network. International Journal of Advanced Research in Computer Engineering & Technology (IJARCET) 3(3)
6. Ayaz M, Ammad-Uddin M, Sharif Z, Mansour A, Aggoune EHM (2019) Internet-of-Things (IoT)-based smart agriculture: toward making the fields talk. IEEE Access 7:129551–129583
7. Kurosu T, Fujita M, Chiba K (1995) Monitoring of rice crop growth from space using the ERS-1 C-band SAR. IEEE Trans Geosci Remote Sens 33(4):1092–1096
8. Chakraborty M, Manjunath KR, Panigrahy S, Kundu N, Parihar JS (2005) Rice crop parameter retrieval using multi-temporal, multi-incidence angle Radarsat SAR data. ISPRS J Photogramm Remote Sens 59(5):310–322

Singular Value Decomposition and Rivest–Shamir–Adleman Algorithm-Based Image Authentication Using Watermarking Technique

Y. Bhavani, Kiran Kumar Bejjanki, and T. Nagasai Anjani kumar

Abstract Digital watermarking is an approach where some information is embedded into the digital data like images, audio, video by its rightful owner. These days, most of the digital data is forged and copied by using various methods. Among this digital data, digital images are acting as the major subject for the modifications and malicious attacks. Digital watermarking can be used as a method to identify the unauthorized data and also for the copyright protection. The information embedded into the digital data by the owner is known as a watermark. Watermarking is helpful in the authentication of the image and to verify if the image has undergone any tampering. The watermark that is embedded needs to be robust and imperceptible against various attacks. In this paper, we are using Singular Value Decomposition technique (SVD), Rivest–Shamir–Adleman (RSA) encryption techniques for generating the robust watermarking technique which tolerates various image manipulation attacks and adds more strength to our authentication system.

Keywords Host image · Digital watermark · Singular Value Decomposition (SVD) · Rivest–Shamir–Adleman algorithm (RSA)

1 Introduction

Nowadays, the availability of digital data like image, audio, video, etc., has increased significantly. This data can be shared among different persons without losing its quality parameters. This exponential development of digital data has additionally led to a number of threats due to multimedia security controls, copyright protection, and critical content verification. In these days, huge amount of digital data is generated in

Y. Bhavani (✉) · K. K. Bejjanki · T. Nagasai Anjani kumar
Department of IT, Kakatiya Institute of Technology and Science, Warangal, India
e-mail: yerram.bh@gmail.com

T. Nagasai Anjani kumar
e-mail: b18it037@kitsw.ac.in

K. A. Reddy et al. (eds.), *Proceedings of Fourth International Conference on Computer and Communication Technologies*, Lecture Notes in Networks and Systems 606,
https://doi.org/10.1007/978-981-19-8563-8_37

real world. So, it is essential to handle issues related to privacy, security, and copyright protection. Copyright protection can be provided using watermarking technique.

Digital watermarking is a technique proposed for protecting the ownership rights of digitalized data by determining the original copyright owners of information contents. Digital watermarking integrates or embeds some information as the owner's name or logo in to a digital media. Thus, watermark information will serve as the identification mark of its owner. With the aid of this embedded watermark, we can suspect whether the data or an image is illegally edited and copied. Digital watermarking is the process of embedding other specific digital data like text, audio, or image into source content. The data embedded into source content is called as a watermark or can be termed as Label.

Digital watermark may be visible or invisible or fragile identification code that is embedded permanently in digital data. This watermark remains in the digital data even after the digital watermark is extracted by means of various decryption algorithms, so that rightful ownership of data remains solid at all times. Visual watermarks are those that are visible to the naked eye and are widely used to show image identity but invisible watermarks are not visible to the human eye.

2 Related Work

Abdulsattar [1] in proposed method gave a short overview of watermarking techniques including conversion techniques widely utilized in computerized signal processing systems. Abdallah et al. [2] had presented a novel process that uses a singular valued algorithm which determines a few features like durability, resistance, and reliability of the watermark image. He additionally delivered a mixture of SVD and a homomorphic approach with a block chain.

Liu and Tan [3] in their proposed work endorsed a watermarking scheme to discover and confine the interfered areas within the image that is extracted. Wong and Memom [4] proposed a scheme that can detect any changes done on the image and location that has been modified using secret key and public key versions.

Byun et al. [5] proposed a fragile watermarking method for image authentication, detection of modification in watermarked images. In this, image integrity is checked using singular values of SVD. Joseph and Anusudha [6] introduced watermarking technique based on DWT–SVD for improving the robustness and imperceptibility.

Prasad and Koliwad [7] elevated the technique with the help of a wavelet decay process referred to as Haar wavelet transform. Oktavia and Lee [8] proposed a technique that is used to detect modifications which are done on watermarked images. If any modifications are done, the extracted watermark will be changed.

Ramos et al. [9] used discrete wavelet transformations and digital signature techniques to add and extract watermark. The digital signature is either LSB of uncompressed image or header of compressed image. Hsu and Wu [10] worked on hidden watermarking images with the help of Discrete Cosine Transmission (DCT).

Kaewkamnerd and Rao [11] have developed a wavelet-based adaptive watermarking scheme. Embedding is done within the high-band, sub-bands of the wavelet transform, although this will obviously change the reliability of the image. Barni et al. [12] improved wave-based watermarking using pixel-wise masking. It is supported by a masking watermark that is compatible with HVS features. The watermark is added according to the most important bands.

Lakshmi Priya and Nelwin Raj [13] advised a brand-new set of rules that incorporates all the exceptional categories of homomorphic filtering, SVD and DWT. Voloshynovskiy et al. [14] presented a novel stochastic approach that involves with computing Noise Visibility Function (NVF) which is based on stationary or a non-stationary Gaussian model of cover image, with various combinations of watermarking techniques.

Parashar and Singh [15] verified the literature on existing digital watermarking techniques, and comparison is done on the basis of the outputs. Karla et al. [16] worked on digital watermarking algorithm involving two encryption methods. Fares et al. [17] suggested a novel substitution method, based on the Fourier transformation, for coloured image watermarking.

Our paper focuses on providing authenticity or ownership to images. We suggest a soft watermarking process using the same values in the image block to validate the image. The safety of this method depends on the double keys utilized in installation technique.

3 Proposed Watermarking Technique

The proposed SVD-based watermarking process is used in extracting effective features of cover image. SVD is a numerical analysis method used for diagonalizing the matrices. SVD is defined in linear algebra, as factorization of matrices into symmetric square matrices, whose diagonal values are Eigen vectors. To decompose a matrix into three different matrices, a mathematical tool SVD is used. The image is represented in the form of a matrix, which include eight-bit numbers of diverse sizes relying on the form of image. For example, a grayscale photo has a size of $1 \times M \times N$ and a colour image has a size of $3 \times M \times N$, wherein M, N represent height (quantity of rows) and width (quantity of columns) of the matrix, respectively. The following features of SVD are used in creating image display:

- Diagonal values in S are very stable, where there is a slight change within the singular values. As there will be no impact on the image pixels that appear, watermark details are regularly embedded by not affecting the host image view.
- Singular values in S are structured such that smaller values are found near the end of the matrix. Adding or updating those small values throughout the restoration stage has little impact on image quality. In addition, including new values to all positions in S additionally has a small impact on image quality.

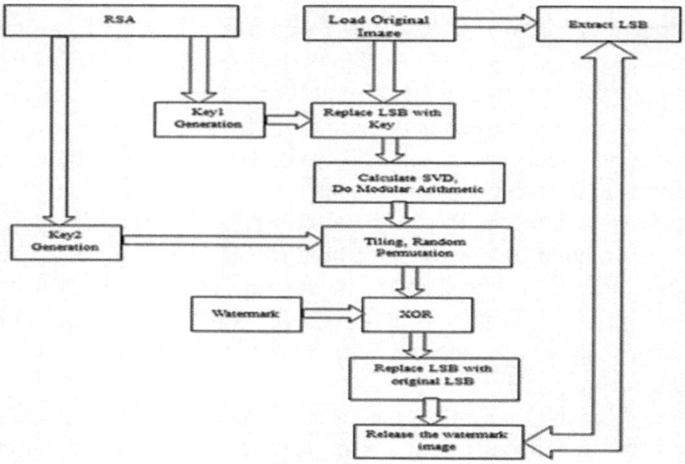

Fig. 1 Schematic representation

The proposed scheme has three phases: key creation, embedding the watermark image, and extraction of the watermark image. As depicted in Fig. 1, two keys (Key1 and Key2) that are generated using RSA algorithm are used in embedding and extracting of watermark image. This secrete key is used for initial conditions and parameters to produce a complex mapping system. This system is used to change the watermark before embedding process. This updated watermark will protect the real watermark from attack.

Authenticated data is generated by applying Exclusive-OR operation on binary bits found on singular image block and watermark image. The two keys generated by RSA algorithm used in embedding process and extraction process enhance the system security. This method improves durability and tolerates various attacks of image deception. The proposed system adds power to the authentication system.

3.1 Embedding of Watermark Image

Consider a picture A and picture W of pixel size n x n as an original and watermark image, respectively. Divide the first image A into blocks of $S \times S$. Each image is named as A_r, where $r = 1, 2, 3, \ldots, S^2$. Similarly, the watermark image is divided into "b" blocks of size $S \times S$ and each image is called W_r. The Key1 will be randomly generated by RSA algorithm whose size is equal to the size of a block pixel. For each A_r block,

1. Remove the LSB from the A_r image and replace LSB of A_r with Key1.
2. Obtain the singular values (SV's) of A_r. If A_r is a $n \times n$ sized matrix then, n singular values will be obtained. Namely S_r (where $r = 1, 2, \ldots, n$).

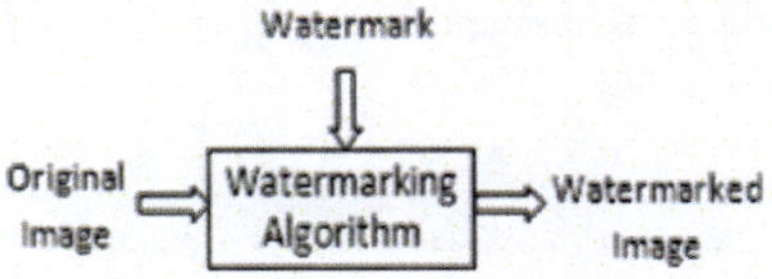

Fig. 2 Embedding watermark

3. Do round off operation after multiplication of S_r with a scalar α and apply modular operation to obtain binary bits ($S_r = \text{floor}(\alpha S_r)$, $B_r = S_r \bmod 2$).
4. Tile these binary bits within the dimensions of the block of image i.e., create a matrix B_r whose row contains B vectors.
5. Enable B_r with irregular permutation primarily dependent on Key2 which is encrypted and then apply XOR among B_r and W_r. ($X_r = B_r$ XOR W_r)
6. Insert validated information (X_r) in LSB of A_r to get block of image consisting of watermark.

By performing the above steps with complete blocks, we obtain an image as shown in Fig. 2.

3.2 Extraction of Watermark

Similar to embedding process, the extraction of the image is performed, besides that final XOR function is performed between B_r and X_r (LSB of the image block including watermark respective block). The detailed procedure is as follows:

Initially, split the image into blocks of 8 × 8 size each. Perform the following steps on each block:

1. Extract LSB of image block with watermark A_r^w (named as X_r^w).
2. Replace LSB of A_r^w with the Key1.
3. Obtain the singular values of block $A_r{}^w$. If A_r^w is $n \times n$ sized matrix then n singular values will be obtained. Namely S_r^w (where $r = 1, 2, \ldots, n$)
4. Do round off operation after multiplication of S_r^w with a scalar α and apply modular operation to obtain binary bits.

$$S_r^w = floor\left(\alpha S_r^w\right) B_r^w = S_r^w \text{mod2}.$$

5. Permutate the image block by random permutation based on Key2.
6. Apply XOR among B_r^w and X_r^w.

$$W_r = B_r^w \text{ XOR } X_r^w$$

By performing above procedure on complete blocks of an image with a watermark, we obtain a watermark image. By examining the extracted watermarked image, one can know whether the watermarked image has been corrupted or not. The extraction

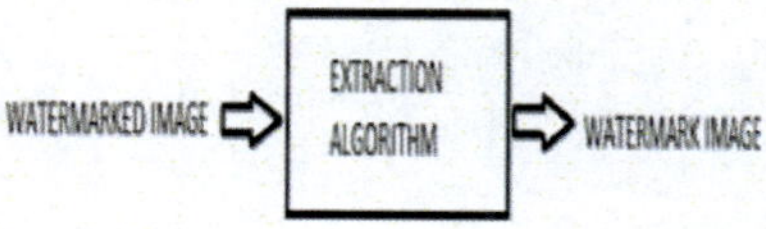

Fig. 3 Watermark image extraction

algorithm generates non-corrupted watermark image on submission of correct keys as shown in Fig. 3.

4 Experimentation and Results

We used MATLAB to implement the proposed algorithm. We considered the images shown in Figs. 4 and 5 as original and watermark images, respectively for experimentation. Each image we considered is of size 512 × 512. This experiment generates a high robust and high imperceptible watermark image.

Figure 6 shows the watermarked image which is obtained after the watermark image is embedded into original image. Figure 7 represents the watermark image after extracting from the image.

Figure 8 represents if the image has undergone any changes. The noise or sound indicates the area where the change was made to an image with a watermark. Therefore, in this case the image is considered uncertain. When the incorrect keys are used,

Fig. 4 Original image

Fig. 5 Watermark image

Fig. 6 Watermarked Image

Fig. 7 Extracted Watermark Image

Fig. 8 Modified watermark image

Fig. 9 Extracted watermark image with incorrect keys

the watermark that is extracted will be an image that represents noise, as shown in Fig. 9.

Gaussian Noise attack is used to compare the efficiency of our work. The proposed SVD–RSA-based watermarking technique is measured by using Normalization Coefficient (NC) given in Eq. 1.

$$NC = \frac{\sum_i \sum_j W_i(\text{i, j}).RW_i(\text{i, j})}{\sqrt{\sum_i \sum_j W_i(\text{i, j})}.\sqrt{\sum_i \sum_j RW_i(\text{i, j})}} \tag{1}$$

where W_i represents original watermark image and RW_i represents extracted watermark from the embedded image

The NC value ranges from 0 to 1, where NC value 0 indicates that the similarity of two images is low and 1 indicates that the similarity of two images is high. The NC values are calculated for different watermarking techniques and compared with our proposed technique. The comparison results are given in Fig. 10.

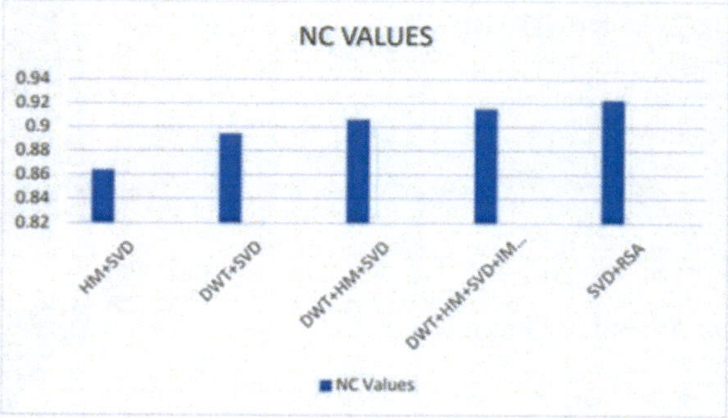

Fig. 10 Comparison bar graph for NC values for various watermarking techniques

4.1 *Security Analysis*

The proposed SVD and RSA based digital watermarking method is resistant to following attacks.

Active Attacks: The hacker attacks intentionally by removing the watermark from the original image or he can make it undetectable. The hacked image is critical for identification of the owner, proof of identity, etc. To overcome this attack, encryption is used in the proposed method.

Passive Attacks: Passive attacks are also intentional and detect the presence of watermark. The hacker hides the watermark without destroying it. In this work, we used SVD to overcome these attacks.

5 Conclusion

In current age of multimedia communication, authentication plays an important role. This is just a way to make sure that the information, which has been received, has been sent by a legitimate source or in other terms a way to exchange data securely with none modification within the contents of the particular information by an unauthorized source. We suggest a delicate watermarking method using SVD of the block of image to validate it. With the aid of this system, we will confirm the image authenticity. The safety of this system depends about the double keys utilized in the inserting procedure. To improve privacy, double-key encryption was performed using the RSA algorithm. This method will find the areas where the changes took place. This technique improves the robustness, tolerates against various image manipulation attacks, and adds strength to the authentication system.

References

1. Abdulsattar FS (2012) Robust digital watermarking technique for satellite images. J Eng Dev 16(2):133–143
2. Abdallah HA, Ghazy RA, Kasban H, Faragallah OS, Shaalan AA, Hadhoud MM (2014) Homomorphic image watermarking with a singular value decomposition algorithm. Inf Process Manage 50(6):909–923
3. Liu R, Tan T (2002) An SVD-based watermarking scheme for protecting rightful ownership. IEEE Trans Multimedia 4(1):121–128
4. Wong P, Memon N (2001) Secret and public key image watermarking schemes for image authentication and ownership verification. IEEE Trans Image Processing 10(10):593–1601
5. Byun S, Lee S, Tewfik A, Ahn B (2002) A SVD-based fragile watermarking scheme for image authentication. In: International workshop on digital watermarking, pp 170–178
6. Joseph A, Anusudha K (2014) Singular value decomposition based wavelet domain watermarking. In: International conference on computer communication and informatics, Coimbatore, pp 1–5
7. Prasad RM, Koliwad S (2010) A robust wavelet-based watermarking scheme for copyright protection of digital images. In: Second international conference on computing, communication and networking technologies, pp 1–9
8. Oktavia V, Lee WH (2004) A fragile watermarking technique for image authentication using singular value decomposition. In: Advances in multimedia information processing. Lecture notes in computer science, pp 42–49
9. Ramos CC, Reyes RR, Miyatake MN, Meana HMP (2011) Watermarking-based image authentication system in the discrete wavelet transform domain. Discrete Wavelet Transforms-Algorithms and Applications
10. Hsu C-T, Wu J-L (1999) Hidden digital watermarks in images. IEEE Trans Image Process 8(1):58–68
11. Kaewamnerd N, Rao KR (2000) Wavelet based image adaptive watermarking scheme. Electron Lett 36(4):312–313
12. Barni M, Bartolini F, Piva A (2001) Improved wavelet-based watermarking through pixel-wise masking. IEEE Trans Image Process 10(5):783–791
13. Lakshmi Priya CV, Nelwin Raj NR (2017) Digital watermarking scheme for image authentication. In: International conference on communication and signal processing (ICCSP), pp 2026–2030
14. Voloshynovskiy S, Herrigel A, Baumgaertner N, Pun T (2000) A stochastic approach to content adaptive digital image watermarking. In: International workshop on information hiding. pp 211–236
15. Parashar P, Singh RK (2014) A survey: digital image watermarking techniques. International Journal of Signal Processing, Image Processing and Pattern Recognition 7(6):111–124
16. Kalra GS, Talwar R, Sadawarti H (2015) Adaptive digital image watermarking for color images in frequency domain. Multimedia Tools and Applications 74(17):6849–6869
17. Fares K, Amine K, Salah E (2020) A robust blind color image watermarking based on Fourier transform domain

Crop Yield Prediction Using Machine Learning Algorithms

Boddu Rama Devi, Prashanth Ragam, Sruthi Priya Godishala, Venkat Sai Kedari Nath Gandham, Ganesh Panuganti, and Sharvani Sharma Annavajjula

Abstract Agriculture is the most crucial aspect in ensuring survival. Climate and other environmental changes have become a significant threat to agriculture. Estimating the crop yield before the harvest would assist farmers in choosing marketing and storage strategies. Machine learning algorithms are used for developing practical and efficient solutions to predict the yield. Historical data, such as rainfall, temperature, fertilizer, and past crop yield data, are used to predict crop yield. This paper focuses mostly on estimating yield by utilizing a variety of machine learning methods. The models utilized here are ensemble XGBoost-RF, gradient boosting, random forest, and XGBoost out of which ensemble XGBoost-RF showed maximum accuracy with the R^2 of 0.976111 and MSE of 0.002163.

Keywords Crop yield · Gradient boosting · Random forest · XGBoost · Ensemble XGBoost-RF

1 Introduction

Agriculture is very significant in India's economy. Advancement in agriculture is essential to fulfilling the demands of a country like India, which has an ever-increasing food demand due to its growing population. In India, agriculture is a primary occupation since antiquity. The agriculture field is slowly declining since the development of new creative technology and procedures. People have been concentrating on producing artificial items, which are hybrid products, which leads to unhealthy living. Farmers used to estimate their crop yield based on the previous year's yield results. Thus, different methodologies or algorithms exist for this type of data analytics in yield prediction, and yield can predict using such algorithms to help farmers. These

B. Rama Devi (✉) · S. P. Godishala · V. S. K. N. Gandham · G. Panuganti · S. S. Annavajjula
Department of ECE, Kakatiya Instistute of Technology and Science, Warangal 506015, Telangana, India
e-mail: brd.ece@kitsw.ac.in

P. Ragam
School of Computer Science and Engineering, VIT-AP University, Vijayawada 522237, Andhra Pradesh, India

K. A. Reddy et al. (eds.), *Proceedings of Fourth International Conference on Computer and Communication Technologies*, Lecture Notes in Networks and Systems 606,
https://doi.org/10.1007/978-981-19-8563-8_38

algorithms help to predict the crop yield, which is a better way than using excessive hybrid products to increase the crop. This work emphasizes crop yield prediction with the help of machine learning (ML) algorithms. It is vital to make efficient use of agricultural land to ensure the food security of the country. So, ML algorithms can be used to predict the yield from the historical data. Various ML algorithms [1] such as random forest, XGBoost, gradient boosting, and ensemble XGBoost-RF are used to predict the yield based on various parameters [2] like rainfall, temperature, fertilizers, etc. By using the above-mentioned algorithms, from the results, it can be concluded that the proposed hybrid model called extreme gradient boosting–random forest gave maximum accuracy.

2 Crop Yield

The quantity of a crop produced per unit of land is referred to as crop yield. It is a crucial measurement to comprehend since it helps us to understand food security. Crop yield is one of the measures used to assess the efficiency of food production. Understanding crop yield and being able to estimate, it is significant for several reasons. First, understanding food security, or the capacity to produce enough food to fulfill human needs soon, requires the ability to estimate crop yield. Second, for each crop, the potential yield should be estimated prior. Finally, crop yields are important because they have a direct impact on how much money people will spend on food. Rainfall, temperature, and fertilizers are the different factors that are important to achieving high yields [2]. In this work, a dataset related to agriculture (shown in Table 1) is used for the analysis. The dataset contains rainfall, fertilizer, temperature, nitrogen, phosphorus, and past crop yield.

Table 1 Minimum, maximum, and standard deviation of the parameters

Parameters	Minimum	Maximum	Standard deviation
Rainfall (in mm)	400	1300	400.0427
Fertilizer (urea) (kg/acre)	50	80	10.0282
Temperature (°C)	24	40	5.42635
Nitrogen (N)	59	80	6.677079
Phosphorus (P)	18	25	1.951695
Potassium (K)	15	22	1.817254
Yield (Q/acre)	5.5	12	1.965902

3 Machine Learning Techniques

The collecting of electronic data has grown more prevalent in most domains of human endeavor because of advances in computer technology over the last several decades. Many organizations need vast volumes of data dating back many years. This information relates to individuals, financial activities, biological data, etc. Simultaneously, data scientists have been working on algorithms which are iterative computer software applications that can look at vast amounts of data, evaluate it, and find patterns and links that people cannot. Analyzing the previous events can reveal a wealth of information on what to expect in future from the same or nearly comparable events. These algorithms may learn from the past and use what they have learned to make better decisions in future. Data analysis is not a novel concept. ML algorithms distinguish themselves from other techniques and can cope with significantly large amounts of data and data with minimal structure. It enables ML algorithms to be effective in a wide range of applications previously thought to be too complicated for conventional learning techniques.

In this current work, developed and applied four ML algorithms: random forest, gradient boosting, XGBoost, and ensemble XGBoost-RF used to predict the crop yield.

3.1 Random Forest

For classification and regression tasks, this is the most common and powerful supervised ML approach. During training and creating class outputs, this technique creates a vast number of decision trees. Random forest (RF) is a bagging technique that employs many decision trees on subsets of a given set of observations and averages the results to improve the dataset's estimated accuracy.

The predictions from each tree are collected by random forest, which then predicts the ultimate output based on the popular vote of predictions. The more trees in the forest, the more accurate it becomes, and the risk of errors is reduced. There are two random factors in a random forest. They are as follows:

1. Random subset of features.
2. Bootstrap samples of data.

A random forest [3] is merely a group of trees, each of which makes a prediction, and they gather from all of them and use the mean, mode, and median of the collection as the forest's prediction, based on the data which may be continuous or categorical. To a greater extent, this appears to be acceptable. But most of the trees may have generated predictions based on random possibilities because each had its own set of conditions.

3.2 Gradient Boosting

Gradient boosting is one of the boosting methods that is used to reduce the bias error of the model. It can be used for predicting continuous target values, i.e., as a regressor. The gradient boosting regressor (GBR) reduces the prediction error and increases the accuracy of the model. GBR is a fully integrated model that offers improved performance and stability. To overcome the regression problem, the GBR method [4] extends the boosting technique. This method makes use of negative gradients of the loss function to solve the minimum value. GBR has been widely utilized in biological research because of its capacity to handle messy and noisy data and has a good predictive ability for non-linear data.

3.3 XGBoost

XGBoost refers to the extreme gradient boost algorithm. It provides a parallel tree boosting that solves the issues in data science fast and accurately. This algorithm performs best on datasets that are well-structured or tabular.

This model uses boosting ensemble learning with the help of decision trees. Gradient boosting is XGBoost's original model that involves iteratively merging weak base learning techniques into a stronger learner. The residual will be utilized to adjust the previous predictor at each iteration of gradient boosting, so that the stated loss function may be improved. Regularization is introduced to the loss function in XGBoost to create the objective function for monitoring model performance, which is represented by

$$J(\varphi) = L(\varphi) + \Omega(\varphi) \tag{1}$$

where φ denotes the parameters trained from the provided dataset; L denotes the training loss function which is a metric for how well a model fits on a training set data.

3.4 Ensemble XGBoost-RF

XGBoost-RF is a hybrid model. The XGBoost library implements gradient boosting in a fast and flexible way to train random forest ensembles. Compared to gradient boosting, random forest is a much simpler technique. The XGBoost library was there to train random forest (RF) models by repurposing and harnessing the library's computational efficiency. The core XGBoost algorithm [5] can also be configured to support different tree ensemble algorithms, such as random forest, in addition to gradient boosting. A random forest algorithm is a collection of decision trees.

The argument 'n'_estimators sets the number of trees used in the ensemble. The XGBoost-RF ensemble is first fitted to the available data, after which the predict function generates predictions on new data.

Gradient boosting is extremely slow to train a model and exacerbated by big datasets. XGBoost solves the speed concerns of gradient boost by incorporating different strategies that drastically speed up the model's training and, in many cases, improve the model's overall performance [6]. The primary advantage of training random forest ensembles with the XGBoost library is to boost the speed.

3.5 Evaluation Metrics

The performance of developed algorithms is measured by mean square error (MSE) and coefficient of determination (R^2). The equations are as follows [7]:

$$\text{MSE} = \frac{\sum_{x=1}^{y}(b^x - b^*)}{y} \tag{2}$$

$$R^2 = 1 - \frac{\sum_{x=1}^{y}(b^x - f(a^x))^2}{\sum_{x=1}^{y}(b^x - b^*)^2} \tag{3}$$

where y refers to the number of target values; $b = (b^1, b^2, \ldots, b^y)^{\mathrm{T}}$; b^* is the prediction value, and $f(a^x)$ denotes the regression function for the feature vector a^x.

4 Results and Discussion

The collection and processing of sample data are the initial step in the construction of a prediction model. To serve as input data, large amount of data must be compiled. To train the algorithm, dataset with different parameters is to be considered. The variables in the dataset are rainfall, temperature, fertilizer, nitrogen, phosphorus, potassium, and yield. After collecting data, apply the four ML algorithms to check the accuracy of each algorithm. In this project, a random forest, gradient boosting, XGBoost, and ensemble XGBoost-RF are implemented using Python on the Jupyter notebook application. Pandas, scikit learn, NumPy, and matplotlib are the main libraries used. The data has two parts as follows: (i) testing and (ii) training (67% for training and 33% for testing). During experimentation with tuning the hyperparameters, model depths had taken the maximum values. Figure 1 shows the correlation between actual crop yield and predicted crop yield of ensemble XGBoost-RF algorithm.

Figure 2 depicts the correlation between actual crop yield and predicted crop yield of XGBoost algorithm. Similarly, Fig. 3 shows the correlation between actual crop

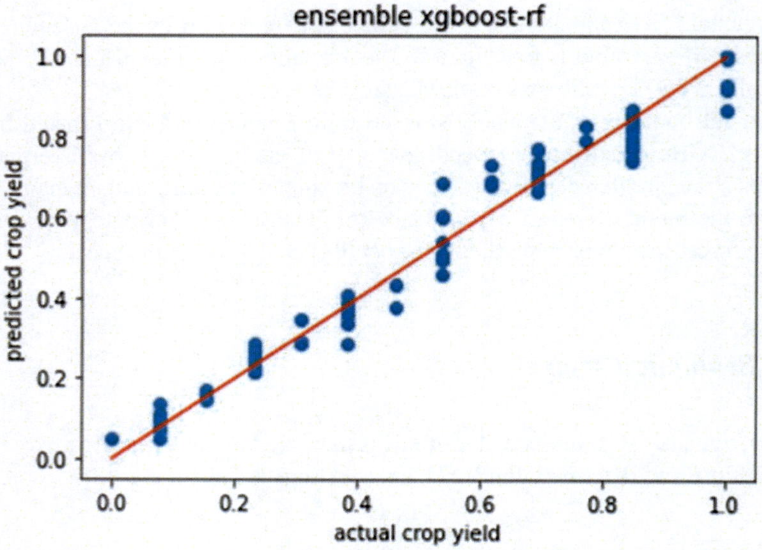

Fig. 1 Plot of measured crop yield versus predicted crop yield of ensemble XGBoost-RF

yield and predicted crop yield of random forest algorithm. Finally, Fig. 4 depicted the correction correlation between actual crop yield and predicted crop yield of gradient boosting algorithm.

The obtained R^2 and MSE values of the four algorithms applied to the collected datasets were shown in Table 2. The R^2 value for gradient boosting is 0.952457, which

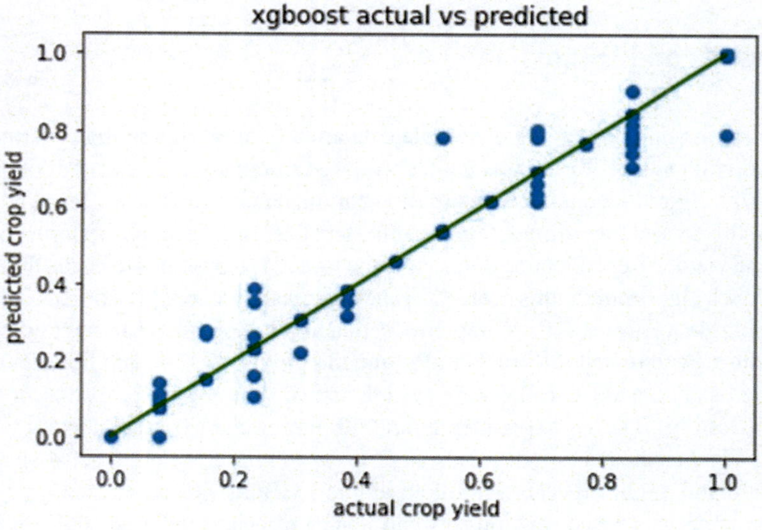

Fig. 2 Plot of measured crop yield versus predicted crop yield of XGBoost

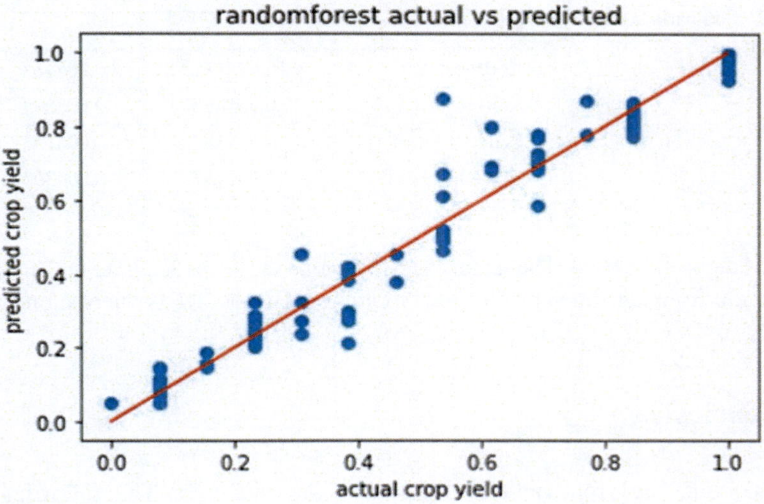

Fig. 3 Plot of measured crop yield versus predicted crop yield of random forest

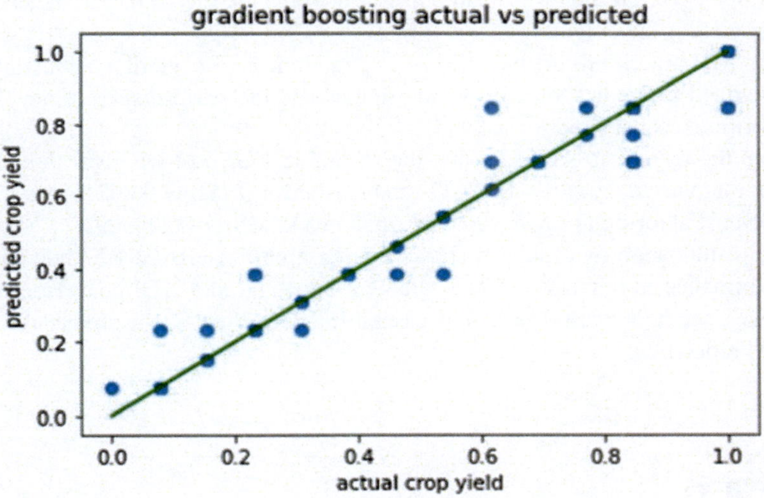

Fig. 4 Plot of measured crop yield versus predicted crop yield of gradient boosting

means the accuracy level for gradient boosting is 95.24%. Likewise, the accuracy level for random forest, XGBoost, and ensemble XGBoost-RF is 95.43%, 96.58%, and 97.61%, respectively. From the above results, an ensemble XGBoost-RF has the highest R^2 value. The higher the R^2 value, the more accurate the algorithm. The value of MSE for gradient boosting is 0.004303, for the XGBoost algorithm, MSE value is 0.003092, for the random forest algorithm, MSE value is 0.004133, and for the ensemble XGBoost-RF algorithm, the MSE value is 0.002163 which is least from all

Table 2 Comparison of R^2 and MSE

Model	R^2	MSE
XGBoost-RF	**0.976111**	**0.002163**
XGBoost	0.965855	0.003092
Random forest	0.954357	0.004133
Gradient boosting	0.952457	0.004303

the four algorithms used. The lesser the MSE value the more accurate the algorithm will be. So from the above results, the ensemble XGBoost-RF is the best one.

5 Conclusion

In this work, the crop yield data (Q/acre) was analysed. The following key observations are observed from the analysis. They are as follows:

- The data used for constructing the model consists of rainfall, temperature, fertilizer, nitrogen, phosphorous, and potassium which are the input parameters, and crop yield is the output.
- Four ML algorithms include XGBoost, random forest, gradient boosting are developed to predict the crop yield. To evaluate the performance of developed algorithms, considered R^2 and MSE.
- From the results, it is evident that the ensemble XGBoost-RF algorithm shows maximum accuracy ($R^2 = 0.97611$) and least error (MSE $= 0.002163$), whereas XGBoost algorithm provides R^2 of 0.965855 and MSE of 0.003092.
- The results analysis (Table 2) shows that ensemble XGBoost-RF shows better performance over random forest, gradient boosting, and XGBoost. Hence, the above specified results show that ensemble XGBoost-RF can predict the crop yield efficiently.

References

1. Raja SP, Sawicka B, Stamenkovic Z, Mariammal G (2022) Crop prediction based on characteristics of the agricultural environment using various feature selection techniques and classifiers. J IEEE Access 10:23625–23641
2. Venugopal A, Aparna S, Mani J, Mathew R, Williams V (2021) Crop yield prediction using machine learning algorithms. Int J Eng Res Technol 9(13):87–91
3. Priya P, Muthaiah U, Balamurugan M (2018) Predicting yield of the crop using machine learning algorithm. Int J Eng Sci Res Technol 7(4):1–7
4. Khan R, Mishra P, Baranidharan B (2020) Crop yield prediction using gradient boosting regression. Int J Technol Exploring Eng 9(3):2293–2297
5. Ravi R, Baranidharan B (2020) Crop yield prediction using XG boost algorithm. Int J Recent Technol Eng 8(5):3516–3520

6. Oikonomidis A, Catal C, Kassahun A (2022) Hybrid deep learning-based models for crop yield prediction. J Applied Artificial Intelligence 36(1)
7. Ragam P, Nimaje DS (2018) Evaluation and prediction of blast-induced peak particle velocity using artificial neural network: a case study. Noise Vib Worldw 49(3)

Analysis of Students' Fitness and Health Using Data Mining

P. Kamakshi, K. Deepika, and G. Sruthi

Abstract The level of physical development of students' has greatly improved as living standards have continued to rise. According to the previous study, students are dealing with a variety of issues such as stress, sadness, anxiety, disinterest, and behavioral issues. These issues may be caused by a lack of physical education as well as the numerous expectations placed on their shoulders. Because of numerous difficulties reported among students, awareness over physical fitness and health management is fast growing. On a daily basis, most modern university campuses save statistical information trails from varied databases addressing many aspects of student life. However, aggregating these data to get a comprehensive picture of a student, using these data to accurately anticipate student health, and applying such predictions to promote positive student involvement with the universities are all complicated tasks. This paper proposes a use case for Internet of things technology, particularly data mining techniques such as the random forest algorithm, decision tree algorithm, voting classifier, and LSTM are implemented to monitor fitness and health management among students. Here, data is processed through random forest because the random forest model has the greater precision, performance, and classification effect. As a result, this paper is beneficial to schools and colleges in terms of mastering students' health of students and scientifically preventing health issues among students. The trials reveal that the AugmentED model has a high level of accuracy in predicting pupils' wellness.

Keywords Data mining · Random forest algorithm · Accuracy · Machine learning (ML) · Voting classifier · Long short-term memory (LSTM)

P. Kamakshi (✉) · K. Deepika · G. Sruthi
Department of Information Technology, Kakatiya Institute of Technology and Science, Warangal, India
e-mail: pk.it@kitsw.ac.in

K. Deepika
e-mail: dr.dk.it@kitsw.ac.in

G. Sruthi
e-mail: b19it004@kitsw.ac.in

K. A. Reddy et al. (eds.), *Proceedings of Fourth International Conference on Computer and Communication Technologies*, Lecture Notes in Networks and Systems 606,
https://doi.org/10.1007/978-981-19-8563-8_39

1 Introduction

Each country's progress requires high-quality education. The amount of data in the education domain is expanding by the day, thanks to admission systems, academic information systems, learning management systems, and e-learning. As a result, using this vast amount of educational data to predict student health is a hot topic. The technique of obtaining useful insights from vast quantities of data is referred as data mining, or knowledge discovery (KDD) in databases. Student health and fitness analysis is a critical topic in the education [1] data mining business since it is a significant step toward personalized education. The following aspects have been shown to have a significant impact on academic performance:

- Personality traits of students (e.g., neurotic tendencies, conscientiousness, and extroversion).
- Personal concerns of students (e.g., age, sex, physical fitness, indifference, emotional stability, stress, mood, panic attacks, activeness, and energy levels).
- Lifestyle behaviors (e.g., nutrition, regular exercise, sleeping habits, social connections, and effective planning); and
- Learning conduct (e.g., presence in class, active participation, and study time).

Many data-driven approaches for predicting health status have been developed by analyzing the impact of various factors on student health and fitness. Despite the development of various health prediction systems for college students, substantial challenges remain, such as acquiring student's whole profile and merging this data to achieve a comprehensive overview. Analyzing the elements that affect students' health and utilizing that data to construct a strong, high-accuracy prediction model, as well as leveraging the statistical model to give individualized support that could assist students improve their behavior and enhance their study-life balance.

2 Literature Survey

Laith et al. presented Aquila-Optimizer (AO), a new population-based optimization technique that they evaluated with four techniques. The test analysis revealed that when compared to the prior meta-heuristic algorithm, the AO method was more favorable and appropriate for the evaluation of health data from university students. Using Internet of things technology, DM techniques, and miscellaneous data processing tools, Wang et al. constructed a big data management system for public healthcare centers and accomplished disease risk early-warning and therapy.

With the national education reformation, physical health of college students has become a focal point of social significant concern. Perspectives on academic achievement for fostering literally the entire quality of students' personal health through sports are included in the document no. 27, and the State Council stated that physical education should be enhanced. Many earlier studies have concentrated solely on selecting an effective [2] classification method, ignoring concerns such as complexity of data and classification inaccuracy that develop during the data mining phases. These flaws lowered the model's accuracy.

The author, Thota et al. [3], proposes a healthcare platform that combines the Internet of things and cloud computing, as well as a centralized fog network for securing the resources delivered. The paper by Raghupathi and Raghupathi [4] "describes the promises and potential of big data analytics while explaining the benefits, organizational framework, methodology, as well as the problems and constraints." The research article [5] supports the IoT combined with big data analytics for monitoring patients in remote regions deals with the analysis of cardiac disease patients. The Design of Distribution Transformer Health Management System Using IoT Sensors [6] by Rajesh is a soft computing paradigm dedicated to the study of soft computing. A machine learning algorithm is proposed for Optimization of Prediction of Kidney Failure by Gosh et al. [7]. Dubey et al. [8], Yassine et al. [9], Suma [10]; the author in the paper claims that combining big data with IoT will enable optimal decision-making in industries, resulting in industry sustainability.

3 Proposed Methodology

The proposed methodology deals with the tree regression, LSTM, KNN, random forest regression, voting classifier for accuracy [11] comparison. The algorithm is trained using a student dataset that contains information on students' health status. Students' data is collected from educational institutes in order to create a health and fitness management [12] system. The most effective and efficient model would be determined by comparing the performance and accuracy of these models (Fig. 1).

3.1 Model Building

KNN stands for "*k*-nearest neighbor." KNN is a classification algorithm, and it requires some reference data for classification. For the KNN algorithm, some reference data is required. It calculates the distance between the specified data item and all of the reference data records then searches the *k*-nearest data items in the reference data. If $k = 5$, then the nearest neighbor class is the data to be checked. The predicted class [3] of algorithms will be the majority class in this group of k data records [6]. The input instance's categorization is determined by the majority of its *k* neighboring training instances.

SVM stands for "support vector machine." SVM is a data mining tool that can be used to tackle a range of problems, including regression (time series inquiry) and also pattern matching. SVM works by creating a classification hyperplane with two margin lines with some distance so that they are linearly separable. We ensure that the two margin lines are parallel to the hyperplane and pass through the nearest points.

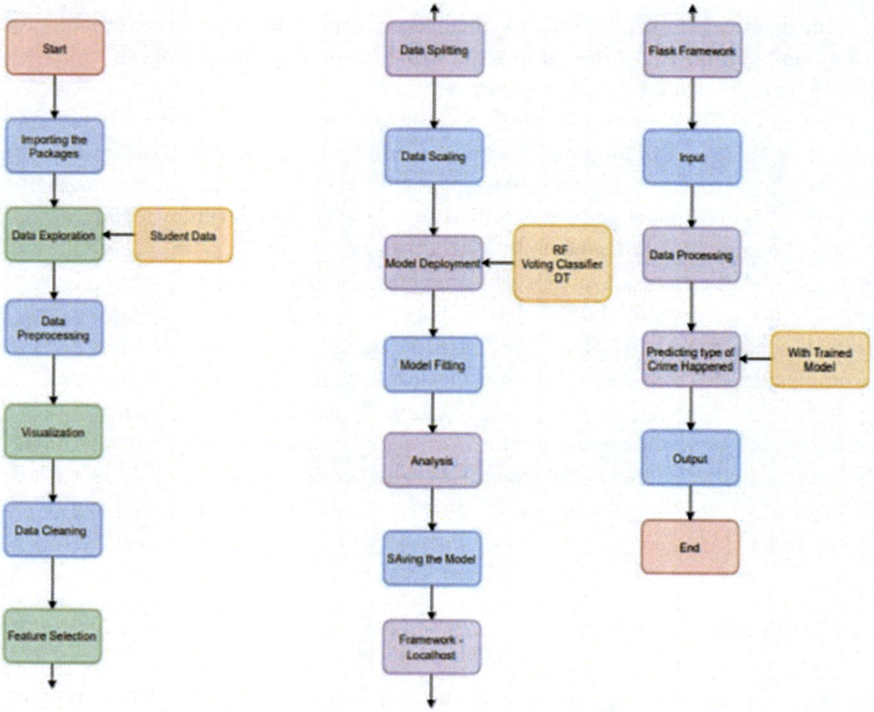

Fig. 1 Flowchart for fitness and health management system

The finest classification of linear separable data can theoretically be accomplished with SVM.

LSTM is a type of RNN-based deep learning model. In order to transfer data, LSTM includes the deployment of a specific unit (memory subsystem) to learn recent data and retrieve related information and rules from the input. Because of the memory module, LSTM is better for deep neural network calculations. There are three gates in each memory module: input gate (it), forget gate (ft), and output gate (ot). As the gradient lowers, they are utilized to systematically recall the correction coefficients of the feedback error function.

A decision tree is a decision-making tool. It divides huge data into subsets using a tree-like paradigm for categorization. There are three basic structures in decision tree: root node, leaf nodes, and internal nodes. It is one way of demonstrating an algorithm comprised solely of conditional control structures. In operations research, decision trees are frequently used to assist in the identification of the finest strategy for achieving a target, specifically in decision analysis.

A voting classifier is a classifier model that trains from a set of models using bagging and boosting techniques. It anticipates an outcome (class) build on the output having the highest probability of being the targeted class. It sums up the outcomes fed into the voting classifier and determines the resultant class with majority votes.

We develop a single model that trains on numerous models and forecasts results depended on the cumulative majority [13] of votes for each resultant class, rather than building separate models and evaluating their performance.

Random forest is a flexible supervised machine learning algorithm which uses bagging techniques to improve the performance. A random forest is a collection of tree classifiers with the parameters $\{h(x, \beta k), k = 1\ldots\}$. The meta classifier $(x, \beta k)$ is a CART-based regression tree with x as the input vector and k as an independent random vector with the same distribution. The forest algorithm's final output is determined by voting. Randomness is determined in two ways: The bagging algorithm is used to choose the training sample set, and the split attribute series is generated at random as well. Considering the classification model has N attributes in all, we set an trait value $S \leq N$ to each intermediate node, select S attributes at random from the N attribute set as the split attribute set, and determine the optimal strategy of splitting for the S attributes. The tree classifier's vote determines the final classification outcome as shown in Fig. 2.

Gini index, Gini(T) is defined as follows:

$$\text{Gini} = 1 - \sum_{i=1}^{c} (p_i)^2 \quad (1)$$

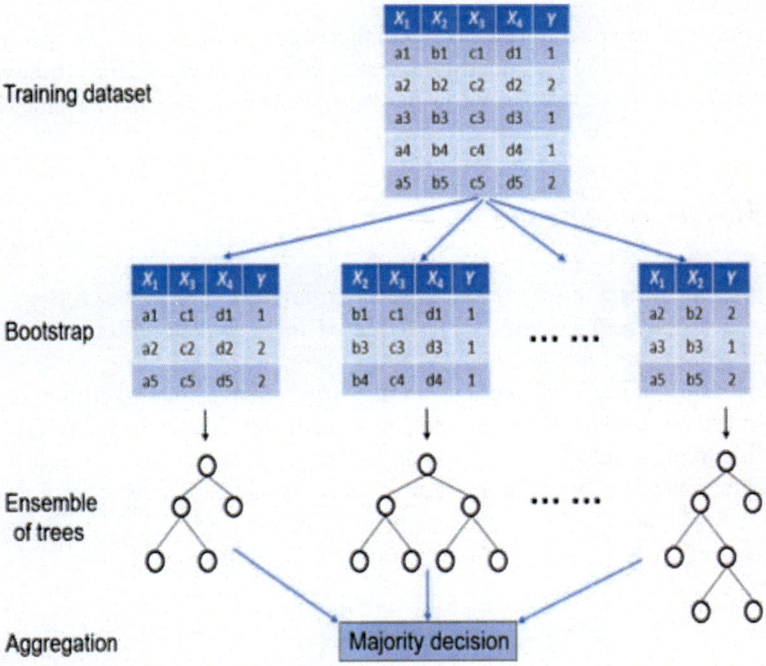

Fig. 2 Random forest classifier

Table 1 Health problems in a college

Serial number	Lack of energy	Bad mood	Emotional instability	Indifference	Personality embrittlement	Weird behavior	Physical quality decline
1	Yes	Yes	No	No	Yes	Yes	No
2	No	Not	No	No	Yes	No	Yes
3	Yes	No	No	No	No	Yes	Yes
4	No	Yes	Yes	Yes	Yes	No	No
5	No	Yes	No	Yes	No	No	Yes
6	Yes	No	No	No	No	Yes	No
7	Yes	No	Not	Yes	Yes	No	No
8	Not	No	Yes	Not	Yes	Yes	Yes
9	No	Yes	Yes	Yes	No	Yes	No
10	No	Yes	Yes	Yes	No	No	Yes

Table 1 shows some of the pupils' health data.

The health analysis data of each student corresponds to a transaction *Ti* (TID) in Table 1. The fitness evaluation tables of all students constitute a transaction set $T = \{T1, T2, T3, \ldots, Ti\}$.

Yes indicates symptomatic, whereas no indicates asymptomatic in the seven health factors that follow.

To apply the discussed models for getting accuracy and a confusion matrix, the 300-student health database is partitioned into different blocks, and the transaction database is transformed into a Boolean matrix using dynamic data memory allocation.

3.2 Model Comparison

In the first experiment, three classification algorithms (random forest, voting classifier, and decision tree) are run on a dataset containing student personal and health details.

The accuracy of Decision tree algorithm, matrix-based Apriori algorithm, support vector machine and the K-nearest neighbor algorithm is between 73% to 76%. According to the graphical representation in Fig. 3, the best accuracy was achieved by random forest (79.8%), which was satisfactory in comparison with prior studies, while the lowest accuracy was achieved by decision tree.

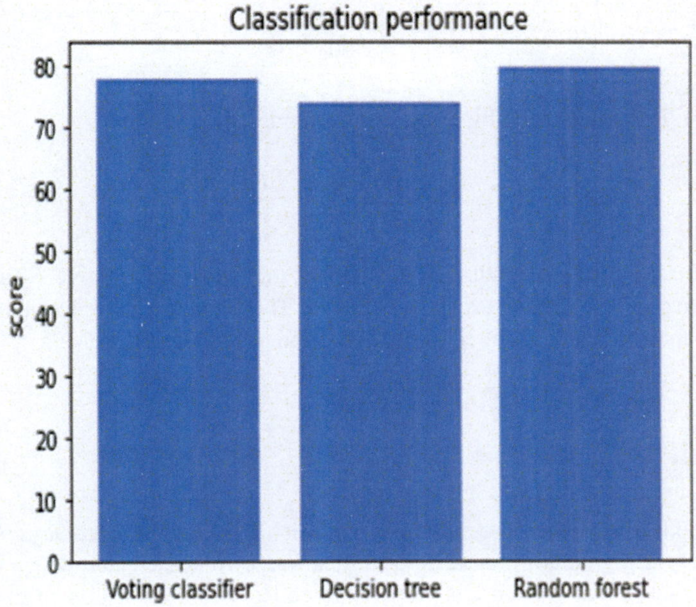

Fig. 3 Comparison graph

3.3 Evaluation Measures

In our study, we evaluate categorization quality using five popular distinct measures. Details are as follows:

Accuracy: It is also abbreviated as CCI (correctly classified instances). It is determined by the formula

$$\text{Accuracy} = \frac{T_p + T_n}{T_p + T_n + F_p + F_n} \tag{2}$$

ICI: obtained by calculating the count of misclassified instances divided by the overall instances.

Precision: The fraction of accurately classified instances among all truly classified instances is represented by the algorithm.

$$\text{Precision} = \frac{T_P}{T_P + F_P} \tag{3}$$

Recall: obtained by calculating the count of accurately labeled instances divided by the overall instances.

$$\text{Recall} = \frac{T_P}{T_p + T_n} \tag{4}$$

F-Measure: It is determined by using Eqs. (3) and (4)

$$F_1 = 2 \times \frac{\text{precision} \times \text{recall}}{\text{precision} + \text{recall}} \tag{5}$$

From Eqs. (2–5), T_p indicates true positive, T_n indicates true negative, F_p indicates false positive, and F_n indicates false negative. These values were derived using a confusion matrix, which resulted in the execution of the algorithm.

4 Results

The best model turned out to be random forest as it has the highest accuracy compared to others. By using random forest algorithm, a framework is created which predicts the score. Now, a questionnaire is prepared to collect details of a student to analyze students' health.

The questionnaire consists of few personal questions such as name, gender, and age and questions related to students' health status mentioned in Table 1. The students answer these questions according to their choice. By analyzing the answers of these questions, we get a numerical score at the end. A framework is developed using random forest algorithm to predict the score based on the inputs collected from the students.

Through this score obtained in Fig. 4, we can conclude that a student having score from.

- 16–20 is healthy and extremely active and attends college regularly.
- 11–15 is experiencing mild stress.
- 6–10 is weak, with emotional instability and apathy. The student must practice self-care and seek professional counseling.
- 1–5 is unhealthy, suffering from serious disorder, stress, and depression. The student should see a doctor and receive treatment.

5 Conclusion

In this paper, data mining methods are used to extract seven health dimensions, resulting in a health and fitness management system. The findings provide a realistic framework for educational institutions to master student health and colleges to scientifically prevent health problems among college students. Every educational institute is in need of an accurate student health and fitness prediction model. However,

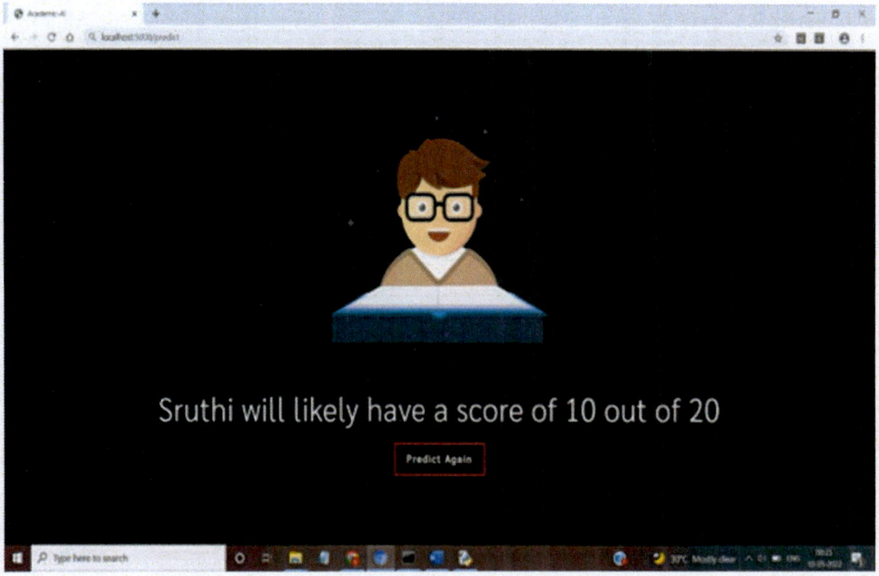

Fig. 4 Predicted result

resolving data quality issues in student health prediction models is sometimes the most difficult task. This research develops a random forest model-based student performance prediction model. Many academics have looked into student health and fitness status prediction as an essential topic in the field of education data mining. However, there are still several hurdles in predicting accuracy and interpretability due to a lack of abundance and diversity in both data sources and characteristics. This system has the potential to lead to extensive investigations. The knowledge gained in this study has the potential to help with related studies among students who are interested in developing a student health management system.

References

1. Abd-Ali RS, Radhi SA, Rasool ZI (2020) A survey: the role of the internet of things in the development of education. Indonesian J Electrical Eng Computer Sci 19(1):215
2. Zhang X, Liu L, Xiao L, Ji J (2020) Comparison of machine learning algorithms for predicting crime hotspots. IEEE Access
3. Thota C, Sundarasekar R, Manogaran G, Varatharajan R, Priyan MK (2018) Centralized fog computing security platform for IoT and cloud in healthcare system. In: Fog computing: breakthroughs in research and practice. IGI global, pp 365–378
4. Raghupathi W, Raghupathi V (2014) Big data analytics in healthcare: promise and potential. Health Information Science and Systems 2(1):3
5. Smys S, Raj JS (2019) Internet of things and big data analytics for health care with cloud computing. J Inf Technol 1(01):9–18

6. Rajesh SR (2021) Design of distribution transformer health management system using IoT sensors. Journal of Soft Computing Paradigm 3(3):192–204
7. Ghosh P, Shamrat FMJM, Shultana S, Afrin A, Anjum A et al (2020) Optimization of prediction method of chronic kidney disease using machine learning algorithm. In: 2020 15th international joint symposium on artificial intelligence and natural language processing (iSAI-NLP), pp 1–6
8. Dubey H, Yang J, Constant N, Amiri AM, Yang Q, Makodiya K (2015) Fog data: enhancing telehealth big data through fog computing. In: Proceedings of the ASE bigdata & social informatics 2015. ACM, p 14
9. Yassine A, Singh S, Hossain MS, Muhammad G (2019) IoT big data analytics for smart homes with fog and cloud computing. Futur Gener Comput Syst 91:563–573
10. Suma V (2019) Towards sustainable industrialization using big data and internet of things. Journal of ISMAC 1(01):24–37
11. Furnham A, Monsen J (2009) Personality traits and intelligence predict academic school grades. Learning and Individual Differences 19(1):0–33
12. Yaacob WFW, Sobri NM, Nasir SAM et al (2020) Predicting student drop-out in higher institution using data mining techniques. J Phys Conf Ser 1496(1):13–15
13. Alam S, Abdullah H, Abdulhaq R et al (2021) A blockchain-based framework for secure educational credentials. Turkish J Comput Math Edu (TURCOMAT) 12(10):5157–5167

Local Agnostic Interpretable Model for Diabetes Prediction with Explanations Using XAI

Vivekanand Aelgani, Suneet K. Gupta, and V. A. Narayana

Abstract Diabetes mellitus is the deadliest disease that affects the production of insulin. Diabetes is a life-taking disease, if it is not detected early in advance. Recently, artificial intelligence-based machine learning (ML) predictive models are predominantly used in sensitive healthcare domain for predicting diseases in advance. Most of these ML models are black-box models which provide approximate explanations of how a model behaves. If the models were interpretable, then domain expert can understand the reasons and modify the model accordingly to get the best results. In this paper, we present an ensemble local explainable agnostic model for predicting diabetes. Our study shows that the ensemble voting classifier produced 81% accuracy on the Pima Indian diabetes dataset as compared to other conventional predictive models. We then applied the explainable AI (XAI) technique which helps the medical experts in understanding the predictions made by the mode.

Keywords Diabetes mellitus · Machine learning · Black-box models · Artificial intelligence · Explainable AI · Pima dataset

1 Introduction

As per the International Diabetes Federation (IDF) [1] report, nearly 537 million people are suffering from diabetes across the world. Every year diabetes causes 6.7 million casualties, and more than a million children and adolescents (0–18 years) are suffering from insulin-dependent diabetes. Every year, more than 21 million children are born with diabetes [2]. About 541 million grown-ups are in danger

V. Aelgani (✉) · V. A. Narayana
CMR College of Engineering & Technology, Kandlakoya, Telangana, India
e-mail: aelgani.vivekanand@gmail.com

V. A. Narayana
e-mail: narayanaphd@gmail.com

S. K. Gupta
Bennett University, Greater Noida, Uttar Pradesh, India
e-mail: suneet.gupta@bennett.edu.in

K. A. Reddy et al. (eds.), *Proceedings of Fourth International Conference on Computer and Communication Technologies*, Lecture Notes in Networks and Systems 606,
https://doi.org/10.1007/978-981-19-8563-8_40

of acquiring adult-onset diabetes. If diabetes is diagnosed early, it is much easier for us to control it. Thus, analyzing and predicting diabetes accurately and swiftly are a topic worth studying. In the last two decades, the area of deep learning has paid enormous attention due to its vast application in real-life applications such as health care, transportation, security surveillance, agriculture [3–9]. Nowadays, many researchers have developed various applications based on deep learning for the identification and classification of human disease by analyzing the different types of data such as X-rays, ultrasound. Explainable AI (XAI) is an innovation in the field of artificial intelligence that focuses on increasing transparency in AI products [8]. It allows to build an AI model that is simple and easy to interpret by humans, ensuring domain experts understand how machines make predictions. XAI has many advantages that include reduction of cost of mistakes, responsibility, accountability, and Code Confidence [9].

In this paper, a Local Agnostic Interpretable model (LIME) for generating local explanations of a complex black-box model has been reviewed with a case study on diabetic prediction. The best-performing classifier on the Pima India diabetes dataset is obtained by comparing the performance of traditional machine learning algorithms and ensemble approaches for diabetic prediction. Our study reveals that the soft voting ensemble classifier outperformed all classifiers on accuracy. We then used Local Interpretable Model-Agnostic Explanations (LIME) to generate the local explanations for instance of interest.

The paper is organized in to five sections. In Sect. 2, there is a brief discussion on existing literature related to diabetes prediction. Section 3 is devoted to the description of the dataset used in the study. An explanation of the proposed model, experimental setup, and performance comparison is provided in Sect. 4. These sections are followed by the conclusion and future scope in Sect. 5.

2 Related Work

In this section, we review a few prominent works done by researchers in predicting diabetes. Hameed et al. [10] proposed the ID3-based method to predict gestational diabetes on the Irvine dataset, and the model is interpretable for the small number of features. Moreover, it becomes more complex and highly non-explainable for a large number of features. Jakka et al. [11] concluded that logistic regression outperformed all classical machine learning algorithms. Kalagotla et al. [12] proposed a stacking technique with a multilayer perceptron, support vector machine, and logistic regression (MLP, SVM, and LR, respectively) for diabetes prediction. Kumari et al. [13] proposed a voting-based ensemble approach for diabetes prediction. They concluded that voting-based ensemble approach gives maximum accuracy. Lu et al. [14] proposed a patient network-based machine learning model for disease prediction. Maniruzzaman et al. [15] used the Gaussian process (GP)-based classification technique using three kernels, namely: linear, polynomial, and radial basis kernel to predict diabetes. They achieved encouraging results. However, their models lack interpretability. Palimkar et al. [16] studied the performance of logistic regression, random forest classifier, support vector machine, decision trees, K-nearest neighbors, Gaussian process classifier, AdaBoost classifier, and Gaussian Naıve Bayes and concluded that random forest outperformed all the models. Prabhu et al. [17] studied the performance of deep neural network models on the Pima Indian diabetes dataset. Sarwar et al. [18] compared the performance of machine learning algorithms based on the accuracy of the models. This review shows that ML and DL models are extensively used for the accurate prediction of diabetes. However, these models are black-box models. They lack explainability and trustworthiness. The user or domain expert does not know why the model failed to predict certain instances. The study also reveals that there are very few attempts made toward the explainability of the predictive model.

3 Materials and Methods

In this section, a discussion on materials and methods used for conducting the study has been presented. This section is divided into three sections, namely Sects. 3.1, 3.2, and 3.3. Section 3.1 describes the dataset used in the study. Section 3.2 describes the problem statement, and finally, Sect. 3.3 explains the explainable model.

3.1 Dataset Description

The ML and DL predictive models were trained and tested on [19]. This dataset was created by the National Institute of Diabetes and Digestive and Kidney Diseases, USA. The dataset consists of 768 diabetic patients from the Pima Indian population near Phoenix, Arizona. Dataset consists of 268 diabetic patients (positive) and 500 non-diabetic patients (negative) with eight different features.

Soft Voting Classifier: The idea behind the soft voting classifier (SVC) is to integrate theoretically diverse predictive models and use a majority result (predicted outcome) or the average predicted probabilities to predict the category labels. A classifier of this type can be useful for a collection of equally well-performing models to compensate for their deficiencies or shortcomings.

Example: *Diabetic prediction is classification task with class label k belonging to {0, 1}.*

0 indicates negative (non-diabetic class) and 1 indicates positive (diabetic class). Sample calculations are shown in Eqs. 1–3.

$$\text{prob}(k0|x) = (0.8 + 0.8 + 0.8 + 0.7 + 0.7)/5 = 0.76; \tag{1}$$

$$\text{prob}(k1|x) = (0.2 + 0.2 + 0.2 + 0.3 + 0.3)/5 = 0.24; \tag{2}$$

$$y = \arg\max_{k}\left[\text{prob}(k_0|x),\ \text{prob}(k_1|x)\right] = 0. \tag{3}$$

3.2 Problem Description

The objective of the study is to build an explainable predictive model to predict whether a patient has diabetes or not, formulated on measurable features of the dataset. It means we need to find a classifier function f such that $Y = f(X_i)$ where $X_i \in R^8$ and $Y \in B = \{0, 1\}$. If $Y = 0$, then patient has no diabetes, else patient has diabetes.

3.3 Interpretable Machine Learning Models

LIME [20] is an explainable model. LIME is an acronym for Local Interpretable Model-Agnostic Explanations. Here, local refers to the scope of the model, i.e., LIME is used to explain a single observation or record. LIME produces human-understandable explanations of the model. Mathematically, LIME can be expressed as shown in Eq. 4, [20].

$$\text{Explainability}(p) = \arg\min_{H \in S} L(X, H, \Pi_p) + \Omega(H) \tag{4}$$

It is a mathematical optimization problem. Here, we want to create a local approximation of our complex model for a specific input p. X is a nonlinear non-interpretable complex model, H is simple interpretable model, $S = \{\text{interpretable models}\} = \{\text{DT, LR}\}$, and πp is local neighborhood of p. $\Omega(H)$ regularization parameter which indicates the measure of complexity of the model. The working of LIME is presented in Algorithm 1.

Algorithm 1 Local Interpretable Model Explainer

Input: p //observation whose explanation sought
Input: n // glass box model Sample Size
Input: m // the number of FEATURES for the simple model
Input: NearestNeighbor. //a distance metric in the INITIAL dataset
Output: Lime predictions of p
$p^{'} \leftarrow H(p)$ //mapping of p in the REDUCED Dataset
for j in range(n) **do**
$A^{'}[j] \leftarrow SyntheticSample(p^{'})$ //PERTURBED DATASET
$B^{'}[j] \leftarrow f(A'[j])$ //prophecy for new observation A′[i]
$D^{'}[j] \leftarrow NearestNeighbor(p', A'[j])$
end for
return *LASSO Regression* $m(B', p', D')$

4 Results and Discussions

This section presents the experimental procedure, results, and analysis. This section is divided into three sections, namely Sects. 4.1, 4.2, and 4.3. Section 4.1 presents the experimental setup of the study, Sect. 4.2 describes results, and Sect. 4.3 describes explanations generated by LIME explainer.

4.1 Experimental Setup

The experimental setup of the suggested method is provided in Fig. 1.

The experiment was carried out on the Pima Indian diabetes dataset. The dataset set contains 768 instances and 9 feature vectors. The dataset is class imbalance as there were only half of the total instances as negative. The dataset is divided into training and test datasets with a 70:30 ratio. To mitigate the class imbalance, SMOTE oversampling technique has been used [21]. On train dataset, predictive models were developed using various conventional machine learning algorithms.

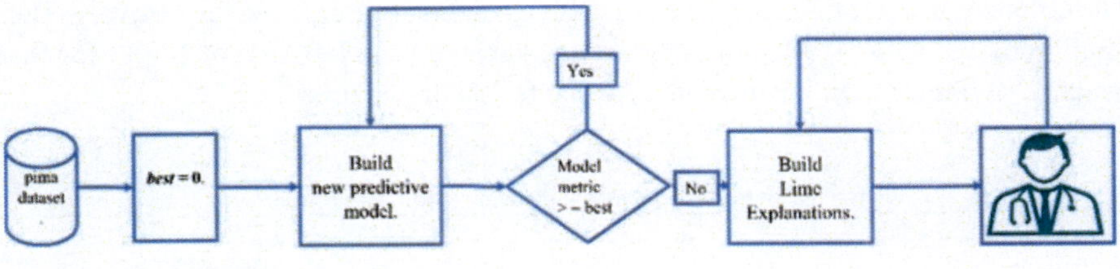

Fig. 1 Experimental setup

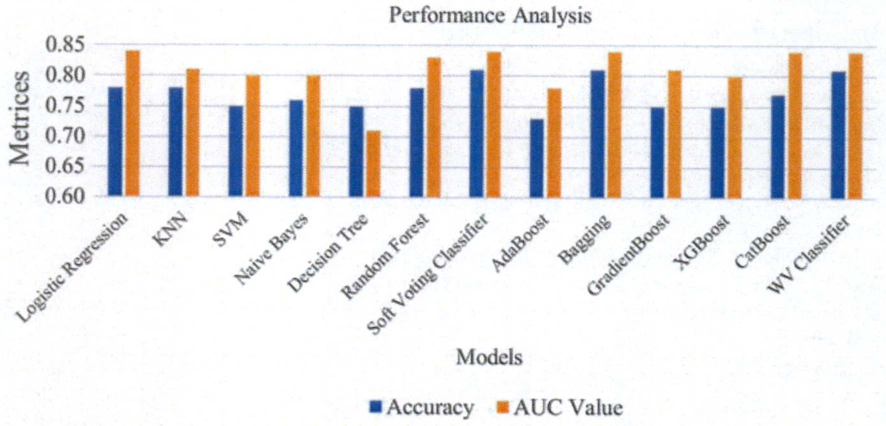

Fig. 2 Performance analysis

4.2 *Results*

As the dataset is imbalanced, accuracy may not be the right metric to select the best-performing model as it misleads the classification decisions. The accuracy and AUC values are popular metric(s) for comparing predictive models on class imbalanced datasets. From performance analysis bar graphs presented in Fig. 2, it is clear that soft voting classifier has the best values for accuracy (81%) and AUC (84%). Therefore, we have selected soft voting classifier as the best-performing complex predictive model to generate explanations for the instance of interest using LIME.

4.3 *LIME Explanations*

An instance of test data is shown in Fig. 3.

The output generated by LIME explainer for the instance of Fig. 3 is shown in Fig. 4. The three numerical values on the top left of Fig. 3 represent the intercept, i.e., constant part of the mathematical representation of the simple model, predicted local prediction of simple model, and right: prediction of complex model, respectively. The colors blue and orange represent negative and positive classes, respectively. The high predicted value 0.90 shown by orange baron on the left side of Fig. 3 for the instance of Table 1 can be attributed for the following factors.

	Num_preg	glucose_conc	diastolic_bp	thickness	insulin	bmi	diab_pred	Age	Actual	Predicted
113	0.046	-1.41	-0.37	-1.29	-0.69	0.25	-0.24	-0.7	0	1

Fig. 3 Single instance of test dataset

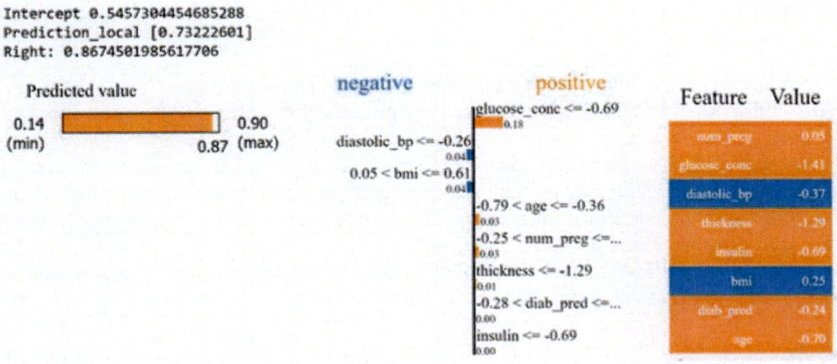

Fig. 4 LIME explanations

- The high value of glucose conc. has a positive effect on prediction.
- The low values of BMI and diastolic BP have a negative effect on prediction.

5 Conclusion and Future Scope

Diabetes mellitus is a long-lasting incurable disease that is more commonly found in humans of all ages, these days. Therefore, the early detection of Diabetes mellitus is the current need. The core objective of this study is to find the best-performing predictive model with local explanations model. Classical ML algorithms that have been employed in the previous decade were compared for accuracy. The models proposed in the last five years were lacking the interpretability of the model. Hence, the authors proposed an explainable ensemble classifier based on voting with a combination of five machine learning algorithms that include LR, KNN, DT, RF, and AdaBoost classifiers. The Pima Indian diabetes dataset has been taken for study. The proposed ensemble classifier has given 81% accuracy when tested on the well-known publicly available Pima Indian diabetes dataset. To generate local explanations, we used the well-known post-hoc XAI method—LIME. In the future work, this model may be enhanced by using deep neural network models and can be extended to generate explanations with counterfactuals.

References

1. World Health Organization et al. (2018) Global report on diabetes, 2016
2. Tao Z, Shi A, Zhao J (2015) Epidemiological perspectives of diabetes. Cell Biochem Biophys 73(1):181–185
3. Saba L, Sanagala SS, Gupta SK, Koppula VK, Laird JR, Viswanathan V, Sanches MJ, Kitas GD, Johri AM, Sharma N et al (2021) A multicenter study on carotid ultrasound plaque tissue characterization and classification using six deep artificial intelligence models: a stroke application. IEEE Trans Instrum Meas 70:1–12
4. Misra D, Mohanty SN, Agarwal M, Gupta SK (2020) Convoluted cosmos: classifying galaxy images using deep learning. In: Data management, analytics and innovation. Springer, pp 569–579
5. Agarwal M, Sinha A, Gupta SK, Mishra D, Mishra R (2020) Potato crop disease classification using convolutional neural network. In: Smart systems and IoT: innovations in computing. Springer, pp 391–400
6. Suri JS, Agarwal S, Pathak R, Ketireddy V, Columbu M, Saba L, Gupta SK, Faa G, Singh IM, Turk M et al (2021) Covlias 1.0: lung segmentation in covid19 computed tomography scans using hybrid deep learning artificial intelligence models. Diagnostics 11(8):1405
7. Agarwal M, Kaliyar RK, Singal G, Gupta SK (2019) Fcnn-lda: a faster convolution neural network model for leaf disease identification on apple's leaf dataset. In: 2019 12th international conference on information & communication technology and system (ICTS). IEEE, pp 246–251
8. Balamurugan D, Aravinth S, Reddy P, Rupani A, Manikandan A (2022) Multiview objects recognition using deep learning-based wrap-cnn with voting scheme. Neural Processing Letters 1–27
9. Shaker Reddy PC, Sureshbabu A (2020) An enhanced multiple linear regression model for seasonal rainfall prediction. International Journal of Sensors Wireless Communications and Control 10(4):473–483
10. Hameed SA (2022) An efficient method of classification the gestational diabetes using id3 classifier. Al-Nahrain Journal of Science 25(1):51–58
11. Jakka A, Vakula Rani J (2019) Performance evaluation of machine learning models for diabetes prediction. International Journal of Innovative Technology and Exploring Engineering (IJITEE) 8(11):1976–1980
12. Kalagotla SK, Gangashetty SV, Giridhar K (2021) A novel stacking technique for prediction of diabetes. Comput Biol Med 135:104554
13. Kumari S, Kumar D, Mittal M (2021) An ensemble approach for classification and prediction of diabetes mellitus using soft voting classifier. International Journal of Cognitive Computing in Engineering 2:40–46
14. Lu H, Uddin S, Hajati F, Moni MA, Khushi M (2022) A patient networkbased machine learning model for disease prediction: the case of type 2 diabetes mellitus. Appl Intell 52(3):2411–2422
15. Maniruzzaman M, Kumar N, Abedin MM, Islam MS, Suri HS, El-Baz AS, Suri JS (2017) Comparative approaches for classification of diabetes mellitus data: machine learning paradigm. Comput Methods Programs Biomed 152:23–34
16. Palimkar P, Shaw RN, Ghosh A (2022) Machine learning technique to prognosis diabetes disease: random forest classifier approach. In: Advanced computing and intelligent technologies. Springer, pp 219–244
17. Prabhu P, Selvabharathi S (2019) Deep belief neural network model for prediction of diabetes mellitus. In: 2019 3rd international conference on imaging, signal processing and communication (ICISPC). IEEE, pp 138–142
18. Sarwar MA, Kamal N, Hamid W, Shah MA (2018) Prediction of diabetes using machine learning algorithms in healthcare. In: 2018 24th international conference on automation and computing (ICAC). IEEE, pp 1–6
19. Find opendatasets and machine learning projects—kaggle. https://www.kaggle.com/datasets. Accessed on 06 June 2022

20. Ribeiro M, Singh S, Guestrin C (2019) Why should I trust you? Explaining the predictions of any classifier. arxiv160204938 cs stat. 2016
21. Vivekanand A, Vadlakonda D, Lendale V (2021) Performance analysis of predictive models on class balanced datasets using oversampling techniques. Soft computing and signal processing. Springer, Singapore, pp 375–383
22. Felzmann H, Fosch-Villaronga E, Lutz C, Tam'o-Larrieux A (2020) Towards transparency by design for artificial intelligence. Science and Engineering Ethics 26(6):3333–3361
23. Langer M, Oster D, Speith T, Hermanns H, Kästner L, Schmidt E, Sesing A, Baum K (2021) What do we want from explainable artificial intelligence (XAI)? A stakeholder perspective on XAI and a conceptual model guiding interdisciplinary XAI research. Artificial Intelligence 296:103473

Exploring the Potential of eXplainable AI in Identifying Errors and Biases

Raman Chahar and Urvi Latnekar

Abstract Artificial intelligence has virtually pervaded every field and its adaptation is a catalyst for organizational growth. However, the potential of artificial intelligence is often associated with a difficulty to understand the logic veiling behind its decision making. This is essentially the premise upon which XAI or eXplainable AI functions. In this field of study, researchers attempt to streamline techniques to provide an explanation for the decisions that the machines make. We endeavor to delve deeper into what explainable means and the repercussions of the lack of definition associated with the term. We intend to show in this paper that an evaluation system based solely on how easy it is to understand an explanation, without taking into account aspects such as fidelity, might produce potentially harmful explanation interfaces.

Keywords XAI · Artificial intelligence · Machine learning · Decision making

1 Introduction

Artificial Intelligence is an umbrella term commonly used to describe a multitude of technologies and approaches. While a precise definition of this term involves many complex aspects, such as defining the meaning of intelligence and its relation with human thinking ([9], Chap. 1.1), in this paper we will focus on those computing techniques currently used to take automated decisions in complex situations, i.e., in which the desired behavior cannot be easily synthesized using simple rules.

Most of these algorithms are in fact built to reason in a statistical way, are composed of a large number of elements (e.g., neurons in DCNNs) and can have very complex structures.

R. Chahar
Delhi Technological University, New Delhi, India

U. Latnekar (✉)
Bennett University, Noida, India
e-mail: urvirl9@gmail.com

K. A. Reddy et al. (eds.), *Proceedings of Fourth International Conference on Computer and Communication Technologies*, Lecture Notes in Networks and Systems 606,
https://doi.org/10.1007/978-981-19-8563-8_41

This aspect represents a huge technical and ethical issue for this field, especially when building autonomous systems that are meant to replace or aid humans in highly impacting decisions. If we can't explain "why" a certain algorithm took a certain decision, how can we trust these systems? How do we ensure that their internal models are not biased or broken? How do we understand when the machine is failing?

The problem of introspection and accountability for these systems is a very serious one. Marvin Minsky et al. raised the issue that AI can function as a form of surveillance, with the biases inherent in surveillance, suggesting Humanistic Intelligence (HI) as a way to create a more fair and balanced "human-in-the-loop" AI [8].

As a natural result of these emerging concerns about AI, the field of Explainable AI (XAI) was born. The goal of this research field is to build systems that can provide humans with a deeper understanding of AI algorithms [5], with the ultimate objective of making errors and biases easier to spot or predict and AI-based systems generally more trustworthy.

In this paper, we will analyze some of the extreme consequences of the lack of such definition, and more generally the lack of a comprehensive way to evaluate AI explanations.

To explain this idea, we will proceed as follows.

Section 2 provides some background on the problems which XAI is trying to solve and a classification of the solutions that are currently being developed. Section 4 introduces the problem of defining interpretability, and proposes a classification of the aspects that define an explanation. Section 5 discusses the idea that explanation interfaces might be able to fool a human user into believing that a specific algorithm is doing the right thing, leveraging his or her own bias. Finally, Sect. 6 contains some concluding thoughts on this subject.

2 Background

2.1 The XAI Approach

The term *Explainable AI* refers to methods and techniques in the application of artificial intelligence that aim at improving the possibility for humans to understand AI algorithms. We can consider [3] as a starting point for modern XAI research. The idea is that we want to be able to better understand when a given AI-based system is doing something wrong, when we can trust it and why an error occurred. The three main aspects that XAI aims to improve are: *accountability*, by enabling vendors, companies and governments to verify the technology they are using or providing, *transparency*, since the reasons for a certain decisions are in theory stated in an understandable way, and *fairness*, since understanding the reasons behind automated decisions also enables us to challenge them.

3 Methods

From a general perspective, [2] identifies two families of approaches to this problem:

Transparent box design, which aims at building algorithms that are more interpretable by design. A transparent box design is a cognitive process to try and simplify things so it is easier for the human brain to understand them. This enables people with cognitive impairments, parents, or carers of young children, among others, as well as scholars. It has been seen that clear visualizations help in various types of user engagement too.

Reverse-engineering approaches, also called *post-hoc interpretability* approaches, which try to provide explanations for already existing algorithms. Post-hoc interpretability refers to the nuanced understanding and prediction of anonymous public data.

Some examples of the latter type are listed in Ref. [4].

Visualization, for instance, focuses on representing visually some key aspects of the model, for example which pixels of an image are important for a classification output.

Approximation consists in using simple models or simplifying already existing models: in *single tree approximation*, for example, the internal structure of an AI algorithm is approximated to a classification tree, shown in Fig. 1.

Causal Models (CAMEL) try to generate causal explanations of Machine Learning operations and present them to the user as intuitive narratives. A scheme of the architecture needed for this approach is illustrated in Fig. 2.

Other approaches include *Learning and Communicating Explainable Representations*, where explanations themselves are learned as a separate part of the training process, and *Explanation by Example*, where the AI is able to provide an example, or a *prototype*, of how it thinks that a typical member of a given class should appear and/or which characteristics should be changed to change the outcome.

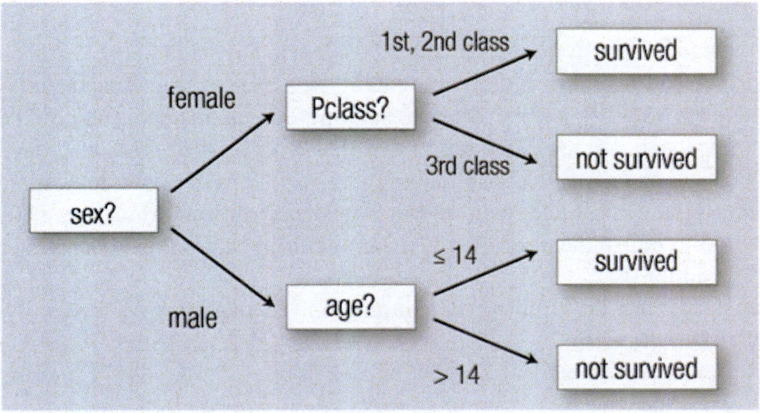

Fig. 1 A simple decision tree

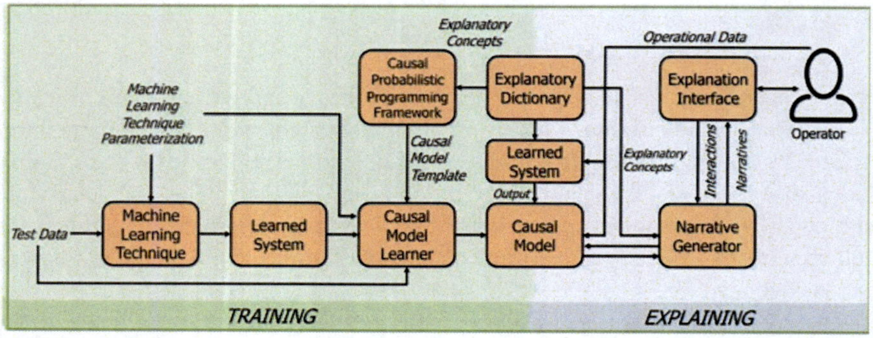

Fig. 2 Architecture of the CAMEL approach

It is important to notice how these approaches differ in how *thick* the explanation interface is, i.e., how many complex manipulations the initial model undergoes before being presented to the user. Intuitively, we can see for example that the visualization approach tries to give a close insight on how the internal elements are activated by a certain picture, while in techniques such as CAMEL and Learned Explanations there is a much more indirect connection between elements of the original model and elements of the explanation, which is also reflected on the increased complexity of the interface itself.

This intuitive idea will be further expanded in Sect. 5.1 using the concept of *fidelity*.

4 Defining Interpretability

4.1 Dimensions of AI Explanation

As anticipated in Sect. 1, one fundamental problem in the field of XAI is that there is no single conventional notion of interpretability. Reference [7] goes as far as considering the term itself ill-defined, therefore stating that claims about interpretability generally have a quasi-scientific nature. Reference [2] on the other hand, considers the lack of a mathematical description as an obstacle for the future development of this field. Reference [3] itself defines the formalization of an evaluation metric for explanations as one of the goals of the XAI program, to be developed in parallel with technical solutions.

When analyzing the problem of defining and evaluating interpretability, two questions naturally arise:

Explainable to whom? The concept of *user* of an AI system is not always well-defined, nor is the concept of user of an explanation. This might include:

- The *developer* of the AI system, as he is only partially in control of what the algorithm does.
- The *operator* of an AI system: many AI algorithms nowadays are being used as an input for a human to make decisions on a certain subject.
- The *end user* which is affected by the decision of an AI.

Explainable for which purpose? Different users have different needs, which may partially overlap. Therefore there is a variety of goals that explanations try to accomplish, which are possibly in contrast with each other. Some of these are:

- *Debugging*: finding errors and backtracking them to a specific reason.
- *Human-in-the-loop*: creating systems where human and AI decisions can co-exist and influence each other.
- *Validation*: understanding if a certain model is good enough to be deployed for a certain task.
- *Failure Prediction*: understanding which are the weaknesses of an AI system and when it is likely to fail.
- *Appeal AI decisions*[1]: giving the right to users and citizens that are affected by AI decisions to know, understand and possibly appeal decisions that are automated with AI systems.

It appears quite evident that different XAI solutions built with different users in mind will have very different notions of what a good explanation is.

4.2 Possible Metrics

Bearing in mind the goals of XAI, there are a number of metrics that can be used to characterize and evaluate a solution:

- *Complexity*: how many elements are there in the explanation?
- *Clearness*: how cognitively hard is the explanation? How difficult is it to understand the correspondence between the elements of the explanation and the information we are trying to gain?
- *Informativeness*: how much information, weighted on how meaningful it is, can be extracted by the explanation? E.g., does the explanation significantly modify the level of uncertainty about the AI behavior?
- *Fidelity*: how closely does the explanation represent the functioning of the system? Are all the facts inferred from the explanation also applicable to the original system?

Clearly, a specific metric will be more or less important depending on the specific user and use-case. There is however a deeper distinction that has to be made, which is related to how these metrics are measured.

[1] This goal is not explicitly listed in the original scope of XAI, but has gained traction recently with the introduction of the concept of *right for an explanation* in Europe's new GDPR [10].

Complexity, for instance, is often measured using a proxy quantity such as the number of elements in the explanation, which can be for example the depth of the decision tree or the number of neurons. On the other hand, clearness and informativeness are more difficult to quantify a-priori, but could be empirically evaluated by providing the explanations to a group of humans and verifying how they respond.

In general, we can identify two ways of evaluating an AI explanation: one is using a direct measurement of some quantity that we can derive directly from the explanation. The second one is considering an explanation itself a black-box, and check if it actually provides a better understanding of the AI model to some selected group of individuals used as a benchmark. While the first method is not always feasible, since choosing which quantity is representative of a certain aspect is in itself a difficult decision to make, the second method clearly presents the same problems of opaqueness and unreliability that AI models themselves have.

5 The Troubles of Explanations

5.1 Measuring Fidelity

Of all the metrics highlighted in Sect. 4.2, fidelity, also called *faithfulness* in literature [1], is probably the most complex to evaluate. On one hand, the maximum fidelity is already represented by the implementation itself, but on the other hand the reason we need explanations is that the implementation itself is not clear enough.

This is particularly important since AI explanations are also targeted to unspecialized users, which need to understand what's happening without necessarily having a solid background on the internal functioning of such systems.

Yet, fidelity plays a fundamental role when we have to evaluate an AI algorithm, as it quantifies the difference between what is being evaluated (the AI model) and the instrument we are using for this evaluation (the AI explanation). This represents in some sense the "measurement error" introduced by the explanation.

Let's take for example the situation depicted in Fig. 3: in this case, a human operator is evaluating an AI model trough an explanation interface.

While this idea might seem easy enough to understand, devising an operational way to measure it is a non-trivial task.

Let's take for example Causal Models: in this case, the explanation and the original model will typically have a very different nature, since the explanation interface produces causal relationships, while the AI model typically reasons in terms of statistical correlation. In this case, how can we measure the fidelity of this interface?

On the other hand, being unable to measure fidelity poses another question: if both the AI and the explanation are treated as black boxes, how can we be sure that evaluating the AI using that explanation interface will effectively improve our understanding of the underlying AI model? Couldn't it be that we just *think* we understand it?

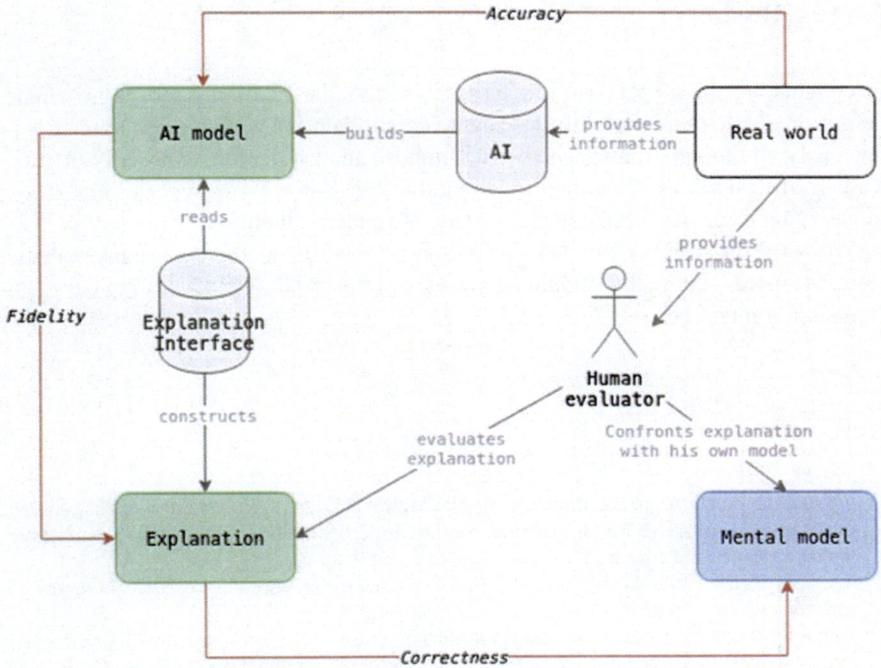

Fig. 3 Evaluation process of an AI through an explanation interfaces

5.2 Decision Making Biases

Human decision making is known to be affected by many cognitive bias, which are deeply rooted in our thinking and are often difficult, if not impossible, to exclude when we make decisions. Recently, [6] studied the consequences of the *framing effect* in the domain of AI, in particular how likely is a person to accept or reject an AI recommendation based on how the output was framed. An interesting result of this research is, for example, that "*perceived reasonableness was significantly higher when the suggestion of AI was provided before the decision is made than after the decision is made when perceived accuracy was controlled*" ([6], page 5).

While this is not a direct study on AI explanation interfaces, it does show how the same local decision of an AI can be judged differently simply varying the timing of the explanation. Similar results have been observed when varying how the explanation is framed (positive or negative sentences, etc.).

This shows how the evaluation of the correctness of an AI model is not only a subjective matter, but can vary in the same individual depending on factors that are external to the AI behavior itself.

6 Conclusions

In conclusion, this paper should have shown how the fact that there is no single definition of what interpretability is and no comprehensive way of evaluating simultaneously all the important aspects that compose an explanation, especially fidelity, leads to the possibility of creating yet another black-box layer over the black-box model, which can accentuate biases instead of reducing them.

While the proposed argument is just a thought experiment, there are many realistic elements in this setting that should warn us about the possibility of creating deceitful explanation interfaces.

References

1. Gilpin LH, Bau D, Yuan BZ, Bajwa A, Specter M, Kagal L (2018) Explaining explanations: an overview of interpretability of machine learning. In: 2018 IEEE 5th international conference on data science and advanced analytics (DSAA), pp 80–89
2. Guidotti R, Monreale A, Ruggieri S, Turini F, Giannotti F, Pedreschi D (2018) A survey of methods for explaining black box models. ACM Comput Surv 51(5)
3. Gunning D (2017) Darpa's explainable artificial intelligence (xai) program. In: Proceedings of the 24th international conference on intelligent user interfaces, IUI'19, page ii, New York, NY, USA. Association for Computing Machinery
4. Gunning D (2018) Xai for nasa
5. Islam MR, Ahmed MU, Barua S, Begum S (2022) A systematic review of explainable artificial intelligence in terms of different application domains and tasks. Appl Sci 12(3):1353
6. Kim T, Song H (2020) The effect of message framing and timing on the acceptance of artificial intelligence's suggestion
7. Lipton Z (2016) The mythos of model interpretability. Commun ACM 61:10
8. Minsky M, Kurzweil R, Mann S (2013) The society of intelligent veillance. In: 2013 IEEE international symposium on technology and society (ISTAS): social implications of wearable computing and augmediated reality in everyday life, pp 13–17
9. Russell S, Norvig P (2009) Artificial intelligence: a modern approach, 3rd edn. Prentice Hall Press, USA
10. Selbst AD, Powles J (2017) Meaningful information and the right to explanation. Int Data Privacy Law 7(4):233–242

Novel Design of Quantum Circuits for Representation of Grayscale Images

Mayukh Sarkar

Abstract The advent of quantum computing has influenced researchers around the world to solve multitudes of computational problems with this promising technology. Feasibility of solutions for computational problems, and representation of various information, may allow quantum computing to replace classical computer in near future. One such challenge is the representation of digital images in quantum computer. Several works have been done to make it possible. One such promising technique, named Quantum Probability Image Encoding (Yao in (Phys Rev X 7(3):031–041) [1], Quantum Edge Detection—QHED Algorithm on Small and Large Images [2]), requires minimal number of qubits, where the intensity of n pixels is represented as the statevector of $\lceil \log_2 n \rceil$ qubits. Though there exist quantum circuit design techniques to obtain arbitrary statevector, they consider statevector in general Hilbert space. But for image data, considering only real vector space is sufficient that may constraint the circuit in smaller gate set, and possibly can reduce number of gates required. In this paper, construction of such quantum circuits has been proposed.

Keywords Quantum computing · Digital image processing · Quantum image representation

1 Introduction

Quantum computing, one of the buzzwords in today's research, started from the idea of quantum mechanical model of Turing machine proposed by Paul Benioff, in 1980 [3]. In 1982, Richard Feynman, ideate the possibility of quantum computer [4], and it became a buzzword when, in 1994, Peter Shor proved its capability by proposing a quantum polynomial time algorithm of integer factorization [5]. Since then, researchers all around the world have been trying to solve multitudes of computational problems using this technology. One of the promising applications is the domain of image processing using this powerful paradigm.

M. Sarkar (✉)
Department of Computer Science and Engineering, Motilal Nehru National Institute of Technology Allahabad, Prayagraj, India
e-mail: mayukh@mnnit.ac.in

K. A. Reddy et al. (eds.), *Proceedings of Fourth International Conference on Computer and Communication Technologies*, Lecture Notes in Networks and Systems 606,
https://doi.org/10.1007/978-981-19-8563-8_42

To implement any image processing algorithm on a quantum computer, it is first necessary to represent the image using qubits. There have been a multitude of techniques devised for representation of an image on a quantum computer, such as Real Ket [6], Qubit Lattice [7], Entangled image [8], Flexible representation of quantum images (FRQI) [9], Novel enhanced quantum representation (NEQR) [10], a novel quantum representation of color digital images (NCQI) [11], quantum probabilistic image encoding (QPIE) [1, 2] etc., and their further improvements. A good overview of major quantum image representations, along with an improvement of NCQI, named INCQI, can be found in Su et al. [12]. Among these techniques, QPIE represents the pixels of a grayscale image, as probability amplitudes of statevector of a possible quantum state. The advantage of this technique is requirement of minimal number of qubits, namely $\lceil \log_2 \mathrm{n} \rceil$ qubits for representation of n pixels. Higher the number of qubits used in a system, it becomes more costly and error-prone. Added advantage is the straightforward implementation of edge detection [1], if pixels are represented using QPIE technique.

But the techniques used in quantum circuit design for arbitrary statevector [13, 14], considers general statevectors in complete Hilbert space. But image data do not require complex numbers, and hence it is sufficient to consider only real vector space. This allows us to confine the gates used in circuit within NOT, CNOT, Toffoli (NCT), and R_y gates, and their controlled counterparts. It has been shown in Sect. 3.2 that this technique also requires lesser number of gates than circuit obtained using generalized techniques [13, 14], at least for small circuits. This paper proposes an algorithm to design quantum circuits consisting of only gates with real matrices, to represent an arbitrary real statevector, which can be utilized to represent pixels of a grayscale image on a quantum computer following QPIE technique.

Section 2 gives a brief overview of QPIE techniques. Section 3 describes the complete proposed work, in which at first, single-qubit, and two-qubits circuits are designed as base cases for smaller statevectors, followed by general recursive algorithm to design quantum circuit for real statevectors of any dimension. Section 4 then finally concludes the work.

2 Background Information

Several representations of digital images have been proposed in quantum computing literature over last decade, some of which has been mentioned in Para 2 of Sect. 1. Among the proposed techniques, the most common ones used in practice are FRQI, NEQR, and QPIE. Yao et al. [1], in the base paper of QPIE, have proposed a quantum image representation technique in which pixel values are encoded as probability amplitudes. In this section, an overview of QPIE technique is being provided.

In QPIE, the 2-dimensional image data is first unfolded into a vector form. If $I = (f_{ij})_{MxL}$ be an image data, where f_{ij} represents the pixel value at position (i, j), M and L are number of rows and number of columns, respectively, it is first unfolded into vector v as,

$$v = \left[f_{11}, f_{21}, \ldots, f_{M1}, f_{21}, f_{22}, \ldots, f_{M2}, \ldots, f_{ij}, \ldots, f_{ML}\right]^T \tag{1}$$

Then this image data is encoded into a quantum state

$$|I\rangle = \sum_{k=0}^{2^n-1} \mathrm{c_k}\,|k\rangle \tag{2}$$

of $n = \lceil \log_2(ML) \rceil$ qubits. $|k\rangle$ represents the computational basis encoding the position (i, j) and $c_k = \frac{f_{ij}}{\sqrt{\sum(f_{ij})^2}}$ represents the pixel values, encoded as probability distribution satisfying $\sum |c_k|^2 = 1$.

Given that, c_k and $\sum |c_k|^2$ can be calculated efficiently, the n-qubit state representing the image data can be created efficiently in $O(poly(n))$ steps, where $poly(n)$ represents some polynomial function of n [1]. Arbitrary state preparation techniques proposed by Grover et al. [13] and Soklakov et al. [14], includes unit vectors in 2^n-dimensional Hilbert space, i.e., vectors may contain complex amplitudes. But pixel data of an image is always real. Though the generalized state preparation techniques can also prepare such states, removing the necessity of handling complex amplitudes has the ability to obtain circuits with smaller subset of gates, such as NCT, and $\mathrm{R_y}$ gates and their controlled counterparts, which keeps the states only in real vector space. Current paper proposes exactly the same, i.e., goal of the current paper is to propose an algorithm to produce a quantum circuit producing state as in Eq. (2), solely for unit vectors in real vector space, such as normalized pixel data of a grayscale image.

Note that, as the pixel data is being represented as probability amplitudes of a quantum statevector, it cannot be used to store the image for further retrieval, as measuring the statevector will collapse the complete quantum state, thereby destroying the complete pixel data. This work is expected to be important in the applications requiring state preparation circuits, where an image needs to be represented using minimal number of qubits, temporarily. These qubits are then further processed via the image processing circuit, performing important image processing applications.

3 Proposed Work

In this section, the technique to generate a quantum circuit with n qubits that will produce an arbitrary unit vector in 2^n-dimensional real vector space, is proposed. The circuit consists of only NCT, and R_y gates and their controlled counterparts. To demonstrate the technique, let us first start with a 2-dimensional unit real vector.

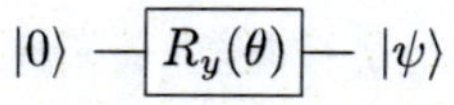

Fig. 1 Quantum circuit to generate arbitrary 2-D real statevector

3.1 *Single-Qubit Circuit Generating 2-Dimensional Arbitrary Real Statevector*

Let us consider an arbitrary real vector $|\psi\rangle = \begin{bmatrix} \alpha_1 \\ \alpha_2 \end{bmatrix}$ with $\alpha_1^2 + \alpha_2^2 = 1$. Thus we can readily consider $\alpha_1 = \cos\frac{\theta}{2}$ and $\alpha_2 = \sin\frac{\theta}{2}$ for certain angle θ, which can be obtained as $\theta = 2\arccos(\alpha_1)$. The following circuit will generate the desired state (Fig. 1).

3.2 *Two-Qubit Circuit Generating 4-Dimensional Arbitrary Real Statevector*

Now, let us consider an arbitrary real vector $|\psi\rangle = \begin{bmatrix} \alpha_1 \\ \alpha_2 \\ \alpha_3 \\ \alpha_4 \end{bmatrix}$ satisfying $\alpha_1^2 + \alpha_2^2 + \alpha_3^2 + \alpha_4^2 = 1$. Thus we can consider three real angles $\theta_1, \theta_2, \theta_3$ such that $\alpha_1 = \cos\frac{\theta_1}{2}$, $\alpha_2 = \sin\frac{\theta_1}{2}\cos\frac{\theta_2}{2}$, $\alpha_3 = \sin\frac{\theta_1}{2}\sin\frac{\theta_2}{2}\cos\frac{\theta_3}{2}$, and $\alpha_4 = \sin\frac{\theta_1}{2}\sin\frac{\theta_2}{2}\sin\frac{\theta_3}{2}$. This is in accordance with the spherical coordinate system.

Now, with initial state of a two-qubit quantum system being $(1, 0, 0, 0)^T$, the circuit generating the desired quantum state can be designed as follows.

(a) Employing $R_y(\theta_1)$ on first qubit yields the state $\left(\cos\frac{\theta_1}{2}, \sin\frac{\theta_1}{2}, 0, 0\right)^T$, following the similar logic in Sect. 3.1.

(b) Employing controlled-$R_y(-\theta_2)$ gate with control on first qubit and target on second qubit performs following operation.

$$\begin{bmatrix} 1 & 0 & 0 & 0 \\ 0 & \cos\left(-\frac{\theta_2}{2}\right) & 0 & -\sin\left(-\frac{\theta_2}{2}\right) \\ 0 & 0 & 1 & 0 \\ 0 & \sin\left(-\frac{\theta_2}{2}\right) & 0 & \cos\left(-\frac{\theta_2}{2}\right) \end{bmatrix} \begin{bmatrix} \cos\frac{\theta_1}{2} \\ \sin\frac{\theta_1}{2} \\ 0 \\ 0 \end{bmatrix} = \begin{bmatrix} \cos\frac{\theta_1}{2} \\ \sin\frac{\theta_1}{2}\cos\frac{\theta_2}{2} \\ 0 \\ -\sin\frac{\theta_1}{2}\sin\frac{\theta_2}{2} \end{bmatrix}.$$

(c) Employing controlled-$R_y(\pi + \theta_3)$ gate with control on second qubit and target on first qubit performs following operation.

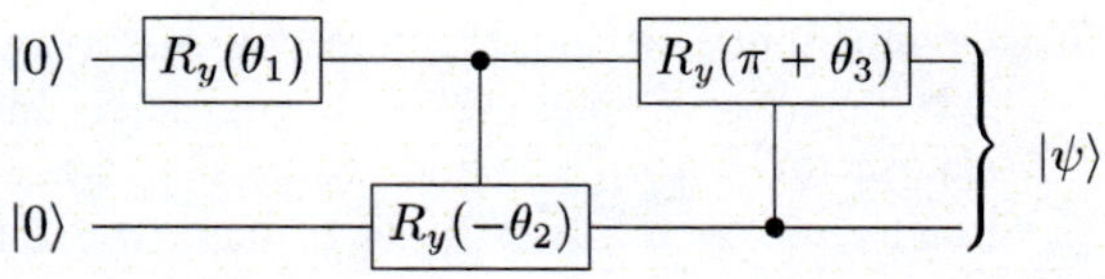

Fig. 2 Quantum circuit to generate arbitrary 4-D real unit vector

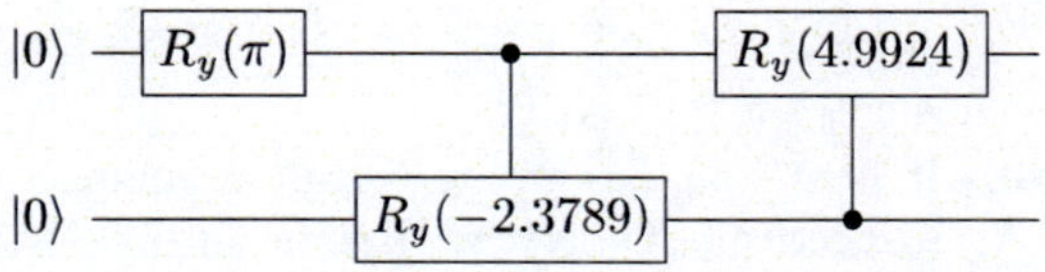

Fig. 3 Quantum circuit to generate statevector corresponding to [0, 128, 192, 255]

$$\begin{bmatrix} 1 & 0 & 0 & 0 \\ 0 & 1 & 0 & 0 \\ 0 & 0 & \cos\left(\frac{\pi}{2}+\frac{\theta_3}{2}\right) & -\sin\left(\frac{\pi}{2}+\frac{\theta_3}{2}\right) \\ 0 & 0 & \sin\left(\frac{\pi}{2}+\frac{\theta_3}{2}\right) & \cos\left(\frac{\pi}{2}+\frac{\theta_3}{2}\right) \end{bmatrix} \begin{bmatrix} \cos\frac{\theta_1}{2} \\ \sin\frac{\theta_1}{2}\cos\frac{\theta_2}{2} \\ 0 \\ -\sin\frac{\theta_1}{2}\sin\frac{\theta_2}{2} \end{bmatrix}$$

$$= \begin{bmatrix} \cos\frac{\theta_1}{2} \\ \sin\frac{\theta_1}{2}\cos\frac{\theta_2}{2} \\ \sin\frac{\theta_1}{2}\sin\frac{\theta_2}{2}\cos\frac{\theta_3}{2} \\ \sin\frac{\theta_1}{2}\sin\frac{\theta_2}{2}\sin\frac{\theta_3}{2} \end{bmatrix}.$$

The output state, as observed, matches with our desired statevector. The circuit thus demonstrated, is shown as in Fig. 2.

The generated circuit has been tested on several randomly generated 4-dimensional real array with elements in range [0, 255], using Qiskit library in Python 3.9.

As an example, when the above-mentioned procedure is employed on the pixel data [0, 128, 192, 255], the following quantum circuit, as shown in Fig. 3, is produced.

Whereas as accessed on the day of this writing, the circuit proposed by the Qiskit tutorial website [2], for the 4-pixel image with pixel values [0, 128, 192, 255], consists of 5 quantum gates.

3.3 Multi-Qubit Circuit Generation for Arbitrary 2^n-Dimensional Real Statevector

Suppose we have been given any arbitrary grayscale image. We can readily pad the image with zeros to make number of pixels as power of 2. Let number of pixels, after padding, turns out to be 2^n. After scaling and converting it into probabilistic amplitudes of a possible quantum system statevector, the n-qubit quantum circuit to generate the arbitrary 2^n-dimensional statevector, can be obtained as follows. Generation of a 3-qubit quantum circuit for 8-dimensional statevector, is being shown as example along with each step.

(a) Obtain spherical angles from the statevector. With 2^n-dimensional statevector, we will obtain $(2^n - 1)$ angles. As an example, for an 8-dimensional statevector $[c_0, c_1, c_2, c_3, c_4, c_5, c_6, c_7]$, we can obtain 7 spherical angles $[\alpha_0, \alpha_1, \alpha_2, \alpha_3, \alpha_4, \alpha_5, \alpha_6]$ such that, the statevector can be represented as $\left[\cos\frac{\alpha_0}{2},\ \sin\frac{\alpha_0}{2}\cos\frac{\alpha_1}{2},\ \sin\frac{\alpha_0}{2}\sin\frac{\alpha_1}{2}\cos\frac{\alpha_2}{2},\ \sin\frac{\alpha_0}{2}\sin\frac{\alpha_1}{2}\sin\frac{\alpha_2}{2}\cos\frac{\alpha_3}{2},\ \sin\frac{\alpha_0}{2}\sin\frac{\alpha_1}{2}\sin\frac{\alpha_2}{2}\sin\frac{\alpha_3}{2}\cos\frac{\alpha_4}{2},\ \sin\frac{\alpha_0}{2}\sin\frac{\alpha_1}{2}\sin\frac{\alpha_2}{2}\sin\frac{\alpha_3}{2}\sin\frac{\alpha_4}{2}\cos\frac{\alpha_5}{2},\ \sin\frac{\alpha_0}{2}\sin\frac{\alpha_1}{2}\sin\frac{\alpha_2}{2}\sin\frac{\alpha_3}{2}\sin\frac{\alpha_4}{2}\sin\frac{\alpha_5}{2}\cos\frac{\alpha_6}{2},\ \sin\frac{\alpha_0}{2}\sin\frac{\alpha_1}{2}\sin\frac{\alpha_2}{2}\sin\frac{\alpha_3}{2}\sin\frac{\alpha_4}{2}\sin\frac{\alpha_5}{2}\sin\frac{\alpha_6}{2}\right]$.

(b) If $n = 1$ or 2, employ design techniques mentioned in Sects. 3.1 and 3.2, respectively. Otherwise, design an $(n-1)$-qubit arbitrary statevector generator circuit, recursively, employing the first $(n-1)$ qubits on the system. This will involve first $(2^{n-1} - 1)$ spherical angles, and will build up first $(2^{n-1} - 1)$ entries of the statevector completely, and 2^{n-1}th entry partially. As an example, for the 3-qubit system, employ the design of Sect. 3.2 with first $(2^{n-1} - 1) = 3$ angles, as shown in Fig. 4. The output of the partial circuit is $\left[\cos\frac{\alpha_0}{2}, \sin\frac{\alpha_0}{2}\cos\frac{\alpha_1}{2}, \sin\frac{\alpha_0}{2}\sin\frac{\alpha_1}{2}\cos\frac{\alpha_2}{2}, \sin\frac{\alpha_0}{2}\sin\frac{\alpha_1}{2}\sin\frac{\alpha_2}{2}, 0, 0, 0, 0\right]$. Observe that c_3 has been created partially.

(c) Employ an $(n-1)$-qubit controlled $R_y(\alpha_{2^{n-1}-1})$ gate, with control on first $(n-1)$ qubits and target on last qubit, where $\alpha_{2^{n-1}-1}$ represents the 2^{n-1}th spherical angle. For the 3-qubit system, this will employ $R_y(\alpha_3)$ on entries $0\underbrace{111\ldots111}_{n-1}$ (2^{n-1}th entry) and $1\underbrace{111\ldots111}_{n-1}$ (2^nth entry). The circuit in Fig. 5 has the output as $\left[\cos\frac{\alpha_0}{2}, \sin\frac{\alpha_0}{2}\cos\frac{\alpha_1}{2}, \sin\frac{\alpha_0}{2}\sin\frac{\alpha_1}{2}\cos\frac{\alpha_2}{2}, \sin\frac{\alpha_0}{2}\sin\frac{\alpha_1}{2}\sin\frac{\alpha_2}{2}\cos\frac{\alpha_3}{2}, 0, 0, 0, \sin\frac{\alpha_0}{2}\sin\frac{\alpha_1}{2}\sin\frac{\alpha_2}{2}\sin\frac{\alpha_3}{2}\right]$.

(d) Employ $(n-1)$ CNOT gates, one by one, on each of the first $(n-1)$ qubits. Each of these CNOT gates have controls on the last qubit. These $(n-1)$ gates will take the entry at $\underbrace{1111\ldots111}_{n}$ (last entry) to $1\underbrace{000\ldots000}_{n-1}$ ($2^{n-1}+1$th entry). As the continuation of the example, for the 3-qubit system, the circuit in Fig. 6 has

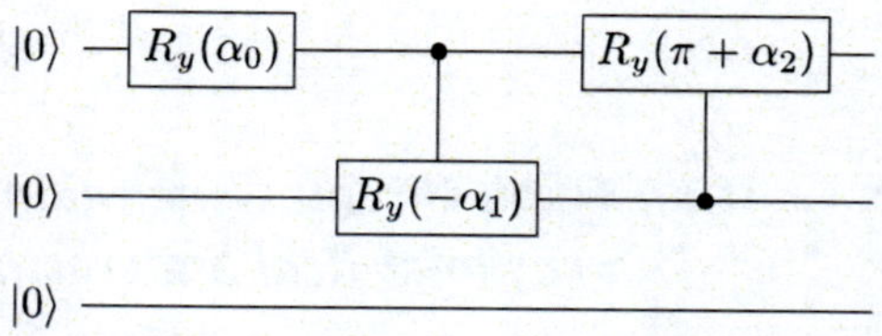

Fig. 4 Partial circuit for arbitrary 8-dimensional real statevector: Building $(n-1)$-qubit subcircuit recursively

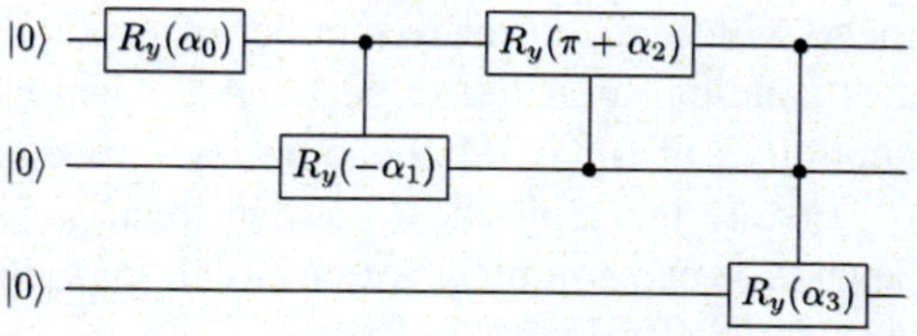

Fig. 5 Partial circuit for arbitrary 8-dimensional real statevector: Adding $(n-1)$ qubit controlled-R_y gate on last qubit

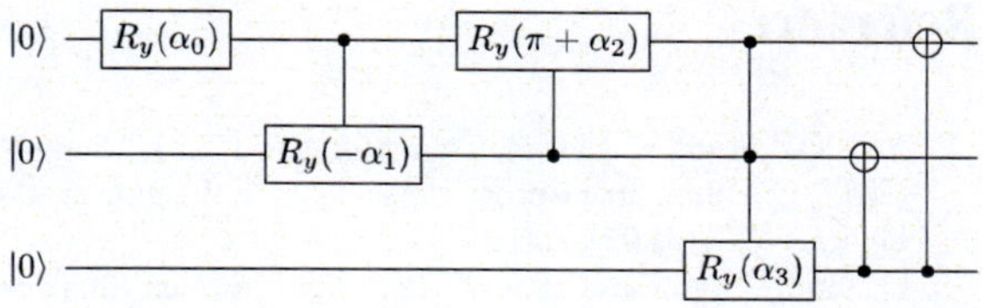

Fig. 6 Partial circuit for arbitrary 8-dimensional real statevector: Adding $(n-1)$ CNOT gates

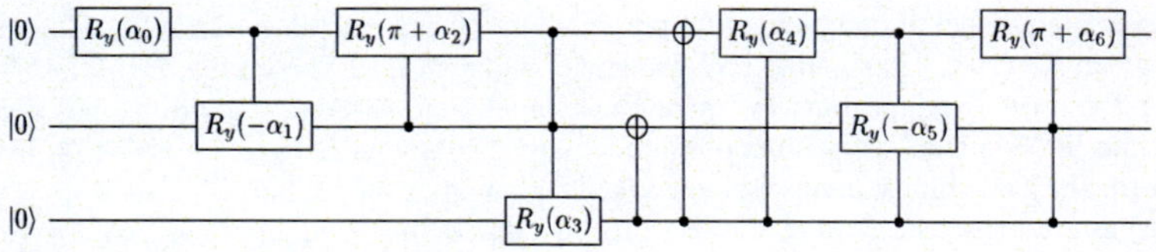

Fig. 7 Complete quantum circuit to generate arbitrary 8-dimensional real statevector

the output $[\cos\frac{\alpha_0}{2}, \sin\frac{\alpha_0}{2}\cos\frac{\alpha_1}{2}, \sin\frac{\alpha_0}{2}\sin\frac{\alpha_1}{2}\cos\frac{\alpha_2}{2}, \sin\frac{\alpha_0}{2}\sin\frac{\alpha_1}{2}\sin\frac{\alpha_2}{2}\cos\frac{\alpha_3}{2}, \sin\frac{\alpha_0}{2}\sin\frac{\alpha_1}{2}\sin\frac{\alpha_2}{2}\sin\frac{\alpha_3}{2}, 0, 0, 0]$.

(e) Employ another $(n-1)$-qubit arbitrary statevector generator circuit with last $(2^{n-1}-1)$ angles, recursively, on first $(n-1)$ qubits. Each gate in this sub-circuit must have additional control from last qubit. The final 3-qubit circuit is shown in Fig. 7. It has 2-qubit arbitrary statevector generator circuit of Fig. 2 with angles $[\alpha_4, \alpha_5, \alpha_6]$, employed after the partial circuit of Fig. 6, each having additional control from last qubit.

The circuit of Fig. 7 eventually has the desired output statevector. The circuit thus designed, has also been verified successfully with Qiskit library in Python 3.9, on several randomly generated 8-dimensional arrays with values in range [0, 255].

4 Conclusion

In this work, a novel technique to design a quantum circuit in order to create an arbitrary real statevector has been proposed. Designed circuit is composed solely of NCT, and R_y gate and its controlled counterparts. The circuit needs $O(poly(n))$ gates, where n is number of pixels in the image. Though the designed circuit may not have minimal number of gates performing the task, number of gates have been found to be lower than circuits designed using generalized techniques [13, 14], at least for small circuits. This work is expected to be important in the applications requiring state preparation circuits, where an image needs to be represented using minimal number of qubits, temporarily. The qubits representing pixel data can then be further processed via an image processing circuit for important and interesting applications.

References

1. Yao XW, Wang H, Liao Z, Chen MC, Pan J, Li J, Zhang K, Lin X, Wang Z, Luo Z, Zheng W (2017) Quantum image processing and its application to edge detection: theory and experiment. Phys Rev X 7(3):031–041
2. Quantum Edge Detection—QHED Algorithm on Small and Large Images. https://qiskit.org/textbook/ch-applications/quantum-edge-detection.html
3. Benioff P (1980) The computer as a physical system: a microscopic quantum mechanical Hamiltonian model of computers as represented by Turing machines. J Stat Phys 22:563–591
4. Feynman RP (1982) Simulating physics with computers. Int J Theor Phys 21(6/7):467–488
5. Shor PW (1994) Algorithms for quantum computation: discrete logarithms and factoring. In: 35th IEEE annual symposium on foundations of computer science, pp 124–134, IEEE
6. Latorre JI (2005) Image compression and entanglement. Comput Sci
7. Venegas-Andraca SE, Bose S (2003) Storing, processing, and retrieving an image using quantum mechanics. In: Proceedings of SPIE—the international society for optical engineering, vol 5105
8. Venegas-Andraca SE, Ball JL (2010) Processing images in entangled quantum systems. Quantum Inf Process 9:1–11
9. Le PQ, Dong F, Hirota K (2011) A flexible representation of quantum images for polynomial preparation, image compression, and processing operations. Quantum Inf Process 10:63–84
10. Zhang Y, Lu K, Gao YH, Wang M (2013) NEQR: a novel enhanced quantum representation of digital images. Quantum Inf Process 12:2833–2860
11. Sang JZ, Wang S, Li Q (2017) A novel quantum representation of color digital images. Quantum Inf Process 16:14
12. Su J, Guo X, Liu C, Lu S, Li L (2021) An improved novel quantum image representation and its experimental test on IBM quantum experience. Sci Rep 11(1):1–13
13. Grover L, Rudolph T (2002) Creating superpositions that correspond to efficiently integrable probability distributions. arXiv preprint quant-ph/0208112
14. Soklakov AN, Schack R (2006) Efficient state preparation for a register of quantum bits. Phys Rev A 73(1):012307

Trajectory Tracking Analysis of Fractional-Order Nonlinear PID Controller for Single Link Robotic Manipulator System

Pragati Tripathi, Jitendra Kumar, and Vinay Kumar Deolia

Abstract Increasing demand for automation is being observed especially during the recent scenarios like the Covid-19 pandemic, wherein direct contact of the healthcare workers with the patients can be life-threatening. The use of robotic manipulators facilitates in minimizing such risky interactions and thereby providing a safe environment. In this research work, a single link robotic manipulator (SLRM) system is taken, which is a nonlinear multi–input–multi–output system. In order to address the limitations like heavy object movements, uncontrolled oscillations in positional movement, and improper link variations, an adaptive fractional-order nonlinear proportional, integral, and derivative (FONPID) controller has been suggested. This aids in the effective trajectory tracking of the performance of the SLRM system under step input response. Further, by tuning the controller gains using genetic algorithm optimization (GA) based on the minimum objective function (J_{IAE}) of the integral of absolute error (IAE) index, the suggested controller has been made more robust for trajectory tracking performance. Finally, the comparative analysis of the simulation results of proportional & integral (PI), proportional, integral, & derivative (PID), fractional-order proportional, integral, & derivative (FOPID), and the suggested FONPID controllers validated that the FONPID controller has performed better in terms of minimum J_{IAE} and lower oscillation amplitude in trajectory tracking of positional movement of SLRM system.

Keywords Single link robotic manipulator · FONPID controller · GA · IAE · PID

P. Tripathi · J. Kumar (✉) · V. K. Deolia
Department of ECE, GLA University, Mathura, UP 281406, India
e-mail: jitendra.kumar@gla.ac.in

P. Tripathi
e-mail: pragati.tripathi_mtec20@gla.ac.in

V. K. Deolia
e-mail: vinay.deolia@gla.ac.in

K. A. Reddy et al. (eds.), *Proceedings of Fourth International Conference on Computer and Communication Technologies*, Lecture Notes in Networks and Systems 606,
https://doi.org/10.1007/978-981-19-8563-8_43

1 Introduction and Literature Surveys

Originally designed to substitute human intervention in high-risk environments such as handling radioactive and bio-hazardous materials, robotic manipulators consist of a series of interconnected linkages. In the recent Covid-19 pandemic situation, wherein several healthcare workers were exposed to unsafe environments, robotic surgeries played an instrumental role by enabling physical distancing and thereby reducing the risk factor.

These systems are susceptible to various process variations and initial conditions, making device management much more difficult [1]. In the domains of mechatronics and automation also, robotic manipulators are widely deployed. Positioning is the most basic goal in motion control and parallelly, one of the most important difficulties in servomechanism application. There has been rigorous research ongoing in the field of control for the proper, controlled, and effective operation of robotic manipulators. In the proportional-integral-derivative (PID) controller, the I term enables strong steady-state tracking of step instructions, while the P and D terms give stability and desired transient behavior making the PID or three-term controllers indisputably the most widely used industrial controller in the last 50 years [2].

Fractional-order calculus dates back to around 300 years. Despite this, its usage in fields like control systems and simulation is still under research [3]. The increase in the degree of freedom is achieved by fractional-order simulation by which models can be made like regular system dynamics [4]. In 1911, Hultmann Ayala, H.V. developed the Integral Order-PID (IO-PID) controller for use in automated ship steering [5]. A 3DOF articulated robotic arm was used to build a robotic system. Kinematics and dynamics of the manipulator played a significant part in the design. By linking the end effector's location and orientation, the kinematic model was created and control action was performed using PID controller [6]. The three-link revolving joint robotic manipulator system used an NPID controller, and the gain parameters were modified using a genetic algorithm (GA) to reduce the weighted sum of the integral of absolute error (IAE) signal. And comparative performance assessment was carried out to show the efficacy of the chosen controller [1].

In this work, a fractional-order nonlinear proportional, integral, and derivative (FONPID) controller [7] has been suggested. This aids in the effective trajectory tracking performance of the single link robotic manipulator (SLRM) system over step input response. Further, by tuning the controller gains using genetic algorithm optimization (GA) based on the minimum objective function (J_{IAE}) of the integral of absolute error (IAE) index, the controller has been made more robust for trajectory tracking performance. The comparative analysis validates the simulation results of proportional & integral (PI), PID, fractional-order proportional, integral, & derivative (FOPID), and the suggested FONPID controllers.

The overall work is structured in five sections beginning with the introduction and literature survey in this section, the modeling of the SLRM system has been shown in Sect. 2. Section 3 talks about the suggested control techniques and tuning. The simulation and investigation of trajectory tracking of the SLRM system with

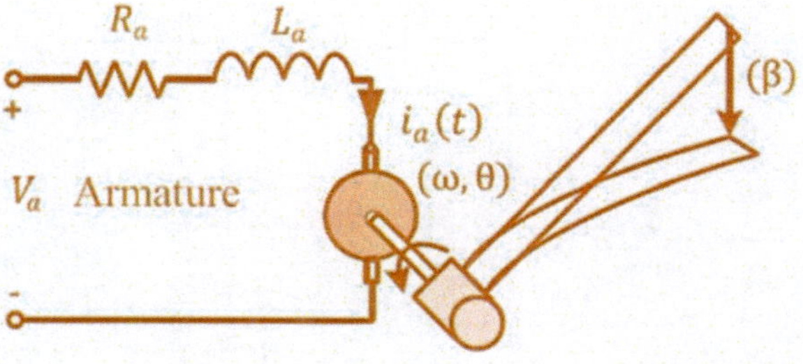

Fig. 1 Robotic manipulator system with DC circuit [8]

the application of various controllers is explored in Sect. 4. Finally, the paper is concluded in Sect. 5 along with the key future trends.

2 Modeling of the SLRM System

The SLRM systems are very flexible in their links due to which positional oscillation and unsustained vibration can be observed. The SLRM system as shown in Fig. 1 comprises of a single link joint, and rotational base, modeled using Euler's Lagrangian technique by assessing the kinetic energy (KE) and potential energy (PE) of the system. The model equations of the SLRM system in the state-space form are described below in Eq. (1) [8, 9].

$$\begin{bmatrix} \dot{\theta} \\ \dot{\beta} \\ \ddot{\theta} \\ \ddot{\beta} \end{bmatrix} = \begin{bmatrix} 0 & 0 & 1 & 0 \\ 0 & 0 & 0 & 1 \\ 0 & \frac{C_s}{J_{\text{eq}}} & -\frac{\eta_g C_t C_m \eta_m N^2 + R_m \gamma}{J_{\text{eq}} R_m} & 0 \\ 0 & -\frac{C_s (J_l + J_{\text{eq}})}{J_{\text{eq}} J_l} & \frac{\eta_g C_t C_m \eta_m N^2 + R_m \gamma}{J_{\text{eq}} R_m} & 0 \end{bmatrix} \begin{bmatrix} \theta \\ \beta \\ \dot{\theta} \\ \dot{\beta} \end{bmatrix} \tag{1}$$

where θ is the angle of rotation; β is the angle of oscillation; $\dot{\theta}$ is the rate of change of angular rotation; $\dot{\beta}$ is the rate of change of oscillational angle; C_s : 1.3792 is the stiffness constant; C_t : 0.0069 is the thermal constant; J_{eq} : 0.00208 kgm^2 is the moment of inertial without load; J_l : 0.000410 kgm^2 is the moment of inertia of link; η_g : 0.90, and η_m : 0.69 is the efficiency of the gearbox and motor, respectively; C_m : 0.0078 v/rad/s is back emf constant; $N = 70$ is the ratio of the gearbox; γ : 0.004 Nm/(rad s) is coefficient of damping; R_m : 2.6 ohm is resistance of armature [8, 9].

3 Suggested Control Technique and Tuning Using GA

The discovery of fractional calculus has enabled in switching from the long-established models and controllers to those based on noninteger order differential

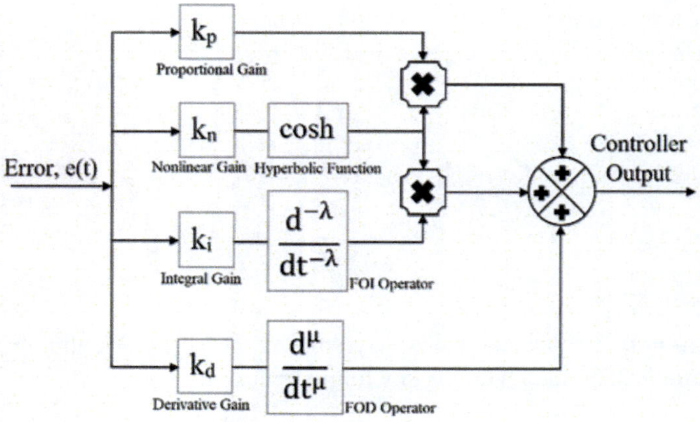

Fig. 2 Structure of FONPID controller [1]

equations. As a result, fractional-order dynamic models and controllers were devised [10–13]. To overcome such complications, the controller built for these systems should rectify its control gains dynamically. The nonlinearity is used to accomplish the objective of producing a self-tuning control signal. It will improve output tracking performance by compensating for quickly fluctuating inputs, reducing overshoot, and compensating for parameter change.

The FONPID controller is presented for redundant robot manipulators in this study, by cascading nonlinear hyperbolic function combinations with the traditional FOPID controller. The control structure of FONPID is represented in Fig. 2. The mathematical expression of the FONPID controller is specified in Eq. (2).

$$u_{\text{FONPID}} = \left[k_p f(h)\mathrm{e}(t) + k_i \frac{\mathrm{d}^{-\lambda}}{\mathrm{d}t^{-\lambda}}\mathrm{e}(t) f(h) + k_d \frac{\mathrm{d}^{\mu}}{\mathrm{d}t^{\mu}}\mathrm{e}(t) \right] \tag{2}$$

where u_{FONPID} is the controller output, i.e., control signal considered in V; k_p, k_i, and k_d are proportional, integral and derivative gains of the suggested FONPID controller; λ, and μ are the fractional-order integral (FOI), and fractional-order derivative (FOD) operators, respectively. $\mathrm{e}(t)$ is the error signal; $f(h) = \cosh(k_n \mathrm{e}(t))$ is a nonlinear hyperbolic gain function for the proportional and integral term; k_n is a positive gain for the nonlinear hyperbolic function.

The logic behind the nonlinear hyperbolic function [14] is that if the total of errors $\mathrm{e}(t)$ and manipulator system output is big, the nonlinear function is considerably high, resulting in greater corrective actions that quickly direct the output toward the intended trajectory. Hence, the incorporation of the FONPID controller [7, 15] has been suggested for such a purpose which lessens the error $\mathrm{e}(t)$ as well as the flexibility to changes in robot output. The combined error signal and the actual robot output are input into a nonlinear function in the supplied loop.

For the implementation of FOI and FOD, the fifth-order of Oustaloup's Recursive Approximation (ORA) filter is considered for the proper distribution of poles and zeros as shown in Eq. (3) having a lower frequency of 0.01 rad/sec and upper frequency of 100 rad/sec.

$$D_t^{\lambda,\mu} = \frac{\mathrm{d}^{\mu}}{\mathrm{d}t^{\mu}} \forall R(\mu) > 0;\ 1 \forall R(\lambda, \mu) = 0;\ \frac{\mathrm{d}^{-\lambda}}{\mathrm{d}t^{-\lambda}} \forall R(\lambda) < 0 \tag{3}$$

The key to building an effective control scheme is to tune a controller. The parameters tuned using a nature-inspired algorithm yield higher performance than parameters tuned with traditional algorithms. Because the system requires precise tracking with little fluctuation in control effort, a machine learning control optimization strategy is required. For the precise and effective positional movement of the link, GA optimization [16] is utilized in order to tune the controller gains based on the minimum objective function (J_{IAE}) of the IAE as given in Eq. (4). The genetic algorithm is the most common type of evolutionary algorithm (EA) that solves optimization problems by maintaining approaches triggered by natural processes including selection, inheritance, mutation, and crossover. The closed-loop control system with a tuned controller structure is showcased in Fig. 3.

$$J_{\mathrm{IAE}} = \int_{0}^{30} |\mathrm{e}(t)|\mathrm{d}t \tag{4}$$

The procedure of GA is mentioned as follows; At first, the FONPID controller gains are generated at random (k_p, k_i, k_d, λ, μ, k_n) for each population. Then each possible set is represented by a matrix of eligible solutions with n number of populations. And the fitness function in the suggested technique is presented in Eqs. (3) for calculating the minimum objective function value for each set of solutions. Further, choosing a smaller population with higher performance and creating a new set of

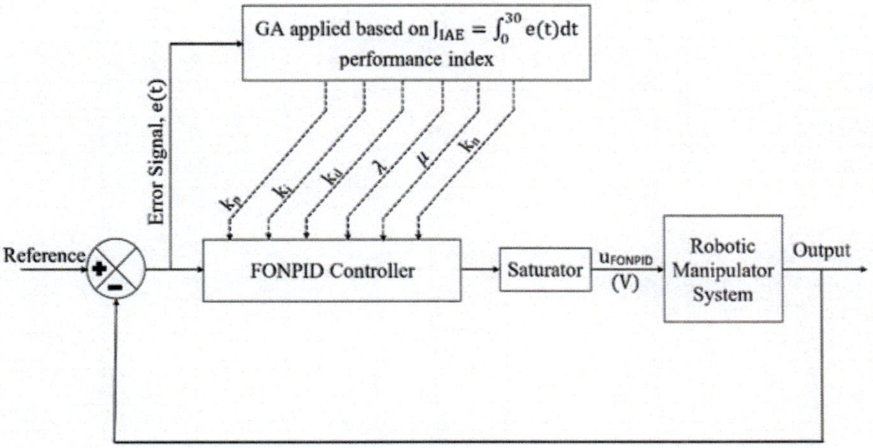

Fig. 3 Closed-loop control configuration of the whole system with tuned FONPID controller

solutions (new—$k_p, k_i, k_d, \lambda, \mu, k_n$) from the best solution found in the previous loops. Later on, using the role of genetic operators (crossover & mutation). By repeating steps 3 through 6 until the best FONPID controller coefficients can be obtained.

4 Simulation and Analysis of Trajectory Tracking

The trajectory tracking of the single link robotic manipulator system using the suggested FONPID controller is simulated and tested in this part for step input when it comes to effective and proper control in the positional movement of the link of the robotic manipulator system. The simulation results are carried out in MATLAB/Simulink R2016b environment. The Runge Kutta method of order four is used for solving the differential equation used in the modeling of this system. The step size is considered as 1 ms at the time of simulation analysis along with the control signal saturation limit ranging between −5 V and 5 V.

When compared to traditional analogues, FOPID, and classical PID, PI controllers, the trajectory tracking the performance of the suggested FONPID controller is studied based on the minimum objective function value of GA optimization considering the IAE performance index. The convergence curve is shown in Fig. 5, and corresponding IAE values are showcased in Table 1 validating the performance of PI, PID, FOPID, and the suggested FONPID controller. The controller gains are optimized using the GA optimization approach as showcased in Table 2. The bar chart in Fig. 4 represents the values of the objective function for PI, PID, FOPID, and FONPID controllers.

Table 1 Gains for PI, PID, FOPID, and FONPID controllers tuned using the IAE performance index

Controller	Controller gains					
	K_P	K_I	K_D	λ	μ	K_0
PID	9.99	0.50	0.48	–	–	–
PI	5.08	−9.60	–	–	–	–
FOPID	9.26	0.50	0.92	0.19	0.91	–
FONPID	4.20	0.50	0.83	0.15	0.90	3.23

Table 2 Values of Objective function for PI, PID, FOPID, and FONPID controllers

Controller	Objective function value (J_{IAE})
PI	0.2226
PID	0.1960
FOPID	0.1519
FONPID	0.1493

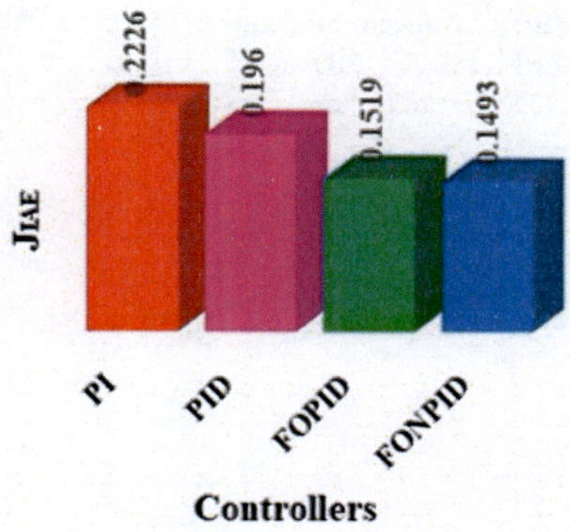

Fig. 4 Bar chart representation of objective function values for PI, PID, FOPID, and FONPID controllers

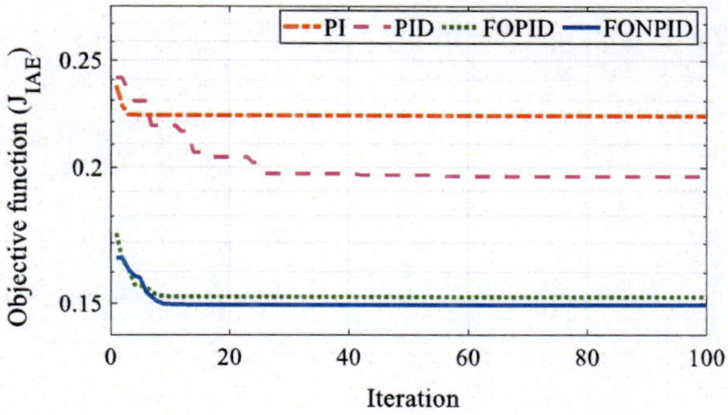

Fig. 5 Objective function versus iteration curves for PI, PID, FOPID, and FONPID controllers

The control goal is to monitor the reference trajectory of the positional movement of a single link robotic manipulator system applied to the various controller such as PI, PID, FOPID, and FONPID as showcased in Fig. 6. The corresponding error signal is generated in Fig. 7. And the control signal, i.e., controller output showcased in Fig. 8. The step response of rate of change of rotational position for PI, PID, FOPID, and FONPID controllers incorporated into the system is showcased in Fig. 9. The step response of oscillational position, as well as the rate of change of oscillational position for PI, PID, FOPID, and FONPID controllers, are showcased in Figs. 10, and 11, respectively.

5 Conclusion and Future Trends

This paper, by deploying a fractional-order nonlinear proportional, integral, and derivative (FONPID) controller in the single link robotic manipulator (SLRM) system, the efficacy of the system was improved by achieving trajectory tracking performance under step input response. By tuning the controller gains using a genetic algorithm (GA) optimization approach based on the minimum objective function (J_{IAE}) of integral of absolute error (IAE) to make the suggested FONPID controller

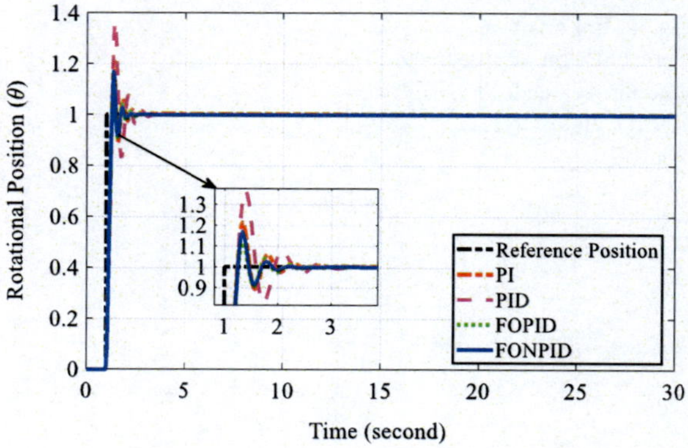

Fig. 6 Trajectory tracking for PI, PID, FOPID, and FONPID controllers

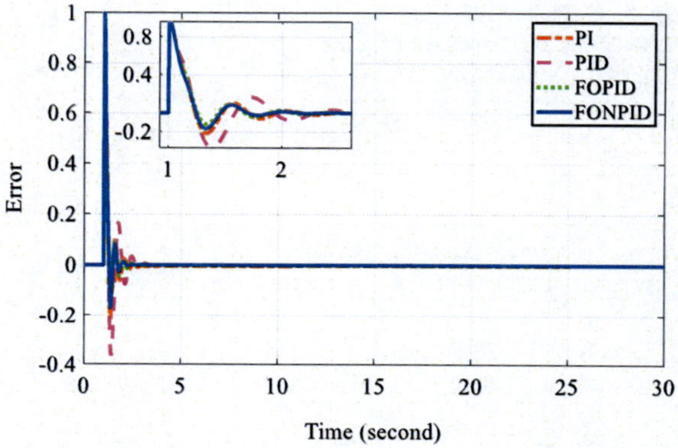

Fig. 7 Error signal for PI, PID, FOPID, and FONPID controllers

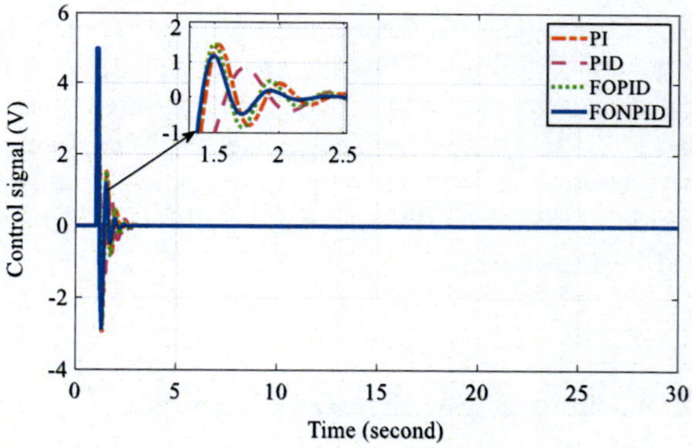

Fig. 8 Controller output for PI, PID, FOPID, and FONPID controllers

adaptive and robust. The suggested FONPID controller observed 32.9%, 23.8%, and 1.7% improvement over proportional & integral (PI), proportional, integral, & derivative (PID), fractional-order proportional, integral, & derivative (FOPID) controllers, respectively. Hence, the suggested FONPID controller showcased more resilient, effective, and better performance over PI, PID, and FOPID controllers based on

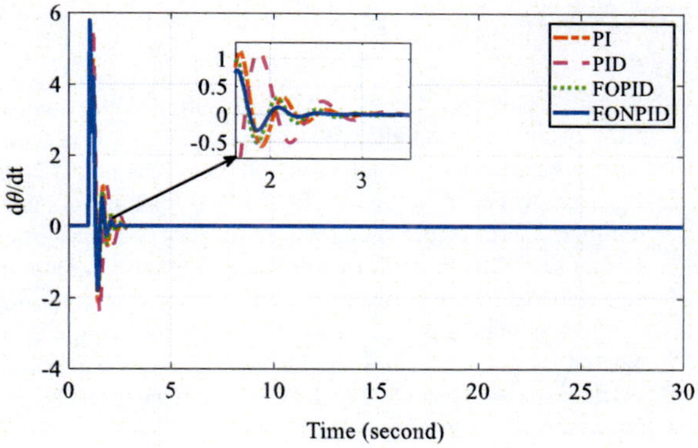

Fig. 9 Response of rate of change of rotational position for PI, PID, FOPID, and FONPID controllers

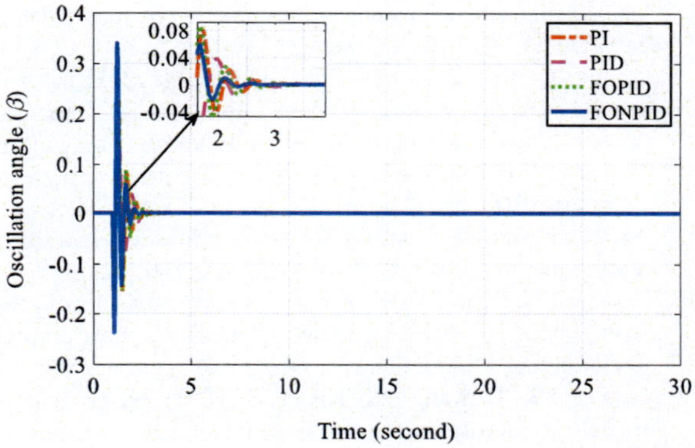

Fig. 10 Response of oscillational position for PI, PID, FOPID, and FONPID controllers

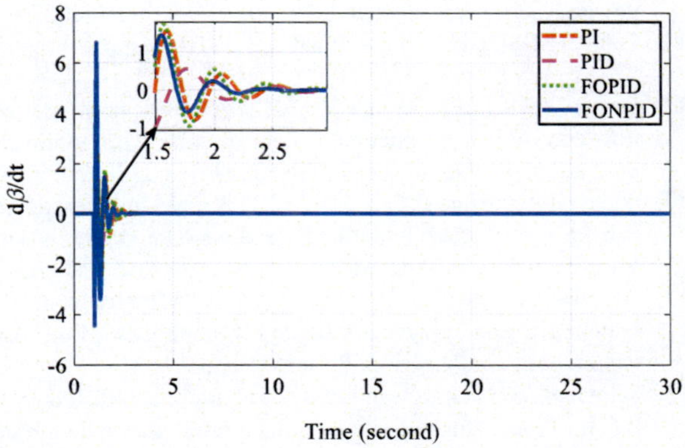

Fig. 11 Response of rate of change of oscillational position for PI, PID, FOPID, and FONPID controllers

minimum J_{IAE} function values. In the future, this work can be further extended by modeling complex multi-link robotic manipulators controlled with various intelligent control techniques such as fuzzy and neural networks.

References

1. Kumar J, Gupta D, Goyal V (2022) Nonlinear PID controller for three-link robotic manipulator system: a comprehensive approach BT. In: Proceedings of ınternational conference on communication and artificial intelligence. Presented at 2022
2. Agrawal A, Goyal V, Mishra P (2021) Comparative study of fuzzy PID and PID controller optimized with spider monkey optimization for a robotic manipulator system. Recent Adv Comput Sci Commun (Formerly Recent Patents Comput Sci) 14
3. Agrawal A (2021) Analysis of efficiency of fractional order technique in a controller for a complex nonlinear control process BT. In: Proceedings of ınternational conference on big data, machine learning and their applications. Presented at 2021
4. Boulkroune A, M'saad M (2012) On the design of observer-based fuzzy adaptive controller for nonlinear systems with unknown control gain sign. Fuzzy Sets Syst 201:71–85. https://doi.org/10.1016/j.fss.2011.12.005
5. Hultmann Ayala HV, dos Santos Coelho L (2012) Tuning of PID controller based on a multi-objective genetic algorithm applied to a robotic manipulator. Expert Syst Appl 39:8968–8974. https://doi.org/10.1016/j.eswa.2012.02.027
6. Renuka K, Bhuvanesh N, Reena Catherine J (2021) Kinematic and dynamic modelling and PID control of three degree-of-freedom robotic arm. In: Kumaresan G, Shanmugam NS, Dhinakaran V (eds) Advances in materials research. Springer Singapore, Singapore, pp 867–882
7. Rawat HK, Goyal V, Kumar J (2022) Comparative performance analysis of fractional-order nonlinear PID controller for NPK model of nuclear reactor. In: 2022 2nd International conference on power electronics & IoT applications in renewable energy and its control (PARC), pp 1–6. https://doi.org/10.1109/PARC52418.2022.9726661
8. Jayaswal K, Palwalia DK, Kumar S (2020) Analysis of robust control method for the flexible manipulator in reliable operation of medical robots during COVID-19 pandemic. Microsyst Technol 9. https://doi.org/10.1007/s00542-020-05028-9
9. Jayaswal K, Palwalia DK, Kumar S (2021) Performance investigation of PID controller in trajectory control of two-link robotic manipulator in medical robots. J Interdiscip Math 24:467–478. https://doi.org/10.1080/09720502.2021.1893444
10. Gupta D, Goyal V, Kumar J. (2019) An optimized fractional order PID controller for ıntegrated power system. Presented at 2019. https://doi.org/10.1007/978-981-13-8461-5_76
11. Faieghi MR (2011) On fractional-order PID design. Presented at 2011. https://doi.org/10.5772/22657
12. Goyal V, Mishra P, Kumar V (2018) A robust fractional order parallel control structure for flow control using a pneumatic control valve with nonlinear and uncertain dynamics. Arab J Sci Eng. https://doi.org/10.1007/s13369-018-3328-6
13. Agarwal A, Mishra P, Goyal V (2021) A novel augmented fractional-order fuzzy controller for enhanced robustness in nonlinear and uncertain systems with optimal actuator exertion. Arab J Sci Eng 46:10185–10204. https://doi.org/10.1007/s13369-021-05508-8
14. Agrawal A, Goyal V, Mishra P (2019) Adaptive control of a nonlinear surge tank-level system using neural network-based PID controller BT. In: Applications of artificial ıntelligence techniques in engineering. Presented at 2019
15. Kumar J (2021) Design and analysis of nonlinear PID controller for complex surge tank system BT. In: Proceedings of ınternational conference on communication and artificial intelligence. Presented at 2021
16. Deb K (1999) An introduction to genetic algorithms. Sadhana 24:293–315. https://doi.org/10.1007/BF02823145

PCA-Based Machine Learning Approach for Exoplanet Detection

Hitesh Kumar Sharma, Bhupesh Kumar Singh, Tanupriya Choudhury, and Sachi Nandan Mohanty

Abstract The search of planets capable of sustaining life has been taken to a whole new level with NASA's Kepler mission. The mission has successfully discovered around 4000 planets, however, the task of manual evaluation of this data is cumbersome and labor intensive, and calls for more efficient methods of discovering exoplanets to remove false positives and errors. The goal of this project is to utilize machine learning algorithms to classify stars as exoplanets through the data collected by the Kepler satellite. To this end, we plan to use preprocessing methods and apply suitable classification algorithms to build an accurate and optimal classifier, increasing the proficiency of the process.

Keywords Exoplanet · Principle Component Analysis (PCA) · Convolutional Neural Network (CNN) · Machine learning · Deep learning

1 Introduction

The everlasting curiosity of humans to know more about the world around them has been a key factor in the advancement of civilization. Since ancient times, humans have wondered where the edge of the world might be, which led pioneers such as Columbus to set sail into an unknown horizon. This human penchant to discover more and

All authors contributed equally.

H. K. Sharma (✉) · T. Choudhury (✉)
School of Computer Science, University of Petroleum and Energy Studies (UPES), Energy Acres, Bidholi, Dehradun, Uttarakhand 248007, India
e-mail: hkshitesh@gmail.com; hksharma@ddn.upes.ac.in

T. Choudhury
e-mail: tanupriya1986@gmail.com; tanupriya@ddn.upes.ac.in

B. K. Singh
B. S. Anangpuria Educational Institutes, Alampur, Ballabgarh-Sohna Major District Road, Faridabad 121004, India

S. N. Mohanty
Department of Computer Science, Singidunum University, Belgrade, Serbia

K. A. Reddy et al. (eds.), *Proceedings of Fourth International Conference on Computer and Communication Technologies*, Lecture Notes in Networks and Systems 606,
https://doi.org/10.1007/978-981-19-8563-8_44

know more has now taken the form of space exploration. Even though the Universe is infinitely vast, we have not yet discovered any other life form in outer space and this seemingly paradoxical situation has baffled scientists and astronomers since decades as they continuously try to look for new worlds where other life forms might be thriving. In this noble quest, one of the most important factors is the discovery of exoplanets, which could possibly host new life forms. The planet which moves around a star beyond our solar system is known as exoplanet. In the hope of finding an exoplanet with similar conditions as that of Earth, ultimately supporting life, humankind took a huge step forward when NASA launched the Kepler Mission. The Kepler Mission was the first of its kind, capable of finding exoplanets smaller or equal to the size of the earth, orbiting around a star. When a planet crosses its star, it momentarily obstructs the light emitted by the star. Thus a depression in the intensity of light of the star is observed, as shown in Fig. 1. This event is known as 'transit' and can be observed from the Earth as well when Venus or Mercury passes the Sun. The Kepler Space Observatory satellite makes use of this transit method by observing a solar system for a long time and looking for variations in the star's flux. It accurately measures the brightness of the star. Astronomers use this data to determine if a regular transit exists, and if it does, then it is evidence that a planet may be orbiting the star. Once a planet is discovered, other aspects like the size of the planet, its orbit and the star are observed and calculated. These values help in knowing whether this newly discovered planet is capable of hosting life forms. Figure 1, shows the light intensity of a planet during its movement.

Manually interpreting this data is a complex, time consuming task and is subject to human error. Moreover, more planet hunting missions like TESS and PLATO are underway and with advanced technology, they provide more comprehensive data. This calls for progressive data analysis methods. Hence, this project aims to simplify and accelerate the process of discovery of exoplanets with the use of machine learning techniques. The data amassed by Kepler for over a decade has been made available by NASA to the public to let researchers carry on discoveries. In this project, we make use of data preprocessing methods like normalization of data and Principal

Fig. 1 Light Intensity of a Planet

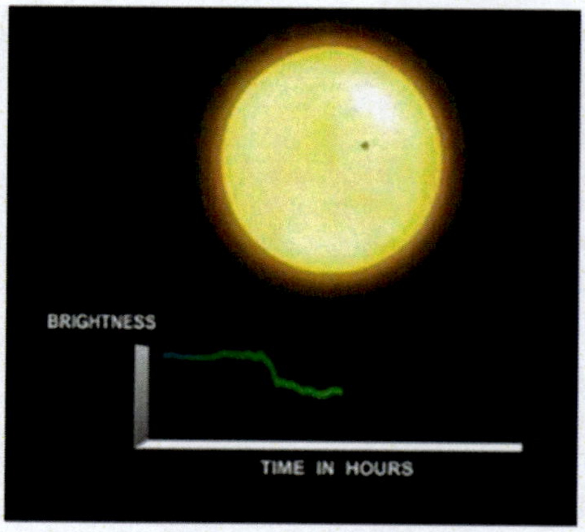

Component Analysis (PCA) on the dataset and apply machine learning models to predict whether an object is an exoplanet or not from the given data points. Through highly efficient prediction models, it will be extremely helpful to determine the general characteristics about exoplanets as recorded by the Kepler and whether the exoplanets confirmed by literature are supported by the measurements of the satellite.

2 Literature Review

Since Exoplanet detection is tedious task to do manually and leaves probability of false positives detections, researchers are trying to automate this process by trying different approaches for, e.g., one of the traditional approach is to detect is by applying a threshold value on all the features through which an exoplanet is detected, this method is good if we are looking for specific type of transit exoplanets but it cannot detect subtle signals of light curve [1]. Machine learning models are being tested on this problem area, many researchers are trying different machine and deep learning models for detecting exoplanets by learning from data collected by Kepler [2], WASP using transit techniques. Some of the methodologies where different machine learning approaches are used to detect planets are given below: Searching for exoplanets using artificial intelligence [3]. In this work, authors have implemented Convolutional neural network (CNN) to detect exoplanets in large datasets. The training of models on self-made dataset and the performance of CNN is validated by kepler's light curves. The proposed CNN is capable of detecting earth like planets even in noisy data and that too with higher accuracy. Overall test accuracy achieved using CCN1D is 88.45%. Transit shapes and maps of self-organizing used as an instrument for grading the planetary candidates: application to Kepler and K2 [4, 5]. In this work, machine learning techniques called self-organizing maps (SOM) are used to detect planets using transit methodology. This work presents a method in which SOM is implemented and planetary candidates are listed according to the possibility of it being a planet. 87% of success rate is achieved using this approach. Apart from this, time to completely run the whole methodology is also very less. SIDRA [6, 7]: a blind algorithm for detection of the signal in photometric surveys [8]: Signal Detection using Random Forest Algorithm (SIDRA), is proposed in this work and is evaluated by applying on the Kepler space mission dataset. They have used 5 classes of simulated light curves, and received promising results with 90% plus accuracy on all the classes on an average. TRANSIT class shows the minimum accuracy of 60% but still is a 7% improved result. Transit clairvoyance: making enhanced the TESS follow-up by the use of artificial neural networks [9, 10]. In this work, ANN is trained on Kepler's catalog and using this neural network they have predicted short period transistors which are accompanied by additional transistors in Transiting exoplanet Survey Satellite (TESS). The developed neural network showed better results than were expected and improved by factor of 2.

Identifying Exoplanets with Deep Learning: A Five-planet Resonant Chain around Kepler-80 and an Eighth Planet around Kepler-90. In this paper deep learning is deployed in searching for optimal exoplanets in kepler's dataset using convolutional neural network (CNN) for this task. The proposed model provides the result with high probability of whether the object is exoplanet or just false positives. The paper concluded by discovering Kepler-80 with 5 planet chains around it and Kepler-90 [10] which is also having 8 planets around it just like our sun.

Transiting Exoplanet Discovery Using Machine Learning Techniques: A Survey. This work summarizes all the machine learning and deep learning algorithms which are being used to analyze light and velocity curve data for detecting and discovering exoplanets. Apart from listed machine learning algorithms along with their properties, comparison is done based on four divisions, i.e., first is preprocessing of light curve, second is detection of exoplanet signal and out of those selected signals making decisions whether signal is or not an exoplanet. Random Forests: Machine Learning: Random Forest as a trademark was registered [10], after they developed an extension of the random decision trees algorithm developed. This work gives a detailed description of the Random Forest classifier as we know it today. Support Vector Networks [7]: this is the original SVM algorithm which was invented. A new way to create nonlinear classifiers by applying kernel tricks [4].

In his work authors has given an ML-based approach for healthcare automation. IoT, ML and Cloud technology has been used in it [7].

3 Methodology

The following are the steps that we have followed for this research work (Fig. 2).

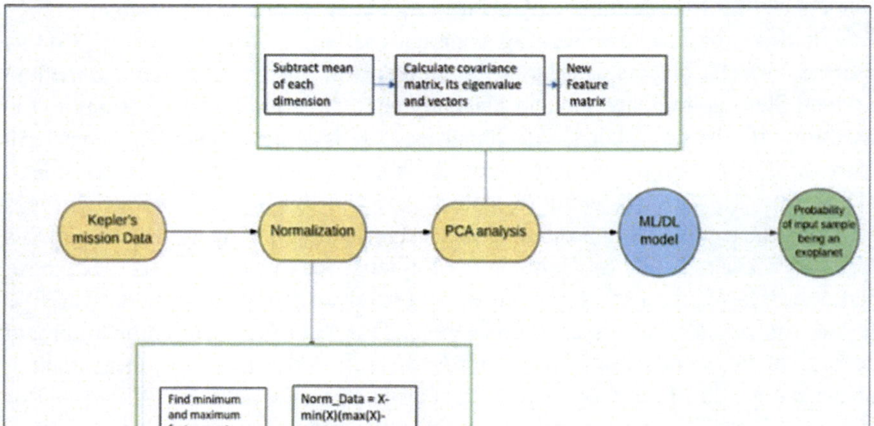

Fig. 2 Methodology flow diagram

- Study of the methods for determining the exoplanet stars.
- In-depth study of the Transient method used to determine the exoplanet stars.
- Selection of appropriate data sets to train our model.
- Preprocessing our dataset.
- Doing a principle component analysis of the dataset.
- Selection of an appropriate classifier to classify the exoplanet stars.
- Training of the models and labeling of clusters.
- Improvement and testing.

4 Implementation

4.1 Normalization

1. Make necessary imports.
2. Load data into the workspace, using pandas library.
3. Find the minimum and maximum feature vectors.
4. Calculate the numerator using the formula below: numerator $= X - \min(X)$.
5. Calculate the denominator using the formula below: denominator $= \max(X) - \min(X)$.
6. Calculate the normalized data points using the formula: normalized_data = numerator/denominator.

4.2 Principal Component Analysis (PCA)

1. Make necessary imports.
2. Load data into the workspace, using pandas library.
3. Find the mean for each dimension and subtract it from the corresponding dimension of data.
4. To calculate the covariance matrix.
5. Finding the eigenvectors and eigenvalues of the covariance matrix.
6. Selecting the components and make new feature vectors.

4.3 Training of Data

1. Make necessary imports.
2. Load the first 37 components of training data obtained from PCA.
3. Load the first 37 components of test data obtained from PCA.
4. Oversample the label '1' data.
5. Train the resultant data using Support Vector Machine (SVC) as the classifier.
6. Train the resultant data using Random Forest as the classifier.

	FLUX.1	FLUX.2	FLUX.3	FLUX.4	FLUX.5	FLUX.6	FLUX.7	FLUX.8 \
0	119.88	100.21	86.46	48.68	46.12	39.39	18.57	6.98
1	5736.59	5699.98	5717.16	5692.73	5663.83	5631.16	5626.39	5569.47
2	844.48	817.49	770.07	675.01	605.52	499.45	440.77	362.95
3	-826.00	-827.31	-846.12	-836.03	-745.50	-784.69	-791.22	-746.50
4	-39.57	-15.88	-9.16	-6.37	-16.13	-24.05	-0.90	-45.20
..	...	...	...	...	...	...	...	...
565	374.46	326.06	319.87	338.23	251.54	209.84	186.35	167.46
566	-0.36	4.96	6.25	4.20	8.26	-9.53	-10.10	-4.54
567	-54.01	-44.13	-41.23	-42.82	-39.47	-24.88	-31.14	-24.71
568	91.36	85.60	48.81	48.69	70.05	22.30	11.63	37.86
569	3071.19	2782.53	2608.69	2325.47	2089.37	1769.56	1421.09	1142.09

	FLUX.9	FLUX.10	...	FLUX.3188	FLUX.3189	FLUX.3190	FLUX.3191 \
0	6.63	-21.97	...	14.52	19.29	14.44	-1.62
1	5550.44	5458.80	...	-581.91	-984.09	-1230.89	-1600.45
2	207.27	150.46	...	17.82	-51.66	-48.29	-59.99
3	-709.53	-679.56	...	122.34	93.03	93.03	68.81
4	-5.04	14.62	...	-37.87	-61.85	-27.15	-21.18
..	...	...	...	...	...	...	...
565	135.45	107.28	...	-123.55	-166.90	-222.44	-209.71
566	-11.55	-10.48	...	-12.40	-5.99	-17.94	-11.96
567	-13.12	-14.78	...	-0.73	-1.64	1.58	-4.82
568	28.27	-4.36	...	2.44	11.53	-16.42	-17.86
569	902.31	714.47	...	695.41	865.97	882.41	1203.06

Fig. 3 Snippet of the original dataset

5 Experimental Results

5.1 *Normalization*

In Fig. 3, the dataset used originally is presented. Few records from the whole dataset is extracted and presented on desktop console.

In Fig. 4, dataset after applying normalization techniques is presented. Dataset is properly transformed into same length for equal contribution. We have again extracted some records and presented from the whole dataset.

5.2 *Principle Components (Two)*

In Fig. 5, we have provided a visualization based on graph that shows the results of the Principle Component Analysis (PCA) applied on provided dataset. Principle Component 1 versus Principle Component 2 has been drawn on min–max scaled form on the graph. The results shows the relations between PCA 1 and PCA 2 and it is much scattered as compare to Fig. 5.

	FLUX.1	FLUX.2	FLUX.3	FLUX.4	FLUX.5	FLUX.6	FLUX.7	\
0	0.313818	0.223566	0.160476	-0.012870	-0.024616	-0.055496	-0.151024	
1	0.654377	0.649262	0.651662	0.648249	0.644212	0.639648	0.638982	
2	1.000000	0.958864	0.886589	0.741705	0.635793	0.474128	0.384692	
3	0.191387	0.190588	0.179124	0.185274	0.240450	0.216565	0.212585	
4	0.686792	0.707634	0.713545	0.716000	0.707414	0.700446	0.720812	
...	...	...	...	...	...	...	...	
565	0.862362	0.828238	0.823874	0.836819	0.775699	0.746299	0.729737	
566	-0.734376	-0.713372	-0.708279	-0.716373	-0.700343	-0.770579	-0.772830	
567	-0.919947	-0.851469	-0.831370	-0.842390	-0.819171	-0.718048	-0.761436	
568	0.537442	0.535440	0.522649	0.522607	0.530033	0.513432	0.509722	
569	-0.370904	-0.391749	-0.404302	-0.424755	-0.441804	-0.464899	-0.490063	

	FLUX.8	FLUX.9	FLUX.10	...	FLUX.3188	FLUX.3189	FLUX.3190	\
0	-0.204203	-0.205809	-0.337035	...	-0.169607	-0.147721	-0.169974	
1	0.631030	0.628371	0.615569	...	-0.228321	-0.284506	-0.318984	
2	0.266083	0.028806	-0.057780	...	-0.259941	-0.365838	-0.360702	
3	0.239841	0.262373	0.280639	...	0.769384	0.751520	0.751520	
4	0.681839	0.717170	0.734466	...	0.688288	0.667192	0.697719	
...	...	...	...	...	...	...	...	
565	0.716419	0.693851	0.673990	...	0.511245	0.480682	0.441524	
566	-0.750878	-0.778554	-0.774330	...	-0.781910	-0.756603	-0.803782	
567	-0.716870	-0.636540	-0.648045	...	-0.550665	-0.556973	-0.534655	
568	0.518842	0.515507	0.504163	...	0.506527	0.509687	0.499970	
569	-0.510210	-0.527526	-0.541090	...	-0.542467	-0.530150	-0.528963	

Fig. 4 Data after performing normalization

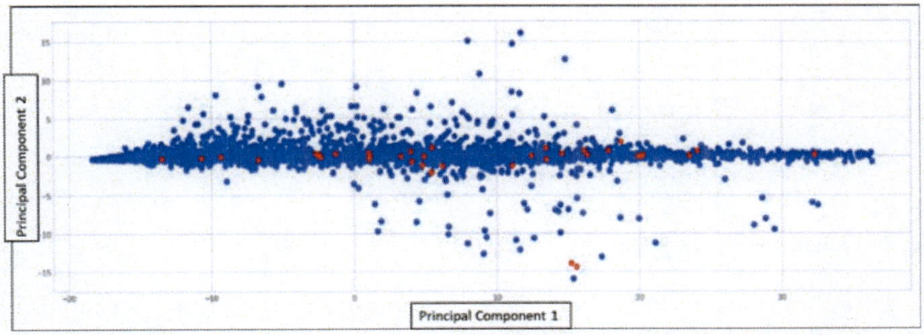

Fig. 5 First two principal components in min–max scaled form

5.3 *Support Vector Machine*

Following are the results (Fig. 6) obtained from training SVM classifiers on the exoplanet data. The output screenshot represents the values of various statistical parameters used for the accuracy measurement of the model performance.

In Fig. 6, we have presented the confusion matrix generated by the model accuracy measurement code module. This is generated by training using SVC.

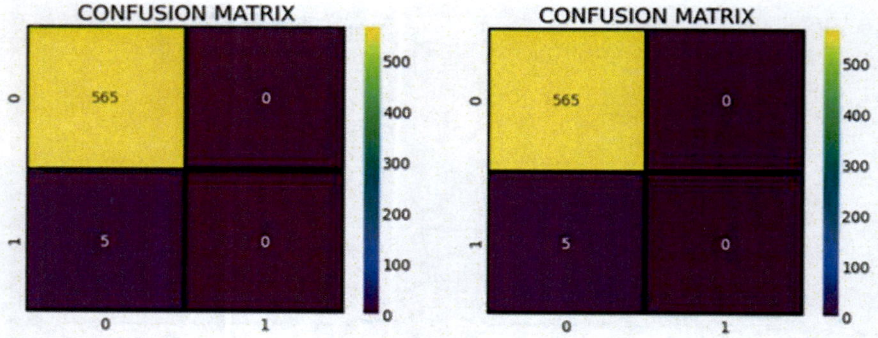

Fig. 6 The confusion matrix of the results obtained by training using SVC

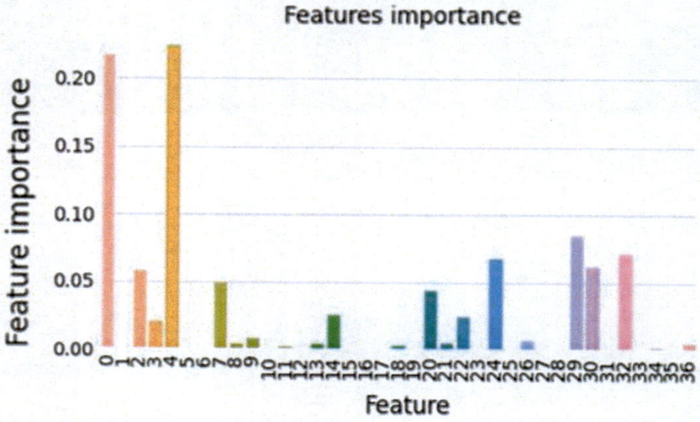

Fig. 7 Feature importance graph

5.4 *Random Forest Classifier*

In Fig. 7, we have presented the confusion matrix generated by the model accuracy measurement code module. This is generated by training using Random Forest Classifier.

5.5 *Feature Importance*

In a given dataset all the features are not equally important, some features have greater importance than others. Figure 7 shows the relative importance of the features obtained using PCA as per the Random Forest Classifier.

6 Conclusion

In this research work, application of two of the most popular ML classifiers—Support Vector Machine and Random Forest Classifier is done, in order to predict if the given flux data belongs to a star with an exoplanet. This work would greatly assist astronomers who have to analyze a huge amount of data to find an exoplanet. In the future, we plan to use deep learning models to further enhance our results. Deep learning has already proven its usability in various fields such as face recognition and speech recognition and we are confident that using techniques involving deep learning would undoubtedly enhance our results.

References

1. Cortes C, Vapnik VN (1995) Support-vector networks. Mach Learn 20(3):273–297. CiteSeerX 10.1.1.15.9362. https://doi.org/10.1007/BF00994018. S2CID 206787478.
2. Khanchi I et al, Automated framework for real-time sentiment analysis. In: International conference on next generation computing technologies (NGCT-2019).
3. Kshitiz K et al (2017) Detecting hate speech and insults on social commentary using nlp and machine learning. Int J Eng Technol Sci Res 4(12):279–285
4. Choudhury T, Kumar V, Nigam D, Mandal B (2016) Intelligent classification of lung & oral cancer through diverse data mining algorithms. In: 2016 International conference on micro-electronics and telecommunication engineering (ICMETE). IEEE, pp 133–138
5. Mittal V, Gupta S, Choudhury T (2018) Comparative analysis of authentication and access control protocols against malicious attacks in wireless sensor networks. In: Satapathy S, Bhateja V, Das S (eds) Smart computing and informatics. Smart innovation, systems and technologies, vol 78. Springer, Singapore. https://doi.org/10.1007/978-981-10-5547-8_27
6. Agarwal A, Gupta S, Choudhury T (2018) Continuous and integrated software development using DevOps. In: 2018 International conference on advances in computing and communication engineering (ICACCE), 2018, pp 290–293. https://doi.org/10.1109/ICACCE.2018.8458052
7. Ahlawat P et al (2020) Sensors based smart healthcare framework using internet of things (IoT). Int J Sci Technol Res 9(2):1228–1234
8. Taneja S et al (2019) I-Doctor: An IoT based self patient's health monitoring system. In: 2019 International conference on innovative sustainable computational technologies, CISCT 2019
9. Pearson KA, Palafox L, Griffith CA (2018) Searching for exoplanets using artificial intelligence, Monthly Notices of the Royal Astronomical Society, vol 474, issue 1, pp 478–491. https://doi.org/10.1093/mnras/stx2761
10. Armstrong DJ, Pollacco D, Santerne A (2017) Transit shapes and self-organizing maps as a tool for ranking planetary candidates: application to Kepler and K2. Mon Not R Astron Soc 465(3):2634–2642. https://doi.org/10.1093/mnras/stw2881

Self-build Deep Convolutional Neural Network Architecture Using Evolutionary Algorithms

Vidyanand Mishra and Lalit Kane

Abstract The convolutional neural network (CNN) architecture has shown remarkable success in image classification and segmentation. Its popularity has increased promptly due to various factors such as exponential growth in computational resources, availability of benchmark datasets, supporting libraries, and open-source software. The efficiency of CNN architecture majorly depends on the complexity of the architecture, availability of datasets, and hyperparameter selection. But, due to the huge number of parameters in CNN architecture, its selection has completely remained ad hoc in the past works. In this article, a novel encoding technique has been proposed that can represent complex CNN architecture effectively. The article defines basic building blocks to represent CNN architecture such as the genesis block, transit block, agile block, and output block. This encoding structure is used to generate dynamic length chromosome structure and initialized using evolutionary algorithms. A comparative analysis is also presented that shows compared its effectiveness with existing encoding representations on the basis of the number of encoding parameters, training cost, and efficiency.

Keywords Evolutionary algorithms · Convolutional neural network · Deep learning · Genetic algorithm

1 Introduction

The convolutional neural network (CNN) is a special type of deep neural network that is specially designed for image classification and segmentation [1]. The CNN architecture is a layered combination of convolutional layer, pooling layer, and fully connected layer. In the convolutional layer, input images are passed pixel by pixel and multiplied with an adaptive weight matrix known as filters. A number of filters have been employed to capture different features of the input dataset. In each convolutional layer, the number of filters and dimensions vary according to the input image size

V. Mishra (✉) · L. Kane
School of Computer Science, University of Petroleum and Energy Studies, Dehradun, Uttarakhand, India
e-mail: mishra.vidyanand28@gmail.com

K. A. Reddy et al. (eds.), *Proceedings of Fourth International Conference on Computer and Communication Technologies*, Lecture Notes in Networks and Systems 606,
https://doi.org/10.1007/978-981-19-8563-8_45

and complexity of the CNN network. The output of the convolutional layer is passed to on next convolutional layer or pooling layer. In the pooling layer, the output of the convolutional layer is optimized by reducing redundant or less useful features using different pooling operations, such as max, mean, and min. After multiple convolutions and pooling operations, data is passed to a fully connected layer. In a fully connected layer, multidimensional data is converted into a one-dimensional layer based on predicted output classes. In the training of CNN architecture, the feed-forward process is followed iteratively with some fixed set of hyperparameters and adaptive learning parameters. In each iteration, the input image is passed, and based on the output, an average learning error is calculated that is passed using a feedback mechanism to adopt filter and weight values. As the number of hyperparameters is numerous and their values are not correlated, we need to check a lot of combinations before predicting suitable CNN architecture. The hyperparameter tuning problem is NP-hard, and finding an optimum solution will take exponential time. Based on the literature, we observed that evolutionary algorithms and RNN methods are helpful to solve that problem. The RNN-based [2] models are efficient to solve hyperparameter tuning problems, but they required huge computation power. Evolutionary-based algorithms [3, 4] are more suitable in a computationally constrained environment with comparative effectiveness. Another limitation of manual selection of architecture is that requires a good amount of knowledge in the CNN design with the problem domain. To solve both problems evolutionary algorithms help to design architecture and hyperparameters selection [5, 6] automatically. In this paper, we proposed an encoding scheme to map the CNN architecture and pass on the evolutionary algorithm as input parameters.

The remaining content of the paper is organized in the following sections. Section 2 represents various existing techniques with their methodology and findings. The proposed encoding scheme is elaborated with the contribution of work in Sect. 3 followed by a discussion of results in Sect. 4. Section 5 concludes the work with the outcome of the reviews.

2 Related Work

To design CNN architecture using an evolutionary algorithm, first, we need to define encoding representation to represent CNN genotype and phenotype. Based on the literature survey, encoding scheme is broadly classified into two categories as follows, one is a fixed-length encoding scheme, and the other is a variable-length. In a fixed-length, encoding scheme depth of architecture is required to define in the beginning. The advantage of fixed-length encoding representation is that it is easy to implement on existing architecture and suitable to define mutation and crossover operators. Genetic CNN [7] and Evo-CNN [3] are used fixed-length encoding to define CNN architecture. In the variable-length encoding scheme, depth of architecture is not restricted which helps to explore the architecture more generously. But it is quite a complex task to define the genetic operation to mutate the architecture. Architecture such as CNN-GA [8], CGP-CNN [9], and AE-CNN [4] is used variable-length encoding representation. The effectiveness of encoding representation is evaluated based on computation cost, accuracy, number of parameters, and adaptability. If encoding schemes are represented with only a few parameters, then it will restrict the exploration of the architecture. If we represent individual training parameters as one unit, then the number of possible chromosomes is too huge which will increase computation cost.

The evolutionary-based algorithms are used to generate improved CNN architecture using an input encoding scheme, and efficiency is compared with benchmark datasets such as CIFAR-10 and CIFAR-100 [10]. Table 1 represents a comparison of the existing encoding schemes with their representation. We compared based on representation, decoded architecture, and accuracy. A comparative analysis of existing encoding techniques is shown in Fig. 1, and methodology with performance on CIFAR-10 dataset using genetic algorithm is presented in Table 1.

3 Proposed Encoding Scheme

This section details the proposed encoding scheme to represent CNN topology. The proposed scheme employs a variable-length encoding scheme that represents the depth as well as the width of the architecture. The scheme comprises four basic building blocks as shown in Fig. 2. A few bit strings represent each building block, and concatenated structure will represent the complete CNN architecture. Genesis block (combination of convolutional block and pooling block) is used to pass input image size. To reduce the feature map and dimension of the input pixel, a transit block is introduced that uses 1×1 convolution and pooling operation. The value of the pooling operation is defined in the range of 0–1. If the value is less than 0.5 means max pool operation is used, else mean pool operation is used. The agile block, working on the concept of dense connection, uses multiple convolutional blocks with the same learning parameters. These convolutional blocks are connected using skip connection to reduce the number of parameters. An agile block is the combination of five elements; operation, size of filters, number of filters, depth, and interconnection of a different convolutional layer. In the end, fully connected blocks are introduced to flatten the layer with one dimension data and convert it into the next output layer with the number of classes.

The main advantage of the proposed encoding scheme is that it can represent architecture with a combination of two different layers. It makes the representation simple, and one can increase the depth of architecture easily. Also, due to fewer parameters, one can define different evolutionary operations like mutation and crossover efficiently. The scheme also supports increasing the complexity within the block. In the agile block, it can generate filer size and depth randomly and thereby increases complexity. The proposed scheme supports a hybrid encoding scheme that utilizes binary as well as decimal representation. The encoding scheme offers the maximum

Table 1 Methodology used to represent CNN architecture in existing encoding techniques

Model/Error	Encoding	Methodology
Genetic CNN, 2017 [7] **Error: 7.10%**	0-01-100	1. Network is represented using binary encoding 2. Architecture is divided into S-stages including input and output. The stages are ordered, and only the lower stage can connect to a higher stage 3. If there is a connecting edge between two stages, we can use bit 1 else 0 4. Encoding '0-01-100'—First '0' indicates that there is no connection between stage 1 and 2. '01' indicates stage 3 is connected with 2 but not 1. '100' means stage 4 which is directly connected with stage1 but not 2 and 3
CGP-CNN, 2017 [9] **Error: 5.98%**	4 [C3 | 1 | 2]	1. Each node is assigned a decimal number in increasing order 2. Each node consists of 3 fields: operation, input 1, and input 2. Operations are defined as convolution, pooling, sum, and output 3. First C3 field represents the convolutional layer, and the second and third integer value represent connectivity from 1st and second layer to layer 4 from lower to higher number block
GACNN, 2019 [11]	[C1 | C2 | L1 | L2]	1. Architecture represents an array of convolution layers and fully connected nodes 2. C1 and C2 are representations of convolution layers with different filter sizes. L1 and L2 represent the number of nodes in a fully connected layer 3. Pooling layer is fixed after each convolution block with max pool operation
CNN-GA, 2020 [8] **Error: 3.22%**	32-64-0.2-64-256	1. A decimal-coded variable-length string represents CNN architecture 2. Each decimal value represents a number of filters in the convolutional block with fixed dimensions 3. Pooling layer is represented by a fractional value. It means if the value is less than 0.5, it uses the max pool else mean pool function will be used

choice of exploration in depth and width as well as faster optimization. We pass our initialized encoding method in evolutionary algorithms to optimize for better architecture. The maximum number of iterations is fixed at 50 as limited computation power is available.

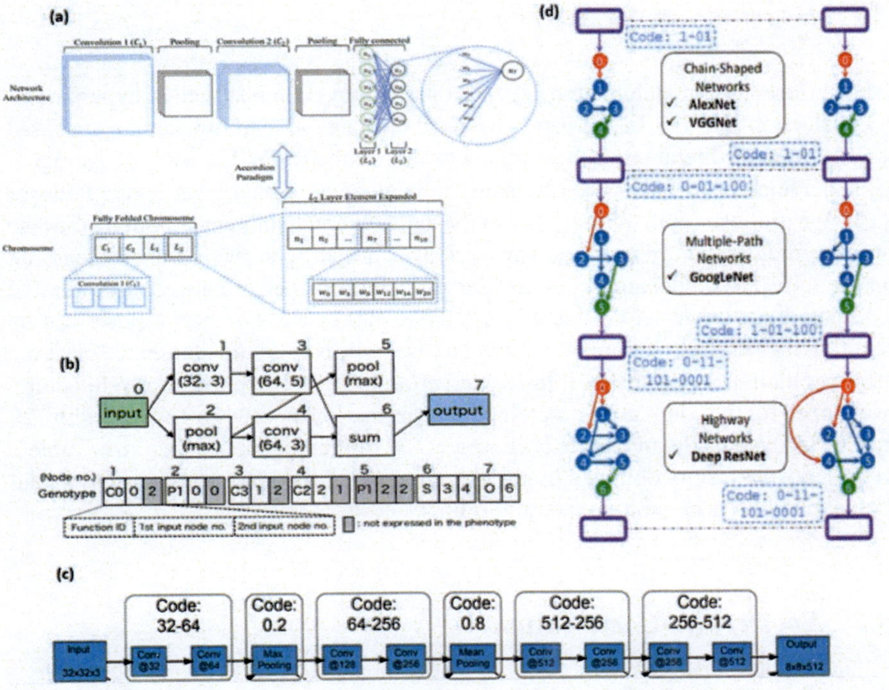

Fig. 1 Block diagram of the encoding representation of CNN architecture in the literature reviewed; **a** GACNN [7], **b** CGP-CNN [8], **c** CNN-GA [2], **d** genetic CNN [5]

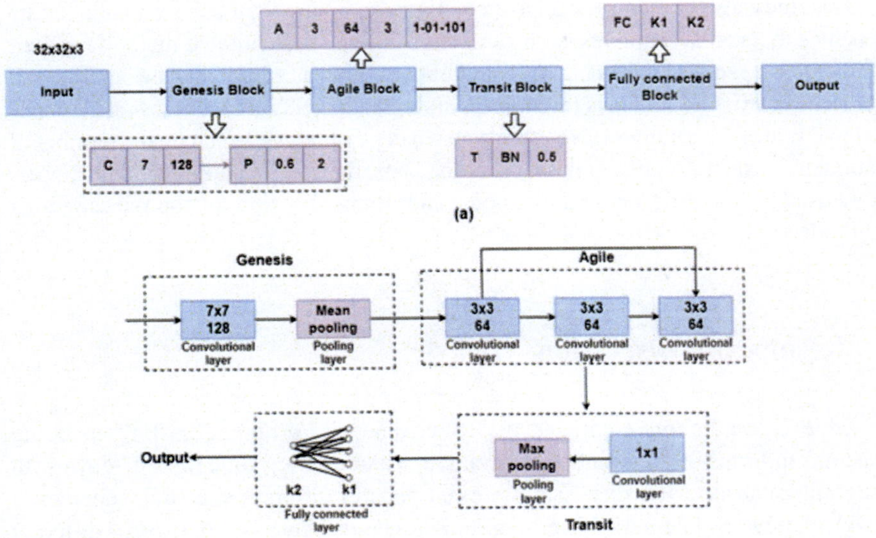

Fig. 2 **a** Block diagram of the proposed encoding representation, **b** CNN architecture of corresponding proposed encoding scheme

3.1 Hyperparameter Selection

The performance of architecture depends on topology design as well as hyperparameter selection [12–14]. In this paper, we fixed the value of learning parameters based on the previous literature such as we chose filter dimensions 1 × 1, 3 × 3, and 5 × 5, and a number of filters 64, 128, and 256 in a convolutional layer and agile layer [15]. We propose depth in range 3–7 in the agile layer to maintain simplicity. For the weight initialization of a fully connected layer, we suggest using transfer learning in the CIFAR-10 benchmark dataset for evaluation of performance as the limited computational power is available else CIFAR-100 or any complex dataset can be used. In the fitness function, we propose to use only 10% of the dataset to select the next population which makes it faster, and after the selection of the top architecture, we trained for the complete dataset. Input image size is proposed as 32 × 32, with 128 batch size, and the learning rate is chosen as 0.1 for homogeneous data size. Table 2 represents the results obtained by proposed encoding methods implementation and compared with peer evolutionary algorithms.

3.2 Novelty and Contribution

A novel encoding scheme for representing CNN architecture is proposed which can effectively be used to represent a complex architecture with a variable number of parameters.

The study also presents a decisive comparison among various existing encoding schemes that can help the researchers in choosing the best suitable method for their application-specific projects. The comparative analysis highlights the merits and demerits of existing schemes through multiple parameters like accuracy and computational power. The authors also represented a depth analysis based on the number of parameters used to represent input chromosomes, their initialization methods, operators used to find different combinations, and fitness function to stop the searching methods.

4 Results and Discussion

In Table 2, the authors compared the effectiveness of different existing encoding schemes in terms of the number of parameters and training time in GPU days with different datasets. After discussing relevant gaps in different encoding schemes, a novel adaptive coding model is proposed. Figure 3 depicts a comparison of different encoding schemes based on architecture using the CIFAR-10 dataset using genetic

Table 2 Comparative analysis of existing encoding techniques

Model	Dataset	Parameters	GPU days	GPU	Relevant findings
Genetic CNN, 2017 [7]	CIFAR-10	0.52 M	17	Titan-X	1. Fixed-length binary encoding scheme is used 2. This encoding is used to represent an existing model 3. All convolutional layers use the same number of filters and filter size
	ILSVRC 2012	–	20		
CGP-CNN, 2017 [9]	CIFAR-100	0.83 M	14	Nvidia Geforce-GTX 1080	1. This representation is suitable to represent basic architecture as only two fields are used for concatenation 2. The limited number of pooling and convolution layer make it restricted to explore 3. Number of nodes is between 10 and 50
CNN-GA, 2020 [8]	CIFAR-10	2.9 M	35	Nvidia Geforce-GTX 1080 Ti	1. Variable-length encoding is used 2. Fixed filter dimension 3X3 and stride value 2 are used in the convolutional layer 3. Fully connected layers are not part of the representation
	CIFAR-100	4.1 M	40		

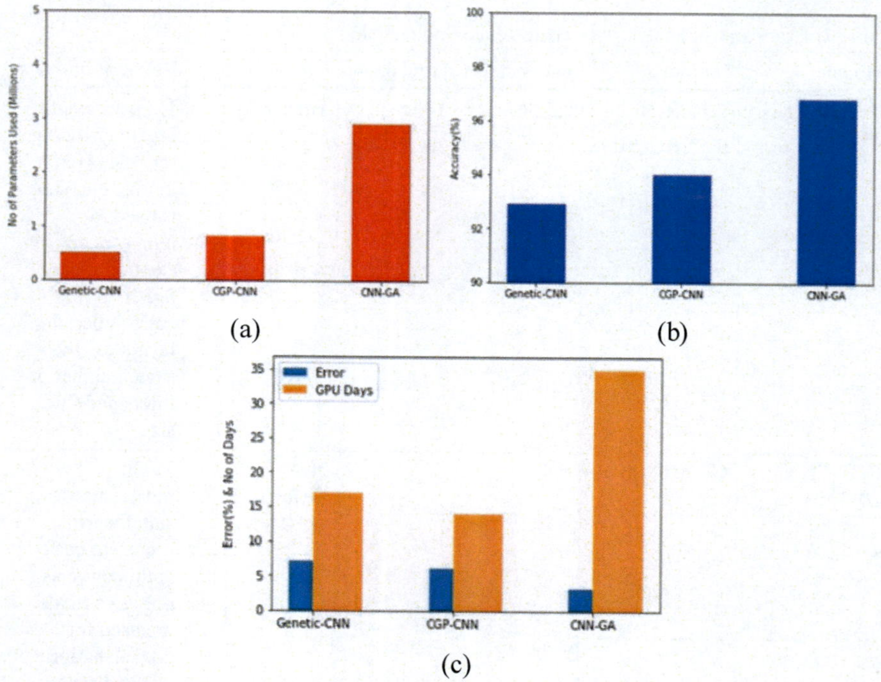

Fig. 3 Comparison of various encoding schemes under training in CIFAR-10 dataset using genetic algorithm; **a** accuracy achieved, **b** number of parameters used, **c** error rate and training cost

algorithms. The proposed scheme is adaptive and versatile in nature. A simple representation offers a better understanding of the complex network. Adaptive behavior scales up the application domain of the proposed scheme.

5 Conclusion

The study proposed a novel encoding method that is used to represent complex CNN architecture. In this encoding, we can represent existing architecture as well as generate new architecture using an available dataset. It covers both the depth and width of architecture that reduces the number of parameters and helps to identify comparable architecture in significant improvement of computation power with comparable accuracy. This encoding scheme is used to pass evolutionary algorithms to design new architecture automatically using different datasets. We can use evolutionary algorithms for hyperparameter tuning and use this encoding representation in future.

References

1. Girshick R (2015) Fast r-cnn. In: Proceedings of the IEEE international conference on computer vision, pp 1440–1448
2. Schmidhuber J (2015) Deep learning in neural networks: an overview. Neural Netw 61:85–117
3. Sun Y, Xue B, Zhang M, Yen GG (2019) Evolving deep convolutional neural networks for image classification. IEEE Trans Evol Comput 24(2):394–407
4. Suganuma M, Kobayashi M, Shirakawa S, Nagao T (2020) Evolution of deep convolutional neural networks using cartesian genetic programming. Evol Comput 28(1):141–163
5. Sinha T, Haidar A, Verma B (2018) Particle swarm optimization based approach for finding optimal values of convolutional neural network parameters. In: 2018 IEEE congress on evolutionary computation (CEC), pp 1–6
6. Serizawa T, Fujita H (2020) Optimization of convolutional neural network using the linearly decreasing weight particle swarm optimization. arXiv preprint arXiv:2001.05670
7. Xie L, Yuille A (2017) Genetic cnn. In: Proceedings of the IEEE international conference on computer vision, pp 1379–1388
8. Sun Y, Xue B, Zhang M, Yen GG, Lv J (2020) Automatically designing CNN architectures using the genetic algorithm for image classification. IEEE Trans Cybern 50(9):3840–3854
9. Suganuma M, Shirakawa S, Nagao T (2017) A genetic programming approach to designing convolutional neural network architectures. In: Proceedings of the genetic and evolutionary computation conference, pp 497–504
10. Krizhevsky A, Hinton G (2009) Learning multiple layers of features from tiny images
11. Esfahanian P, Akhavan M (2019) Gacnn: Training deep convolutional neural networks with genetic algorithm. arXiv preprint arXiv:1909.13354
12. Joshi D, Mishra V, Srivastav H, Goel D (2021) Progressive transfer learning approach for identifying the leaf type by optimizing network parameters. Neural Process Lett 53(5):3653–3676
13. Wang Y, Zhang H, Zhang G (2019) cPSO-CNN: an efficient PSO-based algorithm for fine-tuning hyper-parameters of convolutional neural networks. Swarm Evol Comput 49:114–123
14. Loussaief S, Abdelkrim A (2018) Convolutional neural network hyper-parameters optimization based on genetic algorithms. Int J Adv Comput Sci Appl 9(10):252–266
15. Joshi D, Singh TP, Sharma G (2022) Automatic surface crack detection using segmentation-based deep-learning approach. Eng Fract Mech 268:108467

Bird Species Recognition Using Deep Transfer Learning

K. Reddy Madhavi, Jyothi Jarugula, G. Karuna, Shivaprasad Kaleru, K. Srujan Raju, and Gurram Sunitha

Abstract Birdwatching is a form of wildlife observation of birds which is a recreational activity. Bird watchers make use of bird books to know about the birds. Convolutional neural network (CNN) is utilized to assist users in identifying bird species, making it a useful tool for admiring the beauty of birds. A CNN uses bird imagery to learn how to locate essential attributes in the imagery. To adjust the forms and colors of the object granularities, first, build a bounded area of interest and then balance the distribution of bird species. Then, for feature extraction, CNN is used. Finally, activation function softmax is applied to the resulting class probabilities. Instead of training from scratch, the concept of transfer learning is used in which pre-trained models (VGG16, Xception, ResNet) are used. Here, two of these models are used, namely ResNet50 and VGG16. Among these two models, ResNet50 has provided better accuracy. Transfer learning helps us to increase classification accuracy. As a result of the introduction of deep learning algorithms, very complicated cognitive for computer vision and image recognition has emerged.

Keywords Convolutional neural network (CNN) · Bird species · VGG16 · Xception · ResNet models

K. R. Madhavi (✉) · G. Sunitha
CSE, Sree Vidyanikethan Engineering College, Tirupati, AP, India
e-mail: kreddymadhavi@gmail.com

J. Jarugula
CSE, VNITSW, Guntur, India

G. Karuna
CSE Department, GRIET, Hyderabad, India

S. Kaleru
Juniper Networks Inc, Sunnyvale, CA, USA

K. S. Raju
CSE, CMR Technical Campus, Hyderabad, Telangana, India

K. A. Reddy et al. (eds.), *Proceedings of Fourth International Conference on Computer and Communication Technologies*, Lecture Notes in Networks and Systems 606,
https://doi.org/10.1007/978-981-19-8563-8_46

1 Introduction

The study of birds has contributed much both to the theoretical and practical aspects of biology. In classifying birds, most environmentalists have historically relied upon structural characteristics to infer evolutionary relationships. So to help them in order to recognize the bird species, a machine learning application model is built that would assist them. To classify the bird species with greater accuracy, deep learning with the transfer learning model ResNet50 is used.

To predict the bird species, an interface is developed for extracting information from bird images, and the pre-trained model ResNet50 and a few dense layers are added to it. Number of neurons is proportional to number of bird classes. A dataset of birds was taken from the Kaggle. The model is trained on this dataset. The output of the current machine learning model is an array of class probabilities. The class which has a higher probability will be the output. Image can be taken in various situations like the image can be taken in dull light, the bird might be small in the image. To overcome the problem of building the model from scratch, the concept called transfer learning is used.

The proposed system aims to maximize the accuracy in determining the bird species in an unconstrained environment from the images and to overcome the problem of class imbalance within the bird images.

2 Relevant Study

The existing system [1] was developed using convolutional neural networks with skip connections. These skip connections will provide an output of the previous layers as an input to the current layers. This system has to be built from scratch. In the existing system [2], SVM with decision trees has been used. It suffers from an error accumulation problem. Even this system has to be built from scratch. The existing system [3] converts the grayscale images into autographs and makes predictions based on the score sheet analysis. The existing system [4] uses the color segmentation technique to remove the background elements and locate the bird [5]. Later, they use the histogram bin size to recognize the bird species. They cannot differentiate the minute variations using these histogram bin sizes.

In these existing systems, there was a problem of building a new machine learning architecture. The weights that are chosen are random values that take time to reduce the loss. The proposed technique solves the problem of starting from scratch and using random weights to build a model.

3 Proposed Method

ResNeT

A feed-forward network having a single layer can represent any function if it has enough capacity. The layer can be quite vast, and network might be prone to overfitting the data. As a result, researchers are constantly agreeing that our network design has to become more complicated. ResNet's basic concept is to provide an "identity shortcut link" that bypasses one or more levels, as indicated in the diagram below. These are called skip connections. These skip connections are able to overcome the vanishing gradient problem.

Residual Block

ResNets are made up of residual blocks as shown in Fig. 1. It is noticed that there is a direct link that bypasses certain levels (which may change depending on the model) in between. Next, this term goes through the activation function $f()$ and $H(x)$ is considered as the output.

$$H(x) = f(wx + b)$$

By introducing skip connection, the output is converted to

$$H(x) = f(x) + x$$

The architecture of ResNet is shown in Fig. 2. By permitting the gradient to travel down an extra shortcut channel, ResNet's skip connections resolve the problem of disappearing gradient in deep neural networks. Connections also assist the model learn the identity functions, guaranteeing that the top layer performs otherwise better than, the bottom layer.

Fig. 1 Residual block

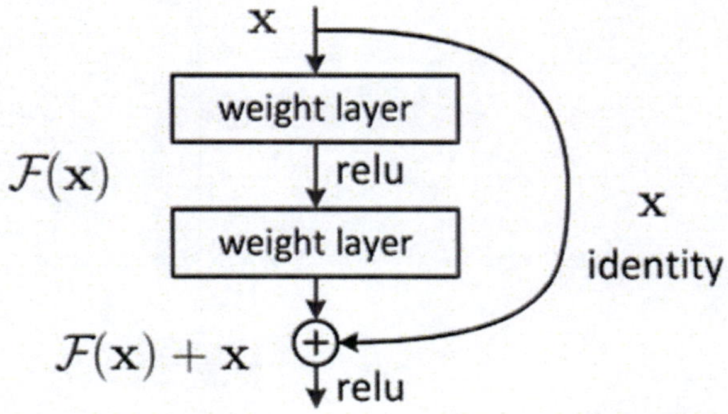

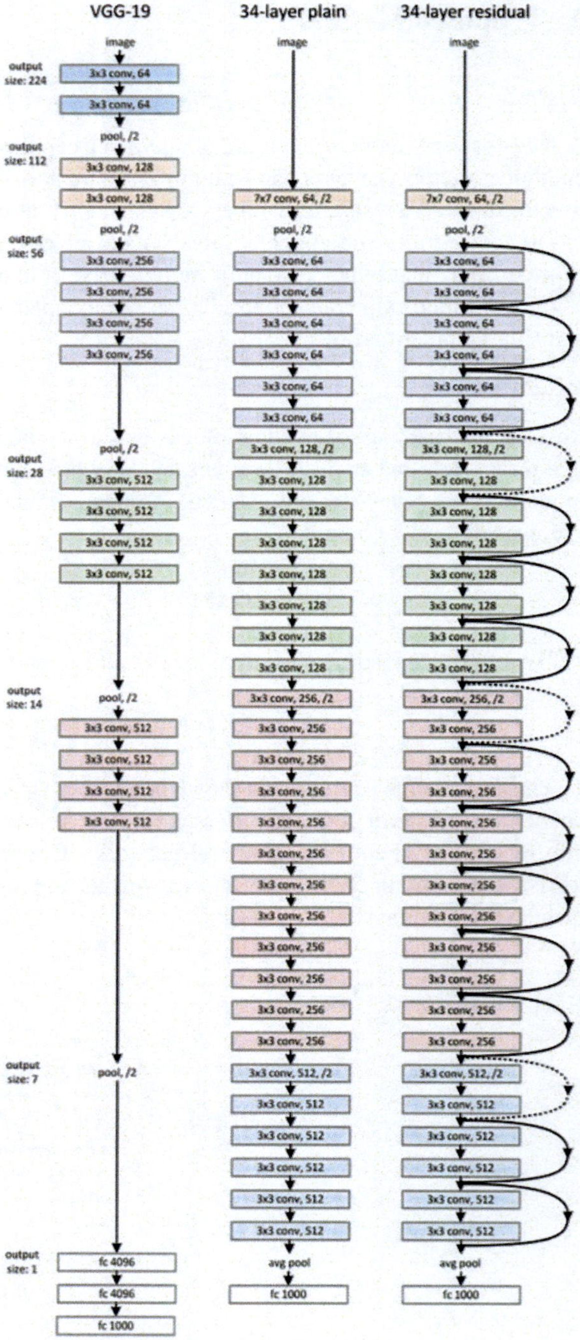

Fig. 2 ResNet architecture

4 Experimental Results

Requirements for implementing the proposed system are Anaconda, GoogleColab, and browser Anaconda for Windows (64-bit) is installed. Google Colab was used to implement the project. It offers notebooks with free GPU support. To complete this project, create a colab notebook and use the code below to mount the GoogleDrive.

The dataset contains train, test, and valid folders which contain bird images. These images are read using ImageDataGenerator. This class helps us to perform data augmentation which is necessary to remove the class imbalance. As a part of data augmentation, different image transformations are applied like rescale, sheer, horizontal flip, and zoom. Two objects are created, namely train_datagen and test_datagen which hold the training and testing/validation data.

Model

Import the model from Keras application and add a few fully connected dense layers to it. The size of the input layer is changed to (224 * 224 * 3). The weights of the pre-trained model are included. Activation functions that are used are ReLU and softmax. Accuracy is used as a metric and categorical gross entropy as loss function.

Training

The training and validation data are used for training. The model is trained for 10 epochs. After each epoch, the model is tested against the validation data.

The user selects the bird image from the local system and uploads the image. The application takes the image provided by the user as input. The image is resized as per the input size of the model and is used for prediction. The model predicts the bird species and provides the result to the user. Input and output were shown in Fig. 3. The application predicts the species based on the image that is provided by the user. The model extracts the features from the image, and based on these features, the model provides an array of class probabilities. From these class probabilities, the output is identified. When the user accesses the application, the following Web page is displayed.

The user selects the choose button to select the image from the local system. The application predicts the bird species and provides the output. The sample image is shown in Fig. 4, and the cropped image was shown in Fig. 4.

Performance Evaluation

The performance of the system is evaluated based on the metric called accuracy. The number of accurately predicted data points out of all the data points is known as accuracy.

$$(\mathrm{TP} + \mathrm{TN})/(\mathrm{TP} + \mathrm{TN} + \mathrm{FP} + \mathrm{FN})$$

Fig. 3 Input and output: indigo bunting

Fig. 4 Input image, Cropped image

TP—"True positives", FP—"False positives", TN—"True negatives", and FN—"False negatives". The training and validation accuracy of the system are 98.73% and 96.11%, respectively, as shown in Table 1.

As a part of the transfer learning concept, the two pre-trained models, namely ResNet50 and VGG16, are used. Among these two models, ResNet50 produced better accuracy.

Table 1 Comparing accuracy values

Model	Training accuracy	Validation accuracy
ResNet50	98.73	96.11
VGG16	87.47	87.89

5 Conclusion

The pre-trained models like ResNet50 and VGG16 are employed as part of the transfer learning process. ResNet50 is able to provide better results than the other model when it comes to predicting bird species. With this model, an accuracy of 98.7% can be achieved. The proposed system outperforms some of the existing systems in predicting the bird species. But whenever a new bird species is included, the model has to be re-trained which is a time-consuming task. The changes can be made in the application such that the images can directly be uploaded from the camera instead of uploading it from the folder. The effectiveness of recognition is hampered by the poor image quality.

References

1. Huang Y-P, Basanta H (2019) Bird image retrieval and recognition using a deep learning platform. IEEE Access 7:66980–66989. https://doi.org/10.1109/ACCESS.2019.2918274
2. Qiao B, Zhou Z, Yang H, Cao J (2017) Bird species recognition based on SVM classifier and decision tree. In: 2017 First international conference on electronic instrumentation & information systems (EIIS)
3. Gavali P, Banu JS (2020) Bird species identification using deep learning on GPU platform. In: 2020 International conference on emerging trends in information technology and engineering (ic-ETITE)
4. Marini A, Facon J, Koerich AL (2013) Bird species classification based on color features. In: 2013 IEEE international conference on systems, man, and cybernetics
5. Cox DTC, Gaston KJ (2015) Likeability of garden birds: importance of species knowledge & richness in connecting people to nature. PloSone 10
6. Ragib KM, Shithi RT, Haq SA, Hasan M, Sakib KM, Farah T (2020) Pakhi Chini: automatic bird species identification using deep learning. In: 2020 Fourth world conference on smart trends in systems, security and sustainability (WorldS4)

CNN-Based Model for Deepfake Video and Image Identification Using GAN

Hitesh Kumar Sharma, Soumya Suvra Khan, Tanupriya Choudhury, and Madhu Khurana

Abstract Deepfakes are the new age tools that automate the syntheses and detection of computer altered videos through GANs. Studies and researches are being done to detect and study the impact of deepfakes on social media and on human lives. In this paper, we will research about the DF technologies such as MTCNN and ResNext-v1 classification models to artificially automate the tasks of deepflakes detection by using datasets from varied sources and having different diversities of people. We also portray another deep learning-based technique that can successfully recognize AI-created counterfeit recordings from genuine recordings. It is inconceivably critical to foster innovation that can spot fakes, so the DF can be recognized and kept from spreading over the Web. Our strategy identifies by looking at the facial zones and their encompassing pixels by parting the video into outlines and separating the highlights with a ResNext-v1 CNN and utilizing the MTCNN catch the transient irregularities between frames presented by GANs during the remaking of the pixels. Our aim is to make an audio-less deepfakes detection system using ML and DL techniques to curb the spread of misinformation.

Keywords Deepfakes · GANs · MTCNN · ResNext-v1 · Artificial intelligence · Machine learning · Deep learning

Hitesh Kumar Sharma and Tanupriya Choudhury contributed equally to the work.

H. K. Sharma (✉) · T. Choudhury (✉)
School of Computer Science, University of Petroleum and Energy Studies (UPES), Energy Acres, Bidholi, Dehradun, Uttarakhand 248007, India
e-mail: hkshitesh@gmail.com; hksharma@ddn.upes.ac.in

T. Choudhury
e-mail: tanupriya@ddn.upes.ac.in

S. S. Khan
Department of CSE, Meghnad Saha Institute of Technology (MSIT), Kolkata, West Bengal 700150, India
e-mail: soumyasuvra@msit.edu.in

M. Khurana
University of Gloucestershire, Cheltenham GL50 2RH, UK
e-mail: mkhurana1@glos.ac.uk

K. A. Reddy et al. (eds.), *Proceedings of Fourth International Conference on Computer and Communication Technologies*, Lecture Notes in Networks and Systems 606,
https://doi.org/10.1007/978-981-19-8563-8_47

1 Introduction

The free and open access to enormous amount of public data through various social media Websites and e commerce Websites, along with the quick advancements of deep learning strategies specifically generative adversarial networks, have prompted the age of deepfakes content in this time of providence of news through social media. Deepfake videos which include biometric information created by digitally manipulating information, with deepfake algorithms, have become matter of grave concern. The well-known term "deepfake" referred to the DL-based technology able to forge synthetic videos by mapping the facial features of a person onto the target person. Human faces are usually preferred to current deep fake algorithms because: In computer vision, augmenting facial details are well researched fields. Faces are the first and most important in human connection because we tend to believe the message if it is coming from a trust worthy faces. These factors stirred consideration around the innovation's disinformation hazards. Be that as it may, an absence of answers to key inquiries has left policymakers and experts without clear direction in creating arrangements to address these hazards. How quickly is the innovation for manufactured media progressing, and what are sensible assumptions around the commodification of these instruments? For what reason would a disinformation crusade decide to disperse deepfakes rather than all the more roughly made phony substance at times similarly as powerful? What sorts of entertainers are probably going to receive these advances for vindictive finishes? How might they utilize them?

Deepfakes offer the online platform an interesting chance to make hoaxed content. ML-driven deception can create strikingly practical portrayals of people furthermore, circumstances. Critically, deepfakes can repeat different unobtrusive subtleties—such as persuading facial spasms or practical shadows for a phony article glued into a picture—that make it trying to distinguish a picture or video as a lie. At the very least, such fakes may plant adequate uncertainty about an objective individual or circumstance to make disarray and doubt. Deepfake innovations are progressively incorporated into programming stages that do not need exceptional specialized mastery. Simple to utilize, ML-driven programming that works with a "face trade"—eliminating one face from a picture or video and embedding another—is progressively accessible for clients with no specialized expertise. Other routine changes of pictures and video controlled by ML are prone to follow. This pattern toward democratization may diminish or successfully take out a significant number of the operational costs in any case making deepfakes an ugly alternative for disinformation culprits. Uncovering reality in such fields along these lines has become increasingly elementary. Nowadays, there are many system independent platforms to create DFs, and anyone can make deepfakes nowadays with little to no knowledge using existing DF models or software. There are many detection models used to detect deepfakes. The majority of them depend on DL, and along these lines, a fight among vindictive and positive employments of DL strategies has been emerging [1]. Taking into consideration the emergence of DF, the US DARPA started an examination conspire in Media Forensics in order to accelerate the advancement of deepfakes techniques. As of late, Facebook Inc. collaborating with Microsoft Corporation and the Partnership on AI alliance have dispatched the deepfake detection challenge to catalyze more examination also, advancement in identifying, and forestalling deepfakes from being utilized to delude watchers.

2 Literature Review

Exposing deepfakes videos by detecting face warping artifacts [2], it is a method to verify whether a videos are DF by analyzing the synthesized resource features by their neighboring pixels using committed CNN [3]. This technique takes into account that this generation DFs can only synthesize the DF resources of restricted goals. Bottlenecks like these have significant impact on DF resources and exploiting these we can detect the deepfakes resources using CNNs and GANs. Others methods use past DF resources to train their CNN models, our strategy need not bother with deepfake created resources, and our model uses modern feature extractions algorithms to detect the deformities in the videos or images. Creating a DF model to generate a negative model is very resource hungry task, this strategy is a time and asset saver.

Exposing AI created fake videos by detecting eye blinking [4] describes another strategy to verify counterfeit resources synthesized with DNN model. The procedure depends on identification of eye gleaming in recordings, a physiological sign that is absent in forged resources. The advancements in DNN [5] have essentially altered the graphics and effectiveness in producing real synthesized facial recordings. In this paper, another strategy to uncover counterfeit face recordings created with deep neural organization models. This technique depends on discovery of eye squinting in the recordings, which is a physiological sign that is not first rate in the combined phony recordings. This strategy is assessed on standards of eyes-flickering location dataset and displays confident execution on distinguishing recordings produced with deep neural network-based programming DF.

Using capsule networks to detect forged images and videos [6] utilizes a strategy that uses a capsule network to distinguish fashioned, controlled pictures and recordings in various situations, similar to replay attack discovery and computer-generated video recognition. Ongoing advances in media age procedures have made it simpler for assailants to make fashioned pictures and recordings. State-of-the-art techniques empower the ongoing production of a fashioned variant of a solitary video got from an informal community. Albeit various techniques have been produced for distinguishing produced pictures and recordings, they are for the most part focused at specific areas and immediately become out of date as new sorts of assaults show up. The technique presented in this paper utilizes a container organization to recognize different sorts of parodies, from replay assaults utilizing printed pictures or filmed recordings to computer synthesized recordings utilizing DNN. It broadens the utilization of case networks past their unique expectation to the taking care of converse designs issues.

Detection of synthetic portrait videos using biological signals [6] technique separates organic signs on face pixels areas on genuine and counterfeit representation resources sets. At that point, the total probabilities to check whether a resource is genuine or not. It is a way to deal with identify manufactured substance in representation recordings, as an obstructive answer for the arising danger of DF [7, 8]. As such, this present a deep phony indicator. We see that identifiers aimlessly using DL are not powerful in getting phony substance, as GAN generates considerably practical outcomes. The key affirmation follows that organic signs covered up in picture recordings can be utilized as a verifiable descriptor of credibility, since they are neither spatially nor transiently safeguarded in counterfeit substance.

3 Proposed CNN Model for COVID X-Ray Detection

There are numerous instruments accessible for making the deepfakes; however for deepfakes detection, there are few apparatus accessible. Our methodology for recognizing the deepfakes will be extraordinary commitment in keeping away from the permeation of the deepfakes over the Internet. In this paper, the expression "deepfakes" allude to the expansive extent of engineered pictures, video, and sound created through ongoing leap forwards in the field of ML [9], explicitly in deep learning. This term is comprehensive of ML procedures that look to adjust some part of a current piece of media, or to create completely new substance. While this paper underlines propels in neural organizations, its examination is pertinent for other strategies in the more extensive field of ML. The expression "deepfakes" reject the wide reach of strategies for controlling media without the utilization of ML, including many existing instruments for "reordering" objects starting with one picture then onto the next. One of the significant goals is to assess its presentation and worthiness as far as security, ease of use, exactness, and unwavering quality. Our technique is zeroing in on distinguishing a wide range of deepfakes. First need to create datasets containing the pictures and videos for both persons that we want to mimic and that person who we want to map that information on. Then, encoder is created to encode the available information on the pictures and videos by using a CNN-based deep learning model [10, 11]. Then, we create a decoder to reenact the image and video information. These autoencoders (the encoder and the decoder) have thousands of pooling layers which is used to extract the image data, reenact them, and argument the image data. Hence, an encoder is required to extract the various facials extracted features to learned the provided input data. To decode the extracted facial maps, we use two separate decoders for both the persons. Encoder and decoder are trained based on backpropagation model, such that output data through the decoder resembles the input data from the encoder (Fig. 1).

After training our model, the video is processed frame-by-frame to map information of one's face to another. Face detection machine learning algorithms are used for face A to identify the feature, then decoder is used from face B for superimpose for GAN-based fake image generation. The dataset is prepared before applying the

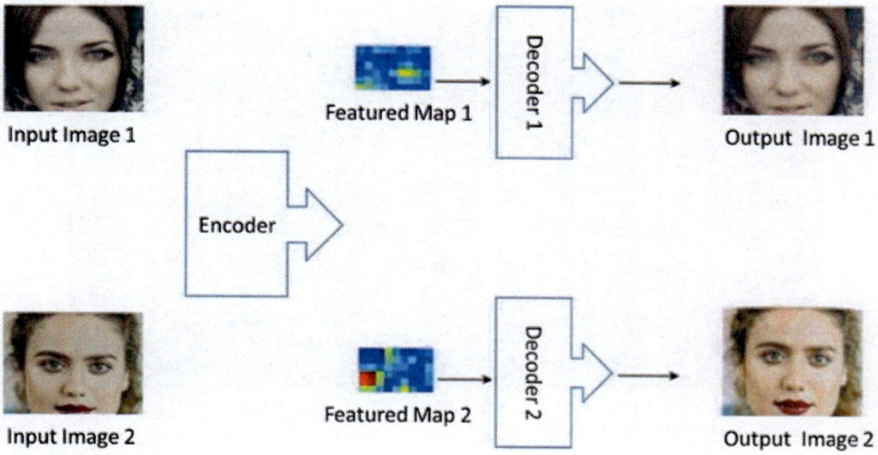

Fig. 1 Encoder and decoder for the images

methodology on subject image. For this work, we have used pre-processed dataset form Kaggle to achieve high accuracy from our algorithm.

- Choose images in datasets that contains only one face.
- There should be lots of videos containing different facial expressions with different angles.
- Remove any bad quality images.

In subject picture, we identified a 5 × 5 network focuses and move them, somewhat far away from their uniquely identified positions. We utilize a straightforward calculation to twist the picture as indicated by those moved matrix focuses. Indeed, even the distorted picture may not look spot on, however, that is the commotion that we need to present. At that point, we utilize a more perplexing calculation to develop an objective picture utilizing the moved framework focuses. We need our made pictures to look as close as the objective image (Fig. 2).

4 Implementation of Proposed CNN Model

The CNN model used for encoder includes 5 CNN layers and 2 dense layers. Dense layers used for fully connected neurons. The CNN layers for decoder consist 4 layers. It reconstructs the 64 × 64 image back. The dimensions up from 16 cross 16 (16 × 16) to 32 cross 32 (32 × 32), a convolution filter (3 cross 3 cross 256 cross 512 filter) to do mapping with (16, 16, 256) layer into (16, 16, 512). Then, we need to reshape it to (32, 32, 128). Face area of fake image is blur, it shows that peoples are using forceful approach for fake image or video generation [1].

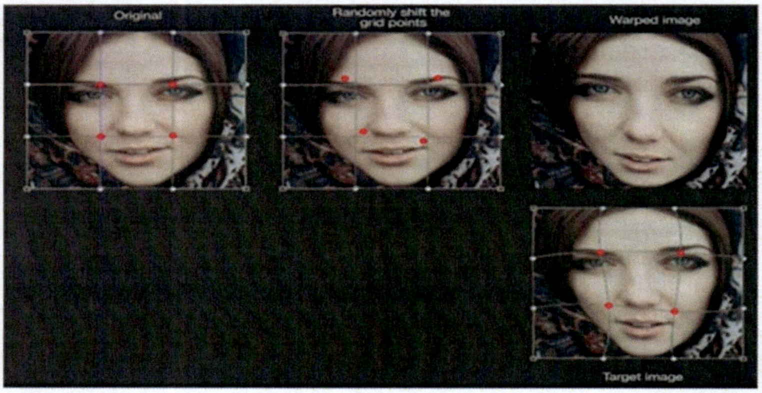

Fig. 2 Subject image original to target

We could understand this by taking in account of deepfake labs, we try to put facial data of image 1 onto the image 2, by cutting the facial image of image 1 and pasting it on image 2, we can clearly see the disorientation of images on one another. But if we change the skin tones around the areas and reduce to opacity in image, we can easily blend in the mage onto one another (Fig. 3).

The mask is created on generated face, and it is blended with the required video. To eliminate further the process of applying Gaussian filter for area boundary, application is configured for compress and expend the mask, and the shape of mask is also controlled. The data size in surely a major factor for good performance. We can add more capacities for betterment of a model. For this purpose, we have used a large dataset with more number of images. We have trained our model with more images. GAN generated images and discriminator is provided for more accuracy. Surely, we can add option misfortune capacities to validate our CNN model. The edge cost is used to estimates that the objective picture and the made picture have a similar edge given at a similar area. A few people likewise investigate the proposed perceptual misfortune. The reproduction cost gages the pixel contrast between given objective picture and the made picture. In any case, this may not be a decent measurement in estimating how our cerebrums see objects. Subsequently, a few people may utilize recognition misfortune to supplant the first reproduction misfortune.

Fig. 3 Stepwise process flow

5 Experimental Results

The work is implemented to achieve multiple goals. Data cleaning, pre-processing, and transformation are done for better results. Real data is extracted and processed into the required format using efficient methodology. ML-based algorithms and functions were used for pre-processing the data based on self-decision approach. In ML measures, data pre-processing is basic to inscribe the dataset in a pipeline that could be deciphered and parsed by the algorithm.

- Pipeline class is initialized to detect the faces in the frames of a video files.
- Then cropping the faces along the frames to reduce the dimensions of our datasets for easier processing.
- Creating a new face cropped video.
- Saving the cropped face videos as our new datasets.

Passing the cropped video datasets in MTCCN model with batch size of 60. From sklearn.model_selection, import train-test-split module with test size of 0.85 and with random state 5. After processing, applying various classification models such as logistic regression, Gaussian Naive Bayes, and SVM to compare the accuracy. In our cloud hosted notebook, we have tested each given packages using deepfakes detection datasets of about 300 images for their processing speeds using GPU. Detection is performed on 1080 × 1920, 720 × 1280, and 540 × 960 FPS. Using cv2's VideoCapture module, we will read the frames of a video. After training our model on pre-processed datasets of 400 images, we will import Scikit-learn Python Library which is used for classification, clustering, and model selection. Then, we will import train-test-split module from sklearn to splitting our training data and testing data in which we will pass arguments (Table 1).

X, Y = parameter is the dataset we are selecting to use, test_size = 0.85 (85% of all test data), random state = 5, importing logistic regression from sklearn library with random state = 0, we get an accuracy score of 0.7891, importing Gaussian Naive Bayesian from sklearn library, we get an accuracy score of 0.6645. Importing SVM from sklearn library, we get an accuracy score of 0.7859. Importing confusion matrix from sklearn (to evaluate the accuracy) with parameters: y_test = Target value. y_pred_lr, y_pred_gnb, y_pred_svm = Estimated target (Figs. 3, 4, and 5).

Table 1 Performance comparison of face detection packages

Packages	FPS (1080 × 1920)	FPS (720 × 1280)	FPS (540 × 960)
FaceNet-PyTorch	12.97	20.32	25.50
dlib	3.80	8.39	14.53
MTCNN	3.04	5.70	8.23
FaceNet-PyTorch (non-batched)	9.75	14.81	19.68

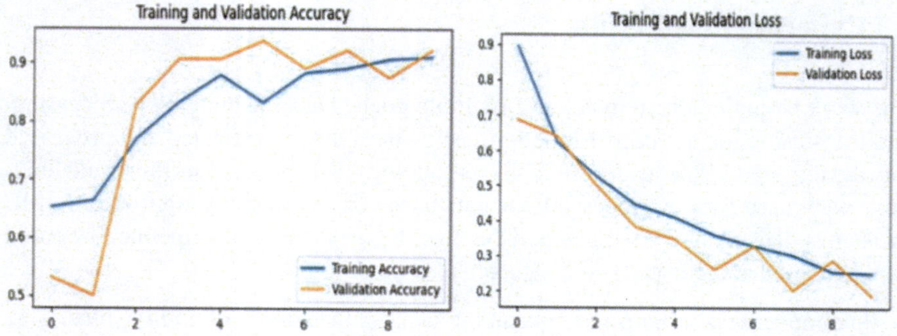

Fig. 4 Training and validation accuracy and loss

Fig. 5 Heat map of confusion matrix for our proposed model

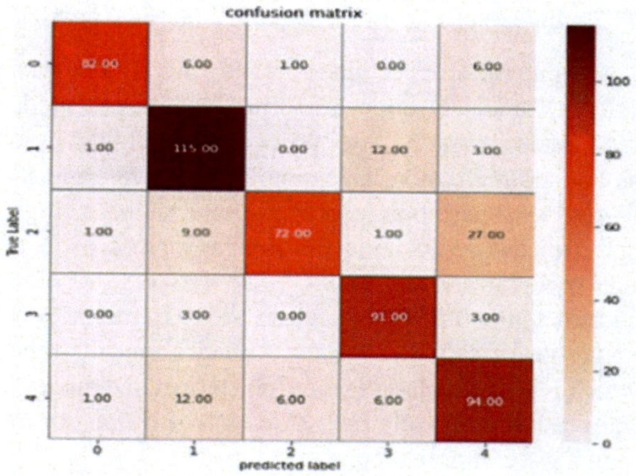

6 Conclusion

We introduce CNN-based model to detect the suspected videos, or image is deepfake or real video, with the accuracy and confidence matrix. This technique draws the inspiration from generation of deepfakes through GANs implementing autoencoders. Our technique does the edge level recognition utilizing Inception ReNet-v1 and video classification utilizing MTCCN. The proposed strategy is fit for predicting whether a videos are real or computer generated. And through tested feedback and rigorous training, it will correctly predict on real-time data. We have not trained on the audio deepfakes and videos; hence, it will not be able to predict or detect the audio deepfakes resources as it requires temporal convolutional network.

References

1. Sharma HK, Khanchi I, Agarwal N, Seth P, Ahlawat P (2019) Real time activity logger: a user activity detection system. Int J Eng Adv Technol 9(1):1991–1994
2. Filali Rotbi M, Motahhir S, El Ghzizal A, Blockchain technology for a Safe and Transparent Covid-19 Vaccination. https://arxiv.org/ftp/arxiv/papers/2104/2104.05428.pdf
3. Choudhury T et al (2022) CNN based facial expression recognition system using deep learning approach. In: Tavares JMRS, Dutta P, Dutta S, Samanta D (eds) Cyber intelligence and information retrieval. Lecture Notes in Networks and Systems, vol 291. Springer, Singapore. https://doi.org/10.1007/978-981-16-4284-5_34
4. Shi F, Wang J, Shi J, Wu Z, Review of artificial intelligence techniques in imaging data acquisition, segmentation, and diagnosis for COVID-19. https://ieeexplore.ieee.org/document/9069255
5. Wang L, Qiu Lin Z, Wong A (2020) COVID-Net: a tailored deep convolutional neural network design for detection of COVID-19 cases from chest X-ray images 19549
6. Chuang M-C, Hwang J-N, Williams K (2016) A feature learning and object recognition framework for underwater fish images. IEEE Trans Image Process 25(4):1862–1872
7. Chuang M-C, Hwang J-N, Williams K (2014) Regulated and unsupervised highlight extraction methods for underwater fish species recognition. In: IEEE Conference Distributions, pp 33–40
8. Kim H, Koo J, Donghoonkim, Jung S, Shin J-U, Lee S, Myung H (2016) Picture based monitoring of jellyfish using deep learning architecture. IEEE Sens Diary 16(8)
9. Sharma HK, Kumar S, Dubey S, Gupta P (2015) Auto-selection and management of dynamic SGA parameters in RDBMS. In: 2015 2nd International conference on computing for sustainable global development (INDIACom). IEEE, pp 1763–1768
10. Khanchi I, Ahmed E, Sharma HK (2020) Automated framework for real-time sentiment analysis. In: 5th International conference on next generation computing technologies (NGCT-2019)
11. Mishra M, Sarkar T, Choudhury T et al (2022) Allergen30: detecting food items with possible allergens using deep learning-based computer vision. Food Anal Methods. https://doi.org/10.1007/s12161-022-02353-9

Comparative Analysis of Signal Strength in 5 LTE Networks Cell in Riobamba-Ecuador with 5 Propagation Models

Kevin Chiguano, Andrea Liseth Coro, Luis Ramirez, Bryan Tite, and Edison Abrigo

Abstract This article analyzes the measured signal strength in 5 LTE network cell located in the central urban area of Riobamba in Ecuador. These measures were done using Network Cell Info Lite and WiFi software for Android systems and 3 campaign were applied getting 50 measured points in each cell. The results were compared with 5 propagation models (log-normal, Okumura-Hata, Cost 231, Walfisch-Bertoni and SUI models) where the SUI and log-normal models fit better for the areas analyzed. The signal strength varies from −70 to −110 dBm. Finally, the model that best fits the real values obtained is determined through the calculation of the quadratic error.

Keywords LTE · Cell · Signal strength · Propagation models

1 Introduction

3G technology is a response to the IMT-2000 specification of the International Telecommunication Union. It takes into account the three main 3G technologies. In Europe, Universal Mobile Telecommunication System (UMTS) provides data transfer up to 2Mbps. For the United States CDMA2000 and for China TDS-CDMA (Time Division Synchronous) was developed. In the fourth generation, mobile and fixed broadband are integrated due to three aspects: network-based entirely on IP technology, use of packet switching and a common service layer for end users [1].

K. Chiguano · A. L. Coro · L. Ramirez · B. Tite · E. Abrigo (✉)
Escuela Superior Polítecnica de Chimborazo, Telecommunications, Riobamba, Ecuador
e-mail: edison.abrigo@espoch.edu.ec

K. Chiguano
e-mail: kevin.chiguano@espoch.edu.ec

A. L. Coro
e-mail: andrea.coro@espoch.edu.ec

L. Ramirez
e-mail: luis.ramirez@espoch.edu.ec

B. Tite
e-mail: bryan.tite@espoch.edu.ec

K. A. Reddy et al. (eds.), *Proceedings of Fourth International Conference on Computer and Communication Technologies*, Lecture Notes in Networks and Systems 606,
https://doi.org/10.1007/978-981-19-8563-8_48

The signal strength and power levels received by mobile devices connected to a cellular network are identified by the parameter received signal code power (RSCP), which indicates the level of signal reception in the UMTS (3G) network. The RSRP parameter (reference signal received power) within the 4G technology measures the signal strength that reaches the mobile from the cell or tower to which it is connected [2]. The mobile operating frequencies in Ecuador for 4G in Movistar operator are 1900 MHz (Band 2), Claro operator 1700/2100 MHz (Band 4). For 3G technology in Movistar operator: 850 MHz or 1900 MHz, in CLARO operator: 850 MHz (Band 5) [3].

Network Cell Info Lite is a monitoring and measurement software tool for 4G long-term monitoring and measurement tool for 4G long-term evolution (LTE), 4G+, wide-band code division multiple access (WCDMA), wideband code division multiple access (CDMA) and GSM [4]. It is dual SIM compatible, except for Android mobile devices below 5.0, due to device/Android limitation. It is capable of measuring the received signal strength in decibels (milliwatts). The application needs to specify the actual network whose signal strength is being measured. This application is available in the Play Store; in addition, there is a more complete paid version, but in this case, the one we have used to make this article is the free one. This app can be very useful to check the mobile network coverage we have at a given time. A limitation of Android devices is GPS: it is recommended, set the GPS mode to "high precision" in the location settings of your device to get the best performance of the application. To approximate the distance between the measurement points and the base station, it was used Google Earth [5].

Empirical propagation models are widely used to calculate path losses in a wireless channel in different types of scenarios, and their results are considered when selecting the location of base stations and planning their coverage area [6]. For 2.1 GHz, propagation models used to estimate signal attenuation in long-term evolution (LTE) mobile communications systems are mainly the Stanford University Interim (SUI) model and Walfisch-Bertoni model, which are models applicable up to 3 GHz. For these models, the equations depend on different variables of the propagation environment (effect of roofs and height of buildings, among others), which makes them precise; however, they are more complex in their calculation [7]. In this article, power reception measurements were taken in 5 strategic points of Ecuador in the city of Riobamba. Using the software Network Cell Info Lite WiFi to obtain the reception power in the coverage area of each of the base stations. We compare graphs obtained using propagation models and the coverage measurements of Movistar and Claro mobile telephony powers in band 2 and band 4, respectively, for suburban environments. Log-normal, Okumura-Hata, Cost 231, Walfisch-Bertoni and SUI models were evaluated in this work in order to estimate a better coupling. Finally, a propagation model is selected that best fits the measurements obtained using error theory to estimate a better analysis. The best fitting models in relation to the measurements were Cost 231, Walfisch-Bertoni and SUI.

Table 1 Values for different areas

Environment	Path loss exponent
Free space	2
Urban area cellular radio	2.7–3.5
Shadowed urban cellular radio	3–5
Inside a building—line of sight	1.6–1.8
Obstructed in building	4–6
Obstructed in factory	4–6

2 Theorical Framework

2.1 Log-Normal Path Loss Model

Log distance path loss model is a generic model and an extension to Friis free space model. It is used to predict the propagation loss for a wide range of environments, whereas the Friis free space model is restricted to unobstructed clear path between the transmitter the receiver. The following equation shows mean path losses

$$PL(d)[dB] = PL(do) + 10n\log\left(\frac{d}{do}\right) + \chi \quad (1)$$

Equation 1 shows mean path losses $\chi =$ A zero-mean Gaussian distributed random variable (in dB) with standard deviation σ. This variable is used only when there is a shadowing effect. If there is no shadowing effect, then this variable is zero. Taking log of the normal (Gaussian)-variable results in the name "log-normal" fading. $n =$ Path loss exponent. See Table 1 that gives the path loss exponent for various environments [8].

2.2 Okumura-Hata Model

This model is considered one the simplest and best in accordance with its precision in path loss calculation and has become the method of mobile system planning in Japan.

The most important result provided by the model is the median value of the basic propagation loss, as a function of frequency, distance and the heights of the base station and mobile antennas. Although it does not include any of the path type correction factors, which are in the Okumura model, the equations proposed by Hata have important practical value [9].

- *f: 150–1500 MHz*
- h_b*: 30–200 m*

- h_m: *1–10 m*
- *d: 1–20 km*

$$L_u = 69.55 + 26.16\log(f) - 13.82\log h_b - \mathrm{CH} + \left[44.9 - 6.55\log(h_b)\right]\log(d) \quad (2)$$

the following equation is used to calculate the correction factor for small cities:

$$CH = 0.8 + (1.1\log(f) - 0.7)h_m - 1.56\log(f) \quad (3)$$

$$L_{su} = L_u - 2\left(\log\left(\frac{f}{28}\right)\right)^2 - 5.4 \quad \text{for sub-urban area} \quad (4)$$

2.3 COST 231 Model

The COST 231 model is a semi-empirical path loss prediction model, resulting from the combination of the Walfisch-Bertoni and Ikegami models. It is recommended for macro-cells in urban and suburban scenarios, with good results of the path loss for transmitting antennas located above the average roof height. However, the error in predictions increases considerably as the transmitter height approaches the rooftop height, with very low efficiency for transmitters below that level [10].

For this analysis, the model is implemented with the following equation:

$$PL = 42.64 + 26\log(d) + 20\log(f) \quad (5)$$

where L_o is the attenuation in free space and is described where is the attenuation as:

$$L_o = 32.45 + 20\log(d) + 20\log(f) \quad (6)$$

L_{RTS} represents diffraction from rooftop to street, and is defined as

$$L_{\mathrm{RTS}} = -16.9 - 10\log(w) + 10\log(f) + 20\log(h_b - h_t) + L_{\mathrm{ORI}} \quad (7)$$

Here, L_{ORI} is a function of the orientation of the antenna relative to the street a (in degrees) and is defined in Table 2.

L_{MSD} represents diffraction loss due to multiple obstacles and is specified as:

$$L_{\mathrm{MSD}} = L_{\mathrm{BSH}} + k_A + k_D\log(d) + k_f\log(f) - 9\log(s_b) \quad (8)$$

where:

Table 2 Equations depending on the range of the angle

L_{ORI}	–
-10 + 0.354a	0 < a < 35
2.5 + 0.075(a-35)	35 < a < 55
4 – 0.114(a-55)	55 < a < 90

Table 3 Formulas according to the range of the height of the base station

L_{BSH}	–
$-18\log(1 + h_t - h_b)$	$h_t > h_b$
$54 + 0.8(h_t - h_b)2d$	$h_t \leq h_b$
–	and $< d <$ 0.5 km

Table 4 Formulas according to the range of the height of the base station

K_D	–
54	$h_t > h_b$
$54 + 0.8(h_t - h_b)$	$h_t \leq h_b$
–	and $< d <$ 0.5km

Table 5 Formulas according to the range of the height of the base station

K_D	–
$18 + 15\frac{(h_t - h_b)}{h_b}$	$h_t > h_b$
18	$h_t \leq h_b$
–	and $< d <$ 0.5 km

$$k_F = -4 + k\frac{(f)}{924} \tag{9}$$

Here, $k = 0.7$ for suburban centers and 1.5 for metropolitan centers (Tables 3, 4 and 5) [11].

2.4 Walfisch-Bertoni Model

This model estimates the influence of building height and ceilings by using diffraction models to predict the average signal power at pavement level [12].

$$L_P = 57.1 + A + \log(f) + 18\log(d) - 18\log(H) - 18\log\left(1 - \frac{d^2}{17H}\right) \tag{10}$$

$$A = 5\log\left[\left(\frac{b}{2}\right)^2 + (h_b - h_r)^2\right] - 9\log(b) + 20\log\left[\arctan\left(\frac{2(h_b - h_r)}{b}\right)\right] \tag{11}$$

Equations should be punctuated in the same way as ordinary text but with a small space before the end punctuation mark.

2.5 SUI Model

SUI is based on the Hata model. It applies to heights from the MS between 2 and 3 m and from the BS between 10 and 80 m. The frequency range for the model is from 0 to 2000 MHz [13].

$$P_L = A + 10\gamma\log\left(\frac{d}{d_o}\right) + X_f + X_h + S \tag{12}$$

$$\gamma = a - bh_b + \frac{c}{h_b} \tag{13}$$

$$A = 20\log\left(\frac{4\pi d_0}{\lambda}\right) \tag{14}$$

$$X_f = 6\log\left(\frac{f}{2000}\right) \tag{15}$$

$$X_h = -10.8\log\left(\frac{h_r}{2000}\right) \tag{16}$$

$$S = 0.65(\log(f))^2 - 1.3\log(f) + \alpha \tag{17}$$

The SUI model groups the propagation scenarios into three different categories, each with its own specific characteristics:

- Category A: mountainous ground with medium and high levels of vegetation, which corresponds to high loss conditions.
- Category B: mountainous ground with low levels of vegetation, or flat areas with medium and high levels of vegetation. Medium level of losses.
- Category C: flat areas with very low or no vegetation density. Corresponds to paths where losses are low.

2.6 Theory of Errors

Errors can be the result of the inaccuracy of the measuring equipment, which are called systematic errors, or caused by external agents or by the operator himself, which are called accidental errors. While the former are repeated in the same sense, whenever the same measuring apparatus is used, the latter vary from one experience to another, both in value and in sign [14]. The absolute error can be conceptualized as the difference between the real value and the value obtained:

$$E_a = X_r - X_o \tag{18}$$

in which a represents the actual value and X_r represents X_o the obtained value. The relative error is the result of the division between the absolute error and the actual value, while the relative percent error is the result of the relative error in percent:

$$E_r = \frac{E_a}{X_r} \quad \text{Relative Error} \tag{19}$$

$$E_r\% = \frac{E_a}{X_r} \times 100\% \quad \text{Relative Percent Error} \tag{20}$$

3 Methodology

3.1 Data Collection

For this study, 5 base stations were selected (Movistar-Claro) located in the downtown area of the city. Claro) that are located in the downtown area of the city of Riobamba; their locations have been selected thanks to the coverage maps provided by the cellular coverage maps provided by the cellular telephone companies (the locations are shown in Fig. 1). The data obtained from the mobile application (Network Cell Info Lite) are collected quantitatively. The sample size is 60 data per base station. The data collection was obtained at different times of the day, in order to observe the influence of variables such as weather, traffic, users, cellular technology, distance between the antenna and the base station, technology, distance between the transmitting antenna and the receiver. To obtain the approximate distance from the point of each measurement and the base station, Google Earth was used to obtain the approximate distance between the point of each measurement and the base station.

Several of the parameters used in the propagation models were obtained from information provided by the country's mobile telephone companies. Each selected base station is located in different environments, and the received power is affected

Fig. 1 Map of cover zone

by several variables such as reflection by buildings, saturation of connected users, street width, attenuation by vegetation (Park Guayaquil base station).

3.2 Parameters to Consider

For each of the propagation models, it was necessary to determine certain parameters as shown in the following tables.

Where it is necessary to consider important parameters, being the operating frequency, the P.I.R.E. and the height of the transmitting antenna as shown in Table 6. It is also necessary to consider that in 4G/LTE, the Movistar operator is working in band 2 with a frequency of 1900 MHz.

Table 7 considers a network cell with similar characteristics to the previous one mentioned, because both base stations belong to the Movistar operator, and their main difference is the height at which it is located at 49 m.

Claro is working in Band 4, at a frequency of 1700/2100 MHz. In addition, this cell is working with a P.I.R.E. of 20 dBm as shown in Table 8. And these are the most important parameters to consider.

Table 9 considers narrow streets and a transmitter height of 49 m. It is also necessary to consider that in the theoretical calculations to determine the P.I.R.E is 19 dBm was obtained; therefore, it is assumed that the P.I.R.E is 20dBm (Table 10).

Table 6 Hotel el Cisne base station

Parameters	Values
Base station transmitter power	30 dBm
Transmitter antenna height	30 m in suburban area
Receiver antenna height	1.5 m
Operating frequency	1900 MHz (Band 2)
Average building height	27 m
Street width	13 m

Table 7 Hotel Zeus base station

Parameters	Values
Base station transmitter power	30 dBm
Transmitter antenna height	49 m in suburban area
Receiver antenna height	1.5 m
Operating frequency	1900 MHz (Band 2)
Average building height	27 m
Street width	13 m

Table 8 Guayaquil Park base station

Parameters	Values
Base station transmitter power	20 dBm
Transmitter antenna height	30 m in suburban area
Receiver antenna height	1.5 m
Operating frequency	1700/2100 MHz (Band 4)
Average building height	27 m
Street width	10 m

Table 9 Pichincha bank base station

Parameters	Values
Base station transmitter power	20 dBm
Transmitter antenna height	49 m in suburban area
Receiver antenna height	1.5 m
Operating frequency	1700/2100 MHz (Band 4)
Average building height	27 m
Street width	6 m

3.3 Application of Propagation Models

Finally, knowing all the parameters required by the propagation models, they were applied to each of the base stations in order to predict the loss curve with respect to

Table 10 Santa Cecilia base station

Parameters	Values
Base station transmitter power	20 dBm
Transmitter antenna height	29 m in suburban area
Receiver antenna height	1.5 m
Operating frequency	1700/2100 MHz (Band 4)
Average building height	27 m
Street width	6 m

distance, in addition to applying the error calculation to determine which model best fits the real values, based on the percentage of error.

4 Results

4.1 El Cisne Base Station

Figure 2 shows the results obtained in the field measurements, providing information about the coverage offered by El Cisne base station located at Daniel Leo´n Borja and Duchicela avenues. In general, the figure shows that in the area near the base station, we were able to receive a signal with an excellent power reaching values of up to −60 dBm; however, at similar distances, we could appreciate a drop that reaches up to −87 dB. Thanks to the field measurements, it was possible to appreciate that because the base station is located in a central area of the city of Riobamba in certain points, it was possible to receive a considerably good power because it was possible to have a direct view to the transmitting antenna; however, in certain areas near the antenna, it was possible to appreciate that there is an abrupt drop in areas where apparently the signal should be good. At this point, we can consider that there is an attenuation due to the infrastructure surrounding the base station. In spite of this, the final result, which is the average of the measurements, indicates that there are no considerable losses, and it can be deduced that in the area of the Hotel Zeus as well as in the Guayaquil Park, the coverage is really good. However, it should be noted that the base stations in an urban area seek to reduce the power and increase more infrastructure is why it was observed that when reaching a distance of 450 m with a drop of −110 dBm the mobile device suddenly changes, connecting to a transmitting antenna that manages to emit a considerably higher power, thus achieving a wider coverage in the urban area.

In Fig. 3, it shows the results obtained by applying the propagation models to the mean of the measurements obtained, which allowed observing the behavior of the logarithmic curve under conditions consistent with the scenario found.

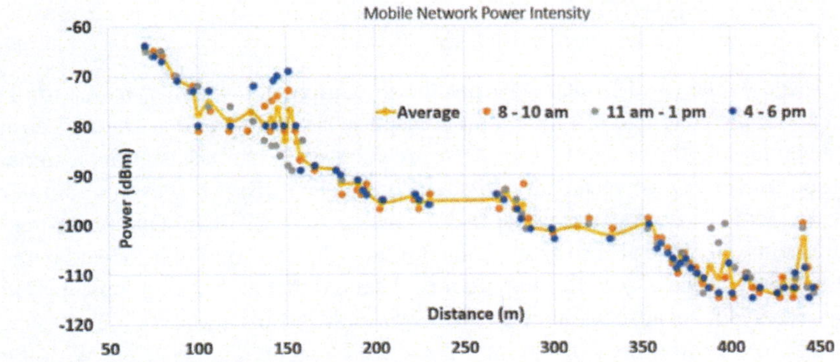

Fig. 2 Signal strength of the base station EL CISNE

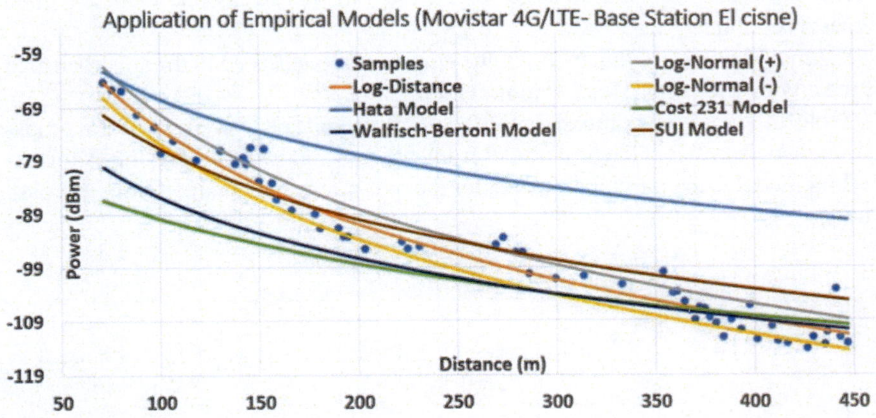

Fig. 3 Propagation models applied at the base station EL CISNE

To determine the model that best fits the measurements obtained, the error theory was applied, giving the results shown in the Table 11, resulting in the Walfisch-Bertoni model showing a smaller error compared to the others. propagation models.

Table 11 Error table

	E_a	E_r	E_r %
Okumura-Hata model	14.7319	0.14,678	15
Cost 231 model	7.2838	0.0884	9
Walfisch-Bertoni model	6.0193	0.0723	7
SUI model	4.7385	0.0473	5

4.2 Zeus Base Station

Figure 4 shows the results obtained from the measurements which gives us information about the coverage offered by the Zeus base station located at Daniel Leo´n Borja Avenue. The figure shows that in the area closest to the base station, a signal with good power is received reaching values up to −65 dBm; however, at similar distances, there is a drop in the signal that reaches up to −89 dBm. Due to the field measurements, we were able to analyze that because the base station is located in the downtown area of the city of Riobamba, it is observed that at certain points, a good signal is received, because we have a close view to the base station.

As it is in a central area of the city, there are elements that interfere with the intensity of the power which will give us an attenuation due to the infrastructure. Finally, the final result shown in Fig. 4 is the average of our three measurements, which shows that there are no large-scale losses, so we can say that there is a good signal reception.

Figure 5 shows the comparison of the empirical models used in the suburban area of the city of Riobamba. Base station (Hotel Zeus).

Table 12 shows the comparison of the absolute and relative errors of the propagation models to select the best model from which it can be noted that the Walfisch-Bertoni model is the most appropriate for the measurements obtained from this base station.

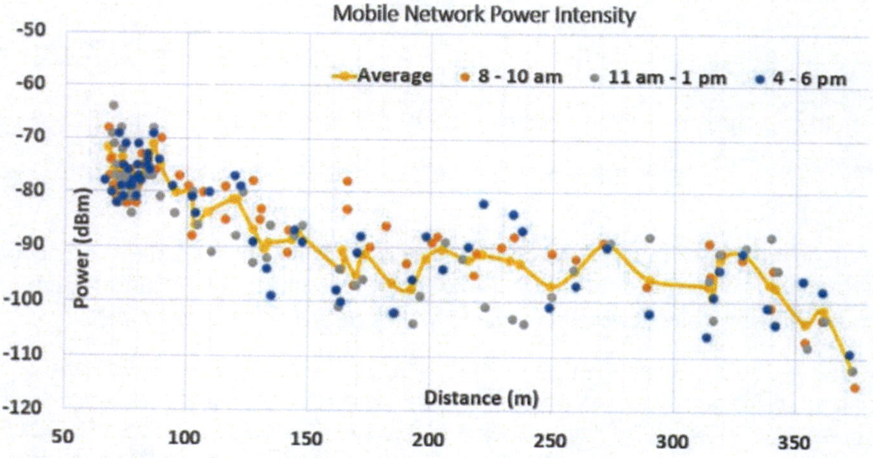

Fig. 4 Signal strength of base station Hotel Zeus

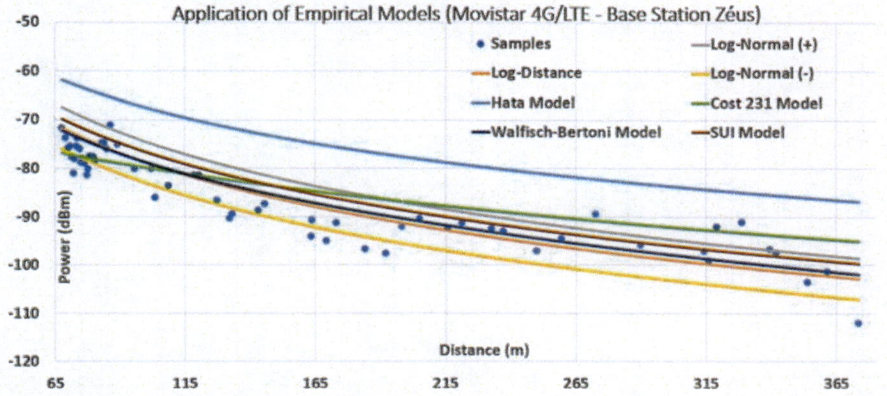

Fig. 5 Propagation models applied at the base station Zeus

Table 12 Error table

	E_a	E_r	E_r %
Okumura-Hata model	15.3769	0.1791	18
Cost 231 model	4.059	0.0454	5
Walfisch-Bertoni model	3.2842	0.0378	4
SUI model	4.4576	0.052	5

4.3 *Guayaquil Park Base Station*

Figure 6 shows the average power level in reception of 150 samples obtained at different times of the day, highlighting that the best time for a good power in reception is from 11:00 to 13:00. On the contrary, it is possible to observe reception values of −100 dBm from 08:00 to 10:00. This is due to many factors such as the number of connected users, the distance from the connection point, the mobile device or the peak hours of the city.

Figure 7 shows the comparison of the empirical models used in the suburban area of the city of Riobamba. Base station (Guayaquil Park).

Table 13 shows the comparison of the absolute and relative errors of the propagation models to select the best model. For the data obtained at this base station, the Walfisch-Bertoni model is the best fit.

4.4 *Santa Cecilia Base Station*

Figure 8 shows the results obtained during measurements at different times in the Santa Cecilia base station providing information about the coverage offered by the operator Claro, sector located between Vicente Rocafuerte and Carabobo streets.

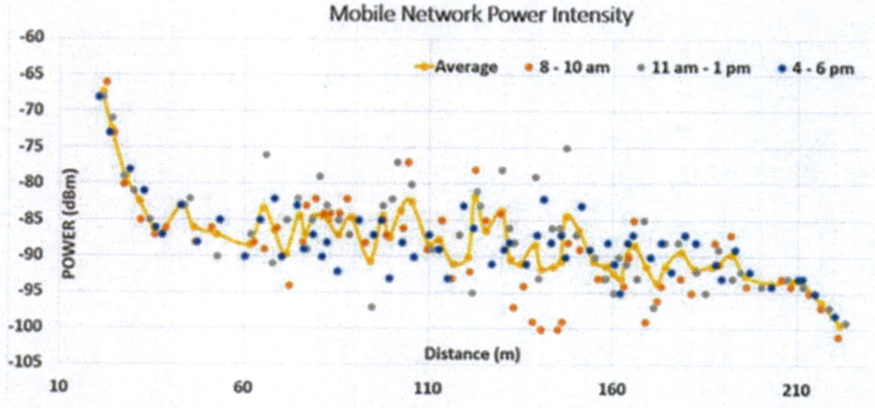

Fig. 6 Signal strength of base station Guayaquil Park

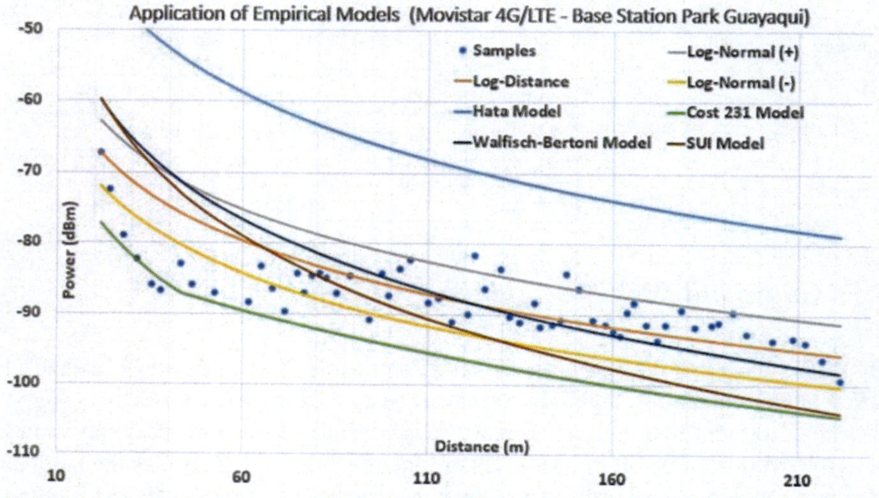

Fig. 7 Propagation models applied at the base station Guayaquil Park

Table 13 Error table

	E_a	E_r	$E_r\%$
Okumura-Hata model	20,3919	0,2345	23%
Cost 231 model	7,8289	0,0894	9%
Walfisch-Bertoni model	4,9579	0,0581	6%
SUI model	6,9139	0,0795	8%

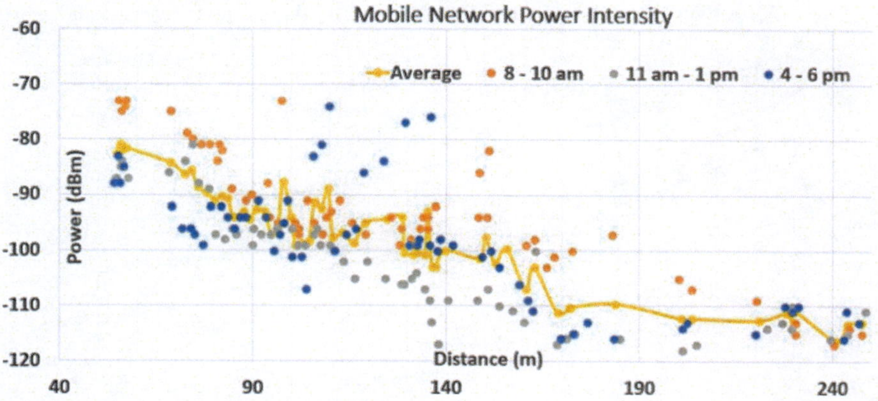

Fig. 8 Signal strength of base station Santa Cecilia

The figure allows to appreciate through an average of the different measurements that has a power range from −80dBm to −112 dBm. It should be noted that when approaching the base station, a signal is received with good power at certain distances, while in other measurements, there is a notable data dispersion due to the different attenuations that exist in the place, where the buildings that are in the place are high due to being a central place there is also traffic of people; therefore, there is a fall of the signal because there are elements that interfere with the intensity of the power.

A result it is obtained that the field measurements were achieved through data collection at different times and distances where the base station located in the downtown area of the city of Riobamba, having greater coverage the base station was able to measure up to a distance of approximately 250 m mentioning that there is a good signal at certain points because we have a close view to the base station and loss of power due to being in the area.

Figure 9 shows that in the "Santa Cecilia" base station, five propagation models have been applied, where the model that best fits the data obtained is the SUI, which is a derivative of Hata, with corrections for frequencies above 1900 MHz. In the application, it has been taken as reference the area as suburban where it is attributed to a terrain where the losses are intermediate of flat relief. And the one that least fits is the Okumura-Hata because this model works at greater distances than those taken in this base station and at lower frequencies.

In Table 14, it can be seen that when calculating the absolute and relative errors of each model in the base radius, it is verified that the best model has to adapt is the SIU due to the characteristics that presents the zone being the lowest error percentage that is 3%, while the model with highest error rate is Okumura-Hata with 9%.

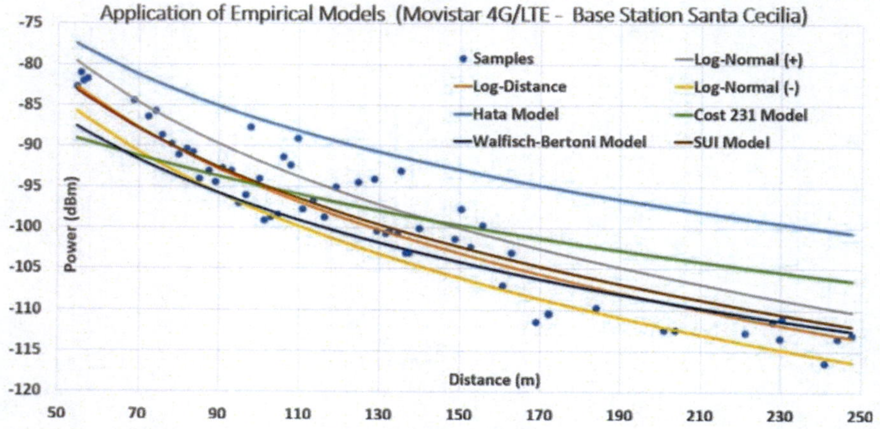

Fig. 9 Propagation models applied at the base station Santa Cecilia

Table 14 Error table

	E_a	E_r	E_r %
Okumura-Hata model	8.6898	0.0859	9
Cost 231 model	4.2978	0.0435	4
Walfisch-Bertoni model	3.5246	0.0376	4
SUI model	2.5125	0.0256	3

4.5 *Pichincha Bank Base Station*

Figure 10 shows the results obtained in the field measurements, providing information on the coverage offered by the Banco Pichicha base station located at Primera Constituente and Pichincha streets. The figure shows that in areas close to the base station, it was possible to receive a signal with excellent power reaching values up to −57dBm; however, at similar distances, it was possible to observe a drop reaching up to −103 dBm.

Through the field measurements, it was possible to appreciate that because the base station is located in a central area of the city of Riobamba in certain points, it was possible to receive a considerably good power; this is because it was possible to have a direct view to the transmitting antenna. While in other data, a notable dispersion in the measurements can be observed due to the different attenuations that exist in the place, which can be noted that the height of the buildings will be an obstacle for the intensity of power that will reach the receiver thanks to the diffraction that occurs in the scenario in where the measurements were collected.

Figure 11 presents the comparison of the empirical models used in the suburban area of the city of Riobamba. Base station (Pichincha bank). Which shows the results obtained by applying the propagation models to average measurements obtained,

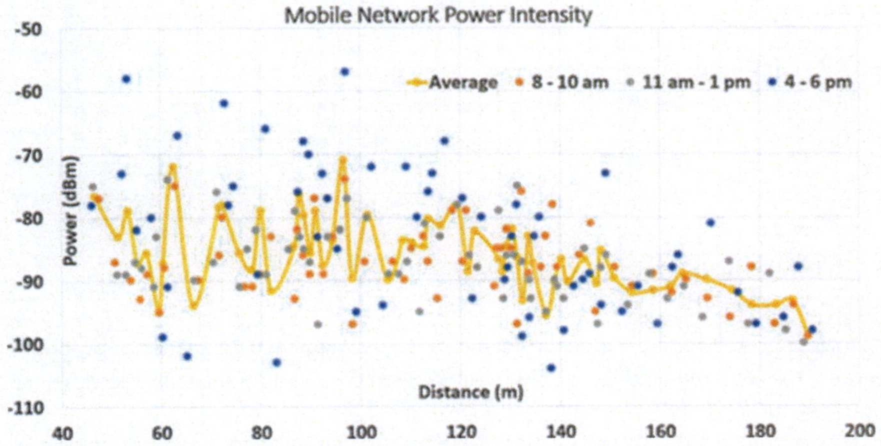

Fig. 10 Signal strength of base station Pichincha bank

which allows observing the behavior of the logarithmic curve under conditions compatible with the results of the scenario found.

Table 15 shows the results of the error estimation for each propagation model applied, of which the one with the highest margin of error is the Okumura-Hata model with 23% and the one with the lowest margin of error is the cost 231 model with 6%, so it can be assumed that the model that best adjusts to the average of the measurements obtained is the COST 231 model.

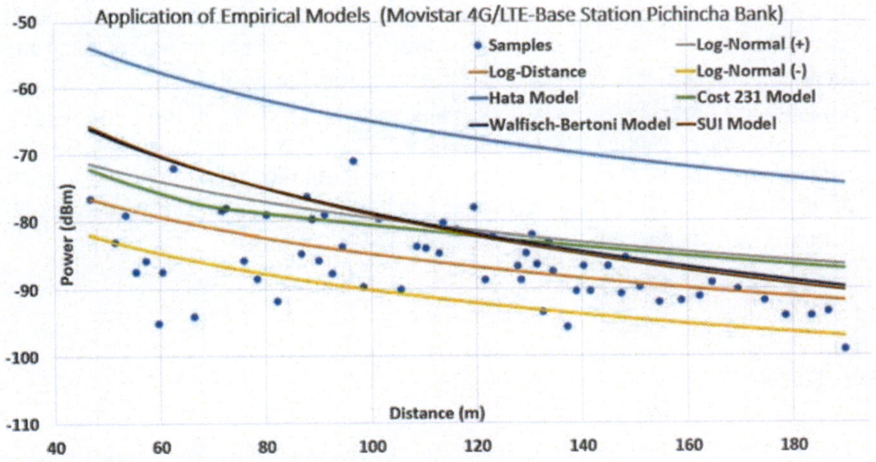

Fig. 11 Propagation models applied at the base station Pichincha bank

Table 15 Error table

	E_a	E_r	E_r %
Okumura-Hata model	19.9758	0.2305	23
Cost 231 model	5.6527	0.0641	6
Walfisch-Bertoni model	6.50	0.0743	7
SUI model	6.3653	0.0728	7

5 Conclusions

- It was observed that the models that most closely adapted to the scenarios proposed were the SUI and COST 231 models, based on the fact that the conditions for the use of these models were met.
- It was concluded that the Okumura-Hata propagation model cannot be used in small cities such as the city of Riobamba since it has a greater error than the other models; in addition, the range for its application of the Hata model is between 200 m and 20 km and that it uses frequencies lower than 1800 MHz.
- According to the observation of the results, 4G technology has a greater stability in its network at a range of up to 250 m or 300 m. But its coverage area is more limited so there are base stations nearby to ensure a permanent connection.
- There is a great influence of the power in reception in relation to the polarity of the antenna located in the base station, although the coverage area acts radially; within the cities, there are antennas directed to cover certain areas. This in most cases is due to the infrastructure of buildings that obstruct the line of sight.
- It was determined that the Walfisch-Bertoni model does not consider the existence of a line of sight between the transmitting antenna and the receiving antenna. It uses diffraction to analyze the losses suffered by the signal before reaching the receiving antenna with respect to the distance from buildings.
- The applied models considered variables such as street width, building height, reflection angles, among other parameters. As could be seen in the results, there is no specific model that fits each base station, since the reception power in each of them is affected by different factors such as weather, traffic, infrastructure and the number of connected users.

References

1. Pavon JO, Jime´nez Motilla J (2017) Evaluación comparativa de redes móviles, Master's thesis, Universidad Politécnica de Madrid
2. Santiago CAO (2020) Diseño e implementación de una celda celular con tecnologías 2g, 3g y lte para la ciudad de san jose de chimbo en la provincia de bolivar (bachelor's thesis). EPN, p 11
3. ARCOTEL (2018) Boletín estadístico. Agencia de regulación y control de las telecomunicaciones

4. I. of GSM Signal Strength in Rural Communities in the South-Eastern Region of Nigeria (2019) Investigation of gsm signal strength in rural communities in the south-eastern region of nigeria," Eur Sci J 15(15):152
5. Luz SD (2017) Network cell info lite: Una aplicacion para ver las estadisticas de tu red movil. Redes Zone
6. Barrios-Ulloa (2021) Comparison of radio wave propagations models of a wireless channel in the urban environment of the city of Barranquilla. J Comput Electron Sci Theory Appl 2(1):31–38
7. López-Bonilla JL, Vidal Beltra´n S, Degollado Rea EA (2017) Simplified propagation model for lte in the frequency of 2.1 ghz. Nova Scientia 9(19)
8. Mathuranathan (2013) Log distance path loss or log normal shadowing model. In: Gaussian waves signal processing for communication systems
9. Okumura TKY, Ohmori E, Fukuda K (1968) Field strength and its variability in the vhf and uhf land mobile radio service. Rev Electr Commun Lab 16(9):825–873
10. Correia LM (2009) A view of the cost 231-bertoni-ikegami model. In: 2009 3rd European conference on antennas and propagation, pp 1681–1685
11. Noman Shabbir HK, Sadiq MT, Ullah R (2011) Comparison of radio propagation models for long term evolution (lte) network. Int J Next Gener Netw (IJNGN) 3(3):5
12. Walfisch J, Bertoni H (1988) A theoretical model of uhf propagation in urban environments. In: IEEE Transactions om Antennas Propagation, pp 1788–1796
13. Sharma D, Sai TV, Sharma PK (2018) Optimization of propagation path loss model in 4g wireless communication systems. In: 2nd International Conference on Inventive System and Control, pp 19–20
14. Torrelavega, TEORIA DE ERRORES. Escuela Politécnica de Ingeniería de Minas y Energía, 2015

Automated System Configuration Using DevOps Approach

Hitesh Kumar Sharma, Hussain Falih Mahdi, Tanupriya Choudhury, and Ahatsham Hayat

Abstract In today's world, a secure and reliable system configuration is very important be it for any service-based or for any product-based company. Through Blaze, we enable the functionality to remotely connect to any device over the network and configure it as per our need. Flexible and scalable are the values that we follow in our roots. Blaze focuses on customer satisfaction with seamless, high-quality support for the users and ensures the correctness of the actions performed through powerful automation scripts to mitigate the human errors. Further, automation is the value that we nourish throughout the project by adding the functionality of a complete system upgrade and including real-time reporting as well for the system to be upgraded. In this project, we aim to develop a command line tool for Service Management and Monitoring that follows cutting-edge automation compliance.

Keywords DevOps · Automation · System provisioning · Service management · Monitoring · Remote report · Command line utility · Ansible · Intelligent system

All authors contributed equally.

H. K. Sharma (✉) · T. Choudhury (✉)
School of Computer Science, University of Petroleum & Energy Studies (UPES), Energy Acres, Bidholi, Dehradun, Uttarakhand 248007, India
e-mail: hkshitesh@gmail.com; hksharma@ddn.upes.ac.in

T. Choudhury
e-mail: tanupriya@ddn.upes.ac.in

H. F. Mahdi
Department of Computer and Software Engineering, University of Diyala, Baquba, Iraq
e-mail: hussain.mahdi@ieee.org

A. Hayat
University of Madeira, Funchal, Portugal
e-mail: ahatsham@iti.larsys.pt

K. A. Reddy et al. (eds.), *Proceedings of Fourth International Conference on Computer and Communication Technologies*, Lecture Notes in Networks and Systems 606,
https://doi.org/10.1007/978-981-19-8563-8_49

1 Introduction

We live in a world where Alexa sets alarms and students study from E-books. Computers have become an inevitable part of our lives, and the ability to connect to and use any service worldwide had completely changed our lives. Coming to the organizations, as an organization grows, workforce, resources, systems, services and infrastructure also tend to grow considerably, and it becomes difficult to maintain each of the system physically. Maintaining system, provisioning them and configuring them are a major concern for most of the IT industries today as we need to ensure that the different system element services are running smoothly with the same configuration to keep the IT services running right. The primary reason is that while using any software, many users notice that there are slight human errors that has degraded the performance or has rendered the system useless. This project focuses on implementing Service Manager and Monitoring System. The project works by using a bunch of automation scripts and roles (modules) for the provisioning and configuration part to mitigate the probability of human errors. The project will use the object-oriented approach as well as handling and processing the data in the database. The entire Service Manager and Monitoring System is working on the real-time data, and we wish to make it an intelligent system using some of the intelligent learning models. Ansible automation engine and various other tools will be utilized to achieve automated [9] using the principles of DevOps (Fig. 1).

The project will be delivered as a command line utility and will be deployed over to a dedicated system (can be both on cloud and on premise as well. Python scripts, In

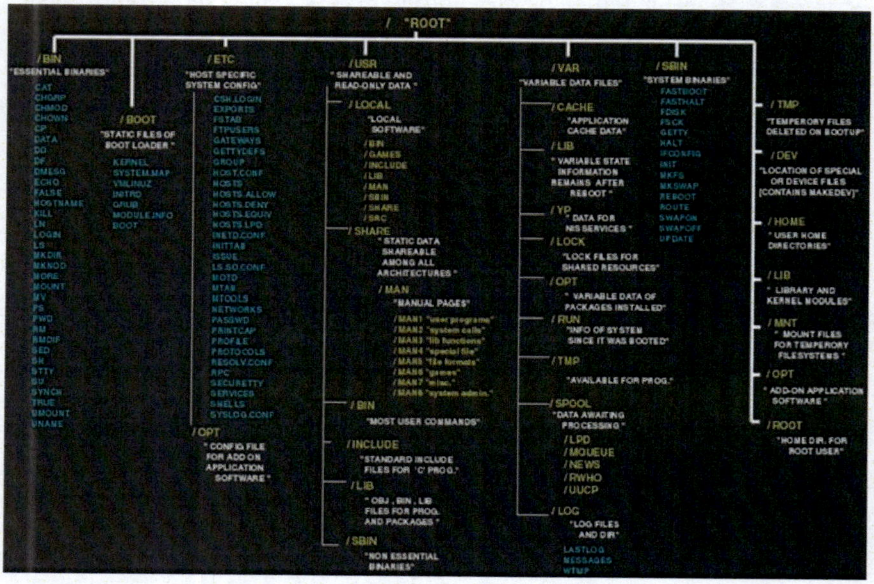

Fig. 1 Linux file system architecture

Linux system architecture, everything is a file and all the files and directories appear under the root (/) directory, even if they are stored on different physical or virtual devices. Each of the files in the file system is isolated and is present in their respective directories, and everything is well organized. The directories are organized in such a way that they provide access to a group of files, e.g., /bin contains essential binaries, /etc. contains the configuration files, /var the variable data files and there are many other directories as well which provides their isolation of files. We will utilize this concept of the Linux architecture and will configure and modify the only file system necessary. Considering the example of system upgrade, we will be modifying only the /bin and /boot directory so that all the other data remains as is.

2 Literature Review

In Refs. [1, 2], the authors have proposed that numerous companies are running distributed operations on their on-premise servers. Still, if load on those servers fluxes suddenly, and also, it becomes tedious to gauge the coffers and requires skillful human authority to address similar situations. It may accelerate the capital expense. Hence, numerous companies have bolted to resettle their on-premise operations to the cloud. This migration of the operations to the cloud is one of the substantial challenges. To setup and address the growing sophisticated architecture after migrating these operations to the cloud are really a time devouring and tedious operation which results in downtime. Hence, we necessitate to automate this contexture. To attain architecture for the distributed systems which advocate security, repetition, trustability and scalability, we require some cloud automation tools. This document summarizes tools like as Terraform and cloud conformation for infrastructure automation and Docker and Habitat for operation automation.

In Ref. [3], the inventors have declaimed about certain embodiments that associations are acquainting in operations to accelerate the pace of their software development procedure and to ameliorate the grade of their software. We also depict the developments of an exploratory interview-predicated study concerning six associations of various sizes that are functional in various diligence. As proportion of our findings, we adhered that all associations were admiring about their guests and only minor challenges were encountered while espousing DevOps.

3 Dataset Characteristics

Here, in our project, the database refers to the list of the devices that is to be provided to the Service Manager and Monitoring System for the configuration purpose [8]. Here, we are using two different kinds of datasets, one being static and other being dynamic in nature.

- The static database is defined in the Blaze inventory and contains the list of IP addresses to be configured.
- The dynamic inventory is used in case of the cloud-based inventories (AWS in our case), where the inventory is parsed dynamically based on the credentials stored in the console.

In case of static inventory, we provide the user with the option to use and configure any Linux distribution according to his/her comfort.

On the other hand, in case of dynamic inventory, we are using RHEL-based OS for making it smooth and easy for user to work and configure the system easily as the user is focused on receiving the working services rather than the underlying architecture [5].

4 Methodology

The objective of the project is to build a command line utility for achieving Service Manager & Monitoring System with all the features of automation [6, 7].

The complete task is further divided into subtasks:

1. Configuring and making resources available on cloud.
2. Managing the workspace and configuring it with the desired toolset.
3. Generating logs for any privilege activity made.
4. Upgrading the system in case the system is old.

The configurations that we want to achieve after the completion of the project are:

1. Installing/updating any package over the remote system.
2. Updating the OS of the remote system from EL6 to EL7 or from EL7 to EL8 [8].
3. Starting/Stopping/Restarting any service over the remote system.
4. Configuring yum repositories.
5. Running any docker image in the remote server.

This project is blending the agile methodology and waterfall methodology of software development as it is rare to find all the qualities in a single software development methodology. There are different ways to implement a waterfall methodology, including iterative waterfall, which still practices the phased approach but delivers in smaller release cycles. The project used the agile methodology for the development in build part of the project to take the advantage of the documentation part of the waterfall methodology as well as utilizing the sprints as a part of agile workflow. Overall, the time for the project is dedicated to an approach where the beginning time is dedicated toward the requirement analysis and the documentation part, and during the implementation part, all the team members are following their dedicated sprints cycles to implement the functionality. After the implementation, testing is to be done for the whole application. Finally, the application is deployed with the documentation.

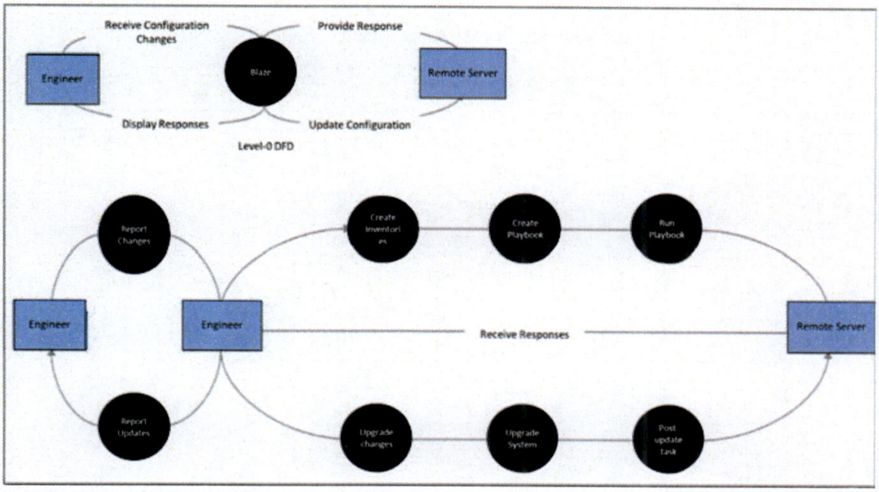

Fig. 2 Data flow diagram

The following data flow diagram depicts how the data is flowing in the application; here in our case, we have three different roes to be considered for the working of the application, i.e., engineer, Blaze and the remote server to be configured (Fig. 2). The engineer is working on his/her machine and wants to configure the remote server as per his needs. First, the user installs and configure the Blaze on his/her system and creates the inventory and playbook for the system to work upon. The playbook syntax is simple and inventory too. Blaze works with Terraform as well as Ansible which it connects in the background; it works as a third-party tool designed to work on the top of existing DevOps tools [4] and utilize their extensive functionality under a single umbrella and make the services available for all the user easily and hassle free. Blaze follows the principle of DevOps and follows the DevOps culture rather than just being a tool for automation. The principle of low-code, idempotence, code-generation, failure detection and many more are used in the project (Fig. 3).

5 Implementation and Results

5.1 sshKeygen

sshKeygen generates a new and a unique key for the authentication purpose and stores it in the keyStorage directory. The status of the sshKeygen is maintained in a file which can be utilized further as required by the program.

- Check if status directory exists,
- If exists, continue.

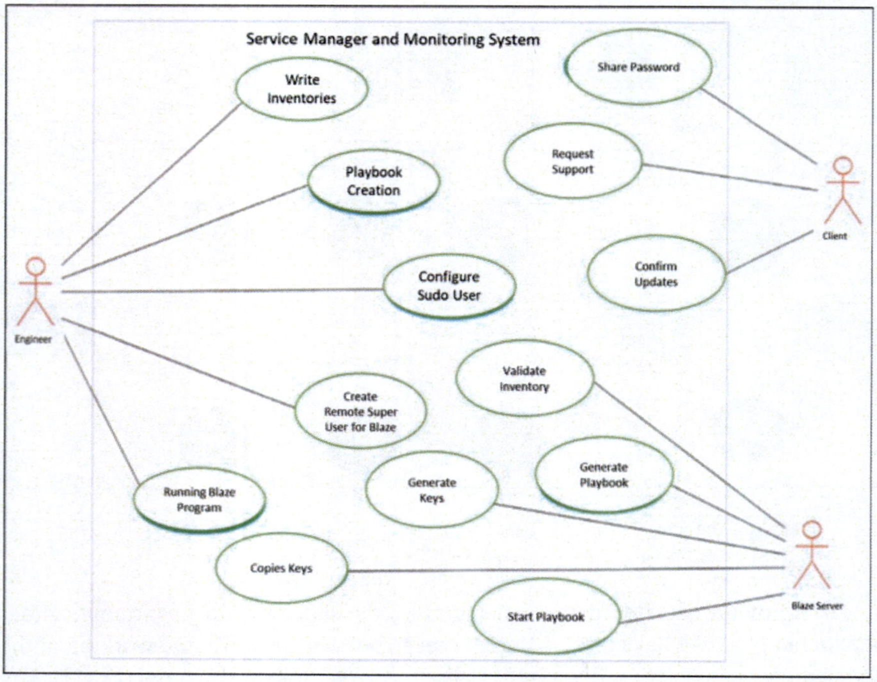

Fig. 3 Use case diagram

- If doesn't exist, create a new directory and continue.
- Check if keyStorage exists,
- If exists, continue.
- If not established, dump the error, update the status and further exit the program.
- Check if any key with the same detail exists,
- If any similar key exists, stop the program and return an error.
- If file doesn't exist, then continue and start the process.
- Generate a new sshKey for the establishing the connection.
- If key is generated successfully, print the status in a file and further exit the program successfully.
- If key is not generated, print the error specified, update the status file and further exit the program with a non-zero exit status.

5.2 *sshCopyId*

sshCopyId utilizes the key generated from the sshKeygen stored in the keyStorage directory and utilizes it to copy it to the remote user so that a password less access can be provided for the remote system. The status of the sshCopyId is maintained in a file which can be utilized further as required by the program (Figs. 4 and 5).

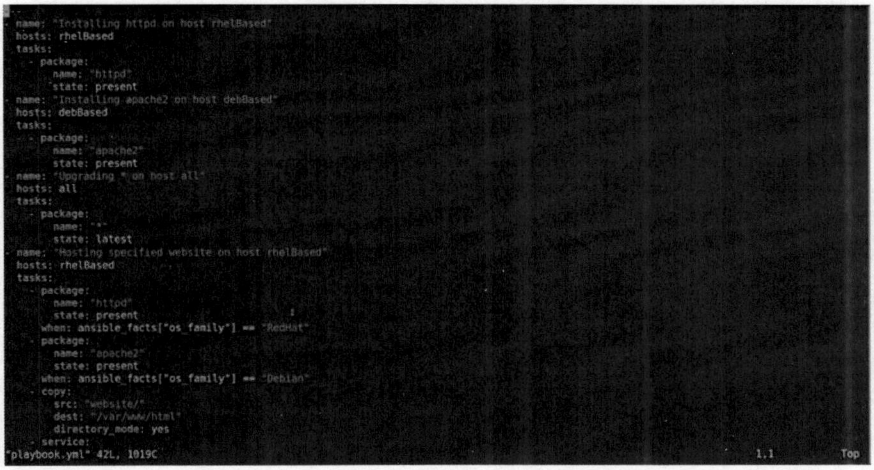

Fig. 4 Generated ansible playbook

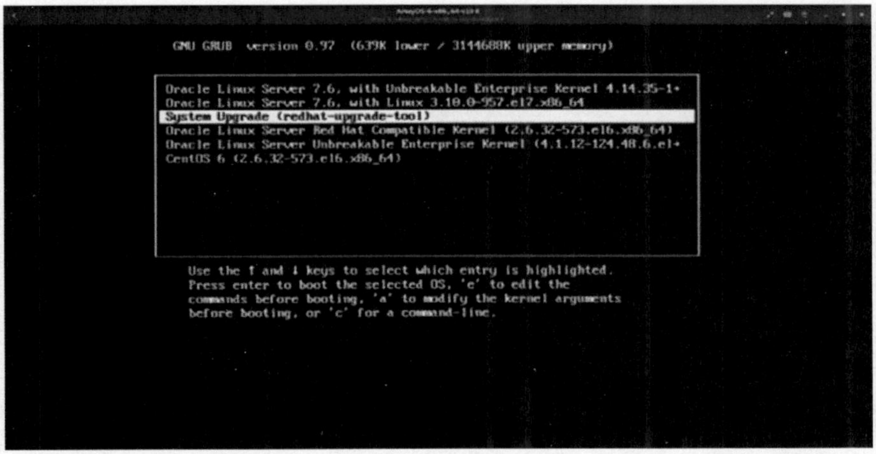

Fig. 5 EL6–EL7 upgrade

- Check if status directory exists,
- If exists, continue.
- If doesn't exist, create a new directory and continue.
- Check if keyStorage exists,
- If exists, continue.
- If not established, dump the error, update the status and further exit the program.
- Parse the command by fetching the details like remoteIp, remotePort and remotePass (remotePass will use the password decrypted from the passwordLoad program).

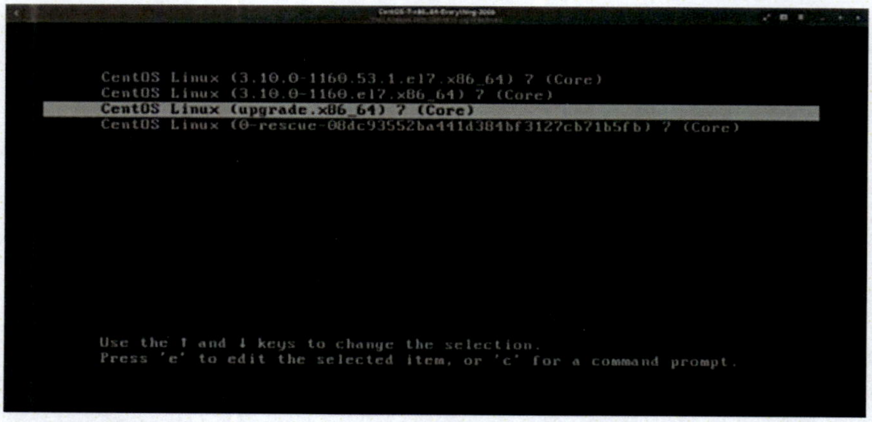

Fig. 6 EL6–EL8 upgrade menu

- Execute the command to copy the sshKey from the server on our end to the server on the client end.
- If not executed successfully, dump the error and exit the program.
- If executed successfully, save the status in the status file (Fig. 6).

5.3 *5x-Automation*

5x-automation is a company-specific (i.e., Ameyo) use case of Blaze, where it works end to end, from provisioning the resources over the AWS Cloud, configuring the systems and even running the softwares on the infrastructure.

5.4 *RPM Packaging*

In the modern world, to install any software, we download the executable file and install it directly rather than downloading the files and placing them in separate directories and then configuring the pre-install, post-install and perform checks for the required packages.

6 Conclusion

A DevOps tool to help and manage other infrastructure provisioning and configuration management tools has been achieved under a single umbrella. Blaze works on all the preferences and is currently under development so much can be expected from it further. In the end, Blaze is able to achieve the following results: configure a remote system according to the provided configuration, upgrade the infrastructure (in-place) without the need to re-configure the system, provide a company-specific full-stack DevOps use case to deploy the infrastructure and configure it.

References

1. Masek P, Štůsek M, Krejčí J, Zeman K, Pokorny J, Kudlacek M (2018) Unleashing full potential of ansible framework: university labs administration. In: Proceedings of the XXth conference of open innovations association FRUCT, p 426. https://doi.org/10.23919/FRUCT.2018.8468270
2. Jayachandran P, Pawar A, Venkataraman N (2017) A review of existing cloud automation tools. Asian J Pharm Clin Res 10.471. https://doi.org/10.22159/ajpcr.2017.v10s1.20519.
3. Erich F, Amrit C, Daneva M (2017) A qualitative study of DevOps usage in practice. J Softw Evol Process. https://doi.org/10.1002/smr.1885
4. Agarwal A, Gupta S, Choudhury T (2018) Continuous and integrated software development using DevOps. In: 2018 International conference on advances in computing and communication engineering (ICACCE), pp 290–293. https://doi.org/10.1109/ICACCE.2018.8458052
5. Kumar S, Dubey S, Gupta P (2015) Auto-selection and management of dynamic SGA parameters in RDBMS. In: 2015 2nd International conference on computing for sustainable global development (INDIACom), pp 1763–1768
6. Tian J, Varga B, Tatrai E, Fanni P, Mark Somfai G, Smiddy WE, Cabrera DeBuc D (2016) Performance evaluation of automated segmentation software on optical coherence tomography volume data. J Biophoton 9(5):478–489
7. Klein R, Klein BEK (2013) The prevalence of age-related eye diseases and visual impairment in aging: current estimates. Invest Ophthalmol Vis Sci 54(14)
8. Biswas R et al (2012) A framework for automated database tuning using dynamic SGA parameters and basic operating system utilities. Database Syst J III(4)
9. Gulia S, Choudhury T (2016) An efficient automated design to generate UML diagram from Natural Language Specifications. In: 2016 6th International conference—cloud system and big data engineering (Confluence), pp 641–648. https://doi.org/10.1109/CONFLUENCE.2016.7508197

Face Mask Detection Using Multi-Task Cascaded Convolutional Neural Networks

Nagaraju Rayapati, K. Reddy Madhavi, V. Anantha Natarajan, Sam Goundar, and Naresh Tangudu

Abstract Detecting faces is a prevalent and substantial technology in current ages. It became interesting with the use of diverse masks and facial variations. The proposed method concentrates on detecting the facial regions in the digital images from real world which contains noisy, occluded faces and finally classification of images. Multi-task cascaded convolutional neural network (MTCNN)—a hybrid model with deep learning and machine learning to facial region detection is proposed. MTCNN has been applied on face detection dataset with mask and without mask images to perform real-time face detection and to build a face mask detector with OpenCV, convolutional neural networks, TensorFlow and Keras. The proposed system can be used as an application in the recent COVID-19 pandemic situations for detecting a person wears mask or not in controlling the spread of COVID-19.

Keywords Face detection · Deep learning · MTCNN · OpenCV

1 Introduction

Advancements in the technology nowadays improved to an extent that a machine can be able to imitate a human brain and its functions. Artificial intelligence made this advancement in technology possible as it is proficient by learning how human brain works, learns and decide and finally outcomes in developing intelligent systems and software. Machine learning is one way to make the machine imitate human by different algorithms. Machine learning is implemented using algorithms and models to make decisions while deep learning structures algorithms to create artificial neural networks that can learn and decide on its own.

N. Rayapati · K. Reddy Madhavi (✉) · V. Anantha Natarajan
CSE, Sree Vidyanikethan Engineering College, Tirupati, AP, India
e-mail: kreddymadhavi@gmail.com

S. Goundar
School of Information Technology, RMIT University, Melbourne, Australia

N. Tangudu
IT Department, Aditya Institute of Technology and Management, Tekkali, Srikakulam, AP, India

K. A. Reddy et al. (eds.), *Proceedings of Fourth International Conference on Computer and Communication Technologies*, Lecture Notes in Networks and Systems 606,
https://doi.org/10.1007/978-981-19-8563-8_50

In these developing technologies, face detection is the most popular and significant technology. In the world that is battling against coronavirus disease COVID-19, the technology is of a great use. Many organizations are getting converted to 'work from home' style as a precaution of this pandemic. As the pandemic effect is slowly getting reduced, many of the workers are now apprehensive about returning to the office 'in-person' work style. During this transformation from work from home (WFH) to in-person style the checking of any violations manually is almost impossible in large premises. Computer vision and artificial intelligence techniques motivated for automatic detection that helps in monitoring, screening society containing coronavirus (COVID-19) pandemic.

2 Related Work

In this section, related works are reviewed that are done in this domain, the methodologies and algorithms that are designated solely for face mask detection are inadequate. The Viola–Jones method for face detection [1] uses Haar features for extracting facial features. A novel detection framework [2] is wearing face mask for helping control the spread of COVID-19. Jignesh et al. [3] proposed a detector using SRCNet classification network. Another model is developed using SRCNet in Ref. [4] with good performance. With deep learning advancements, neural networks learn features without prior knowledge in forming feature extractors like You Only Look Once (YOLO) algorithm [5, 6]. Face detection models developed not only use CNN and pre-trained models but also include independent techniques, classifiers such as support vector machine, softmax and optimizers such as Adam optimizer for better accuracy and efficient classification. A model proposed by Preeti et al. [7] SSDMobileNetV2 proposed face mask detection using OpenCV. Some of the earlier developed systems include pre-trained models called as MobileNet as the main component or as the backbone along with other models with fine-tuning. Some of the models developed [8], which has high computational efficiency and is easy to set up model for embedded systems.

Different pre-trained deep convolutional neural networks (CNN) extract deep features [9] from images of faces. Extracted features are further processed using machine learning classifiers. To overcome the weak generalization ability problem, Qi et al. [10] proposed a network that takes the input images of variable size which is the extension of existing system. Chaves et al. in Ref. [4] evaluate the speed–accuracy trade-off of three popular models. Different face detection methods using deep learning [11, 12] and image processing technologies are presented in Ref. [13]. Receptive field enhanced multi-task cascaded CNN (RFE-MTCNN) is proposed by Xiaochao et al. in Refs. [14, 15]. The existing models take the fixed size input images, which makes the model generalization ability weak.

3 Proposed Method

Proposed approach has been demonstrated in Fig. 1 and aims to detect whether the people wearing a face mask or not. The proposed approach consists of two stages, detection of faces from images and prediction of face mask-wearing conditions.

A. ***MTCNN***

MTCNN is comprised of three deep neural networks—P-Net, R-Net and O-Net, respectively. Hence, MTCNN is called as three-staged neural network. Primary stage is to resize input images into different scales in order to make an image pyramid (Figs. 2 and 3).

Stage-1: Proposal Network (P-Net)

Proposal network is the first stage of MTCNN, which is fully connected network (FCN) as it doesn't have dense layers in the architecture. This network proposes the Region of Interests which will be considered as candidate proposals for the facial detection. These proposals are then processed through non-maximum suppression and bounding box regression. Its architecture is represented in Fig. 1b.

Stage-2: The Refine Network (R-Net)

All the candidate proposals that are obtained from previous network (P-Net) are fed into refine network (R-Net). It has mostly similar structure to the P-Net, but with even more layers. R-Net is a CNN not an FCN due to the fact that R-Net structure consists of dense layer as the last layer of the architecture. The candidate proposals are further refined by reducing number of candidates. Architecture of R-Net is represented at Fig. 1c.

The new bounding box coordinates and confidence in each bounding box after NMS and Bounding box regression of P-Net is shown in Fig. 4, where as for R-Net and O-Net are shown in Figs. 5 and 6 respectively.

Stage-3: Output Network (O-Net)

The output network takes the bounding box coordinates of facial areas from the R-Net and is responsible for giving facial landmarks from the candidate regions. Output Network (O-Net) outputs the probability whether the face is in the box or not, if present the coordinates of bounding box, the coordinates of five facial landmarks (coordinates of the eyes, nose and mouth). Architecture of O-Net is represented at Fig. 1d.

B. ***Face Mask detection***

For the task of feature extraction and face mask detection, a novel architecture is constructed with convolutional, max-pooling, dense and flatten layers. A sequential model is used as it is a perfect for a serial stack of layers where each of the layers has exactly one input tensor and one output tensor. The architecture of the feature extractor convolutional network is illustrated in Fig. 7. The architecture

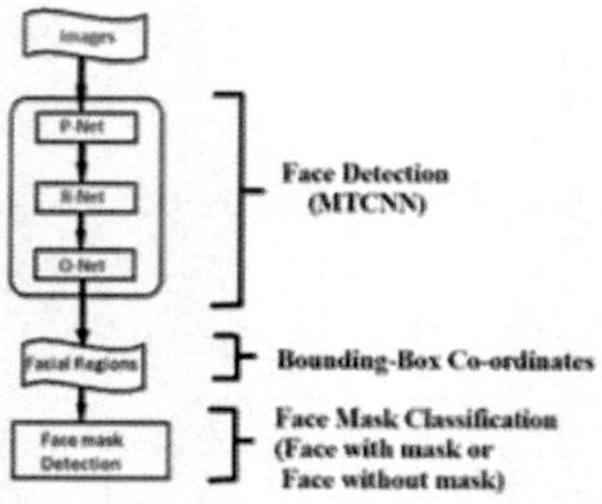

a. Proposed Network with MTCNN Face detector

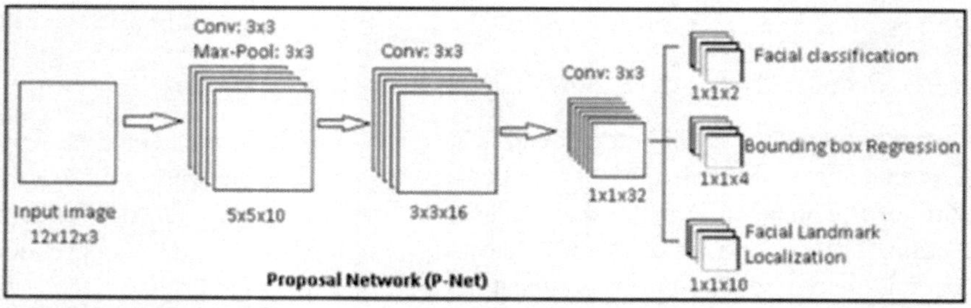

b. Architecture of P-net

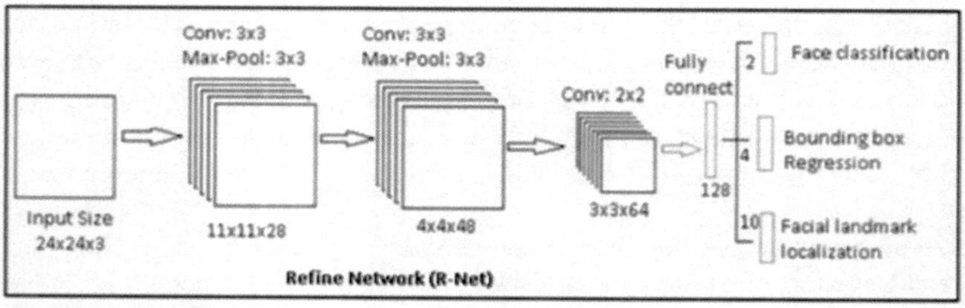

c. Architecture of R-net

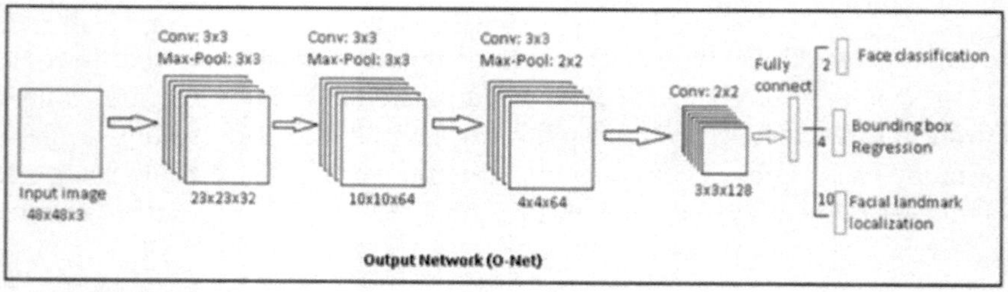

d. Architecture of O-net

Fig. 1 Architecture of MTCNN

Fig. 2 Image pyramid of different scales

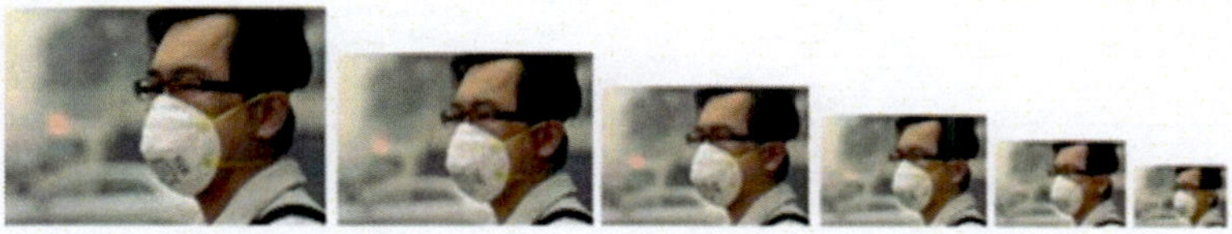

Fig. 3 Images of different scales in image pyramid

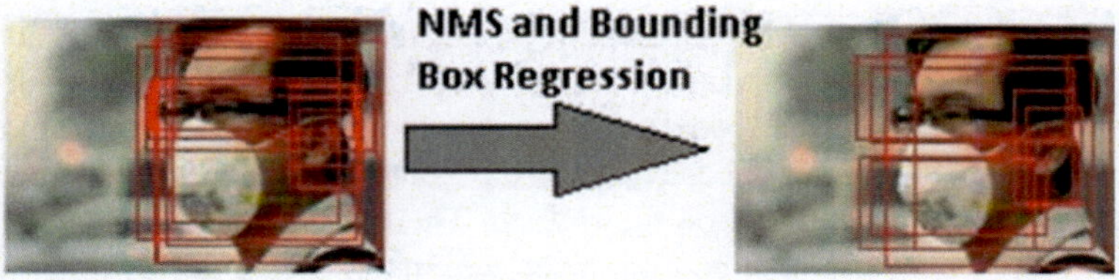

Fig. 4 NMS and bounding box regression in P-Net

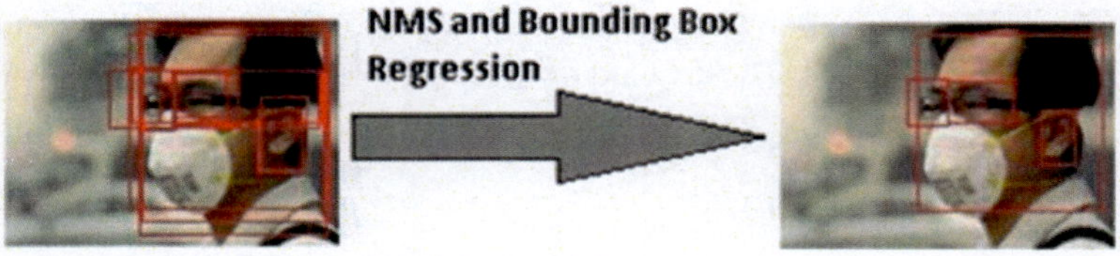

Fig. 5 NMS and bounding box regression in R-Net

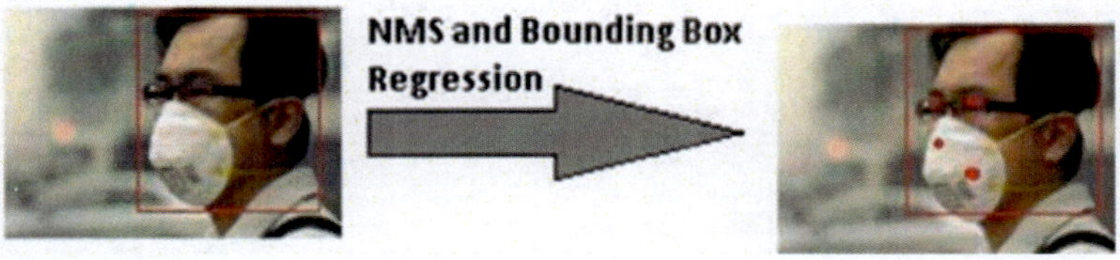

Fig. 6 NMS and bounding box regression in O-Net

consists of 2 convolutional layers, two max-pooling layers, flatten layer, dropout layer and finally, two dense layers.

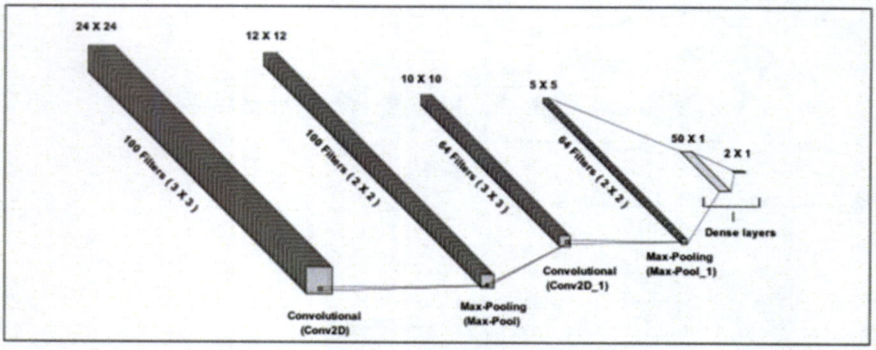

Fig. 7 Architecture of feature extractor

The detailed representation of the shapes of the feature vectors outputted by every layer and parameters in every layer are illustrated in Table 1. Size of the input image fed to the first layer (Conv2d_1) has dimensional size 50 × 50 × 1. After processing, every layer outputs feature vectors of different dimensions. This model consists of a total of 138,816 total parameters which are all trainable with no non-trainable parameters. The architecture contains 2 pairs of Conv2d and max-pooling layers having 100 and 64 filters each. The Conv2d layer consists of 3 × 3 dimensional filters and max-pooling layer consists of 2 × 2 dimensional filters. The feature vectors, processed through 2 pairs of Conv2d and max-Pooling layers fed into a flatten layer, which flattens the feature vector matrix into a column vector having 1600 × 1 dimensions. The flattened vector is then fed to dense layer, which has ReLU activation function (fully connected) having 50 nodes, further to another dense layer having 2 nodes. The detailed representation of dense layers is illustrated in Fig. 8. Final layer in the architecture is incorporated with a classifier called softmax classifier.

Table 1 Summary of network model

	Output	Param#
	24 × 24	1000
	12 × 12	0
	10 × 10	57,664
	5 × 5	0
	1600	0
Dense (Dense)	50	80,050
Dropout (Dropout)	50	0
Dense (Dense)	2	102

4 Experiments

A. ***Dataset***

Dataset used for training is face mask detection (https://www.kaggle.com/wobotintelligence/face-mask-detection-dataset) [16]. The dataset contains a total of 5933 images. The dataset contains a total of 5933 images.

The flow of the proposed approach is illustrated in Fig. 9, showing training phase and testing phase. The proposed model has been trained for 30 epochs. As Adam optimizer is the simple, time-efficient optimizer and best replacement for stochastic gradient descent for training deep learning models. Thus, the Adam optimizer is considered of the proposed model. Hyperparameter settings of the proposed model are listed in Table 2. The training images are annotated by human labelling for the face coordinates in the images for both single face images and multiple face images. The facial coordinates, corresponding image source and class label are loaded from a comma-separated values (CSV) file. During the training phase, the model converts the images into grayscale using OpenCV module. By using the facial coordinates from the CSV file, all the faces are cropped from the images and are resized into 50 × 50 dimensions as a part of image post-processing. Finally, by making the necessary transformations and normalizations on the cropped images and features, they are fed to the model.

Testing phase contains 1698 images. The final result of the testing phase is to classify the images into face_with_mask and face_no_mask. Testing phase consists of two steps, detection faces from the test images and classification of face mask-wearing conditions. As described in the Sect. 3A, MTCNN algorithm is used for the face detection. All the test images are fed to the MTCNN detector which outputs the bounding box coordinates of all the faces in the image. These coordinates are used to crop faces from the images. The cropped image further undergoes image

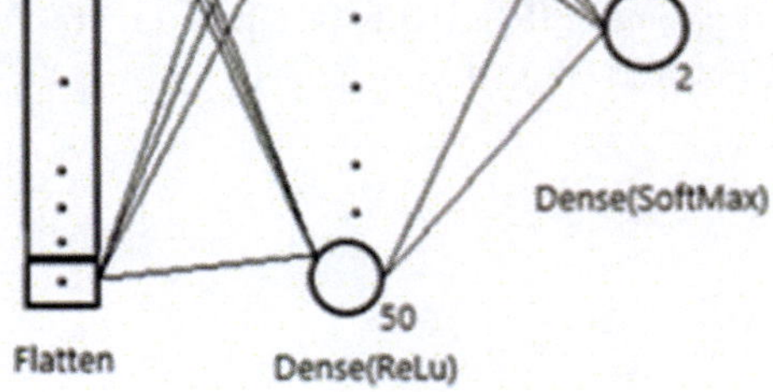

Fig. 8 Dense layers architecture

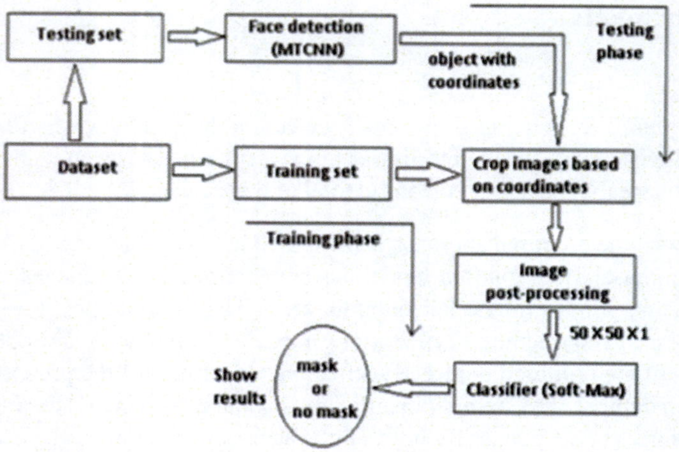

Fig. 9 Phases of proposed approach

Table 2 Hyperparameters of the proposed model

Hyperparameters	Value
Epochs	30
Batch size	5
Optimizer	Adam
Learning rate	0.001
Decay rate	1e-5

post-processing such as conversion into grayscale images, resizing into 50 × 50 × 1 dimensional size. After the image post-processing, the images are fed to the trained model. The faces are further classified and displayed along with the labels of corresponding related class.

The experiments for the proposed system are conducted in Google Colab environment. Accuracy was used as a performance metric to evaluate the model. Given the cascade structure of MTCNN face detector, the facial regions in the images are detected efficiently compared to state-of-the-art detectors. After feeding the post-processed images to the training model, the detailed information of the 30 epochs is represented in Table 2. Table 2 contains cross-entropy loss and accuracy after epochs. The final accuracy after the training of the model reached to 99.53% and cross-entropy loss is 0.14%. Graph 1 represents the accuracy and loss of model during 30 epochs (Fig. 10).

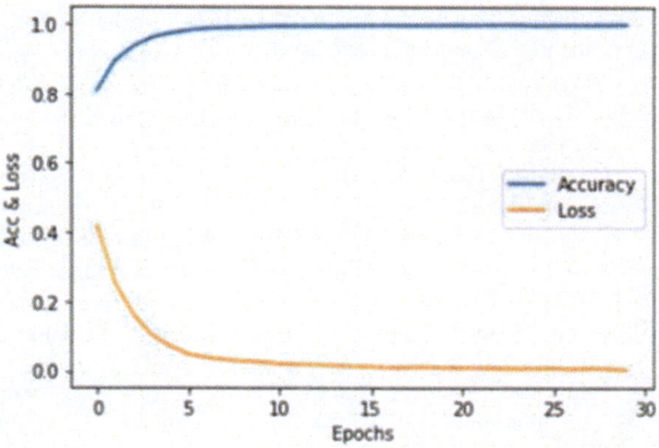

Fig. 10 Accuracy and loss graph

5 Conclusion

During these COVID-19 pandemic times, people are forced to wear a face mask at all the public places such as markets and offices, but manually checking about the conditions of wearing face mask of every person is not achievable. Thus, researches are motivated to develop automatic facial mask detection system. In this paper, a model is proposed for the detection of face mask-wearing conditions. The model proposed accommodates MTCNN algorithm, a hybrid model for the efficient facial region detection in unconstrained environment, which outstands among the existing detectors. For the purpose of face mask detection, CNN architecture is developed that extracts features. Finally, a softmax classifier is used for the binary classification, which classifies faces in the images into two classes, namely face_with_mask and face_no_mask. The model is evaluated on the face mask detection dataset available at kaggle website. This model achieved 99.53% accuracy and 0.14% loss after 30 epochs. Further, the proposed model outperforms several of the existing models in face mask detection area of research.

References

1. Huang J, Shang Y, Chen H (2019) Improved Viola-Jones face detection algorithm based on HoloLens. Springer Access
2. Zhang J, Han F, Chun Y, Chen W (2021) A novel detection framework about conditions of wearing face mask for helping control the spread of COVID-19. IEEE Access 9:42975–42984
3. Jignesh Chowdary G, Punn NS, Sonbhadra SK, Agarwal S (2021) Face mask detection using transfer learning of InceptionV3. Springer access
4. Chaves D, Fidalgo E, Alegre E, Alaiz Rodríguez R, Jáñez-Martino F, Azzopardi G (2020) Assessment and estimation of face detection performance based on deep learning for forensic applications. Sensors
5. Loeya M, Manogaran G, Hamed M, Tahad Nour N, Khalifa EM (2020) Fighting against COVID-19: a novel deep learning model based on YOLO-v2 with ResNet-50 for medical face mask detection. Elsevier

6. Kumar A, Kalia A, Verma K, Sharma A, Kaushal M (2021) Scaling up face masks detection with YOLO on a novel dataset. Elsevier Access
7. Nagrath P, Jain R, Madan A, Arora R, Kataria P, Hemanth J (2020) SSDMNV2: a real time DNN-based face mask detection system using single shot multibox detector and MobileNetV2. Elsevier
8. Qin B, Li D (2020) Identifying facemask-wearing condition using image super-resolution with classification network to prevent COVID-19. Springer access
9. Loey M, Manogaran G, Taha MHN, Khalifa NEM (2021) A hybrid deep transfer learning model with machine learning methods for face mask detection in the era of the COVID-19 pandemic. Elsevier
10. Qi R, Jia R-S, Mao Q-C, Sun H-M, Zuo L-Q (2019) Face detection method based on cascaded convolutional networks. IEEE Access 7:110740–110748
11. Pei Z, Xu H, Zhang Y, Guo M, Yang Y-H (2019) Face recognition via deep learning using data augmentation based on orthogonal experiments. Electronics
12. Zheng G, Xu Y (2021) Efficient face detection and tracking in video sequences based on deep learning. Elsevier
13. Liu Q, Peng H, Chen J, Yang S (2020) Face detection based on open Cl design and image processing technology. Elsevier
14. Li X, Yang Z, Wu H (2020) Face detection based on receptive field enhanced multi-task cascaded convolutional neural networks. IEEE Access 8:174922–174930
15. Zhang K, Zhang Z, Li Z, Qiao Y (2016) Joint face detection and alignment using multitask cascaded convolutional networks. IEEE Signal Process Lett 23(10):1499–1503
16. Cabania A, Hammoudi K, Benhabiles H, Melkemi M (2020) MaskedFace-Net—A Dataset of correctly/incorrectly masked face images in the context of COVID-19. Elsevier

Empirical Study on Categorized Deep Learning Frameworks for Segmentation of Brain Tumor

Roohi Sille, Tanupriya Choudhury, Piyush Chauhan, Hussain Falih Mehdi, and Durgansh Sharma

Abstract In the medical image segmentation field, automation is a vital step toward illness detection and thus prevention. Once the segmentation is completed, brain tumors are easily detectable. Automated segmentation of brain tumor is an important research field for assisting radiologists in effectively diagnosing brain tumors. Many deep learning techniques like convolutional neural networks, deep belief networks, and others have been proposed for the automated brain tumor segmentation. The latest deep learning models are discussed in this study based on their performance, dice score, accuracy, sensitivity, and specificity. It also emphasizes the uniqueness of each model, as well as its benefits and drawbacks. This review also looks at some of the most prevalent concerns about utilizing this sort of classifier, as well as some of the most notable changes in regularly used MRI modalities for brain tumor diagnosis. Furthermore, this research establishes limitations, remedies, and future trends or offers up advanced challenges for researchers to produce an efficient

Authors Roohi Sille and Tanupriya Choudhury contributed equally and all are the first author.

R. Sille (✉)
Systemics Cluster, University of Petroleum and Energy Studies (UPES), Dehradun, Uttarakhand 248007, India
e-mail: rsille@ddn.upes.ac.in

T. Choudhury (✉) · P. Chauhan
Informatics Cluster, University of Petroleum and Energy Studies (UPES), Dehradun, Uttarakhand 248007, India
e-mail: tanupriya@ddn.upes.ac.in

P. Chauhan
e-mail: pchauhan@ddn.upes.ac.in

H. F. Mehdi
Department of Computer and Software Engineering, University of Diyala, Baquba, Iraq
e-mail: hussain.mahdi@ieee.org

D. Sharma
School of Business and Management, Christ University, Delhi NCR Campus, Mariam Nagar, Meerut Road, Delhi NCR, Ghaziabad 201003, India
e-mail: durgansh.sharma@christuniversity.in

K. A. Reddy et al. (eds.), *Proceedings of Fourth International Conference on Computer and Communication Technologies*, Lecture Notes in Networks and Systems 606,
https://doi.org/10.1007/978-981-19-8563-8_51

system with clinically acceptable accuracy that aids radiologists in determining the prognosis of brain tumors.

Keywords MRI · CNN · Deep learning · Cascaded · Fusion · Brain tumor

1 Introduction

For analyzing data about the biological human body, medical imaging is crucial. Various imaging modalities, like computed tomography scans, X-rays, magnetic resonance imaging, and so on, are utilized in order to identify various illnesses. For frequently using in the diagnosis and clinical research, the computed tomography scans and the magnetic resonance imaging are done. The medical imaging approaches presented have certain advantages and disadvantages. MRI has the following advantages over other imaging modalities: high resolution, high signal-to-noise ratio, and soft tissue imaging capability [1]. Segmentation of medical pictures aids in the separation of distinct objects included in a medical image for better brain MRI analysis. The segmentation of brain MRI data has been successfully automated using a variety of deep learning techniques because hand segmentation is laborious and has poor reproducibility [2]. Deep learning algorithms have made transfer learning possible without a vast amount of data or handmade features. It has the ability to extract the features of specific brain MRI tissues automatically. Because of the intensity inhomogeneities in brain MRI images, preprocessing is required before using a deep learning model to process the image. Preprocessing improves the texture quality of images, allowing deep learning approaches to do more accurate segmentation. Computerized diagnostics are necessary to aid radiologists in clinical diagnosis. This allows a large number of instances to be processed with the same precision and in less time. It has also been noticed that due to the overlap in intensity between the two groups, distinguishing non-healthy tissues from healthy tissues is challenging. Recent studies have used deep neural networks or convolutional neural networks to separate brain MRI data.

Different categories of brain datasets are publicly available for research work in object detection, separation of gray matter, white matter, tumor segmentation, and cerebrospinal fluid, among other things, which is due to the extensive research being done in the segmentation of automated brain tumor. Researchers have largely used the BraTS dataset from 2015 to 2021, according to the literature survey undertaken for this publication. T1-weighted CE-MRI, MRBrainS, and iSEG-2017 were among the other datasets examined.

Focusing on qualitative measures like accuracy, specificity, sensitivity, and precision while using quantitative parameters like entropy (a measure of system disorder), peak signal-to-noise ratio, and root mean square error (RMSE)/mean square error (MSE) and can help improve the efficiency of segmentation algorithms [2]. The above-mentioned datasets were used to train various DL models, and their performance was measured using dice score, mean IoU, Hausdorff distance, and other metrics. The majority of the time, dice scores are employed as an evaluation criterion.

In this paper, the benefits and drawbacks of various brain MR segmentation algorithms are explored, with a focus on performance evaluation. DL algorithms have been demonstrated to provide predicted outcomes for brain tumor segmentation when compared to ML methods. When applied to medical image segmentation, DL has a number of advantages over ML techniques [3]. Single-path and multi-path CNN, cascaded CNN, fully convolutional networks (FCN), and fusion approaches are four types of deep learning algorithms. The literature review is covered in these sections. Medical imaging developments have made real-time segmentation of medical images possible, providing for real-time feedback on therapeutic decisions. This study examines the deep learning techniques used to improve the computational speed and efficiency of medical picture segmentation in a real-time setting, in order to address all of the problems mentioned above at once.

2 Related Works

In the subject of medical picture segmentation, extensive research is being conducted in order to speed up the automation of this field. Brain tumors have been shown to be fatal to human health [2]. To achieve high dice score coefficients, the CNNs have been shown which have a direct effect on the performance of deep learning model [4, 5]. Several deep learning models are being developed that use multimodal MRI as inputs. The focus of this research was on deep learning models proposed for brain tumor segmentation between 2020 and 2022. In the clinical therapy of brain tumors, precise segmentation is essential. The endeavor, however, remains difficult due to not only substantial differences in the sizes and shapes of brain tumors but also large differences among individuals. This research focuses on four types of convolutional neural networks (CNNs), as indicated in Fig. 1.

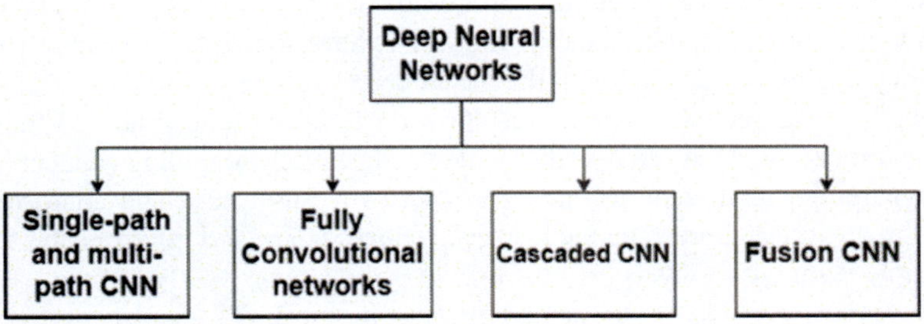

Fig. 1 Categorized deep learning frameworks

2.1 Single and Multi-Path CNN

For multimodal medical picture segmentation, the HybridCTrm network was presented, and it outperforms HyperDenseNet in terms of maximum performance criteria. The proposed network is a hybrid of transformers and CNNs that achieves high generalization and segmentation performance [6]. Traditional deep learning approaches, which result in long-range dependencies and poor generalizations, are replaced by this method. At the time of testing, the proposed model had a mean of 87.16, which increased to 89.75 for CSF, 86.40 for GM, and 85.34 for WM.

When U-net is merged with residual block attention network atrous convolution, a new study area for medical picture segmentation develops. The residual path known as multipath residual attention block (MRAB) is merged with attention unit and the processing block is replaced with two atrous convolution sequences representing the architecture. This architecture attained high dice scores. Dice scores for the ensembled model for the enhanced tumor (ET), core tumor (TC), and the complete tumor (WT) validation datasets were 77.71%, 79.77%, and 89.59% for BraTS2018, 74.91%, 80.98%, 88.48% for BraTS2019, 72.91%, 80.19%, 88.57% for BraTS2020, and 77.73%, 82.19%, 89% [7].

An innovative strategy based on modality pairing learning is proposed to increase segmentation accuracy. In order to capture complex interactions and a wealth of information between modalities, a series of layer connections is used. Paralleled branches are then created to take use of various modalities' unique characteristics. To lessen the variation in predictions between the two losses, the consistency loss is applied. The dice scores which came out as an outcome are 0.842, 0.891, and 0.816 for the CT, WT, and ET, respectively [8].

2.2 Fully Convolutional Neural Networks

The tumor segmentation is a vital and critical step to detect and manage the disease cancer. However, accurately segmenting tumors is an interesting research work because of the characteristics of brain tumors and the noise device. The brain tumor segmentation approaches based on the fully CNN have shown brightly and received a growing amount of attention with the recent deep learning success.

F2 FCN is offered as a way to cut down on CNN training time and improve segmentation accuracy. It is a new distributed and parallel computing concept based on a hypergraph membrane technology. It has a feature reuse and conformance module that extracts more valuable features, reduces noise, and improves the fusion of multiple feature map levels [9].

nnU-net has been updated to include the most recent BraTS team suggestions for post-processing, region-based training, and more aggressive augmentation methods. Based on dice scores and Hausdorff's distance, the nnU-net modification achieved excellent performance results [10]. Following DSC and HD95 are attained with proposed methodology (Table 1).

Table 1 Results of DSC and HD95 [10]

	WT	CT	ET
DSC	88.95	85.06	82.03
HD95	8.498	17.337	17.805

Table 2 Results of DSC and HD95 [12]

	WT	CT	ET
DSC	0.8858	0.8297	0.7900
HD95	5.32	22.32	20.44

A new FCN-based framework is created that concurrently includes dilated convolution and convolution which is separable into the architecture of U-Net so as to fight for the most prevalent and treacherous type of primary brain tumor and glioma. The receptive field is considerably increased and superior global and local feature representations are obtained by combining this architecture with a dilated block. This method contains geographical data as well as pixel-level depth details for more precise and efficient segmentation of brain tumors [11].

(a) **Cascaded CNN**
For brain tumor segmentation, a three-layered deep neural architecture cascade has been proposed. The response obtained from each stage is elaborated using feature maps and prior stage probabilities. The experiment was run using the BraTS 2020 dataset and yielded the following results (Table 2) [12].

Recent breakthroughs in the integration of deep learning and IoT have hastened the expansion of medical automation. IoT devices collect medical data from the human body, which is then analyzed using deep learning algorithms to forecast various abnormalities and diseases. The data that has been accepted can be in the form of text, photographs, or videos. CNN, in collaboration with IoT, has shown remarkable and accurate results in the segmentation of brain tumors. Tumor extraction was the most difficult and time-consuming task due to the wide range of tumor sizes.

It is suggested to use a novel cascaded architecture combining CNN and handcrafted features for automated brain tumor segmentation. Support vector machine classifies the pixels attaining confidence surface modality (CSM). After the mean intensity, the local binary pattern (LBP), and the histogram of oriented gradients (HOG) computations, the CSM is performed then. CSM along with the provided MRI, merged into a special three-path architecture of CNN. The proposed methods were trained on BraTS 2015 dataset and attained good results (Table 3) [13].

When it comes to brain tumor segmentation, semantic segmentation of brain MRI is critical [14]. For semantic segmentation of brain tumors, a multiscale network

Table 3 Results of DSC [13]

	WT	CT	AT
DSC	0.81	0.76	0.73

based on LinkNet is proposed. The precision of brain tumor segmentation has improved because of LinkNet's two-scale technique. The cascade strategy was utilized in this case so that the second network may benefit from the first network's learning. This method yields a dice score of 0.8003 and a mean IoU of 0.9074 [15].

(b) **Fusion CNN**
The fusion or hybrid technique has shown outstanding results in terms of segmentation accuracy. The primary unsolved issue is how to integrate or fuse low-level semantic data with high-level semantic features effectively. The need to run high-end CNN models on low-power computers with limited memory remains a challenge.

To develop a novel multi-path adaptive fusion network, we have to overcome these conditions. "Skip connections" are combined with ResNets in this method to successfully reserve and proliferate more visual characteristics of lower level. By accepting the direction connections from the state of the previous dense block to all levels of the current dense block in the network, an establishment of contiguous memory mechanism was done. Further in the phase of up-sampling, a multi-path adaptive fusion dense block was applied to adaptively change the low-level visual feature and then combine it with high-level semantic data. The layers 84,122,142,194 were used to train this model on BraTS 2015. The values were 0.83 for WT, 0.65 for CT, and 0.62 for ET at 194 layers [16].

Fusion of multipath CNN with FCN improves segmentation precision for multi-modal brain tumor segmentation. This aids in the extraction of useful characteristics from the data of multi-modal MRI. In order to extract various feature receptive fields, 3D dilated convolution is used in each pathway. The trained model yielded the following dice scores on BraTS 2019 and BraTS 2018 datasets (Table 4) [17].

For some CNN models, there are two distinct properties suggested. In order to extract complete feature information from the multi-modality input for mono-modality, paired-modality, and cross-modality data, respectively, the three paths of the feature extraction block were created. In a three-branch classification block, pixels belonging to four separate groups are recognized. Since each branch is trained independently, it is possible to update the parameters using ground truth information from a particular tumor location. There are only 61,843 parameters in the configuration because the system's convolution layers are customized for particular purposes. The BraTS2018 and BraTS2019 datasets have been used to properly test the suggested technique. The mean Dice scores for the enlarged tumor, whole tumor and tumor core from ten investigations on the samples validation for BraTS2018 are 0.787, 0.886, and 0.801, respectively, and for BraTS2019, they are 0.751, 0.885, and 0.776 (Table 5) [18].

Table 4 Results of DSC on BraTS 2018 and 2019 [17]

	WT	CT	ET
BraTS 2019	0.89	0.78	0.76
BraTS 2018	0.90	0.79	0.77

Table 5 Comparison between deep learning frameworks for brain tumor segmentation

S. No	Paper	Dataset	CNN architecture	Input	Performance Parameter
1	[19]	MRBrainS iSEG-2017	Single and Multipath CNN	35 × 35 × 35	Mean IoU: 87.16
2	[20]	BraTS 2018,2019,2020,2021	Single and Multipath CNN	72 × 72 × 72	Dice scores: BraTS 2018: 77.71%, 79.77%, 89.59% BraTS 2019: 74.91%, 80.98%, 88.48% BraTS2020: 72.91%, 80.19%, 88.57% BraTS2021: 77.73%, 82.19%, 89.33%
3	[21]	BraTS 2020	Two Path CNN	2 × 128 × 128 × 128	0.891, 0.842, 0.816
4	[9]	BraTS 2020	Hybrid FCN	4 × 128 × 128 × 128	Dice scores 0.78, 0.91, 0.85 HD 26.57, 4.18 and 4.97
5	[10]	BraTS 2020	U-Net (FCN-Based)	32 × 128 × 128 × 128	DSC: 88.95, 85.06 and 82.03 HD95: 8.498, 17.337 and 17.805
6	[11]	BraTS 2017 and 2018	U-Net (FCN-Based)	128 × 128 × 4	NR (Not Reported)
7	[12]	BraTS 2020	Cascaded DNN	Patches of size 120 × 120	DSC: 0.8858, 0.8297, 0.7900 HD:5.32 mm, 22.32 mm, 20.44 mm
8	[13]	BraTS 2015	Cascaded CNN	NR (Not Reported)	DSC: 0.81, 0.76, 0.73
9	[15]	T1-weighted CE-MRI	Cascaded LinkNet	512 × 512 and 256 × 256	Dice: 0.8003 Mean IOU: 0.9074
10	[16]	BraTS 2015	Skip Connections with ResNets	120 × 120	0.83, 0.65, 0.62
11	[17]	BraTS 2018 and 2019	multipath CNN with FCN	44 × 192 × 192	DSC: BraTS 2019: (0.89, 0.78, 0.76) BraTS 2018: 0.90, 0.79, 0.77

(continued)

Table 5 (continued)

S. No	Paper	Dataset	CNN architecture	Input	Performance Parameter
12	[18]	BraTS 2018 and 2019	FCN and Cascaded	200 × 168	BraTS 2018 0.787, 0.886, 0.801 BraTS 2019 0.751, 0.885, 0.776

The above-mentioned literature survey in Table 1 indicates that various deep learning models are categorized into multipath CNN, cascaded CNN, fusion CNN and FCN. All the mentioned categories have been tested on different performance evaluation parameters.

3 Conclusion

Despite the fact that several deep learning models have been trained on a variety of datasets, brain tumor segmentation remains a difficult task. The CNN models can't be trained using all of the trainable parameters connected to the impacted tumor because of the insufficient datasets available. The segmentation findings are inaccurate as a result. In brain MRI imaging, data imbalance happens as a result of the diminished volume of the tumor or lesion regions. The possibility of inaccurate segmentation due to biased prediction exists as a result of the hand annotation. These reasons allow for the use of generative adversarial networks (GAN) or adversarial learning to replace CNN [22–25] models. GAN have the capability to annotate the images required for training the models, and it is also used to segment the brain tumors from different image modality scans.

References

1. Isa IS, Sulaiman SN, Mustapha M, Karim NKA (2017) Automatic contrast enhancement of brain MR images using Average Intensity Replacement based on Adaptive Histogram Equalization (AIR-AHE). Biocybern Biomed Eng 37(1):24–34
2. Battalapalli D, Rao BP, Yogeeswari P, Kesavadas C, Rajagopalan V (2022) An optimal brain tumor segmentation algorithm for clinical MRI dataset with low resolution and non-contiguous slices. BMC Med Imaging 22(1):1–12
3. LeCun Y, Bengio Y, Hinton G (2015) Deep learning. Nature 521(7553):436–444
4. LeCun Y, Bottou L, Bengio Y, Haffner P (1998) Gradient-based learning applied to document recognition. Proc IEEE 86(11):2278–2324
5. Matsugu M, Mori K, Mitari Y, Kaneda Y (2003) Subject independent facial expression recognition with robust face detection using a convolutional neural network. Neural Netw 16(5–6):555–559

6. Simonyan K, Zisserman A (2014) Very deep convolutional networks for large-scale image recognition. In: 3rd International conference on learning representations, ICLR 2015—conference track proceedings, arXiv preprint arXiv:1409.1556
7. He K, Zhang X, Ren S, Sun J (2016) Deep residual learning for image recognition. In: Proceedings of the IEEE conference on computer vision and pattern recognition, pp 770–778
8. Huang G, Liu Z, Van Der Maaten L, Weinberger KQ (2017) Densely connected convolutional networks. In: Proceedings of the IEEE conference on computer vision and pattern recognition, pp 4700–4708
9. Jia H, Cai W, Huang H, Xia Y (2020) H 2 NF-Net for brain tumor segmentation using multimodal mr imaging: 2nd place solution to BraTS challenge 2020 segmentation task. In: Brainlesion: Glioma, Multiple Sclerosis, Stroke and Traumatic Brain Injuries, pp 58–68
10. Isensee F, Jäger PF, Full PM, Vollmuth P, Maier-Hein KH (2020) nnU-net for brain tumor segmentation. In: International MICCAI brainlesion workshop. Springer, Cham, pp 118–132
11. Zhang J, Lv X, Sun Q, Zhang Q, Wei X, Liu B (2020) SDResU-net: separable and dilated residual U-net for MRI brain tumor segmentation. Curr Med Imaging 16(6):720–728
12. Silva CA, Pinto A, Pereira S, Lopes A (2020) Multi-stage deep layer aggregation for brain tumor segmentation. In: International MICCAI brainlesion workshop. Springer, Cham, pp 179–188
13. Khan H, Shah PM, Shah MA, ul Islam S, Rodrigues JJ (2020) Cascading handcrafted features and Convolutional Neural Network for IoT-enabled brain tumor segmentation. Comput Commun 153:196–207
14. Chaurasia A, Culurciello E (2017) Linknet: Exploiting encoder representations for efficient semantic segmentation. In: 2017 IEEE visual communications and image processing (VCIP). IEEE, pp 1–4
15. Sobhaninia Z, Rezaei S, Karimi N, Emami A, Samavi S (2020) Brain tumor segmentation by cascaded deep neural networks using multiple image scales. In: 2020 28th Iranian conference on electrical engineering (ICEE). IEEE, pp 1–4
16. Ding Y, Gong L, Zhang M, Li C, Qin Z (2020) A multi-path adaptive fusion network for multimodal brain tumor segmentation. Neurocomputing 412:19–30
17. Sun J, Peng Y, Guo Y, Li D (2021) Segmentation of the multimodal brain tumor image used the multi-pathway architecture method based on 3D FCN. Neurocomputing 423:34–45
18. Tong J, Wang C (2022) A performance-consistent and computation-efficient CNN system for high-quality automated brain tumor segmentation. arXiv preprint arXiv:2205.01239
19. Sun Q, Fang N, Liu Z, Zhao L, Wen Y, Lin H (2021) HybridCTrm: Bridging CNN and transformer for multimodal brain image segmentation. J Healthc Eng
20. Akbar AS, Fatichah C, Suciati N (2022) Single level UNet3D with m ultipath residual attention block for brain tumor segmentation. J King Saud Unive Comput Inf Sci
21. Wang Y, Zhang Y, Hou F, Liu Y, Tian J, Zhong C, … He Z (2020) Modality-pairing learning for brain tumor segmentation. In: International MICCAI brainlesion workshop. Springer, Cham, pp 230–240
22. Mishra M, Sarkar T, Choudhury T et al (2022) Allergen30: detecting food items with possible allergens using deep learning-based computer vision. Food Anal Methods. https://doi.org/10.1007/s12161-022-02353-9
23. Choudhury T et al (2022) Quality evaluation in guavas using deep learning architectures: an experimental review. In: 2022 International congress on human-computer interaction, optimization and robotic applications (HORA), pp 1–6. https://doi.org/10.1109/HORA55278.2022.9799824
24. Arunachalaeshwaran VR, Mahdi HF, Choudhury T, Sarkar T, Bhuyan BP (2022) Freshness classification of hog plum fruit using deep learning. In: 2022 International congress on human-computer interaction, optimization and robotic applications (HORA), pp 1–6. https://doi.org/10.1109/HORA55278.2022.9799897
25. Khanna A, Sah A, Choudhury T (2020) Intelligent mobile edge computing: a deep learning based approach. In: Singh M, Gupta P, Tyagi V, Flusser J, Ören T, Valentino G (eds) Advances in computing and data sciences. ICACDS 2020. In: Communications in computer and information science, vol 1244. Springer, Singapore. https://doi.org/10.1007/978-981-15-6634-9_11

An Extensive Survey on Sentiment Analysis and Opinion Mining: A Software Engineering Perspective

S. Vikram Sindhu, Neelamadhab Padhy, and Mohamed Ghouse Shukur

Abstract Context—The authors have analyzed the opinion mining and sentiments related to software Engineering and what are the sentimental issues software engineers are facing in the current scenario. Objective—The authors have obtained the overall solutions to research issue and finding what are the research challenges and gaps related to sentiments and opinion. Conclusion—The authors of current paper, have analyzed the work done in various research papers on sentimental analysis related to software engineering. In software engineering process, the authors include a process where authors analyze and classify the positive, negative and neutral polarities of the opinions and reviews. This process is called sentiment analysis in software engineering. The authors give systematic and extensive survey on sentiment analysis and opinion mining.

Keywords Software testing · Sentiment analysis · Opinion mining · Software engineering

1 Introduction

Analyzing sentiment is important in obtaining opinion about product and services. Analyzing the sentiments of product and services has given rise to following applications.

- Product reviews
- Opinion polls

S. Vikram Sindhu (✉) · N. Padhy
Department of Computer Science and Engineering, School of Engineering and Technology, GIET University, Gunupur, Odisha, India
e-mail: vikram@giet.edu

N. Padhy
e-mail: dr.neelamadhab@giet.edu

M. G. Shukur
Department of Computer Science, King Khalid University, Abha, Saudi Arabia
e-mail: mghoth@kku.edu.sa

K. A. Reddy et al. (eds.), *Proceedings of Fourth International Conference on Computer and Communication Technologies*, Lecture Notes in Networks and Systems 606,
https://doi.org/10.1007/978-981-19-8563-8_52

- Movie review on YouTube.
- New video analysis
- Stress and depression analysis.

Now a day's public is more eager to share their views in social media. Now a day's public are recording their views through audio or video and posting it in the social media. This kind of multi-mode sentimental analysis is better than only posting the text. Video-based analysis is better than text analysis because it can explore the emotions in an appropriate manner [1]. Has done work on multi-mode sentimental analysis which aims for the following

- Present the existing reviews to the researchers so that they can explore it and express them in multi-mode sentimental analysis.
- combine video, audio and text analysis, which gives an idea about existing sentiment analysis.

Sentiment analysis has been introduced in many software engineering scenarios. Studies shows the sentiment analysis tools are unreliable when they are used improperly as they are not designed to process software engineering data sets. It is used to analyze whether customer's review is positive, neutral or negative. It is used to recognize the negative opinion expressed about APIs.

In the existence of web 2.0, most of the people are interested to share their opinion on their activities in day to day life and on global issues. Sentiment analysis helps in analyzing public mood and opinion regarding political movement and market intelligence.

In recent days most customers depend on reviews posted by existing customers in order to analyze the product or service.

1.1 Critical Contributions

1. To write this article, authors have gone through papers from IEEE, Springer, ACM, GS, Elsevier, MDPI, Wiley etc., authors have shortlisted 200 papers which are relevant to the topic. Out of which authors have filtered 59 papers based on abstract and contents. Out of which authors have identified relevant works and finally selected 45 papers.
2. By analyzing work in all papers, authors found 10 research questions which have been mentioned in the section VII.
3. Authors have planned to carryout research in analyzing sentiments and opinion related to software engineering and find necessary solutions to existing challenges.
4. Authors have carried out systematic literature review in order to bring this paper.

Online media provides platform for response and opinion/feedback on global issues. 255 million people login to twitter in a month. It accommodates 500 million tweets per day. Therefore, it is possible to extract various opinion of the people from different backgrounds and may help to improve the services and products.

2 Literature Review

In Marks et al. [2], the author introduced the model that explains the relation-ship between actors in a sentence and thereby obtains the attitude of the actor. This work explains about categorization of opinion mining and sentimental analysis. A more detailed model is introduced by Strapparava and Valitutti [3, 4] who developed Wordnet-Affect that explains direct synsets that elaborates emotions and indirect synsets which include emotion carrier. Khairullah Khan et al. [1] obtained sentiment analysis in sentence level. Naive Bayesian classifier has been used for word level extraction of feature. Individual sentences semantic orientation is obtained from contextual information. This method claims an accuracy rate of 83% on an average. Guzman et al. [5] verified the opinions and sentiments of commit comments which are available in GitHub and given evidence that works having more teams will have higher positive opinions as a result. Authors also expressed that the same comments which are written on Monday will have more negative opinion.

A study conducted by Sinha et al. [6] on 28,466 projects and these projects analysis were done within 7 years of time. This study revealed that most of sentiments which are supposed to be neutral would be negative on Tuesdays. In Bo Pang et al. [7] analyzed the classification of positive and negative tags. Document classification has been done on the topic basis. The work analyses that if we use same techniques of topic-based classification then sentimental analysis will fail. Therefore, more number of techniques are be utilized in solving opinion mining and sentiment analysis problem. Jongeling et al. [8] has compared the four sentiment analysis techniques and they are SentiStrength, NLTK, Stanford CoreNLP and AlchemyAPI. They evaluated their performance also in the paper. They found that all four sentiment analysis techniques failed to provide 100% accuracy. Also they concluded that there are disagreement among tools.

In Fig. 1, authors describe the procedure used to write this paper. To write this article, authors have gone through papers from IEEE, Springer, ACM, GS, Elsevier, MDPI, Wiley etc., and we have shortlisted 200 papers which are relevant to the topic. Out of which authors have filtered 59 papers based on abstract and contents. Out of which authors have identified relevant works and finally selected 45 papers.

3 Sentiment Analysis

Sentiment analysis is a platform for many organizations to get sentiments from unstructured text which are obtained from web sources. Rule based and Hybrid methods are introduced to process data. Rule-based systems are used in analyzing

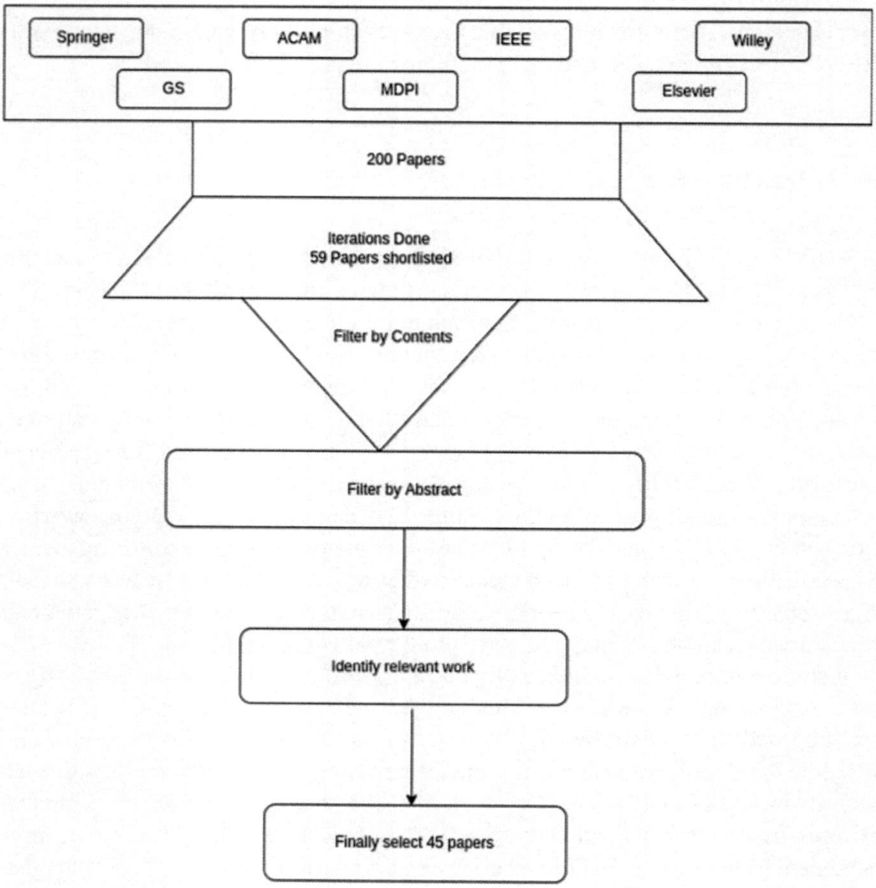

Fig. 1 Literature criteria

lexicon based and predefined rules. Automatic systems are used to learn from techniques used in machine learning. A hybrid sentiment analysis uses both.

Apart from identifying sentiments, opinion mining will obtain polarity which is nothing but the amount of positivity and negativity. Moreover, sentiment analysis are applied to documents, paragraphs, sentences and sub-sentences.

4 Steps Involved in Sentimental Analysis

Figure 2 shows the steps are involved in Sentimental Analysis and Opinion Mining.

First step is setting goal. Then the text goes through prepossessing stage. Text will be read and organized in convenient way for compiler. The next step is parsing

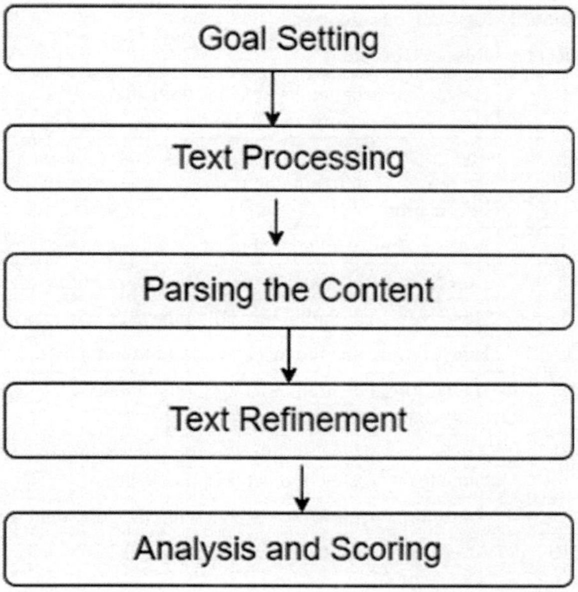

Fig. 2 Steps in sentimental analysis

where the text is divided into tokens. Then the text is refined as per the regulations. Final step is filtered tokens are analyzed and they will be scored.

5 Applications of Sentimental Analysis

The following are applications of sentimental analysis.

1. Applications in Business: Sentiments are easily connected to social discussions. We should not use the estimation of notice. When anyone is qualifying the company's new product notice, one can expect the increase in the notice. This means that the product is accepted by the customer. Some notices may be negative, but can be ignored if they are less in number. This kind of sentiment analysis helps organizations to identify whether their product is reaching the customers or not.
2. Applications in Sociology: It allows us to gather the opinion of the global population in the political and social perspective. Thoughts obtained from gathering public opinions is also important consideration in social and political sentiment analysis.
3. Applications based on reviews in website: The audit of opinions and sentiments of the customers obtained through web portals and social media are very important in analyzing sentiments. Bridging the user surveys to the product is very important aspect in sentimental analysis. Also one has to carefully measure the

Table 1 Research questions

RQ No	Research question
1	How to construct multi-modal sentimental analysis and gather multiple inputs and improve the performance of sentimental analysis?
2	The difficulty in natural language can make information extraction very tough to obtain the information in opinion text. How to overcome this and smoothly extract information?
3	How to identify all textual mentions of the named entities in a text piece?
4	How to make NLP totally capable of representing the meaning from unrestricted text?
5	How to extract relation by finding the syntactic relation between words in a sentence?
6	How to overcome the inaccuracies of training model?
7	How to avoid contradiction in the statements made by the engineers or public regarding their sentiments or opinions?
8	Are sentiments involved in executing the commands are related to programming language where the project is implemented?
9	Are sentiments involved in executing the command is related to approval of project?
10	Are software engineer's emotions are related to project affect the ethical issues of the organization?
11	To what level the sentiment analysis tools are accepted by software developers?
12	To what level results obtained in different sentimental analysis tools are different from each other?

reviews through public portals and social media as the reviews are ambiguous and controversial.

6 Research Challenges and Issues

Sentiment analysis is major research challenge because it creates an ambiguity while analyzing the sentiments. NLP challenges in sentiment analysis is one of the main research topics now a days. Also identifying software developers sentiments in development process is a tedious process. After analyzing sentiments of software engineers, it becomes a time consuming process to overcome those drawbacks. Authors of current paper found research questions (Given in Table 1) in the study carried out related to sentiments and opinions in software engineering.

7 Sentiment Classification Techniques

Sentiment Classification are usually one of following three algorithms.

1. Rule-Based Systems: In Rule-Based Systems, researchers apply number of hand - crafted rules to obtain a pattern for every tag. Rule-based systems rely on

lexicon for sentiment classification problem which are a list of positive terms like beautiful, good useful etc., and list of negative terms like bad, uncomfortable, ugly and frustrated etc.,

When given a piece of text, the model counts the number of positive and negative tokens and assigns the related sentiments. If the input text contains more number of positive terms than negative terms, it will be tagged as positive. If the input contains more number of negative terms it will be tagged as negative.

This technique has limitations. The problem is those words which do not appear in lexicon will not be recognized. It will isolate the unrecognized words from the context.

2. Automated Systems (Based on Machine Learning): Machine learning algorithms are used by automated systems which can predict sentiments from past observations. In this approach researchers need a data set with included tags. This is termed as training data. While carrying out training process, text data will be converted into vectors and patterns are identified so that vectors are associated with predefined tags ("Positive", "Negative" and "Neutral"). Once the related data is fed, the automated system starts obtaining own predictions which classifies the unseen data. Like this one can improve the accuracy of such models with more number of tagged examples.
3. Hybrid Systems: Hybrid systems merge both rule-based and machine learning-based approaches. Hybrid systems tries to learn and detect sentiments from tagged examples and then it verifies the results with lexicon which will improve accuracy. The main goal is to get best possible outcome and to over-come all limitations of this approach.

Figure 3 shows sentiment analysis classification. Data is initially fed to the system and will go through pre-processing stage, then lexicon-based approach has been applied to pre-processed data. The analyzed data is fed to sentiment classifier and opinions and sentiments are extracted. The opinions and sentiments are extracted without using sentiment classifier on some occasions. Then finally classified data will be obtained in the last step.

8 Sentiment Analysis and Software Engineering

ResearcherS of software engineering are interested to consider sentiments and opinions of software engineers now a days. Figure 4 shows how the sentiment analysis can be applied in software engineering. The requirements are collected as per the process of software engineering and the data is pre-processed by applying software engineering techniques and on the other hand the reviews from the users regarding the requirements and the product are taken and pre-processed using sentiment analysis methods and the results extracted from both are analyzed and authors conclude that applying sentiment analysis to software engineering enhances the software product quality.

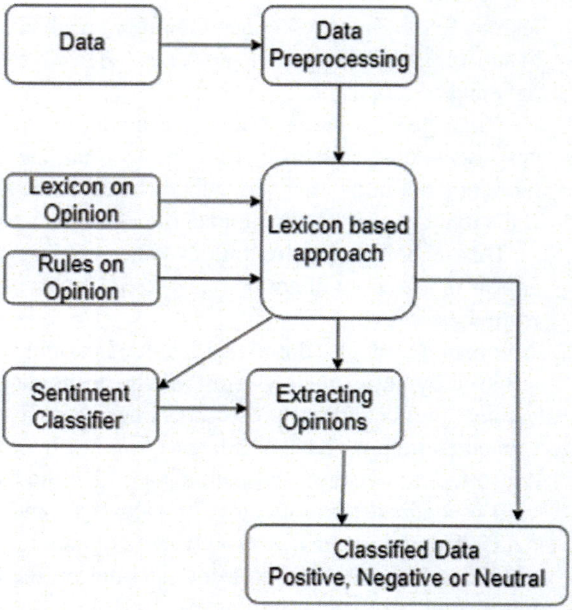

Fig. 3 Sentiment classification

9 Sentiment Analysis and Refactoring

Refactoring needs to have a purpose. The following scenarios leads to refactoring.

(a) Difficulty in understanding program code.
(b) Modules are heavily dependent on each other.
(c) Complicated debugging
(d) Complicated testing.

It is necessary for software professionals to invest their time in resolving issues of refactoring. Refactoring enhances flexibility of the program. We can add the new features easily if we use refactoring and internal structure also will not be complicated. Refactoring reduces code size and avoids duplicate code. Most of the refactoring are done manually. Automatic code smell detection should be incorporated in refactoring. After refactoring, a software engineer has to ensure the correctness in the process and it is a tedious task. Identifying code smell is also an important challenge in refactoring [9].

10 Conclusion

Sentiment analysis purely belongs to machine learning problem and many researchers are interested to carryout research in this area. In this literature survey authors have highlighted the work done to solve the sentiment analysis problems in machine

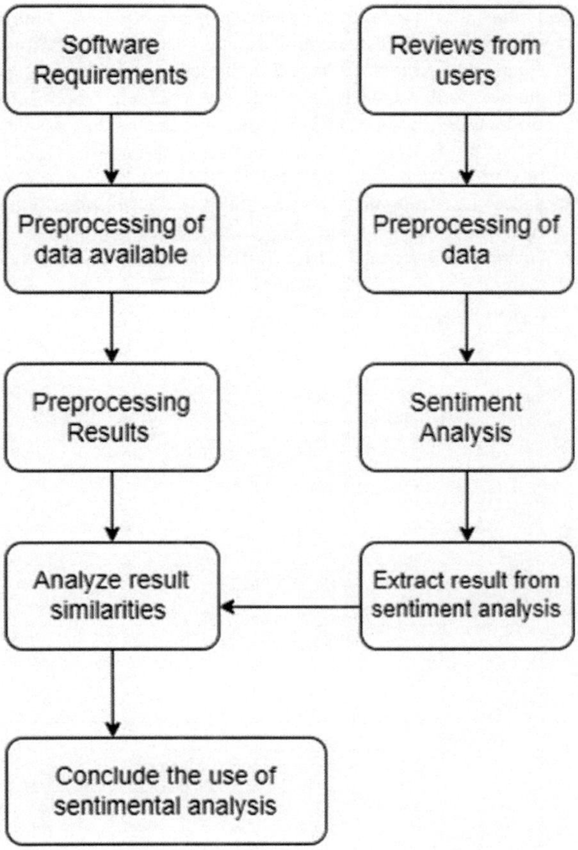

Fig. 4 Applying sentimental analysis in software engineering

learning and it can be studies. However notable work have been done in this field, complete automated systems have not been introduced till now. It may be due to the unstructured nature of natural language. Authors also like to conclude that the opinions and sentiments are controversial and ambiguous and thereby causing more complexity to sentimental analysis.

References

1. Khairullah Khan B, Khan A (2010) Sentence based sentiment classification from online customer reviews. In: ACM, 2010
2. Maks I, Vossen P (2012) A lexicon model for deep sentiment analysis and opinion mining applications. Decis Support Syst 53(4):680–688
3. Strapparava C, Valitutti SA (2004) WordNet-affect: an affective extension of WordNet. In: Proceedings LREC 2004, Lisbon, Portugal, 2004
4. Valitutti A, Strapparava C (2010) Interfacing wordnet-affect with OCC model of emotions. In: Proceedings of EMOTION-2010, Valletta, Malta, 2010

5. Sinha V, Lazar A, Sharif B (2016) Analyzing developer sentiment in commit logs. In: Proceedings of MSR 2016 (13th international conference on mining software repositories). ACM, pp 520–523
6. Jongeling R, Sarkar P, Datta S, Serebrenik A (2017) On negative results when using sentiment analysis tools for software engineering research. Empir Softw Eng 2017:1–42
7. Bo Pang SV, Lee L (2002) Thumbs up? Sentiment classification using machine learning techniques. In: Proceedings of the conference on empirical methods in nat ural language processing (EMNLP), ACL, July 2002, pp 79–86
8. Padhy N, Panigrahi R, Satapathy SC (2019) Identifying the reusable components from component-based system: proposed metrics and model. Springer, pp 89–99
9. Guzman E, Az´ocar D, Li Y (2014) Sentiment analysis of commit comments in GitHub: an empirical study. In: Proceedings of MSR 2014 (11th working conference on mining software repositories). ACM, pp 352–355

Feature Enhancement-Based Stock Prediction Strategy to Forecast the Fiscal Market

Dushmanta Kumar Padhi, Neelamadhab Padhy, and Akash Kumar Bhoi

Abstract According to consensus, the stock market can be viewed as a complex nonlinear dynamic system influenced by numerous factors. Traditional stock market research and forecasting techniques do not correctly disclose the fundamental pattern of the stock market. Researchers have lately applied a range of machine learning techniques to estimate future stock market values with greater accuracy and precision. The literature indicates that researchers have not been interested in feature engineering for stock price prediction. Consequently, the purpose of this work is to present a unique technique to feature engineering for predicting stock values using historical data. So far we have used the ITC stock for our practical experiment purposes. More importantly, the addition of feature engineering techniques to identify the potential features may improve the accuracy of the forecasted model. We have developed eight forecasted models for comparison purposes and found a simple machine learning algorithm even works well when we provide appropriate features for training the model.

Keywords Stock market · VIF · Forecasting · Machine learning

D. K. Padhi · N. Padhy (✉)
Department of Computer Science and Engineering, School of Engineering and Technology, GIET University, Gunupur, India
e-mail: dr.neelamadhab@giet.edu

A. K. Bhoi
Sikkim Manipal University, Gangtok, Sikkim 737102, India

KIET Group of Institutions, Delhi-NCR, Ghaziabad 201206, India

AB-Tech eResearch (ABTeR), Sambalpur, Burla 768018, India

Victoria University, Melbourne, Australia

Research Associate at IIST, National Research Council (ISTI-CNR), Pisa, Italy

K. A. Reddy et al. (eds.), *Proceedings of Fourth International Conference on Computer and Communication Technologies*, Lecture Notes in Networks and Systems 606,
https://doi.org/10.1007/978-981-19-8563-8_53

1 Introduction

The forecast of the stock market has piqued the interest of both academics and those in finance. The problem persists. “To what degree can the price history of a common stock be utilized to generate reliable forecasts about the stock’s future price?” [1]. Earlier research on forecasting relied on the Efficient Market Hypothesis and the random walk hypothesis [1, 2]. These older models said that stock markets couldn’t be anticipated because they are affected by news rather than current market prices. Because of this, stock prices will move in a way that is hard to predict with more than 50% accuracy [3]. In contrast, a growing number of studies [4–14] present data that contradicts the EMH and random walk hypotheses.

Stock market forecasting is critical in the financial business because a reasonably accurate assessment can make a lot of money and protect against market risks [7, 8, 12]. Regardless of how predictable the stock market is, it’s still hard to predict how the price of stocks will move. This is because the financial sector is an extremely complicated, emerging, and highly nonlinear system that comes into contact with political trends, the financial environment, and stockholders’ assumptions [12]. It’s still very important to be able to accurately predict stock prices in the short term and long term because this is the most interesting and important research topic in the investment field. People who are excited about making inaccurate predictions are encouraged to come up with new and better tools and techniques. In the broad sense, there seem to be two ways to figure out how the stock market will go. These two methods are called “fundamental analysis” and “technical analysis.” The first look at economic factors to figure out how much a stock is worth, while the second looks at stock prices in the past to figure out how much a stock is worth. It’s a huge field, and new techniques are being developed each day, notably in the field of automatic feature learning, which is a lot of work.

Information and a framework are the two main parts of ML. At the time of extracting hidden features [15], it is always good practice to select only those features whose results have some contributive meaning. Because potential features give more accuracy to a model during model building. The prime objective of engineering the feature is to not only reduce the dimension but also to find the potential features for the predictive model. Researchers in [16] used machine learning to make predictions about the stock and were pleased with the results.

In addition, there have been a lot of studies done that used feature engineering, but none of them had anything to do with stock prediction. Scholars in [17] used feature extraction to figure out what was wrong with induction motors. Researchers in [18] came up with a semantic feature framework for concurrent engineering that they used. Another study used gradient boosting to make new features for energy theft detection and discovered useful pairings from the original features [19]. The authors of [20] came up with a way to use AETA data to predict short term earthquakes. In [21], researchers looked into how to make search ads easier to recognize. Based on prior investigation, it can be seen that there aren’t a lot of studies that used feature extraction to predict the stock price. So, this study is trying to come up with a new way to figure out how to predict stock prices daily. It's important to point out that our study was the first to look at and use feature extraction for stock prediction with ensemble methods.

The remainder of the article is organized as in Sect. 2 our research framework, Sect. 3 model implementation, Sect. 4 discussion and finally section four consists of the conclusion part.

2 Research Framework

Our research plan is made up of five main steps, such as collecting datasets, preprocessing data, designing features, making a model, and evaluating the model (Fig. 1).

2.1 Collection of Raw Dataset

For our practical experiment, we have collected a 5-year dataset of ITC stock prices for each trading day, which is downloaded from the publicly available website yahoo/finance. The period of the dataset was from April 27th 2017 to April 26th 2022. Our original dataset (ITC) recorded 1235 records of daily transaction historical data. Each record has six pre-existing features.

a. Date—Represents the each trading day's date.
b. Close—It denotes the final stock movement value of each trading days.
c. Volume—It is the total number of shares buy and sell in the particular trading days.
d. Open—It is the Opening price of a stock on each trading days.
e. High—It is the highest value of the stock on that particular trading days.

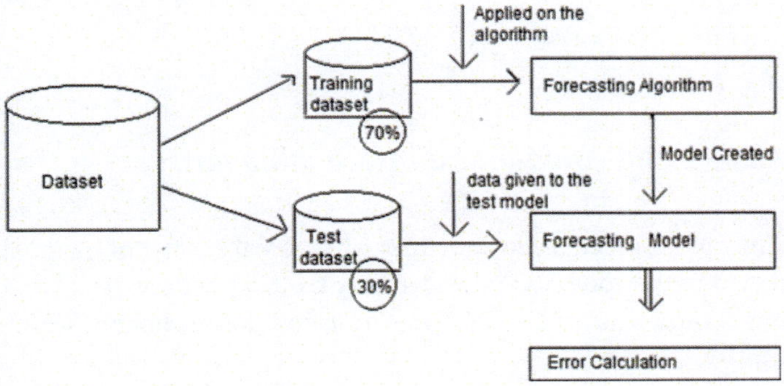

Fig. 1 Sample ML model for forecasting the stock

f. Low—It is the lowest value of the stock on that particular trading days.

2.2 Data Preprocessing

Data preprocessing is the act of converting raw data into a standard format that machine learning models can easily learn from. Next, after receiving the pre-existing historical data, we have to find the missing values and clean up the existing records.

Feature engineering is an essential step to identify the potential features and hidden behavior of our dataset. So, in this segment, we will find out the real features that are essential and fruitful for our experiment. So during our feature enhancement phase, we found the pre-existing features 'Open', 'Volume', 'Close', 'High', and 'Low' are to be used as independent variables, whereas 'Close' will be treated as a dependent variable. During the feature building phase, we found the independent features which are selected by us suffered from multicollinearity issues.

When the independent variables in a regression model are linked to each other, this is called multicollinearity. This is problematic since the independent variables should be kept distinct. If the correlation between variables is high enough, it can be hard to fit the model and figure out what the results mean. It can be hard to tell which independent variables have an effect on the dependent variable if they are all in the same place in a regression model.

The concept is that we may alter the value of one independent variable while leaving the rest unchanged. On the other hand, the correlation between independent variables suggests that changes in one variable are connected with changes in another. Correlation strength reveals how hard it is to change one variable without influencing another. Since the independent variables tend to move together, it's hard for the model to figure out how each independent variable affects the dependent variable on its own.

Multicollinearity is classified into two types:

Structural multicollinearity: This happens when we build new features from the data itself instead of the data that was sampled. **Data multicollinearity**: This is already part of the feature of ones dataframe, and it's much more difficult to see. In this case, this type of multicollinearity is found in the data itself, not because of our framework.

2.3 Detection and Avoidance of Multicollinearity

If we will find out the independent variables which are the cause of multicollinearity and the strength of the correlation then we may fix this issue.

There are certain approaches are there by using that technique we may avoid multicollinearity.

1. Correlation coefficient (Heat map).
2. Variance Inflation Factors (VIF) (Fig. 2)

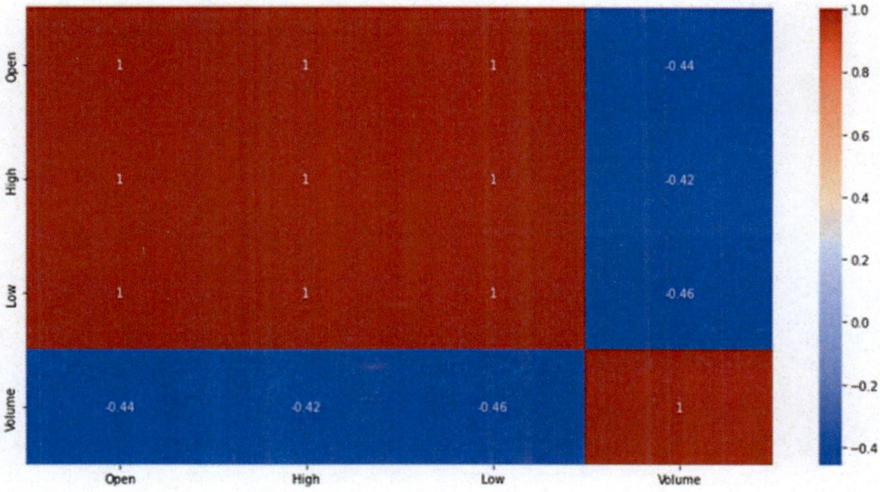

Fig. 2 Heat map of ITC index with pre-existing features

By looking at the heat map we found the features which are selected by us from the pre-existing dataset are highly correlated with each other (Covariance values are 1) except volume. So we must avoid this issue before modeling.

By looking at Table 1 the VIF of all features are more than 10 except Volume feature and VIF value of more than 10 are considered to be the worst one which should be avoided. That means the above experiment clearly gratifies that the combination of features which are selected by us are not suitable. So we have again extracted two more hidden features which are known as technical indicators are shown below.

A. **Triple exponential average (TRX)**—It is a momentum indicator that professional traders use. It shows the percentage change in a triple exponentially smoothed moving average. Its main job is to filter out price changes that don't matter.

B. **Percentage price oscillator (PPO)**—It is a technical momentum indicator that displays the percentage connection between two moving averages, including 26-period and 12-period exponential moving average comparison periods (EMA).

Table 1 VIF values of each pre-existing feature of ITC stock

Variables	VIF
Open	15,305.104990
Volume	3.372619
High	16,512.111440
Low	12,690.170031

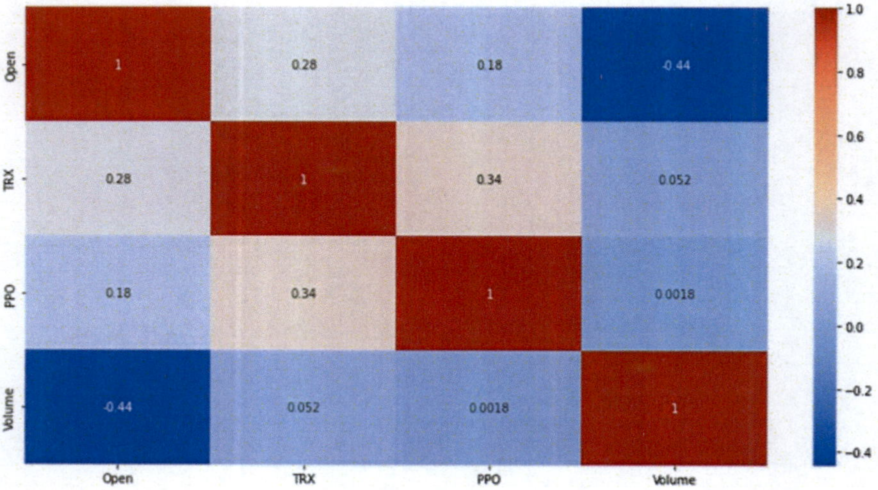

Fig. 3 Heat map of two pre-existing features and two derived features of ITC stock

The website www.ta-lib.org is where we may obtain the Ta-Lib, which is an open-source collection of statistical features commonly used by traders for trading strategies of the stock market [22].

After a certain random process, we found a combination which is most suitable for independent features, and there is no multicollinearity between them (Fig. 3).

3 Model Implementation

Here in this phase, we have selected some base machine learning models with some ensemble models. The list of algorithms are listed below:

1. Linear Regression
2. Lasso Regression
3. SVR
4. KNN
5. GradientBoostingRegressor
6. BaggingRegressor
7. HistGradientBoostingRegressor
8. LGBMRegressor

3.1 Performance Evaluation

After the success of making models that can predict the future, now we have to find the best model among them. For finding the best model, we have used the root mean square error (RMSE) and coefficient of determination (R2). The forecasting model is evaluated in terms of root mean squared error, which is expressed in the following equation.

$$\text{RMSE} = \sqrt{\frac{\sum_{t=1}^{n} \left(\overline{Y}_t - Y_t\right)^2}{n}} \quad (1)$$

where the $\overline{Y_t}$ is the predicted value and Y_t is the actual value, n represents the total number of sample that has been predicted, t represents the time period.

4 Results and Discussion

In this section, we will discuss our findings related to our experiment. Our initial thought process started from the raw dataset whose detailed status has been given in the Sect. 2.1. After getting the raw dataset, we go through the feature engineering process to find the best combination of features to forecast the next day's closing price. But really, it is a tedious job to find out the best features for a certain combination. So to find the best features, we have applied the technique, i.e., multicollinearity and VIF for finding the best combination that has some impotence during the model building process. After successfully engineering the features, our final combination of features is shown in Table 2. For finding the best predictive model and for comparison purposes, we use eight different algorithms. Four of them are base level machine learning algorithms, i.e., Linear Regression, Lasso Regression SVR, KNN, and the other four are based on ensemble techniques, i.e., GBR, BR, and HGBR.

By looking at Table 3 after the successful training of our above said models with 70% data and testing with 30% data we found the linear regression got the R square value is 99.18 and RMSE value 3.451, Lasso regression got the R square value is 99.17 and RMSE value 3.459, SVR got the R square value is 99.23 and RMSE value 3.429, KNN got the R square value is 99.12 and RMSE value 5.859, GBR got the R square value is 99.10 and RMSE value 3.53, Bagging regressor got the R square

Table 2 VIF values of two pre-existing feature and two derived features of ITC stock

Variables	VIF
Open	2.118192
TRX	1.140893
PPO	1.131528
Volume	2.107209

Table 3 Performance evaluation of all models of ITC stock

Models	*R* square	RMSE
Linear regression	99.18	3.451
Lasso regression	99.17	3.459
SVR	99.23	3.429
KNN	99.12	5.859
GradientBoostingRegressor	99.10	3.53
BaggingRegressor	99.08	3.60
HistGradientBoostingRegressor	98.76	3.72
LGBMRegressor	98.63	3.81

value is 99.08 and RMSE value 3.60, HGBR got the *R* square value is 98.76 and RMSE value 3.72 and finally LGBMRegressor got the *R* square value is 98.63 and RMSE value 3.81. If we will consider the best-performing model then we say SVR doing slightly better than other models.

5 Conclusion and Future Scope

This study's goal was to assess machine learning-based forecasting skills to identify difficulties related to intraday trading transactions. As shown in Table 3, while the RMSE for each model is increasing, the advanced approach (ensemble techniques) is unable to give an appropriate improvement over traditional procedures. So it is clearly understood that if we properly find out the best combination of features that impact our machine learning model, even if baseline machine learning models also perform well. So in our future research, we will expand this study in terms of feature engineering to find the best combination of features that really impact a heterogeneous dataset with a heterogeneous combination of algorithms.

References

1. Fama EF (1965) The behavior of stock-market prices. J Bus 38(1):34–105
2. Fama EF, Fisher L, Jensen MC, Roll R (1969) The adjustment of stock prices to new information. Int Econ Rev (Philadelphia) 10(1):1–21
3. Bollen J, Mao H, Zeng X-J Twitter mood predicts the stock market
4. Ballings M, Van den Poel D, Hespeels N, Gryp R (2015) Evaluating multiple classifiers for stock price direction prediction. Expert Syst Appl 42(20):7046–7056
5. Chen Y, Hao Y (2017) A feature weighted support vector machine and K-nearest neighbor algorithm for stock market indices prediction. Expert Syst Appl 80:340–355
6. Rashid TA (2015) Improvement on classification models of multiple classes through effectual processes. Int J Adv Comput Sci Appl 6(7)
7. Chong E, Han C, Park FC (2017) Deep learning networks for stock market analysis and prediction: Methodology, data representations, and case studies. Expert Syst Appl 83:187–205

8. Farias Nazário RT, e Silva JL, Sobreiro VA, Kimura H (2017) A literature review of technical analysis on stock markets. Q Rev Econ Financ 66:115–126
9. Haixiang G, Yijing L, Shang J, Mingyun G, Yuanyue H, Bing G (2017) Learning from class-imbalanced data: review of methods and applications. Expert Syst Appl 73:220–239
10. Wang L, Wang Z, Zhao S, Tan S (2015) Stock market trend prediction using dynamical Bayesian factor graph. Expert Syst Appl 42(15):6267–6275
11. Moghaddam AH, Moghaddam MH, Esfandyari M (2016) Stock market index prediction using artificial neural network. J Econ Financ Adm Sci 21(41):89–93
12. Nayak A, Pai MMM, Pai RM (2016) Prediction models for Indian stock market. Proc Comput Sci 89:441–449
13. Weng B, Ahmed MA, Megahed FM (2017) Stock market one-day ahead movement prediction using disparate data sources. Expert Syst Appl 79:153–163
14. Zhao Y, Li J, Yu L (2017) A deep learning ensemble approach for crude oil price forecasting. Energy Econ. 66:9–16
15. Khurana U, Turaga D, Samulowitz H, Parthasrathy S (2016) Cognito: automated feature engineering for supervised learning. In: 2016 IEEE 16th International conference on data mining workshops (ICDMW), 2016, pp 1304–1307
16. Long W, Lu Z, Cui L (2019) Deep learning-based feature engineering for stock price movement prediction. Knowledge-Based Syst 164:163–173
17. Panigrahy PS, Santra D, Chattopadhyay P (2017) Feature engineering in fault diagnosis of induction motor. In: 2017 3rd International conference on condition assessment techniques in electrical systems, CATCON 2017—Proceedings, 2018, vol 2018, Janua, pp 306–310
18. Liu YJ, Lai KL, Dai G, Yuen MMF (2010) A semantic feature model in concurrent engineering. IEEE Trans Autom Sci Eng 7(3):659–665
19. Punmiya R, Choe S (2019) Energy theft detection using gradient boosting theft detector with feature engineering-based preprocessing. IEEE Trans Smart Grid 10(2):2326–2329
20. Huang J, Wang X, Yong S, Feng Y (2019) A feature enginering framework for short-term earthquake prediction based on AETA data. In: Proceedings of 2019 IEEE 8th joint international information technology and artificial intelligence conference, ITAIC 2019, 2019, pp 563–566
21. Sun Y, Yang G (2019) Feature engineering for search advertising recognition. In: Proceedings of 2019 IEEE 3rd information technology, networking, electronic and automation control conference, ITNEC 2019, 2019, pp 1859–1864
22. TA-LIB: Technical analysis library. Available online: www.ta-lib.org. Accessed on 10 Jan 2022

Early Prediction of Diabetes Mellitus Using Intensive Care Data to Improve Clinical Decisions

Chandrasekhar Uddagiri, Thumuluri Sai Sarika, and Kunamsetti Vaishnavi

Abstract Insulin deficiency causes diabetes mellitus (DM), which can lead to multi-organ failure in patients. Insufficient data is always a threat for detection and diagnosis of health disorders. Intensive Care Units (ICUs) do not have verified medical histories of their patients. This paper presents and elaborates the most significant analysis done during WIDS Datathon 2021. Data from the first 24 h of ICU admission was used to build a model that can identify if a patient has been diagnosed with a particular type of diabetes during admission to an ICU. The work focuses on Diabetes Mellitus type, and to discover a competent classifier to obtain the most accurate result, particularly in comparison to clinical outcomes. For analytic and comparative purposes, four algorithms were used. The optimum result was obtained using the LGBM classifier with roc_auc_score of 0.871. The evaluation is done using Stratified threefold cross-validation and the predictions for the test set won accolades in the Kaggle hackathon.

Keywords Diabetes mellitus · Exploratory data analysis · Missing values · Feature engineering · Light GBM classifier · Stratified K-fold

1 Introduction

During the COVID-19 pandemic, monitoring the overall health of public has become very critical. It paved way to digitalization in healthcare, to obtain faster health analytics. The advances in AI/ML came handy at the right time. It is now possible to detect and there by cure diabetes during early stages through automated techniques. The patient may not be always able to provide other information about his chronic ailments such as injuries, heart diseases. The medical records can take many days to be transferred from another medical service provider. The clinical decisions can

C. Uddagiri · T. S. Sarika (✉) · K. Vaishnavi
Department of Computer Science and Engineering, BVRIT HYDERABAD College of Engineering for Women, Hyderabad 500090, India
e-mail: sarikatumuluri@gmail.com

C. Uddagiri
e-mail: chandrasekhar.u@bvrithyderabad.edu.in

K. A. Reddy et al. (eds.), *Proceedings of Fourth International Conference on Computer and Communication Technologies*, Lecture Notes in Networks and Systems 606,
https://doi.org/10.1007/978-981-19-8563-8_54

be improved with sufficient knowledge about patients' chronic conditions such as diabetes. Hence, it is important, as well as challenging, to detect Diabetes Mellitus from the first 24 h intensive care data.

The MIT-based GOSSIS Community Initiative has supplied a dataset for more than 130,000 patient visits by the intensive care unit (ICU) during a one-year time-frame with a Harvard Privacy Lab privacy certification. A consortium of countries such as Brazil, Sri Lanka, New Zealand, and Australia spanning more than 200 hospitals in the US have contributed to the dataset.

1.1 Features Overview

The dataset includes the following elements:

- Over 130,000 records were used for training and over 10,000 records were used for testing.
- There are 180 observable parameters (Input features), which can be categorized into seven groups, APACHE co morbidity, APACHE covariate, demographic, identifier, labs, labs blood gas, vitals.
- The binary target feature is diabetes_mellitus.

The "Acute Physiology And Chronic Health Evaluation II" (APACHE II) disease severity categorization system is one of the ICU grading methodologies. Within 24 h of a patient being admitted to an intensive care unit (ICU), it is used to produce an integer score ranging from 0 to 71 based on several parameters. Higher scores indicate severe disease and hence a higher probability of death.

Each patient's APACHE III scores were determined using information gathered within the first 24 h of ICU admission. The APACHE III score considers age, sex, race, past co morbidities, and location prior to ICU admission, as well as the major cause for ICU admission. The APACHE III score can be anything between 0 to 299 points.

2 Literature Survey

The following are various works done in this area which are found useful for designing our approach.

Diabetes mellitus (DM), which is brought on by uncontrolled diabetes, can result in multi-organ failure in people. It is now possible to identify and diagnose diabetes in its early stages using an automated method that is more effective than manual diagnosis, thanks to advancements in AIML [1]. In order to perform and analyze tasks effectively, data must be structured. Data was checked for missing information, and diabetes cases are represented by a 1 or a 0. Through the course of the data analysis, it was discovered that there were a fair number of instances with a zero value. Data imputing was used to address missing or zero values in the dataset [2]. The proposed Logistic Regression model has an AROC of 84.0% and a sensitivity of 73.4% compared to the suggested GBM model's 84.7% AROC and 71.6% sensitivity. GBM and Logistic Regression models perform worse than Random Forest and Decision Tree models [3]. Missing data can be handled in a variety of ways, as detailed in a large body of research. There are three approaches to dealing with missing data. The first is based on strategies for disregarding missing data. Imputation of missing data is the second option. Missing data-based modeling is the third option. Usage of missing data imputation methods is concentrated more in this study [4]. XGBoost and pGBRT are two helpful versions of the well-known machine learning algorithm known as the Gradient Boosting Decision Tree (GBDT). Experiments on several publicly available datasets show that Light GBM speeds up the training of traditional GBDT by up to 20 times while keeping roughly the same accuracy [5]. The suggested approach analyses the features in the dataset and picks up the fittest features choosing on correlation values [6]. Random Forest is a dimensionality reduction approach that consists of a few decision trees. It is a classification, regression, and other words ensemble approach. It's a tool for ranking the significance of factors [7]. XGBoost is a type of boosting-based ensemble learning method. The idea behind XGBoost is to use an iterative computation of the CART decision tree classifier to get accurate prediction results quickly. By fusing a linear model with a tree learning model, XGBoost is an optimization model that improves the gradient boost technique. It is highly precise and utilized to solve a number of real-world problems [8]. The Random Forest model that was built might be used to help doctors diagnose diabetes. Other measures, such as classification time, might also be employed to assess the present research's performance. Given the positive outcomes gained with Random Forests, this method can be used to help with pediatric emergency management [9]. Feature selection has already been used to improve classification performance in a variety of medical scenarios. In the current study, the technique for determining the contribution of each characteristic based on its relevance is also important. The most common method for locating such obscured patterns in data is correlation analysis. Although correlation was not included in the prediction challenge, the ranking correlations can be used to enhance our findings at a later stage of model development [10].

3 Exploratory Data Analysis

To begin, an in-depth investigation was conducted by using summary statistics and data visualization techniques to uncover trends, identify anomalies, test hypotheses, and check assumptions. A skewed or unbalanced distribution of samples among the classes gives rise to an unbalanced categorization problem, i.e., one of the two classes has more samples than the other class. Diabetes diagnosis is complicated by this unequal distribution of social classes.

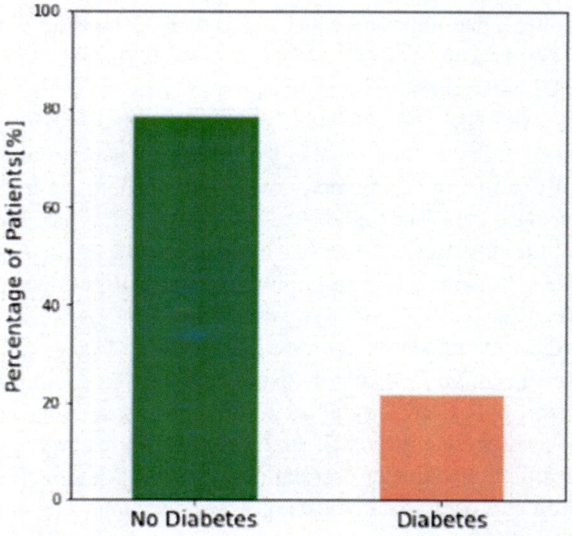

Fig. 1 Percentage of target class types

Table 1 Overview of feature types

float64	157
int64	17
object	6

In the provided training data as shown in Fig. 1, 22% of patients are diagnosed with diabetes.

3.1 Identifying Column Types

The Dtype of 180 columns were obtained which helped in further calculations and analysis. The count of each type is shown in Table 1.

The number of unique classes in each data column are found for investigating the correlations and balancing between the attributes. There are 6 ethnicity classes, 2 gender classes, 15 hospital_admit_source classes, 5 icu_admit_source classes, 3 icu_stay_type classes, and 8 icu_type classes.

3.2 Diabetes Status by Age and Gender

The box plot shown in Fig. 2 depicts the age distribution for men and women, excluding diabetic patients.

Fig. 2 Gender distribution by age

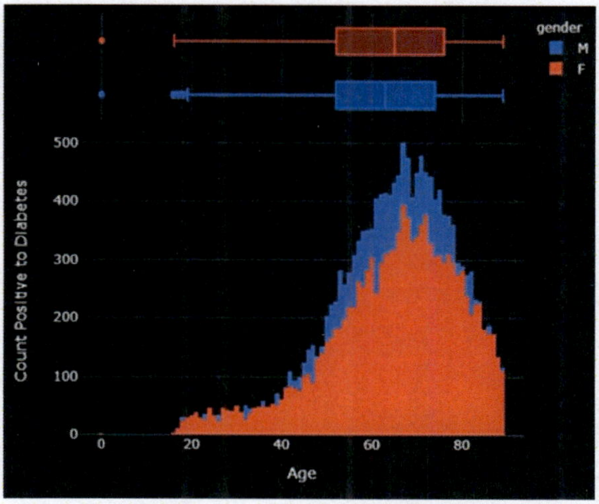

1. The majority of persons diagnosed with diabetes are between the ages of 60 and 70.
2. When it comes to positive instances, males have the most.
3. There is one age 0 value that is an anomaly.

3.3 *Ethnicity Distribution*

This visualization in Fig 3, shows that African Americans have the largest number of diabetes positives cases.

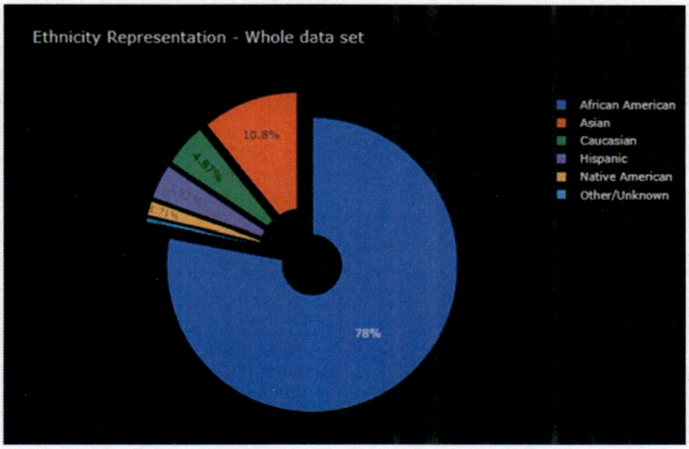

Fig. 3 Ethnicity share

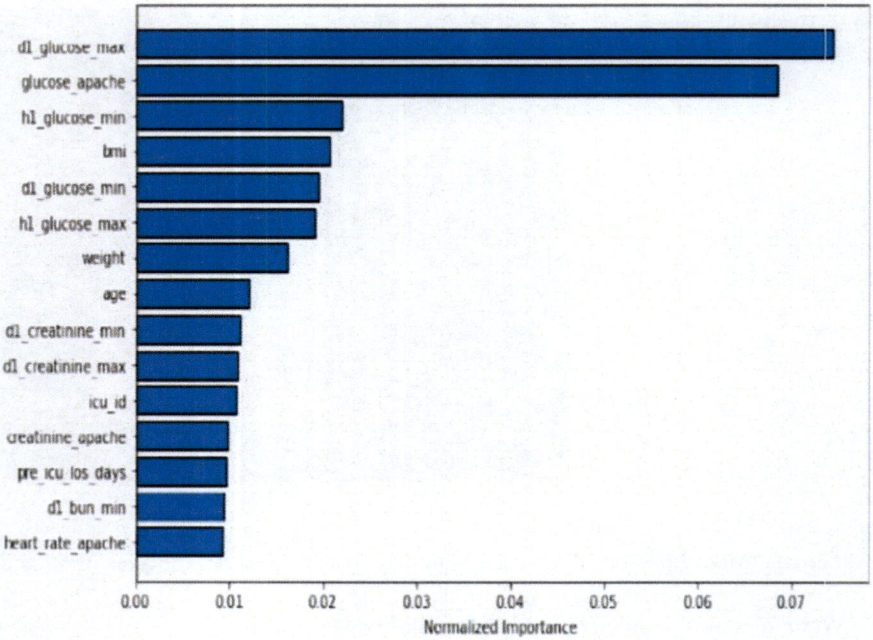

Fig. 4 Feature importance

3.4 Correlation Between Variables

The initial goal of our data analysis was to find the more relevant features in order to focus our preprocessing on them. The more important features are usually those with a higher correlation with the target variable of interest. Figure 4 shows a ranked histogram that shows how the 15 most essential features influence class prediction. Correlation matrices help to discover answers fast. It's used to examine the interdependence of numerous variables at once and to determine which variables in a data table are the most related. The value of the correlation coefficients is shaded in Fig. 5.

4 Preprocessing

4.1 Handling Missing Values

The crucial aspect in this dataset is the amount of missing values for some variables. There are 160 columns that have missing values. The irrelevant columns and the columns with high rate of missing values are also dropped as they could introduce noise into the dataset.

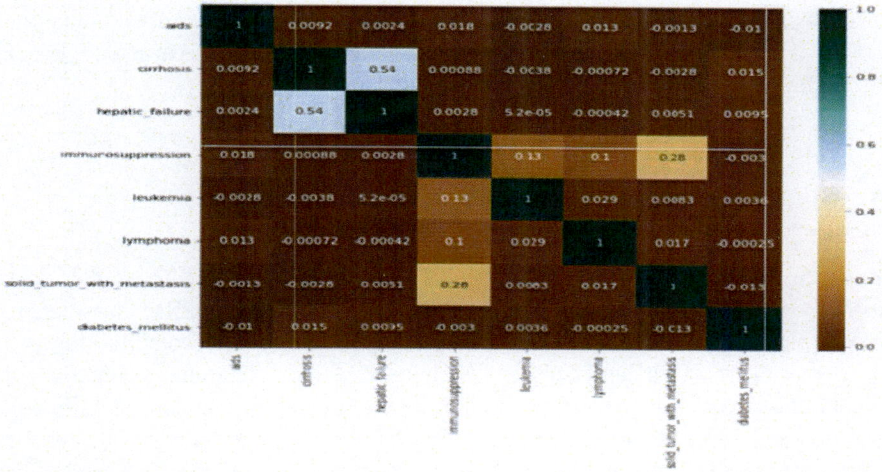

Fig. 5 Heat map among variables—correlation matrix of labs group of feature

The gender data that were missing were predicted using logistic regression using the patient's age, height, and weight. The mean values for the weight and height are aggregated by gender, ethnicity, and age to produce lookup tables. The lookup tables were used to fill in the missing values for height and weight.

4.2 Data Cleaning

Duplicate columns or functionally similar columns in the dataset are removed. The invasive and noninvasive variables are removed from the dataset as they are redundant with respect to diasbp, sysbp, and mpb, and they have a high rate of missing values. Columns like hospital id and encounter id are no longer needed. Because there were no repeat patient visits, the encounter id was irrelevant to our models. The hospitals in the annotated dataset and the hospitals in the unlabeled dataset do not overlap. Furthermore, the readmission_status column is also dropped because it only has one unique value of 0. The dataset relates to young adults and adults aging 16 and older. However, there are 30 data points with age = 0 in the training data. For the initial analysis purposes, these data points are dropped. However, these data points only account for 0.02% of data loss in addition to the earlier argument.

4.3 Encoding Categorical Variables

This dataset has categorical columns such as 'ethnicity' and 'gender' and 'hospital admit source,' 'icu admit source' as well. Those functions are encoded with the categorical data encryption, one hot encoding, when their functions are nominal (do not contain any order). In one hot encryption, a new variable is constructed for each level of a category feature.

5 Methodology

Based on data from the first 24 h of critical care, a model was created in this work using Logistic Regression, Random Forest, XGBoost, and Light GBM classifiers to predict the likelihood that a patient has been diagnosed with Diabetes Mellitus.

One of the most fundamental and widely used Machine Learning techniques for binary classification is logistic regression. The link between one dependent binary variable and independent variables is described and approximated using logistic regression. The Logistic Regression model's roc auc score is 0.638. The most often used classification technique is Random Forest. Random Forest builds several decision trees and combines them to get a more precise and trustworthy prediction. It takes less training time as compared to other algorithms. The roc_auc_score of the Random Forest model is 0.824. Therefore, Random Forest can be considered as a good predictor for diabetes. XGBoost is a distributed gradient boosting library built with efficiency, versatility, and portability in mind. XGBoost efficiently handles the missing value and has in-built cross-validation capability which makes it a great choice for large datasets and classification problems. XGBoost performed very well with roc_auc_score of 0.848.

A decision tree-based gradient boosting system called Light GBM can be utilized for a variety of machine learning applications, including ranking and classification. The execution time for model training differs significantly when Light GBM is used in place of XGBOOST, despite the fact that accuracy and auc score only slightly improve. For handling enormous datasets, Light GBM is a considerably superior approach that is roughly seven times faster than XGBOOST. When working on enormous datasets in a short amount of time, this proves to be a great benefit. By comparing all these models, it is evident that Light GBM performs best and has the highest roc_auc_score of 0.871.

5.1 Hyper Parameter Tuning

Light GBM uses the leaf-wise tree growth algorithm instead of the depth-wise tree development method, which is used by many other widely used approaches. In comparison to the depth-wise technique, the leaf-wise algorithm can converge much more quickly. However, the leaf-wise growth may be over-fitting if the appropriate parameters are not used.

Some important parameters are:

- objective: binary as the problem is binary classification
- metric: auc as the competition metric
- num_leaves and max_depth: This is the major parameter for controlling the tree model's complexity.
- num_leaves = 2^(max_depth)
- It's best to utilize a lower learning rate with a higher number of iterations. Also, if greater num_iterationsaredesired,early_stopping_rounds should be utilized to terminate the training when it isn't learning anything meaningful. The learning rate is initially set to 0.1 and then decreased during the hyper parameter adjustment.
- sample_pos_weight = number of negative samples/number of positive samples.

5.2 *Evaluation Metrics*

The area under the receiver operating characteristic (ROC) curve between the predicted and observed goal (diabetes mellitus diagnosis) was used to evaluate while submitting for the leaderboard. The graph in Fig. 6 shows the performance of a classification model at all classification levels in a ROC curve (receiver operating characteristic curve).

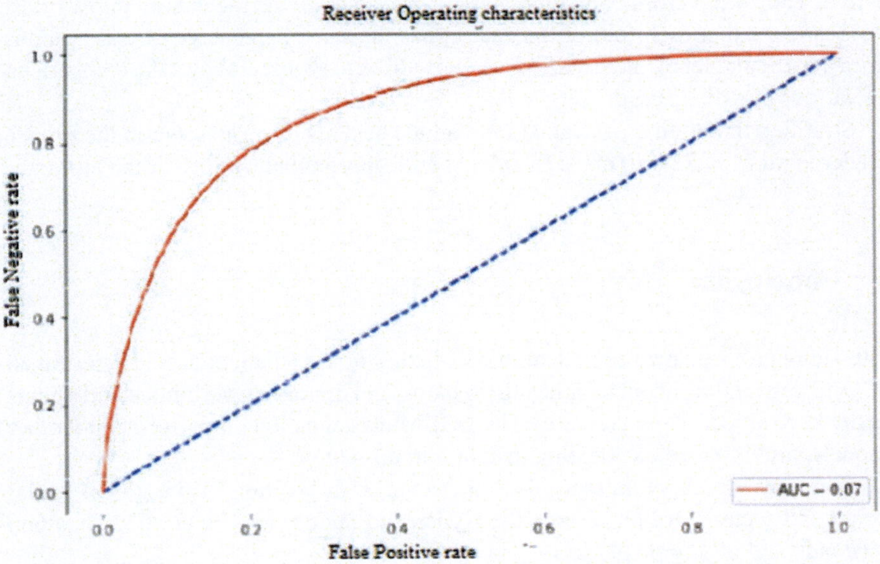

Fig. 6 ROC of LGBM model

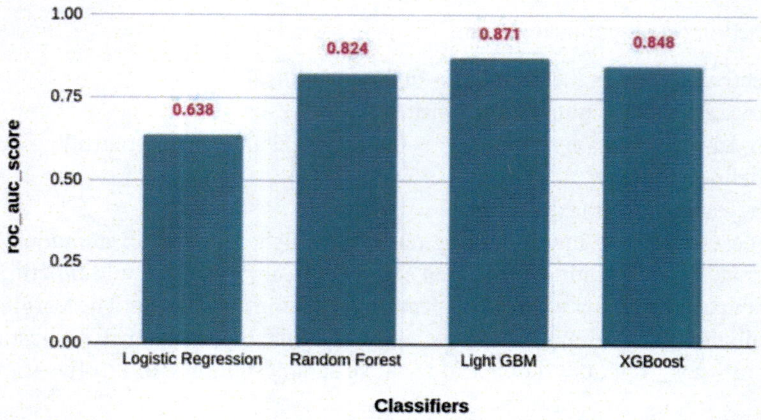

Fig. 7 Comparison of classifiers

6 Results and Discussion

Initially a validation set was created from the training set with test_size = 0.20. Then, the roc_auc_score of Logistic Regression, Random Forest, Light GBM, XGBoost were 0.649, 0.829, 0.844, 0.848, respectively. Light GBM parameters were tuned for obtaining higher results. There are two types of cross-validation: *k*-fold and stratified *k*-fold. The *k*-Fold cross-validation method is used to divide the dataset into *k* folds. To guarantee that each fold of the dataset has the same percentage of observations with a particular label, the stratified *k*-fold is utilized. Figure 7 shows the comparison of various classifiers used.

Stratified threefold cross-validation yielded best results. The score in the private leaderboard is 0.87278. This is calculated with approximately 30% of the test data.

7 Conclusion

Our study provides medics to determine whether or not a patient has diabetes. In an ICU, understanding about their health problems and improving the clinical judgments is the key subject of our research. The LGBM classification outperformed all other models, providing an outstanding insight into this study.

Our team which submitted this work stood 85th position in the global leader board and secured 3rd position in the Hyderabad region, for WiDS 2020 Datathon on conducted on kaggle platform.

References

1. Chaki J, Ganesh ST, Cidham SK, Theertanb SA. Machine learning and artificial intelligence based diabetes mellitus detection and self-management: a systematic review
2. Sarwar MA, Kamal N, Hamid W, Shah MA. Prediction of diabetes using machine learning algorithms in healthcare
3. Lai H, Huang H, Keshavjee K, Guergachi A, Gao X. Predictive models for diabetes mellitus using machine learning techniques
4. Houari R, Bounceur A, Tari AK, Kecha MT (2014) Handling missing data problems with sampling methods. In: 2014 International conference on advanced networking distributed systems and applications, pp 99–104
5. Ke G, Meng Q, Finley T, Wang T, Chen W, Ma W, Ye Q, Liu F. LightGBM: a highly efficient gradient boosting decision tree. In: NIPS
6. Sneha N, Gangil T (2019) Analysis of diabetes mellitus for early prediction using optimal features selection. J Big Data 6(13)
7. Choudhury A, Gupta D (2019) A survey on medical diagnosis of diabetes using machine learning techniques. In: Springer recent developments in machine learning and data analytics, pp 67–78
8. Xu Z, Wang Z (2019) A risk prediction model for type 2 diabetes based on weighted feature selection of random forest and XGBoost ensemble classifier. In: 2019 IEEE eleventh international conference on advanced computational intelligence (I (Houari) (Ke) (Michelle L. Griffith) (Daniel J. Rubin) (N. Sneha1) (Choudhury) (Xu)CACI), pp 278–283
9. Benbelkacem S, Atmani B (2019) Random forests for diabetes diagnosis. In: IEEE 2019 International conference on computer and information sciences (ICCIS), pp 1–4
10. Ahmad HF, Mukhtar H, Alaqail H, Seliaman M, Alhumam A. Investigating health-related features and their impact on the prediction of diabetes using machine learnings

Authorship Identification Through Stylometry Analysis Using Text Processing and Machine Learning Algorithms

Chandrasekhar Uddagiri and M. Shanmuga Sundari

Abstract The project aims to detect the identity of an anonymous author of a defamatory blog post or comment. A dataset samples containing list of authors is acquired, and then predict the anonymous author by using a custom machine learning model. The main task in this proposal is to build an authorship analysis model that will match a sample to the defamatory BlogSpot and reveal the anonymous author. Text preprocessing methods along with a combination of machine learning algorithms such as SDG classifier are employed. Stylometry analysis gives the clarity about the text information like text length, vocabulary and style of text. By this we can use this technique for authorization purpose. The project consists of building a model that can learn authorship style and then scale the model to handle hundreds of such cases. Stylometry analysis plays a major role in this project. An accuracy of 79% is obtained with 40 classes, which was found improving with lesser number of classes.

Keywords Authorship identification · Text processing · Stochastic gradient descent · Stylometry

1 Introduction

The aim of the work is to identify the true author of an anonymous post, which could possibly be defamatory blog or a comment. The sample dataset contains list of existing authors and we try to detect the closest matching author of the post from this sample dataset through a tested machine learning model. Authorship analysis comes under text mining. This work will follow the approach of breaking the text into useful tokens and build predictive models to classify new text. Authorship identification uses a different approach to deal with text. There is a need to perform content analysis

C. Uddagiri (✉) · M. Shanmuga Sundari
Department of Computer Science and Engineering, BVRIT HYDERABAD College of Engineering for Women, Hyderabad, India
e-mail: chandrasekhar.u@bvrithyderabad.edu.in

M. Shanmuga Sundari
e-mail: sundari.m@bvrithyderabad.edu.in

K. A. Reddy et al. (eds.), *Proceedings of Fourth International Conference on Computer and Communication Technologies*, Lecture Notes in Networks and Systems 606,
https://doi.org/10.1007/978-981-19-8563-8_55

and writing style, etc. first. The author is tried to identify irrespective of the content, which is called "Stylometry Analysis."

Stylometric analysis is the study of linguistic style. It can be applied to written texts, music and fine arts. It's based on the intuition that authors have a consistent style of writing such as usage of vocabulary, punctuations. This can be analyzed statistically. The analysis uses input features such as frequency distribution, word-length, *n*-grams (word and character), sentence length, Parts of speech tags, content words, function words.

For example:

- Every human has their own style of writing, which is visible in their vocabulary (rich or poor). The quality of vocabulary is usually associated with their count, which may not be always the case. Noble prize winner for Literature in 1954, Mr. Ernest Hemingway is well-known for his smaller number of words in his writing.
- The length of sentences using clauses varies among the authors.
- No two people use the punctuation the same way.

The goal of optimization is to solve a real-world problem, which includes obtaining the best possible result. The optimization procedure in machine learning is different. Generally, we alter the data features while optimizing and locate the most efficient dataset in the process. In machine learning, we optimize the training data and compare it to new validation data to see how well it performs. As a result, gradient descent is the most widely utilized optimization technique in machine learning.

1.1 Gradient Descent Algorithm

Stochastic Gradient Descent (SGD) algorithm is used to fit the linear classifiers and repressors with convex loss functions. SGD is mainly used on large scale datasets, such as the one in this project, i.e., text mining. SGD helps to find the classifiers in the text with linear function. This mainly used to create matrix array for authors similarity with the linear function. So SGD is the better technique in this project.

Gradient descent runs slow on large datasets. It needs training dataset for prediction. Hence SDG, a variant of this algorithm is used. Randomly select few samples from the entire dataset for each iteration. This is called "batch". If the dataset is redundant, the gradient on first half is identical to second half. Computing the gradients for several samples simultaneously requires matrix multiplications. So GPUs are preferred to improve the efficiency.

2 Literature Survey

Authorship identification is the process of ascribing authorship to an unknown piece of literature based on stylistic similarities between the author's known works and the unknown one. It is concerned with classification issues. From a list of candidate authors, the AI will select the most likely author of a disputed or anonymous document [1]. Main motivation of this research is usage of Stylometry in forensic reports and detective departments.

Starting with data preprocessing, feature engineering (extraction), and modeling document/text as a characteristic vector, AI is often considered a text classification challenge. Feature engineering (FE) is an important aspect of the machine learning process. This is the process of using data domain knowledge to create vector representations of raw data and features that allow machine learning algorithms to function [2].

Everyone's writing style is different, and even if the person intentionally tries to modify it, some identity-related indicators remain. This problem was investigated in the context of authorship recognition, which aims to identify a piece of text's author from a list of candidate authors whose writings are accessible for supervised classifier training [3].

Attempts to determine the author of a book based on writing style predate computers by a long time. Mendenhall, a meteorologist, presented the word-length distribution as an author-invariant feature in 1877, which was the first quantitative technique. In 1901, he used this method to address the Shakespeare–Bacon debate. Mosteller and Wallace, who used function words and Bayesian analysis, pioneered the statistical technique to Stylometry in the computer era in 1964 [4].

The features generated by each post are reduced and supplied to the classifier through an unclear text. They did not utilize a "bag of words" or any other qualities that can distinguish different content in the context, unlike previous research on author identification. The word-based features are limited to a small number of function words that have little to do with the topic of conversation (e.g., "the," "in," and so on). They employ single character frequencies but remove bigrams and trigrams, which could have a substantial impact due to certain words [5]. We are handed a note from an anonymous source. The source states they don't know who the person is that wrote the defaming blog post [6], but they do know it was one of a specific group of people.

To begin, author identification entails determining the author of a document or post in issue based on samples of various authors' writings. The basic goal is to determine which author created the document or post in question. One of those whose samples were presented would be the author's reliability. Load the dataset and convert it to a suitable format with this in mind. After that, run the data through a regular text mining pipeline to generate a classification report, which you can use to evaluate how effective this model will be [7].

3 Proposed System

In Fig. 1. We propose the architecture/flowchart of this project. We preprocessed the dataset and feature extraction is done. After that we applied the algorithm and analyzed the author's signature to identify the ownership of the signature. Many text mining and NLP techniques were used for good word representation. The word2vec (W2V) is a one of the useful tool for better result outcome. W2V is a collection of patterns for generating word embedding [8] (WE). WE is the most prevalent text mining approach in vector representations of words presently.

We designed the system to implement the prediction [9] of author signature. Figure 1 shows the flow of our implementation process. We made an attempt to detect these author signatures using word2vec method. The dataset gives the large volume of author signature. Attribute selection helps to identify the relevant features to support author signature research. After dataset collection, the preprocessing technique is followed to clean our dataset. Fake text detection [10] was done using deep learning in small set of database.

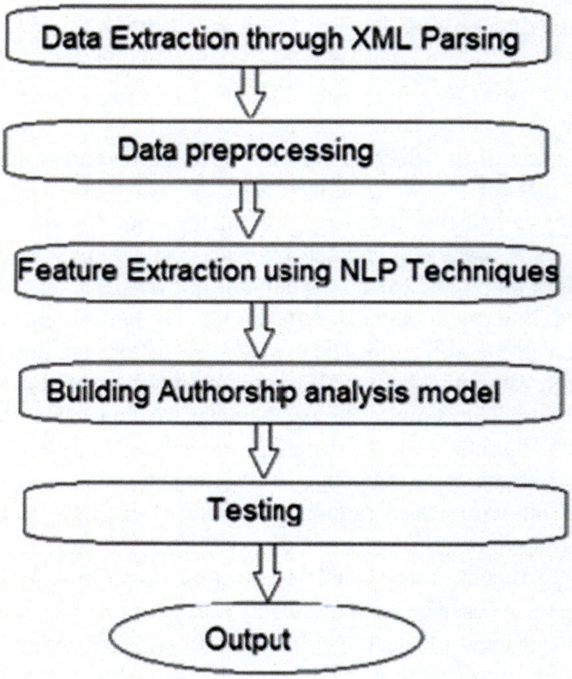

Fig. 1 Flowchart/architecture

4 Preprocessing

4.1 Dataset Preparation

"The Blog Authorship Corpus" dataset was used from bloggers.com website. The corpus consisted of 681,288 posts and over 140 million words, which came up to approximately 7 k words per person. This is a real dataset from genuine blogs. The dataset has required attributes [11] which need not do the preprocessing techniques. They have been labeled with anonymous author ids. There are other features attached to the documents such as age, gender, astrological sign and industry.

4.2 Advantages of Using This Dataset

- These classifiers outperform when using a simple datasets.
- They are usually simpler and easier to implement than complex classifiers [12] and understand the various implementation of a problem.

4.3 Loading the Dataset

To load the XML from the dataset folder, we used the glob library which helps in loading the data into the list with regex expression provided along with the file path in the runtime. The dataset contains all posts in XML files each for an author. We have multiple authors leading this to be multi-classification problem. Loading the dataset is a common step for all the models. However, the steps succeeding these steps vary basing on the model type.

4.4 Preprocessing Step

The files are in XML format which made preprocessing a bit challenging than anticipated. The XML files are first acquired through glob library and made as a blogs folder. Then there are different encodings present throughout the files. Some files with improper encoding are removed. Beautiful soup library is used for parsing the XML files and obtain the posts as strings. Then these strings are being preprocessed.

The Stylometric analysis is way of understanding the author's style of writing [13]. So, in this project we cannot focus much on preprocessing since we may lose author style of writing. But there is some unnecessary text from each of files to remove. Some of the files and posts contained unnecessary information such a url links. We scanned through each file and removed these url links.

Then the strings after preprocessed they are made as posts using post class functions which are created to make post. All the posts are then converted to the post objects. But the problem is we cannot use the post objects for classification directly. So, the posts objects are then converted to data frame. Handling large data frame will be time consuming which led us to create a compressed version of data frame.

5 Methodology

5.1 Classification Using Stochastic Gradient Descent Algorithm

It is one of the linear classifiers (SVM, logistic regression, etc.) with SGD training. The k value is the number of iteration and θ is the constant value. X value is the author's signature list. Y is the linear function value.

SGD Algorithm: update at *k*th iteration.

1. Learning rate $\boldsymbol{\varepsilon}_k$
2. Initial parameter $\boldsymbol{\theta}$
3. **While** stopping criteria is not met do…

   i. Sample a mini batch of m samples from the training set $\{x(^{1}),..x^{(m)}\}$ with
   ii. Corresponding targets $y^{(i)}$

$$\hat{g} \leftarrow +\frac{1}{m}\nabla_{\boldsymbol{\theta}}\sum_{i} L(f(\boldsymbol{x}^{(i)};\boldsymbol{\theta}),\boldsymbol{y}^{(i)})$$

   iii. Compute gradient estimate

$$\boldsymbol{\theta} \leftarrow \boldsymbol{\theta} - \in \breve{a}$$

   iv. Apply update

4. **End While**

The gradient of the loss is calculated for each sample and the model is updated with decreasing learning rate. To obtain good results using default learning rate, the data is expected to have zero mean and unit variance. The feature values are represented as sparse or dense arrays of float points. As shown in Fig. 2 the model it fits can be controlled with the loss parameter.

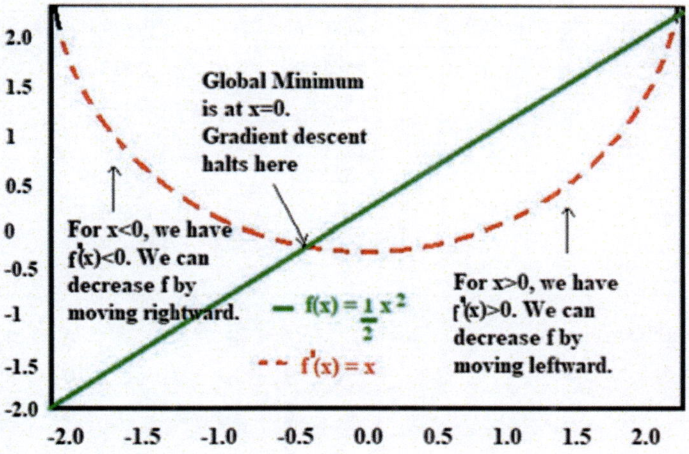

Fig. 2 SGD classifier

6 Result Analysis

Table 1 Displays the accuracy according to class size. If the class size is small the accuracy is high and if the class size is large the accuracy value is reduces. The reason for this reduction can be attributed to the subjective nature of the analysis. The misclassification error is bound to rise due to similarity in writing styles of few authors. If the variance in writing style is high the accuracy was found to improve. Ex: bundle the authors according to country/region.

Table 2 shows the prediction between the real and predicted author ids. So our research gives the maximum correct prediction of authors using stochastic gradient descent. This table also proves that the predicted author ID's writing style is very close to the real author ID. This was manually verified.

Figure 3 shows the confusion matrix value for 10 authors. This interpret the author's identification using array that is given below

Table 1 Accuracy obtained for different numbers of classes

S.No	Classes size	Accuracy (%)
1	10	90
2	50	56
3	90	44
4	150	34
5	200	32
6	500	29
7	40(modified)	79

Table 2 Real author id versus predicted author id

S.No	Real author ID	Predicted author ID
1	3585348	3585348
2	883178	883178
3	1596188	1596,88
4	3522724	3523319
5	1662633	1662633
6	734562	589736

```
array([[ 3,  0,  0,  0,  0,  0,  0],
       [ 0,  4,  0,  0,  0,  0,  0],
       [ 0,  0,  2,  0,  0,  0,  0],
       [ 0,  0,  0, 21,  0,  0,  0],
       [ 0,  0,  0,  1,  2,  0,  0],
       [ 0,  0,  0,  2,  0,  3,  0],
       [ 0,  0,  0,  1,  0,  0,  0]])
```

Fig. 3 Confusion matrix for 10 authors

7 Conclusion and Future Work

The model is used to predict the author of any defamatory blog posts or comments. The current accuracy obtained only gives a lead and cannot completely rely. Also there can be new authors coming up who were not part of the trained model. Author's writing styles are vastly subjective. So, some kinds of filtering techniques may be useful. Bagging and boosting techniques may help if the datasets are bundled according to their features. The future enhancement of this application can be.

- To test it with a larger and complex dataset.
- Bundling the data on combination of features to get more accuracy.

References

1. Benzebouchi NE, Azizi N, Hammami NE, Schwab D, Khelaifia MCE, Aldwairi M (2019) Authors' writing styles based authorship identification system using the text representation vector. In: 2019 16th International multi-conference on systems, signals & devices (SSD), 2019, pp 371–376
2. RamakrishnaMurty M, Murthy JVR, Prasad Reddy PVGD, Satapaty S (2012) Statistical approach based keyword extraction aid dimensionality reduction. In: International conference information systems design and intelligent application—2012, vol 132. Springer—AISC (indexed by SCOPUS, ISI proceeding DBLP etc), pp 445–454. ISBN 978-3-642-27443-5
3. Kuzu RS, Balci K, Salah AA (2016) Authorship recognition in a multiparty chat scenario. In: 2016 4th International conference on biometrics and forensics (IWBF), 2016, pp 1–6

4. Guzmán-Cabrera R, Montes-y-Gómez M, Rosso P, Villaseñor-Pineda L (2008) A web-based self-training approach for authorship attribution. In: International conference on natural language processing, 2008, pp 160–168
5. Zhang W, Hu L, Park J (2022) Politics go "viral": a computational text analysis of the public attribution and attitude regarding the COVID-19 crisis and governmental responses on Twitter. Soc Sci Comput Rev. 08944393211053743
6. Ramnial H, Panchoo S, Pudaruth S Authorship attribution using stylometry and machine learning techniques. In: Intelligent systems technologies and applications. Springer, pp 113–125
7. Singh PK, Vivek KS, Kodimala S (2017) Stylometric analysis of E-mail content for author identification. In: Proceedings of the 1st international conference on internet of things and machine learning, 2017, pp 1–8
8. Narayanan A et al (2012) On the feasibility of internet-scale author identification. In: 2012 IEEE Symposium on security and privacy, 2012, pp 300–314
9. Hossain A, Wahab JA, Khan MSR (2022) A computer-based text analysis of Al Jazeera, BBC, and CNN News shares on facebook: framing analysis on Covid-19 issues. SAGE Open 12(1):21582440211068496
10. Hossain E, Kaysar N, Joy JU, Md AZ, Rahman M, Rahman W (2022) A study towards bangla fake news detection using machine learning and deep learning. In: Sentimental analysis and deep learning. Springer, Singapore, pp 79–95
11. Yafooz W, Emara AHM, Lahby M (2022) Detecting fake news on COVID-19 vaccine from youtube videos using advanced machine learning approaches. In: Combating fake news with computational intelligence techniques. Springer, Cham, pp 421–435
12. An Q, Li R, Gu L, Zhang H, Chen Q, Lu Z, Zhu Y (2022) A privacy-preserving unsupervised domain adaptation framework for clinical text analysis. arXiv preprint arXiv:2201.07317
13. Rebora S (2022) Stylometry and reader response. An experiment with Harry Potter fanfiction. Copyright© 2022 AIUCD Associazione per l'Informatica Umanistica e la Cultura Digitale, 30

Music Genre Classification Using Librosa Implementation in Convolutional Neural Network

M. Shanmuga Sundari, Kamuju Sri Satya Priya, Nandula Haripriya, and Vedaraju Nithya Sree

Abstract Nowadays, numerous songs are available on the Internet and other front-line streaming media. It makes it hard to find the genre to listen to. Quick classification eliminates the idea of searching for music in a specific genre. The classification of music into its respective genre emerges from traditionally extracting the features from time-series data. Another efficient way to classify music into different genres would be to apply a convolutional neural network. Since CNN gives promising results, we have built a CNN model to classify. We used the Librosa library to extract Mel frequencies. This library uses to understand the data using the Mel spectrum. We used the GTZAN data set having ten different genres and 10,000 different audio files (.wav). The classification of our music genre is different zones.

Keywords Convolution neural network · Feature extraction · Librosa · Music classification

1 Introduction

The steps used to classify an audio file into its respective genre are as follows: The first step is feature extraction from the audio file, and the second step would be to build a classifier using these features. Feature extraction [1] depends on one factor, which is Mel-frequency cepstral coefficients (MFCC). We use the Libros library to understand the audio file, its parameters, and the most significant contributing

M. Shanmuga Sundari (✉) · K. S. S. Priya · N. Haripriya · V. N. Sree
BVRIT HYDERABAD College of Engineering for Women, Bachupally, Hyderabad, India
e-mail: sundari.m@bvrithyderabad.edu.in

K. S. S. Priya
e-mail: 18wh1a0559@bvrithyderabad.edu.in

N. Haripriya
e-mail: 18wh1a0530@bvrithyderabad.edu.in

V. N. Sree
e-mail: 18wh1a0548@bvrithyderabad.edu.in

K. A. Reddy et al. (eds.), *Proceedings of Fourth International Conference on Computer and Communication Technologies*, Lecture Notes in Networks and Systems 606,
https://doi.org/10.1007/978-981-19-8563-8_56

factor that helps in classification [2]. We then use the Librosa library to extract Mel-frequency cepstral coefficients from the given audio files in the data set and store them in a .json file. We extract 13 Mel-frequency cepstral coefficients from each audio sample in the data set. GTZAN consists of ten different genres. So, we create ten labels and assign them to each genre correspondingly. As mentioned, each genre consists of 100 audio files that make 10,000 audio files in total. We take each audio file and extract 13 MFCCs from it using the Librosa library. We assign each audio file a label to define which genre the audio file is from. Then we use the extracted features to build a classifier. We load the JSON file data into two different vectors. One consists of extracted MFCCs, and the other consists of labels. The convolutional neural network model is then built with three convolutional layers using Keras. The classifier is trained with a 70:30 ratio of training and test data. The model has been trained, and it can now predict the type of music. The accuracy of the classifier was 77.5%. We take ten random samples from the given data set and predict the genre of each audio file. We now take the input from the user, a .wav audio file, and process it by extracting its 13 MFCCs. We now assign labels to each MFCC and display the most occurring label as the predicted genre.

2 Literature Survey

Global Layer Regularization (GLR) technique is used with CNN and RNN model for the evaluation of training and accuracy [3]. Many music classification techniques use acoustic features for the comparison in terms of the audio signal like Mel-frequency cepstral coefficients (MFCCs). The range bins with the mono-component linear frequency-modulated (LFM) signal model [4]. The music genre recognition was captured using CNN with NetVLAD [5] and aggregated high-level features. It created the proper feature selection to capture musical information across different levels. However, traditional feature coding methods are unsupervised clustering-based approaches that may create chaos to classify the given tasks. EEG signals [6] are tested using many emotions recognition abilities; without traditional utilization strategies, EEG signals are considered the most reliable technique for emotions recognition because of their noninvasive nature. The error rates are extracted using conjecturing that can be the cause of varying noise tag-wise performance differences. Therefore, this research connects to music tagging [7] and neural networks. Music process mining [8] is the technique to categorize the flow of music. The music prediction can also be done with the gender [9] based on the voice frequency. Effective music classification is useful to enhance the level of the music zonal in terms of different platforms [10].

3 Proposed System

As we have mentioned earlier, feature extraction is a crucial step in extracting the MFCCs from the given audio files. The extracted MFCCs are stored in .json file. The .json file is loaded into two vectors having labels and MFCCs. We split the data into training and testing data 70% and 30%, respectively. We build a classifier using the convolutional neural network algorithm having three conventional layers. We use training data and train the classifier to predict the genre. We use the Adam optimizer [11] for controlling the learning rate. It can update neural network weights and optimize the objective function iteratively based on training data. Thus, we used Adam in our architecture. After building and training the classifier, we predict genre using test data. The next step is to take the audio file from the user and extract the MFCCs from it and predict the genre. The process flow of the suggested music classification model is shown in Fig. 1.

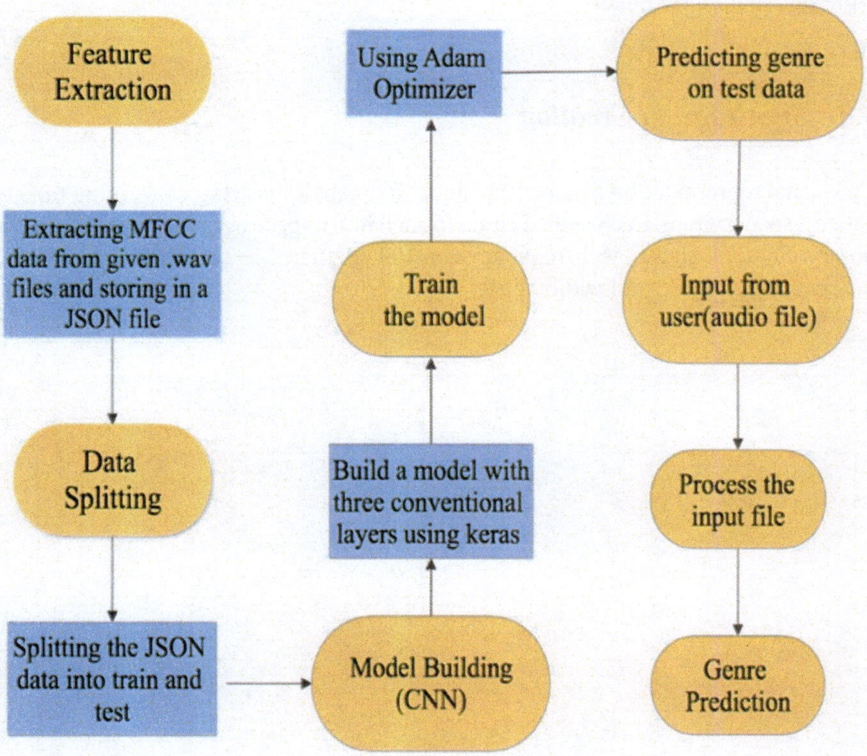

Fig. 1 Proposed model for music classification

4 Implementation

4.1 Data Set

We collected a data set from Kaggle [12] website for different music genres. The data set consists of ten genres and 10,000 audio files. Each genre has 100 different sample audio files. Blues, Classical, Country, Disco, Hip-hop, Jazz, Metal, Pop, Reggae, and Rock are the genres in the data set.

4.2 Data Preprocessing

There are four steps in data preprocessing [13] that will extract the noiseless data from our original data set. Data transformation: The final data set is completely built with the required attributes.

4.3 Spectrogram Creation

The visual representation of spectrum signal frequencies varying concerning time is called a spectrogram. Each audio is transformed into a spectrogram using the Librosa library. Figure 2 shows Mel-frequency cepstral coefficients (MFCCs). It represents the spectrum to spectrum audio clips.

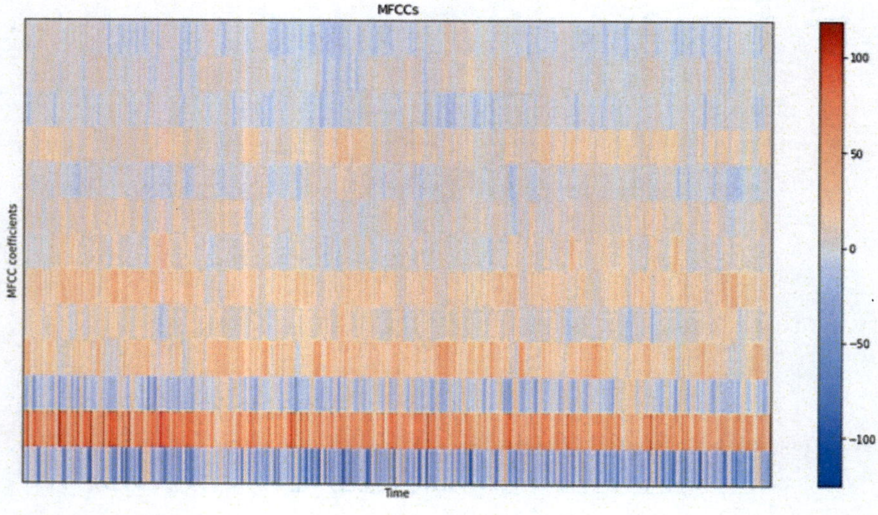

Fig. 2 Spectrogram creation

5 Algorithm

5.1 Convolutional Neural Networks (CNN)

The convolutional layer, pooling layer, and fully connected layer are the major layers of a convolutional neural network. The convolutional layer works on the principle of obtaining the attributes of the image or audio through a fixed-length window (convolutional kernel) by sliding up and down. The feature map which is generated by the activation function is given as the input for the next layer. The pooling layer works to retain the salient attributes, minimize the dimension of each feature map, and lessen the size of the input image or audio file. After obtaining the feature information from the previous convolutional and pooling layers, the fully connected layer acts as a general neural network that is used for classification. The fully connected layer's neurons are only linked to the pixels of the kernel's previous layer and are shared in the same layer with the same weights on each link.

Figure 1 shows the architecture of the convolutional neural network. The pooling layer works on a principle to reduce the attributes to lessen the time and computing resources. Max pooling and average pooling are the two methods of pooling. We have used a max pooling method in our pooling layer that helps choose a maximum value from a matrix and lessen the data of the matrix.

Figure 2 is the illustration of max pooling. Over-fitting is the most occurring issue in a neural network, where the model, when learning features, matches a specific data set too precisely which leads to the model not being generic. It results in the outcomes being more specific to the training data set and low accuracy. The dropout layer eradicates over-fitting and improves generalization. The most used deep learning technique to reduce over-fitting is dropout. Neurons are randomly disconnected during learning by the neural network when dropout is invoked. In current training, these disconnected neurons are not allowed to participate. A sub-network is fabricated from the original neural network after random sampling. This sub-network structure when compared to the original network structure is different. The different machine learning algorithms [14] also can use to predict along with the CNN algorithm for analysis. The sequence of process mining [15] will regulate the activities to get the accurate performance.

Figure 3 shows the architecture of the neural network that contains different layers. The input audio file is sent to the convolution layer. The given audio file is transferred into the pooling layers that are used to reduce the feature dimensions in the input image. Figure 4 shows the feature reduction has done in the previous layer, and the attributes are sent to the CNN layer to get the output.

Figure 5 shows the attribute values after coming across the pooling layer. The attribute selection and reduction will give the better accuracy to find the values.

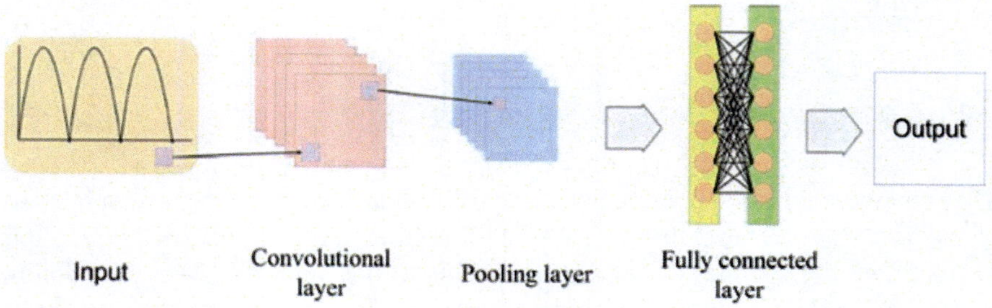

Fig. 3 Neural network architecture

Fig. 4 Attribute reduction in CNN layer

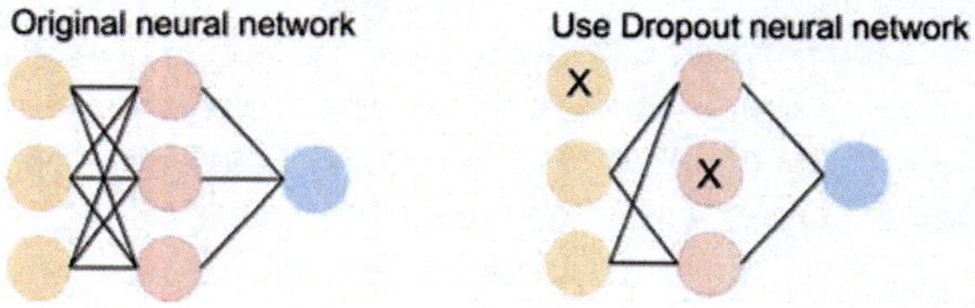

Fig. 5 Feature maps in pooling layer

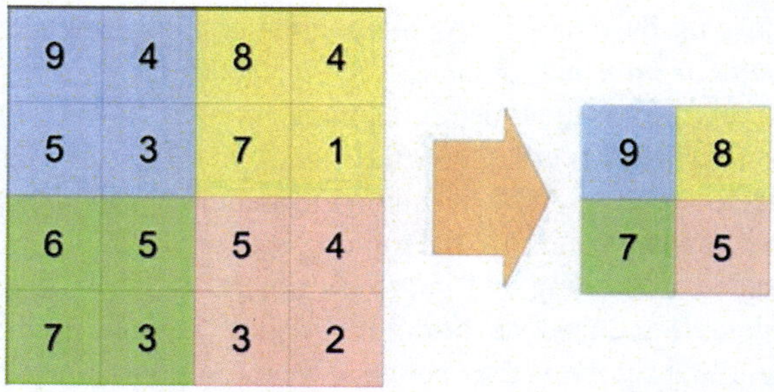

5.2 *Librosa*

The library function is used to load audio from various sources. It will compute spectrogram representations and analyze audio files. Some of the matrix decomposition methods are harmonic-percussive source separation (HPSS) and generic spectrogram decomposition. Time-domain audio processing is also done with these library functions such as pitch shifting and time stretching. Low-level feature extraction is done using feature extraction and manipulation. Various spectral and rhythmic features provided manipulation methods that will help the delta features and memory embedding.

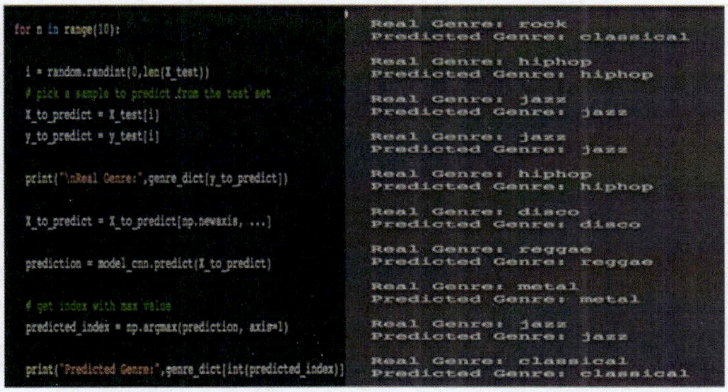

Fig. 6 Result analysis in music genre

6 Result

As mentioned earlier, we predict the genre for randomly chosen audio files from the testing data set. We take ten randomly chosen audio files from the testing data set and pass them to the model. The model then processes the audio files. Processing includes extracting the features from the test sample. Feature extraction entails extracting and storing the 13 MFCCs and assigning a label for the MFCC based on its features that are nearest to the MFCCs retrieved from the audio files of the data set in the preprocessing step. We then check for the most occurring label and predict the genre as the output (Fig. 6).

Figure 7 shows the graph of the accuracy of our CNN model. This model is trained with 50 epochs using Adam optimizer at a learning rate of 0.001. We calculate the loss function and categorize cross-entropy. This also shows validation losses.

7 Conclusion

Music genre classification tool is a time-saving method. Users can find their audio files classified into respective genres within no time. Our proposed model of using the convolutional neural network algorithm gives the highest accuracy of 77.50% which will help in the future work of music genre classification. Music genre classification system can be integrated with Music Recommendation System to recommend more accurate and artist favorite music. Accuracy can be improved by using different models or different optimizers. In the future, our research will lead to finding the gender identification in a large volume of the data set.

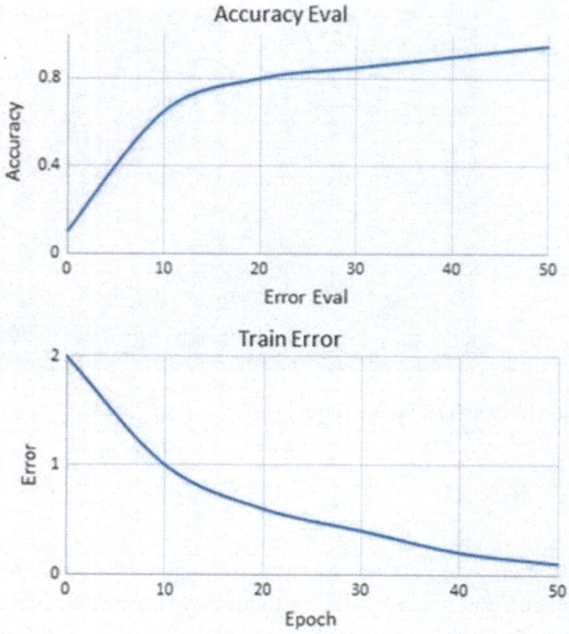

Fig. 7 Accuracy graph

References

1. Sharma AK, Aggarwal G (2021) Classification of Indian classical music with time-series matching deep learning approach. IEEE Access 9:102041–102052. https://doi.org/10.1109/ACCESS.2021.3093911
2. Chen J et al. (2020) An automatic method to develop music with music segment and long short term memory for tinnitus music therapy, vol 8. https://doi.org/10.1109/ACCESS.2020.3013339
3. Ahmad F, Abid F (2020) A globally regularized joint neural architecture for music classification, vol 8
4. Qiuchen LIU, Yong W, Qingxiang Z (2020) ISAR cross-range scaling based on the MUSIC technique, vol 31(5):928–938. https://doi.org/10.23919/JSEE.2020.000070
5. Ng WWY, Member S, Zeng W (2020) Multi-level local feature coding fusion for music genre recognition, pp 152713–152727. https://doi.org/10.1109/ACCESS.2020.3017661
6. Sheykhivand S, Mousavi Z, Rezaii TY, Farzamnia ALI, Member S (2020) Recognizing emotions evoked by music using CNN-LSTM networks on EEG signals, vol 8. https://doi.org/10.1109/ACCESS.2020.3011882
7. Choi K (2018) The effects of noisy labels on deep convolutional neural networks for music tagging, vol 2(2):139–149. https://doi.org/10.1109/TETCI.2017.2771298
8. Sundari MS, Nayak RK (2021) Efficient tracing and detection of activity deviation in event log using ProM in health care industry. In: 2021 Fifth international conference on I-SMAC (IoT in social, mobile, analytics and cloud) (I-SMAC), 2021, pp 1238–1245
9. Reddy RR, Ramadevi Y, Sunitha KVN (2017) Enhanced anomaly detection using ensemble support vector machine. In: 2017 International conference on big data analytics and computational intelligence (ICBDAC), March 2017. IEEE, pp 107–111
10. Zhu Y, Member S, Liu J, Member S, Mathiak K (2020) Deriving electrophysiological brain network connectivity via tensor component analysis during freely listening to music. IEEE Trans Neural Syst Rehabil Eng 28(2):409–418. https://doi.org/10.1109/TNSRE.2019.2953971

11. Castillo JR, Flores MJ (2021) Web-based music genre classification for timeline song visualization and analysis, vol 9:18801–18816. https://doi.org/10.1109/ACCESS.2021.3053864
12. www.kaggle.com/andradaolteanu/gtzan-dataset-music-genre-classification
13. Padmaja B, Prasad VVR, Sunitha KVN, Reddy NCS, Anil CH (2019) Detectstress: a novel stress detection system based on smartphone and wireless physical activity tracker. In: Advances in intelligent systems and computing, 2019, vol 815. https://doi.org/10.1007/978-981-13-1580-0_7
14. Kania D, Kania P, Łukaszewicz T (2021) Trajectory of fifths in music data mining. https://doi.org/10.1109/ACCESS.2021.3049266
15. Sundari MS, Nayak RK (2020) Process mining in healthcare systems: a critical review and its future. Int J Emerg Trends Eng Res 8(9):5197–5208. https://doi.org/10.30534/ijeter/2020/50892020

Web Application for Solar Data Monitoring Using IoT Technology

B. Sujatha, G. Kavya, M. Sudheer Kumar, V. Ruchitha, B. Pujitha, and M. Jaya Pooja Sri

Abstract This paper presents the implementation of a web application for solar data monitoring using IoT technology. A prototype of a solar photovoltaic (PV) panel supporting an electrical load is built. The voltage and current produced by the solar PV panel is continuously sensed and sent to Arduino microcontroller for energy consumption calculations. The computed information is then transferred to the Ethernet shield server and will be stored in firebase. The graphs depict the solar power generation's intermittent behaviour under various weather conditions. Data in the firebase can be accessed through a user interface, which can display the present and past data based on the given inputs like date and time. With this application, one can track the present and historic data given by a solar PV panel. Having first hand information about generation and consumption, the user will enable to optimize their use of generated electrical energy.

Keywords Renewable energy sources · PV · EMS · DER · Ethernet shield server · Firebase · User interface · Microgrid

B. Sujatha (✉) · G. Kavya · V. Ruchitha · B. Pujitha · M. Jaya Pooja Sri
Electrical & Electronics Engineering Department, BVRIT HYDERABAD College of Engineering for Women, Hyderabad, Telangana, India
e-mail: sujatha.b@bvrithyderabad.edu.in

G. Kavya
e-mail: 17wh1a0246@bvrithyderabad.edu.in

V. Ruchitha
e-mail: 17wh1a0259@bvrithyderabad.edu.in

B. Pujitha
e-mail: 17wh1a0250@bvrithyderabad.edu.in

M. Jaya Pooja Sri
e-mail: 17wh1a0258@bvrithyderabad.edu.in

M. Sudheer Kumar
KPIT Technologies, Bangalore, Karnataka, India

K. A. Reddy et al. (eds.), *Proceedings of Fourth International Conference on Computer and Communication Technologies*, Lecture Notes in Networks and Systems 606,
https://doi.org/10.1007/978-981-19-8563-8_57

1 Introduction

Electrical energy plays a significant role in the economic growth of a country. With growing population, the demand for electricity is also increasing rapidly [1]. Renewable energy sources are currently being deployed on a large scale not only to meet the increasing demand for electrical energy but also to mitigate the environmental pollutants and achieve socio-economic benefits for sustainable development [2, 3]. The power produced by these renewable sources can be integrated to the grid. In this regard, microgrid provides low-cost electrical energy. Microgrid is basically a small grid with one or more renewable and/or conventional energy sources (addressed as distributed energy resources (DER) integrated and supplying a small cluster of load) [4]. As the energy is pooled from more than one resource and supplied to different loads, a microgrid needs an energy management system (EMS), which can optimize the use of energy in the most smart, safe, and reliable way. Though EMS was initially developed for management on demand side, emergence of Internet of Things (IoT) paved wave to a better management on both supply and demand sides of the grid. In this paper, a prototype of the microgrid with one DER is made. A solar PV panel is connected to a load through voltage and current sensors. The sensors read the voltage and current every second and send the data to Arduino microcontroller, which calculates the energy consumption. This data is being continuously transferred to firebase using Ethernet shield. This data can be accessed by the user interface application built on Java Script [5, 6].

The importance of an energy management system in a microgrid is discussed in Sect. 1. The relevance of an energy management system in a microgrid is discussed in Sect. 2. Methodology and implementation for the proposed system's hardware and software design is discussed in Sect. 3. Section 4 contains the real-time set-up for the proposed project. The solar data and its outcomes in various weather scenarios are discussed in Sect. 5. The conclusion of the intended research as well as the work's future scope is presented in Sect. 6.

2 Role of an Energy Management System in a Microgrid

In a power distribution network, a microgrid is tailored by integrating distributed energy resources and programmable loads. Currently, energy generation is evolving from a microgrid, which combines numerous energy resources to ensure that clean energy is generated, that operational processes are reliable, and that energy supervision and management arrangements are improved [6].

Microgrid is unable to meet supply demand if energy generation sources are insufficient to feed the requested load. A proper energy management system (EMS) is crucial to avoid this problem. The energy management system for microgrid integrates supply and demand side management while meeting system limits, allowing for optimal implementation with distributed energy sources (DERs). An energy

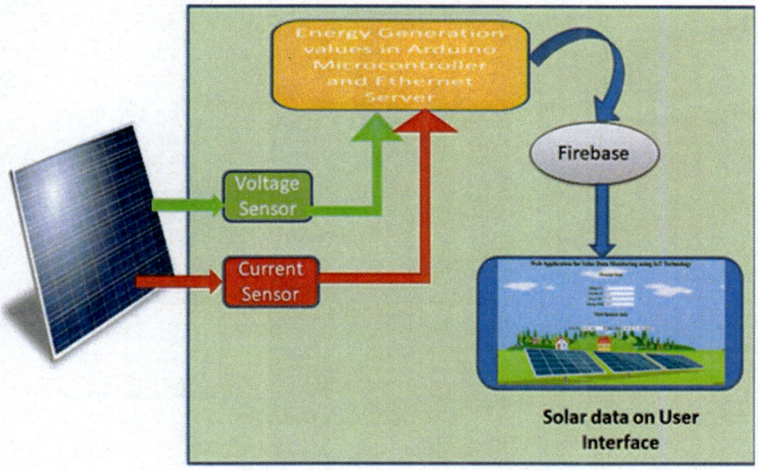

Fig. 1 Block diagram of microgrid EMS and web application for solar data using IoT technology

management system's primary functions are to determine how much energy is created or used, how it is used, and when it is utilized [7]. Microgrid energy management system is implemented with software and hardware integration. Figure 1 shows a microgrid energy management system (EMS) with one distributed energy resource (solar PV panel) and a web application for solar data monitoring using IoT technology [8].

3 Methodology and Implementation for the Proposed Set-up

To develop the proposed system, solar PV panel, Arduino Mega2560 controller, Ethernet shield server, voltage and current sensor, and load are required. The readings from the sensors are given to Ardunio Mega 2560 microcontroller. The serial monitor displays the voltage and current data from the sensors, as well as the computed power and energy values [9].

The Ethernet shield allows an Arduino board to connect to the Internet and read and write to an SD card using the Ethernet and SD libraries, respectively. Data is uploaded into the firebase via Ethernet shield server. Data storage and retrieval are simple with Firebase [9]. A new Firebase project was formed with a real-time database to store the sensed data of components by enabling test mode. JSON data format was retrieved when add.json to the end of the database URL in your browser [10]. A good user interface gives the user a "user-friendly" experience by allowing them to interact with software or hardware in a natural and straightforward manner [11]. The suggested hardware system is depicted in Fig. 2 as a block diagram.

Figure 3 depicts the proposed system's flow chart.

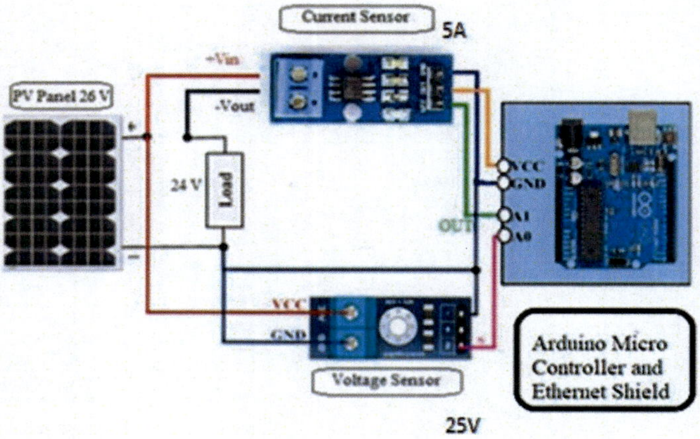

Fig. 2 Block diagram of the proposed hardware system

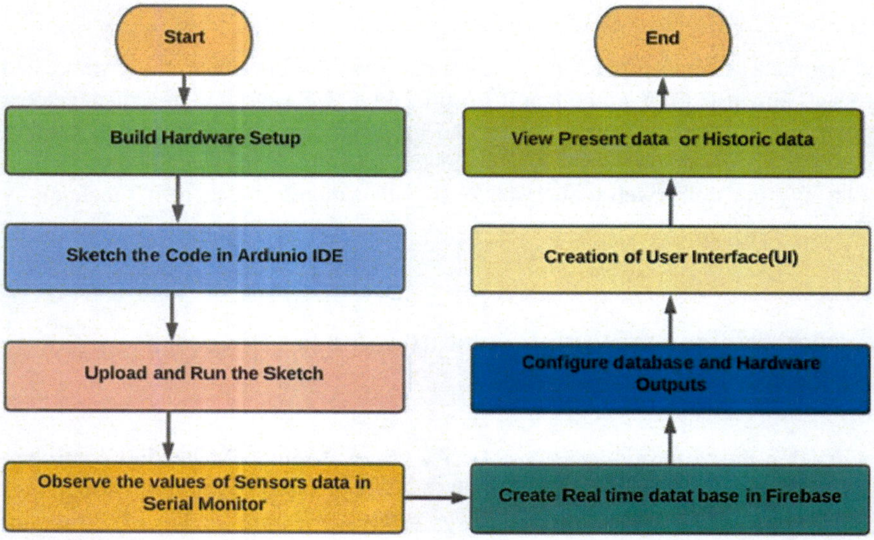

Fig.3 Flow chart of the proposed system

4 Real-Time Set-up for the Proposed Project

Figure 4 shows the proposed system's real-time set-up, which includes a solar PV panel, an Arduino Mega2560 controller, an Ethernet shield server, a voltage and current sensor, and a load. A user interface was built that displays voltage, current, power, and energy values based on the user's preferences. Present data will be shown automatically on the user interface. If user wishes to view previous data, they must enter a date and then click the submit button, after which the UI will display them historic data. If user wants to view present data again, he or she must click the Back Button [12–15].

Figure 5 represents present data and historic data of a solar PV panel on the user interface. The user receives a pop up warning message in case of missing inputs.

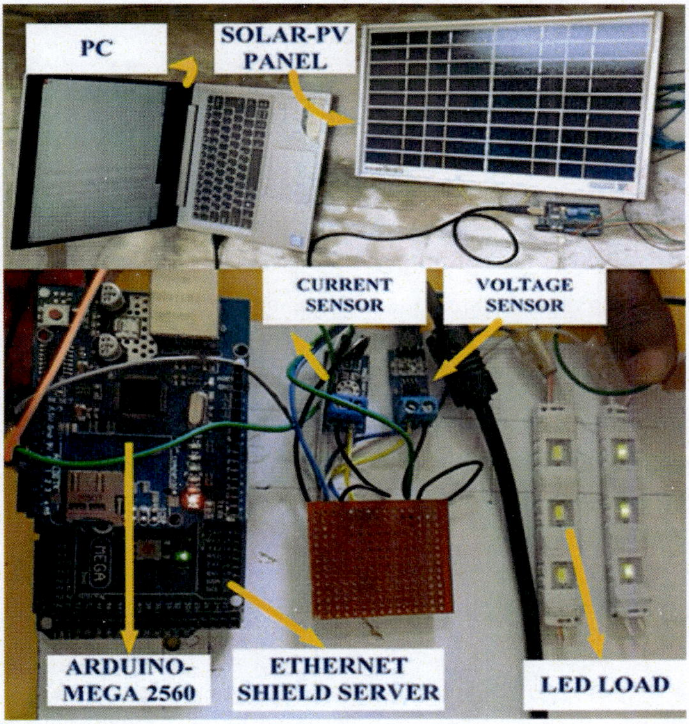

Fig. 4 Real-time set-up for the proposed system

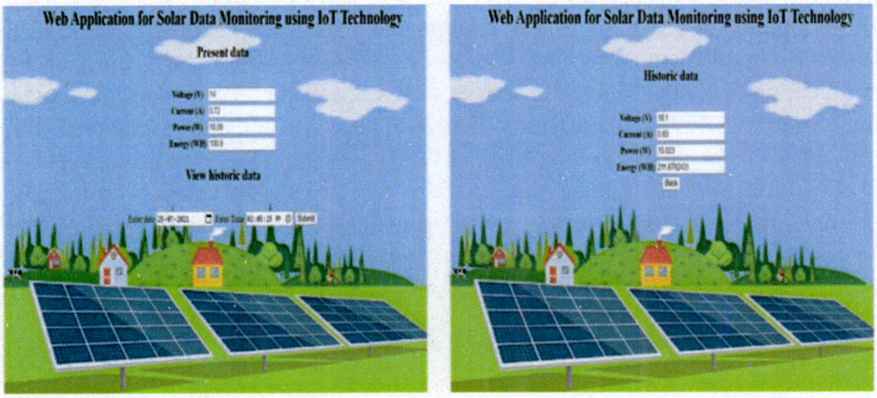

Fig. 5 Present data and historic data of a solar PV panel on UI

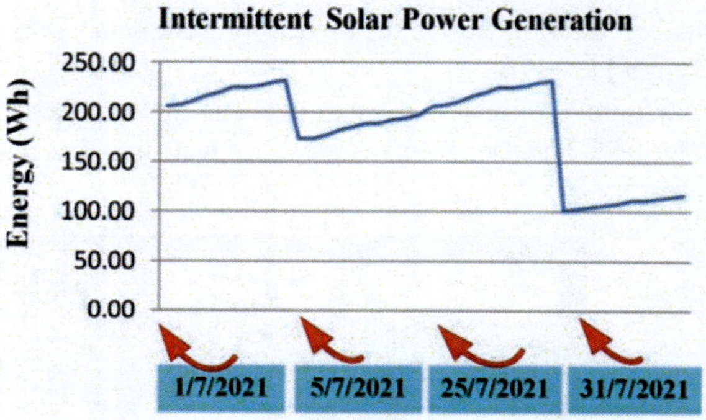

Fig. 6 Intermittent behaviour of solar power generation under a variety of weather conditions

5 Results and Discussion

The table and graph show the intermittent behaviour of solar power generation under a variety of weather conditions, including sunny day with more light intensity, normal sunny day, cloudy day, and rainy day. An energy management system (EMS) is required for a microgrid in order to optimize energy usage in the most intelligent, safe, and reliable manner possible. Though EMS was originally designed for demand management, the advent of the Internet of Things (IoT) has paved the way for better grid management on both the supply and demand sides (Fig. 6; Table 1).

6 Conclusion and Future Work

The proposed system with a solar PV panel supplying an electric load is implemented. The data from the sensors is transferred to the Arduino microcontroller device for computation of the energy consumption. The computed information is then transferred to the Ethernet shield server which directly communicates with the user interface. User interface displays real-time and historical power and energy consumption data according to user requirements. It is useful for industries, which can also be implemented in homes to reduce the cost of energy consumption by coordinating distributed energy resources. The proposed system will be effective by implementing energy management system in a microgrid and can be extended with grid-connected mode of a microgrid using DERs. The data should be protected from all cyber attacks. The user interface can also be equipped with advanced features like sending alerts in case of low energy generation of a source and high energy consumption of the consumer. The system can be optimized by automatic switching to another source in case of low power generation from the existing source.

Table 1 Solar power generation values under a variety of weather conditions

Date	Time	Voltage (V)	Current (A)	Power (W)	Energy (WH)
1/7/2021 (sunny day with more light intensity)	10:00:00	23	0.88	20.24	263.12
	10:05:05	23.02	0.89	20.49	268.08
	10:05:25	23.1	0.91	21.02	275.17
	10:05:35	23.3	0.95	22.14	289.88
	10:05:45	23.25	0.92	21.39	280.06
5/7/2021 (normal sunny day)	10:00:00	16	0.72	11.52	172.8
	10:05:05	16.02	0.72	11.53	173.99
	10:05:25	16.1	0.73	11.75	177.36
	10:05:35	16.25	0.74	12.03	181.49
	10:05:45	16.3	0.75	12.23	184.55
25/7/2021 (cloudy day)	10:00:00	18	0.82	14.76	206.64
	10:05:05	18.02	0.82	14.78	208.12
	10:05:25	18.1	0.83	15.02	211.68
	10:05:35	18.25	0.84	15.33	216.05
	10:05:45	18.3	0.85	15.56	219.26
31/7/2021 (rainy day)	10:00:00	14	0.72	10.08	100.8
	10:05:05	14.02	0.72	10.09	101.8
	10:05:25	14.1	0.73	10.29	103.86
	10:05:35	14.25	0.74	10.55	106.43
	10:05:45	14.3	0.75	10.73	108.28

References

1. Khaparde SA, Mukerjee A (2018) Infrastructure for sustainable renewable energy in India: a case study of solar PV installation. In: IEEE power and energy society general meeting—conversion and delivery of electrical energy in the 21st century, pp 1–7
2. Punna S, Manthati UB, Chirayarukil Raveendran A (2021) Modeling, analysis, and design of novel control scheme for two-input bidirectional DC-DC converter for HESS in DC microgrid applications. Int Trans Electr Energy Syst e12774
3. Punn S, Manthati UB (2020) Optimum design and analysis of a dynamic energy management scheme for HESS in renewable power generation applications. SN Appl Sci 1–13
4. Nayanatara C, Divya S, Mahalakshmi EK (2018) Micro-grid management strategy with the integration of renewable energy using IoT. In: International conference on computation of power, energy, information and communication (ICCPEIC), pp 160–165
5. Arun J, Manivannan D (2016) Smart energy management and scheduling using internet of things. Indian J Sci Technol 9(48)
6. Legha MM, Farjah E (2018) Implementation of energy management of a microgrid using HMAS. In: IEEE smart grid conference (SGC)
7. Hosseinzadeh N, Mousavi A, Teirab A, Varzandeh S, Al-Hinai A (2019) Real-time monitoring and control of. a microgrid - pilot project: hardware and software. In: 29th Australian universities power engineering conference (AUPEC), pp 1–6

8. Legha MM, Rashidifard S (2021) Energy management in multiple micro-grids considering uncertainties of load using hierarchical multi-agent system. Jordan J Electr Eng 7(2)
9. Patil SM, Vijayalashmi M, Tapaskar R (2017) IoT based solar energy monitoring system. In: International conference on energy, communication, data analytics and soft computing (ICECDS), pp 1574–1579
10. Seman LO, Koehler LA, Bezerra EA, Hausmann R (2017) MPPTjs: a JavaScript simulator for PV panels used in a PBL application. Energy Proc 107:109–115
11. Al Kindhi B, Pramudijanto J, Pratama IS, Rahayu LP, Adhi FI, Susila J (2021) Solar cell based integrated sensor system monitoring on smart IoT. In: IEEE International conference on communication, networks and satellite (COMNETSAT), pp 213–218
12. Hadi MS, Maulana MR, Mizar MA, Zaeni IAE, Afandi AN, Irvan M (2019) Self energy management system for battery operated data logger device based on IoT. In: International conference on electrical, electronics and information engineering (ICEEIE), pp 133–138
13. Murdan AP, Joyram A (2021) An IoT based solar powered aquaponics system. In: 13th International conference on electronics, computers and artificial intelligence (ECAI), pp 1–6
14. Mudaliar MD, ivakumar N (2020) IoT based real time energy monitoring system using Raspberry Pi. Internet Things 12:100292
15. Shrihariprasath B, Rathinasabapathy V (2016) A smart IoT system for monitoring solar PV power conditioning unit. In: World conference on futuristic trends in research and innovation for social welfare (startup conclave), pp 1–5

Power Quality Enhancement in Distribution System Using Ultracapacitor Integrated Power Conditioner

K. Bhavya, B. Sujatha, and P. Subhashitha

Abstract With the increased usage of DER like solar and wind energies, power quality (PQ) has become serious concern, in distribution systems and industries. This study suggests utilizing an ultracapacitor (UCAP) at the DC link of the power conditioner employing a bidirectional DC-to-DC converter (BDC) to lessen a variety of power quality issues. The ultracapacitor will improve active power transfer capability and will also reduce the voltage sag and voltage swell problems. UCAP's low energy density, high power density, and quick charging and discharging rates will help to address distribution system power quality problems. The performance of ultracapacitor along with bidirectional DC-DC converter configuration is studied using MATLAB/SIMULINK software. The effectiveness of the UCAP using PID controller and fuzzy controller control (FLC) algorithm is compared.

Keywords Distributed energy resources (DER) · Ultracapacitor (UCAP) · Power quality · Fuzzy logic controller · Bidirectional DC-to-DC converter (BDC) · Energy storage integrator

1 Introduction

Power quality is the term which was given more importance for power engineers nowadays. Quality of power is measured by which parameters like voltage, current, etc., deviate from the given standards. In the distribution systems, power quality (PQ) has become a research in the present scenario due to vast growth in customers on distribution side. The major concern in this is variations in different parameters

K. Bhavya (✉) · B. Sujatha · P. Subhashitha
Electrical and Electronics Engineering Department, BVRIT HYDERABAD College of Engineering for Women, Hyderabad, Telangana, India
e-mail: bhavya.k@bvrithyderabad.edu.in

B. Sujatha
e-mail: sujatha.b@bvrithyderabadd.edu.in

P. Subhashitha
e-mail: subhashitha.p@bvrithyderabad.edu.in

K. A. Reddy et al. (eds.), *Proceedings of Fourth International Conference on Computer and Communication Technologies*, Lecture Notes in Networks and Systems 606,
https://doi.org/10.1007/978-981-19-8563-8_58

like voltage, current, real and reactive power at different customer premises. The main causes for power quality issues are external issues like lightning strikes, motor starting, load variations, nonlinear loads, and arc furnaces. These will lead to various electrical disturbances like voltage sag, voltage swell, harmonic distortions, interruptions, and flicker [1]. This power quality issue will be considered majorly from customer side and is also useful from utility side. Due to the increasing expansion of DG technologies including fuel cells (FC), photovoltaic (PV), wind turbines (WT), small-scale hydro plants (ES), energy storage (ES), and power quality issues arise.

Conventionally, shunt capacitors were used to improve power factor as one of the reactive power compensation techniques. But, sizing and optimal location of the capacitor are the major concerns in the radial system.

Due to development of different custom power devices, the power quality problems are reducing nowadays. These devices are DVR, DS-TATCOM, UPQC, IPFC, etc., are used for mitigation of various power quality issues both in transmission and distribution systems. The D-STATCOM is a shunt controller that produces or consumes reactive power at PCC, allowing for the maintenance of power quality. DVR is a series connected device which will inject three-phase voltage at the same frequency in series to the network voltages to compensate the disturbances.

Advantages of integrating series controllers such as DVR and APF through a converter architecture, which was dubbed UPQC, for power quality enhancement in distribution systems in [2]. The main goal of the conventional UPQC is to recover power quality in distribution systems. Energy storage integration proposed in this paper improves active power capability along with mitigation of voltage sag, swell, and harmonics.

2 3-Ø Converter

2.1 At Supply Point

The planned system is shown in Fig. 1. It contains converter configurations such as DVR and shunt APF which are connected via DC-link capacitor. The DC-link capacitor is supplied over UCAP through BDC. The compensation for voltage sags and swells is provided by series inerter and active, and reactive power support was provided by shunt inverter.

The primary objective of this effort is to enable active power transmission to the load. The proposed topology will compensate.

(1) Voltage sag between 0.1 and 0.9 p.u. and voltage swell between 1.1 and 1.2 p.u., during 3 s–1 min.
(2) Active power or reactive power support and renewable intermittency smoothing.

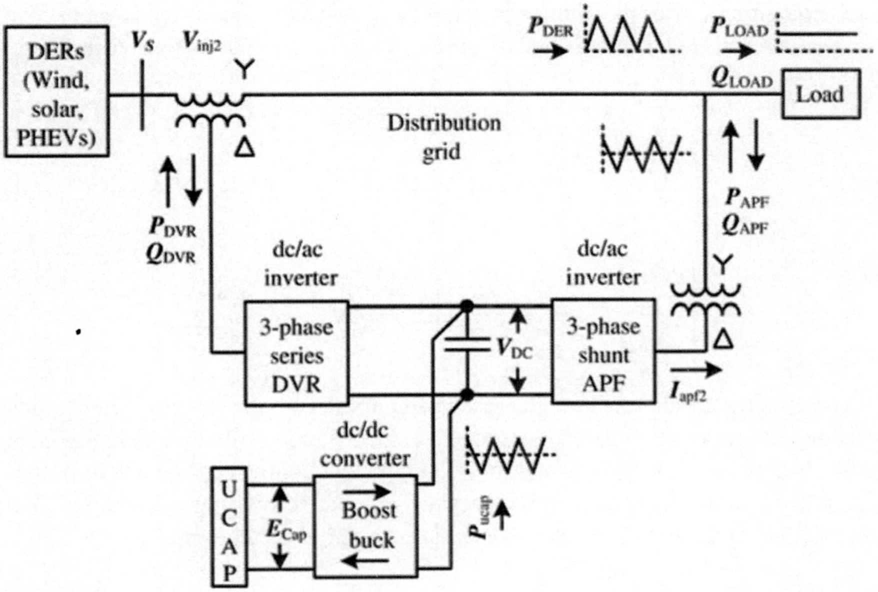

Fig. 1 Block diagram

2.2 Controller Design

In-phase compensation method was implemented for series inverter, which necessitates the use of a PLL to estimate Θ. Based on θ and the L-L source, line voltages are converted to dq coordinates and the L-N components of the source voltage will be calculated by using following equations.

$$\begin{bmatrix} V_{sa} \\ V_{sb} \\ V_{sc} \end{bmatrix} = \begin{bmatrix} 1 & 0 \\ \frac{-1}{2} & \frac{\sqrt{3}}{2} \\ \frac{-1}{2} & -\frac{\sqrt{3}}{2} \end{bmatrix} \begin{bmatrix} \cos\left(\theta - \frac{\pi}{6}\right) & \sin\left(\theta - \frac{\pi}{6}\right) \\ -\sin\left(\theta - \frac{\pi}{6}\right) & \cos\left(\theta - \frac{\pi}{6}\right) \end{bmatrix} \begin{bmatrix} \frac{V_d}{\sqrt{3}} \\ \frac{V_q}{\sqrt{3}} \end{bmatrix} \tag{1}$$

$$\begin{bmatrix} V_{refa} \\ V_{refb} \\ V_{refc} \end{bmatrix} = m * \begin{bmatrix} \left(\sin\theta - \frac{V_{sa}}{169.7}\right) \\ \left(\sin\left(\theta - \frac{2\pi}{3}\right) - \frac{V_{sb}}{169.7}\right) \\ \left(\sin\left(\theta - \frac{2\pi}{3}\right) - \frac{V_{sc}}{169.7}\right) \end{bmatrix} \tag{2}$$

$$\begin{aligned} P_{dvr} &= 3V_{inj2a(rms)} I_{La(rms)\cos\varphi} \\ Q_{dvr} &= 3V_{inj2a(rms)} I_{La(rms)\sin\varphi} \end{aligned} \tag{3}$$

The voltages are kept at normal sine waves as 415 V_{rms} and compared with unit sine. V_{ref} is the voltage needed to have a stable voltage at the load. The DVR injects an equivalent voltage V_{inj2} in-phase in case of mentioned disturbances in the supply, and UCAP is employed to compensate and maintain the promised voltage V_L to the load. Equation (3) uses the voltage injected V_{inj2a} and load current I_{La} to find active power and reactive power delivered by series inverter, where ϕ is difference in-phase.

$$P_{\text{ref}} = -\frac{3}{2} V_{\text{sq}} i_{\text{qref}}$$
$$Q_{\text{ref}} = -\frac{3}{2} V_{\text{sq}} i_{\text{dref}} \tag{4}$$

$$\begin{bmatrix} i_{\text{refa}} \\ i_{\text{refb}} \\ i_{\text{refc}} \end{bmatrix} = \begin{bmatrix} 1 & 0 \\ \frac{-1}{2} & \frac{\sqrt{3}}{2} \\ \frac{-1}{2} & -\frac{\sqrt{3}}{2} \end{bmatrix} \begin{bmatrix} \cos(\theta) & \sin(\theta) \\ -\sin(\theta) & \cos(\theta) \end{bmatrix} \begin{matrix} i_{\text{dref}} \\ i_{\text{qref}} \end{matrix} \tag{5}$$

The id-iq technique was used to create the controller for the shunt inverter, which delivers active power and reactive power compensation, with the id component controlling the reactive power and the iq component controlling the active power. The active and reactive power references are calculated by using i_{qref} and i_{dref}, i.e. from Eq. (4). Equation (5) is used to calculate the reference currents.

3 Description of BDC

The buck-boost converter used as an edge between the UCAP and the DC link in this BDC. This converter will offer active/reactive power assistance as well as voltage sag correction, while the UCAP is in discharge mode. During intermittency smoothing, this converter functions bidirectionally, charging, or absorbing power from the grid. When discharging power from the ultracapacitor, the proposed buck-boost DC-to-DC converter functioned as boost converter, and when charging the ultracapacitor from the grid, it operated as buck converter.

The output voltage of this DC-to-DC converter is controlled using average current mode control. When used in conjunction with other tactics like voltage mode control and peak current mode control, this strategy works better.

4 UCAP

UCAP can transport extremely high power within a short time. When compared to Li-ion batteries, ultracapacitors have lower energy density and higher power density. UCAPs feature superior power density, more lifespan cycles of charge and discharge, and higher terminal voltages for each module than conventional batteries. These are suitable properties for delivering active and reactive power assistance to the distribution system in a short period of time. The terminal voltage at the UCAP, the

DC-link voltage, and the grid voltages on the distribution side all affects how many ultracapacitors are required for grid support.

The UCAP bank has three modules that are practical and economical for a 260-V DC-link voltage. The UCAP bank discharging process can be calculated as follows

$$\text{EUCAP} = \frac{1}{2} * C * \frac{V_{\text{uc,ini}}^2 - V_{\text{uc,fin}}^2}{60} \text{Wmin} \tag{6}$$

$$\begin{aligned}\text{EUCAP} &= \frac{1}{2} * \frac{165}{3} * \frac{144^2 - 72^2}{60} \text{Wmin} \\ &= 7218\,\text{Wmin}\end{aligned} \tag{7}$$

5 Fuzzy Controller

The control action of a fuzzy controller is calculated by the assessment of a set of linguistic rules. A detailed examination of input supply variations on both the supply side and the load side is one of the objectives. Because it is full and constructed sophistically without the use of extrapolation, the rule foundation developed is consistent. The switching instant of the controlled devices in the UPQC's inverter is controlled using fuzzy control in this study.

The fuzzy logic controller-based UPQC in this suggested model includes two inputs: ΔV, ΔI. The input values will first be fuzzified, which is the process of converting them to fuzzy variables. The fuzzy inputs are then transferred to the rule base, and the outputs are defuzzed to compute the final outputs. This process is as shown in Fig. 2.

In the development of fuzzy controller, 7 fuzzy subsets are used for two inputs.

They are zero (ZE), negative small (NS), negative medium (NM), negative big (NB), positive big, positive medium, and positive small. We created 49 control rules using Gaussian membership functions [3], which are displayed in Table 1.

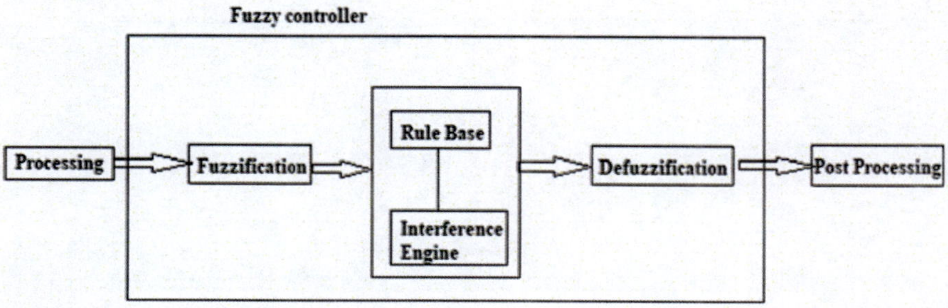

Fig. 2 Fuzzy logic control process

Table 1 Fuzzy controller rules

$\Delta V/\Delta I$	NB	NM	NS	ZE	PS	PM	PB
NB	NB	NB	NB	NB	NM	NS	ZE
NM	NB	NB	NM	NM	NS	ZE	PS
NS	NB	NM	NS	NS	ZE	PS	PM
ZE	NM	NM	NS	ZE	PS	PM	PB
PS	NM	NS	ZE	PS	PS	PM	PB
PM	NS	ZE	PS	PM	PM	PB	PB
PB	ZE	PS	PM	PB	PB	PB	PB

The inputs are modelled as linguistic variables in fuzzification. It is the first controller block and transforms each input value into a degree of membership function. It compares the input data with rule conditions and finds the level of match between the former and latter.

The membership functions of source voltage and change in source voltages are shown in Figs. 3 and 4. The number of levels, which is implemented in this paper, is not fixed, and they depends on the application. If the number of fuzzy levels is more, input will have more resolution. The sinusoidal fuzzy set values are used in fuzzy control implementation.

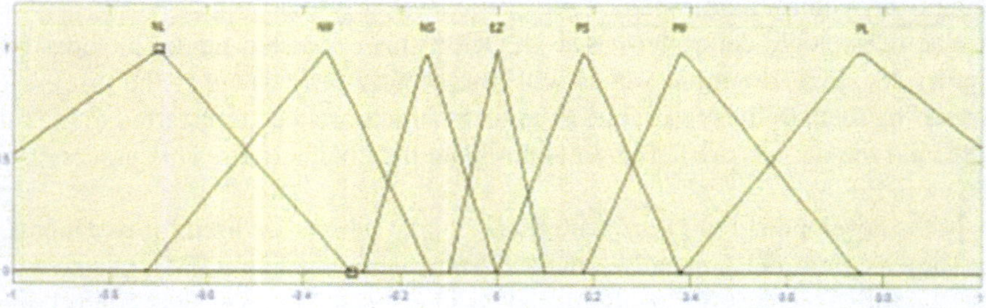

Fig. 3 Member function for source voltage

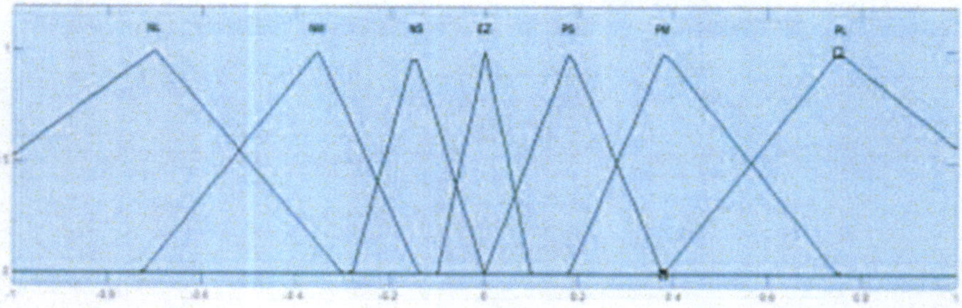

Fig. 4 Membership function for change in voltage ΔV

The procedure of adapting fuzzy data into a single data is known as defuzzification. An aggregate of all the rule outputs is determined in this method. The aggregate indicates the required change in switching instant. Particular change in the triggering angle is calculated which results in change in grid currents to enhance power quality.

6 Simulation Results

See Figs. 5, 6, 7, 8, 9, 10, 11, 12, 13, 14, 15, 16, and 17.

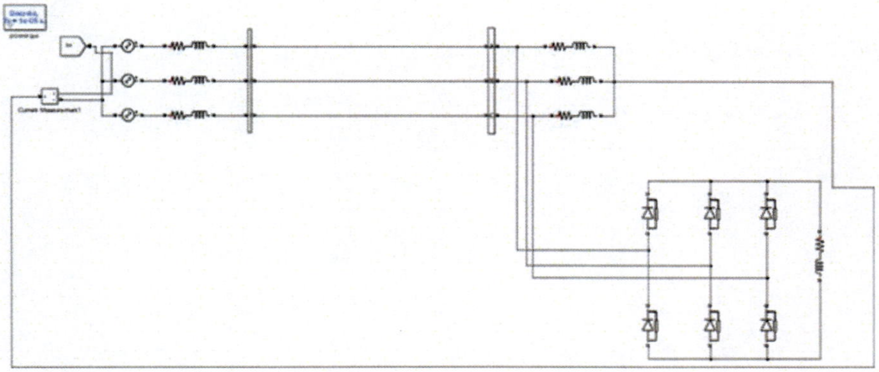

Fig. 5 Simulink model without compensation

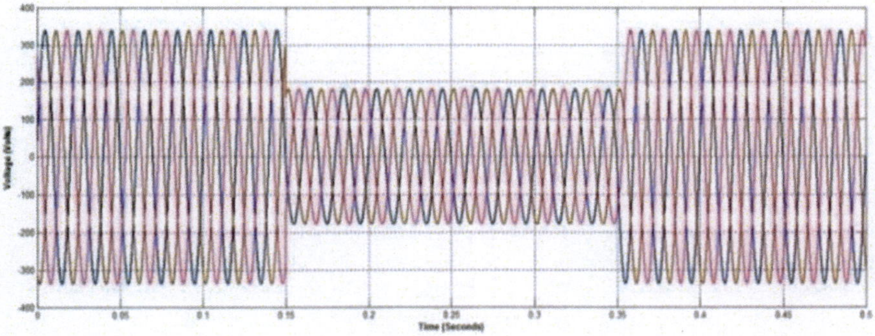

Fig. 6 Voltage of source during voltage sag

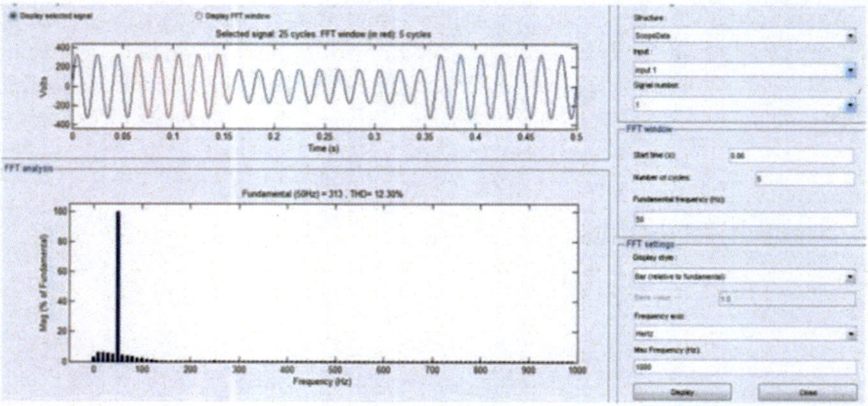

Fig. 7 THD of load voltage before compensation

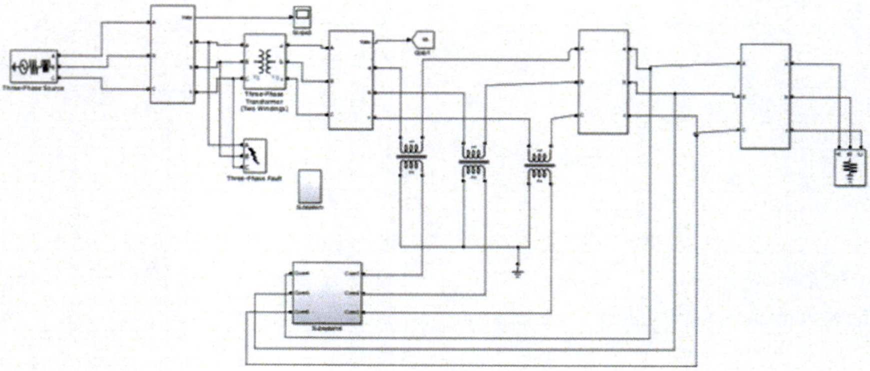

Fig. 8 Simulink model with compensation using PI controller

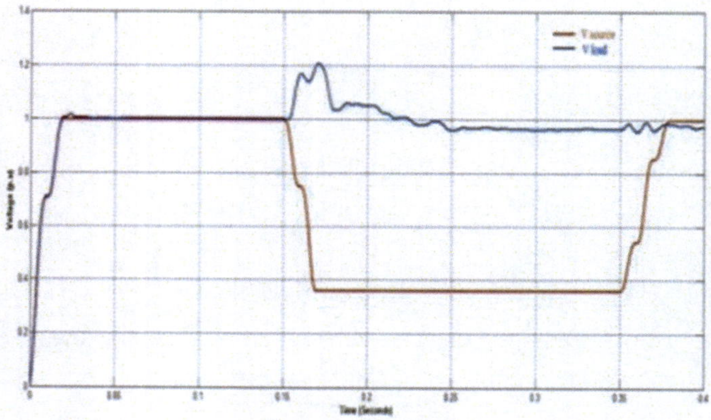

Fig. 9 Source and load RMS voltage

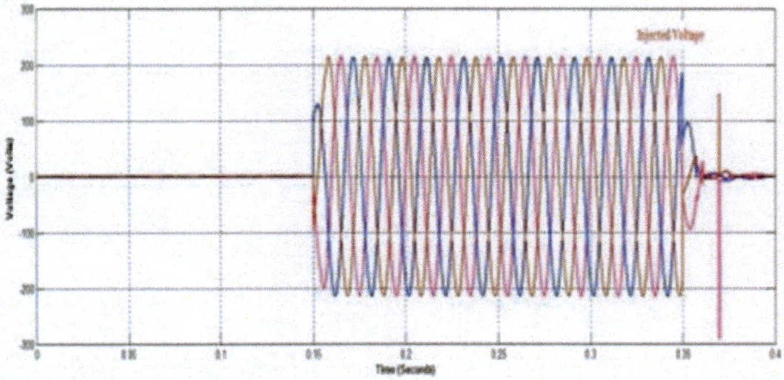

Fig. 10 Injected voltage

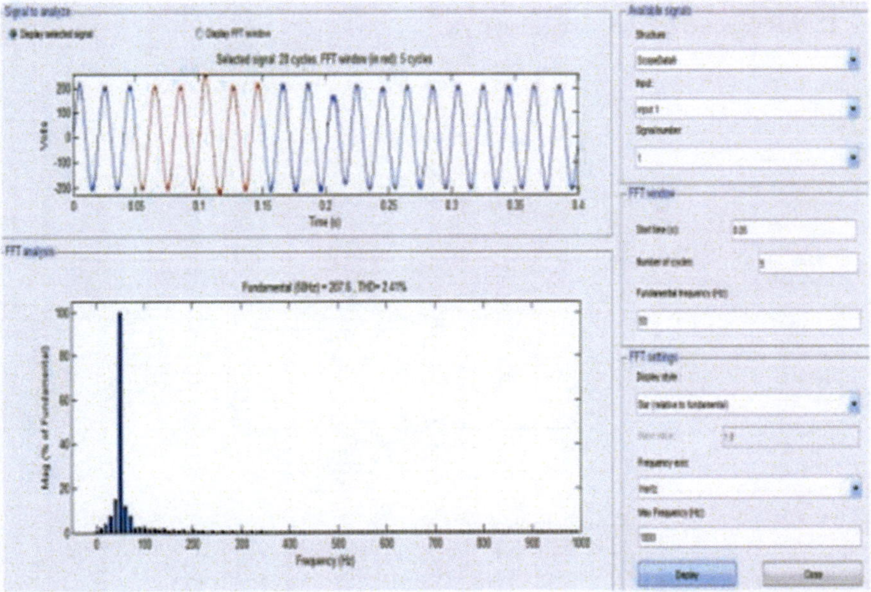

Fig. 11 THD of load voltage after compensation

7 Conclusion

This study describes a power conditioner system for distribution grids that employs UCAP-based energy storage with rechargeable capacity. With the proposed configuration, the active power filter will be able to deliver the distribution grid with active and reactive power, and the DVR component will be able to independently adjust for voltage peaks and valleys. Shunt inverter control techniques and series inverter

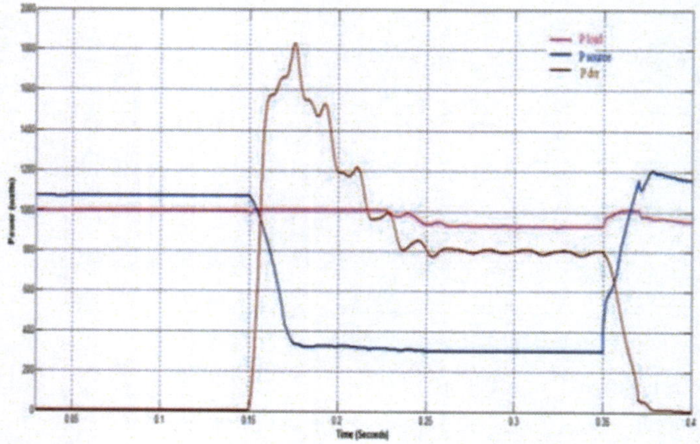

Fig. 12 Active power of load, source, and DVR

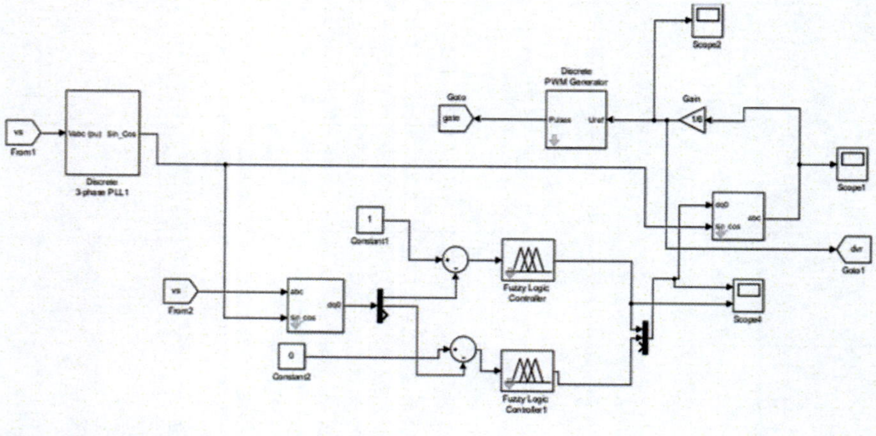

Fig. 13 Control strategy with fuzzy controller

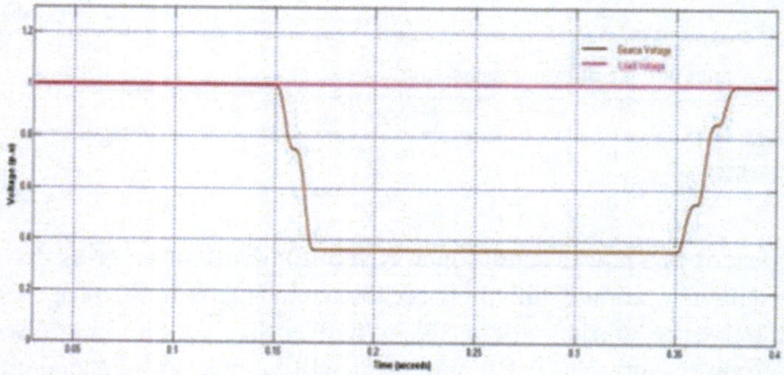

Fig. 14 Source and load RMS voltage

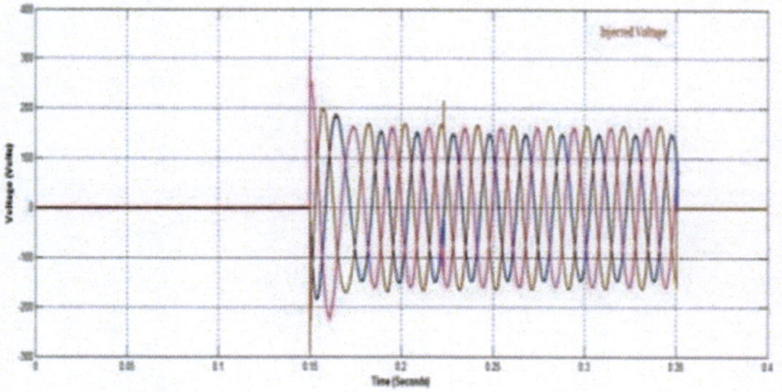

Fig. 15 Injected voltage

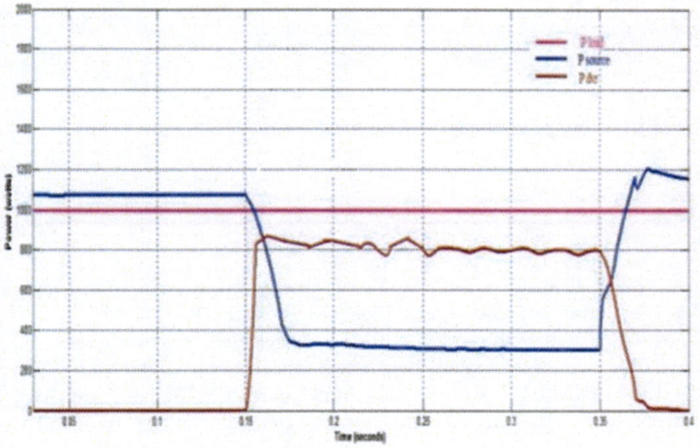

Fig. 16 Active power of load, source, and DVR

(DVR) control strategies both rely on the id/iq methodology and in-phase compensation, respectively. Performance of the UCAP is tracked using a PID controller and fuzzy logic controller. The system's power quality was improved because the fuzzy logic controller responded more effectively in lowering the sags and swells.

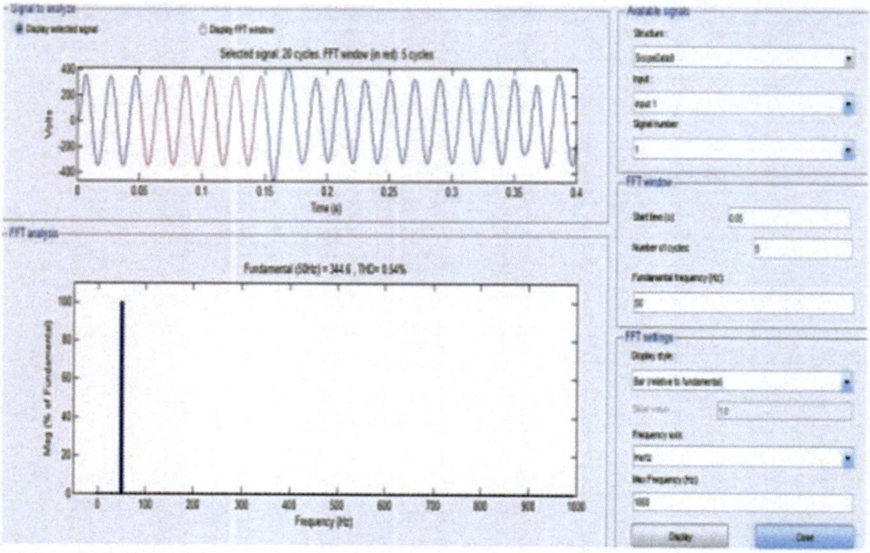

Fig. 17 THD of load voltage after compensation using fuzzy controller

References

1. Elsherif A, Fetouh T, Shaaban H (2016) Power quality investigation of distribution networks embedded wind turbines. J Wind Energy 2016(7820825):1–18
2. Akhil G, Srinivasa Rao D, Bhavya K (2016) Enhancement of power quality in distribution system by ultra capacitor integrated power conditioner. Journal 2(5):99–110 (2016)
3. Arsoy AB, Liu Y, Ribeiro PF, Wang F (2003) StatCom-SMES. IEEE Ind Appl Mag 9(2):21–28
4. Somayajula D, Crow ML (2014) An ultracapacitor integrated power conditioner for intermittency smoothing and improving power quality of distribution grid. IEEE Trans Sustain Energy 5(4)
5. Rittershausen J, McDonagh M (2014) Moving energy storage from concept to reality: Southern California Edison's approach to evaluating energy storage (Online). Available http://www.edison.com/content/dam/eix/documents/innovation/smart-grids/Energy Storage-Concept-toReality-Edison.pdf. Accessed on 15 Jul 2014
6. Li W, Joos G, Belanger J (2010) Real-time simulation of a wind turbine generator coupled with a battery supercapacitor energy storage system. IEEE Trans Ind Electron 57(4):1137–1145
7. Thounthong P, Luksanasakul A, Koseeyaporn P, Davat B (2013) Intelligent modelbased control of a standalone photovoltaic/fuel cell power plant with supercapacitor energy storage. IEEE Trans Sustain Energy 4(1):240–249
8. Ru Y, Kleissl J, Martinez S (2013) Storage size determination for gridconnected photovoltaic systems. IEEE Trans Sustain Energy 4(1):68–81
9. Brekken TKA et al (2011) Optimal energy storage sizing and control for wind power applications. IEEE Trans Sustain Energy 2(1):69–77
10. Santoso S, Mc Granaghan MF, Dugan RC, Beaty HW (2012) Electrical power systems quality, 3rd edn. McGraw-Hill, New York, NY, USAs

CNN-Based Breast Cancer Detection

N. M. Sai Krishna, R. Priyakanth, Mahesh Babu Katta, Kacham Akanksha, and Naga Yamini Anche

Abstract One in eight females globally suffers from breast cancer. It is identified by recognizing the malevolence of breast tissue cells. Current medical image processing techniques use histopathological pictures recorded by a microscope to examine them employing various procedures and methodologies. Nowadays machine learning algorithms are widely used for the interpretation of medical-related images and tools related to pathology. As identifying cancer cells manually is time consuming and may be prone to errors in operation, computer-aided processes are used to get better outcomes than manual pathological detection methods. This is often accomplished in deep learning by the process of feature extraction fully aided by a convolutional neural network (CNN) followed by the classification process via a fully connected network. Deep learning is widely used in medical imaging since it does not need prior knowledge in a related discipline. The process involved in the current study is training a CNN and accomplished prediction accuracy of up to 86.9%.

Keywords Cancer cells · Medical imaging · CNN · ResNet50 · Confusion matrix

1 Introduction

India has seen 30% of the situations of bust cancer cells throughout the last couple of years and it's most likely to enhance [1]. Among all kinds of cancer in ladies,

N. M. Sai Krishna · R. Priyakanth (✉) · M. B. Katta · K. Akanksha
BVRIT HYDERABAD College of Engineering for Women, Hyderabad, Telangana, India
e-mail: priyakanth.r@bvrithyderabad.edu.in

N. M. Sai Krishna
e-mail: saikrishnakumar.n@bvrithyderabad.edu.in

M. B. Katta
e-mail: maheshbabu.k@bvrithyderabad.edu.in

K. Akanksha
e-mail: 19wh1a0470@bvrithyderabad.edu.in

N. Y. Anche
Qualcomm India Pvt. Ltd, Hyderabad, India

K. A. Reddy et al. (eds.), *Proceedings of Fourth International Conference on Computer and Communication Technologies*, Lecture Notes in Networks and Systems 606,
https://doi.org/10.1007/978-981-19-8563-8_59

carcinoma is possible. Carcinoma has the second-highest fatality rate when respiratory organ and cartilaginous tube cancer, and concerning half-hour of freshly diagnosed cases area unit of carcinoma solely. Progressing the combat to eradicate cancer needs an initial outcome that might solely be attainable through an associate in nursing economical detection systems. Techniques are developed to notice carcinoma, together with the processing of medical images and digital pathologies. Pictures are unit tested by histopathology that typically contains diagnostic tests of the contrived tissues. These matters are wracked by the growth area unit deracinated by the medical specialist and are a combination of hematoxylin and eosin stain. They are examined beneath a magnifier for carcinogenic cells. The infinitesimal pictures hence accumulated are utilized for creating computer-aided cancer discovery systems. Performing the detection process manually could be monotonous and possibly encompass human-caused errors, as maximum components in the cell area unit are oftentimes a portion unevenly random. The objective is to find out the nature of the growth of tumors as benign or malicious because of the presence of more hitches in the case of tumors being malicious. Briefly, this is a binary classification problem and has a higher possibility to get solved by widely ranged machine learning processes. It has been proved in the part that using machine learning algorithms in diagnosing various diseases gives additional results when compared to the diagnosis by a human medical expert. Based on the study conducted by Phillips (Europe), a wide assortment of computer-based procedures operated on side of breast pictures have provided additional correct information in the process of detection. This additional information associated with good quality pictures has an opportunity in improvising the performance and accuracy toward detecting cancer [2].

2 Background Study

In the process of diagnosis and recognition of breast cancer, CNN plays a realistic role to offer high accuracy compared to the multilayer perceptron method [3]. A CNN training task involves a huge volume of data, which lack in the medical domain, particularly in breast cancer [4]. Some of the ways to detect breast cancer are mammography, nuclear imaging, computed tomography (CT) scans [5], magnetic resonance imaging, etc. However, there is a limitation on how accurately to detect breast cancer from these techniques. On the other hand, histopathology tests are tissue-based, where cell structures along with additional external elements are stained and captured in high resolution for pathologic analysis. These images give a high level of details to accurately diagnose cancer, but the identification is also equally difficult due to various factors such as multiple appearances of cancer cells, intra-observer variation, due to common hypothermic features, and area of selection of the tissue, so that the selected area is in tumor periphery. Hence, this issue can easily be resolved by deep learning models.

Deep learning is a subsection of machine learning that has resulted in high success rates and accuracies. The deep learning neural networks are inspired and modeled around the human brain to analyze unstructured patterns. These techniques are used to extract and analyze the features at each layer of the neural network and therefore improve the prediction of tumors [6]. There are existing deep learning models designed exclusively for image classification with high accuracies such as VGG19, AlexNet, Mobile Net [7], etc. One can easily use these existing models or design a new model to solve the problem at hand.

As stated above, there are various networks designed aiming to classify breast cancer, such as Artificial Neural Networks based on Maximum Likelihood Estimation (MLE), GRU–SVM model based on Recurrent Neural Networks (RNNs), Gated Recurrent Unit (GRU) reinforced by Support Vector Machine (SVM), and many more. Neural networks together with Multi-variate Adaptive Regression Splines (MARS) can likewise be used in identifying tumor growth. The BreakHis dataset [8] published in 2015 has been used by Fabio A. Spanhol, who described the constraints along with the neural-net system obtaining an accuracy between 80 and 85%.

Arpit B and Aruna T came up with a Genetically Optimized Neural Network [9] (GONN) for breast cancer identification as benevolent or malignant. The neural network architecture has been optimized by presenting state-of-the-art crossover and mutation operators [10]. This had been evaluated by using WBCD and comparing the classification accuracy confusion matrix, sensitivity, specificity, and the receiver operating characteristic curves of GONN by means classical backpropagation model [11]. This technique presented an acceptable accuracy in the classifying process. However, there is a scope for improvement by using a larger dataset. Ashraf O I and Siti M S have given out a computer-based process to classify breast cancer using a multilayer perceptron (MLP) neural network centered around the concept of improved non-dominated sorting genetic algorithm (NSGA-II) for the optimization of the accuracy and network structure.

3 Architecture

Convolutional neural network (CNN): This network is a multilayer neural network that can be used in detecting the different complex features in the images [12]. CNN works on the elementary principle of convolution. This CNN consists of five layers as shown in Fig. 1.

Convolution: In convolution, it involves the integration of two functions in which one function modifies the other function. The convolution process comprises three important items. They are input images, feature detectors, and feature maps. The image considered for detection purposes is called input. A feature indicator is a kernel or matrix. The input represented in the form of a matrix is multiplied element wise with a feature indicator forming the feature map. In this step, some information may be lost but the key features of the image are retained [13].

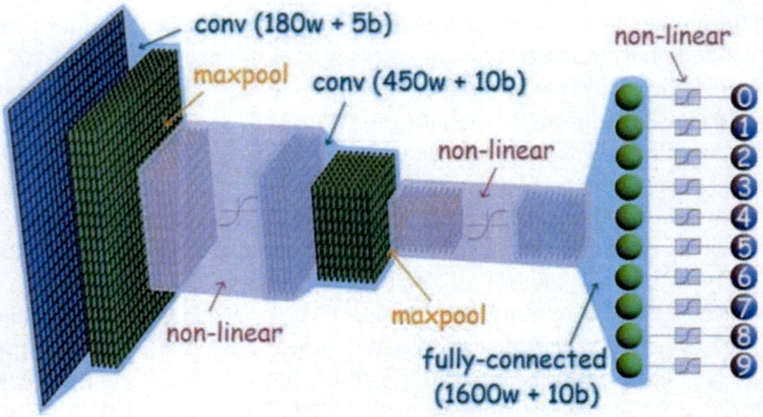

Fig. 1 Layers in CNN

Applying Nonlinear function: From the convolution section, a feature map is obtained. So here, a rectifier function is applied to the feature map to get the nonlinearity of the CNN. Generally, images can be composed of diverse things which are nonlinear in reciprocal. So, the non-application of the rectifier operation leads to glitches in linearity.

Pooling: Max and min pooling techniques are available. Working of max pool layer places a feature map into a matrix and then picks up the highest value in the resultant matrix. Left to right movement is followed by the matrix to pick all the largest values in each pass [14].

Flattening: This technique transforms the whole pooled feature map into one column which in turn is given as input to the neural network.

Full Connection: This mainly consists of three steps: input layer, fully connected layer, and output layer. In the output layer, the prediction classes are obtained. So, after passing into the neural network the error of prediction is calculated. After that, the error is propagated back for an increase in prediction.

ResNet 50: ResNet 50 Architecture [15], as in Fig. 2, has 48 convolution layers with 1 max pool with 1 average pool layer. It is also having 3.8×10^9 radix points function running customary pre-owner ResNet model. ResNet 50 is converse 18 and 34-layer ResNet which gives persisting mapping which is not shown for simplicity. For the ResNet 50, there was a tiny alteration made; previously, the shortcut connections skipped two layers, but now they skip three layers and additional 1×1 convolution layers have been added. The steps in ResNet are shown in Fig. 3.

- Convolution along with kernel size of 7×7 also 64 of diverse kernels all also step size 2 given one layer.
- Following step is max pooling, also with the step size 2.

layer name	output size	18-layer	34-layer	50-layer	101-layer	152-layer
conv1	112×112	7×7, 64, stride 2				
conv2_x	56×56	3×3 max pool, stride 2				
		[3×3, 64; 3×3, 64] ×2	[3×3, 64; 3×3, 64] ×3	[1×1, 64; 3×3, 64; 1×1, 256] ×3	[1×1, 64; 3×3, 64; 1×1, 256] ×3	[1×1, 64; 3×3, 64; 1×1, 256] ×3
conv3_x	28×28	[3×3, 128; 3×3, 128] ×2	[3×3, 128; 3×3, 128] ×4	[1×1, 128; 3×3, 128; 1×1, 512] ×4	[1×1, 128; 3×3, 128; 1×1, 512] ×4	[1×1, 128; 3×3, 128; 1×1, 512] ×8
conv4_x	14×14	[3×3, 256; 3×3, 256] ×2	[3×3, 256; 3×3, 256] ×6	[1×1, 256; 3×3, 256; 1×1, 1024] ×6	[1×1, 256; 3×3, 256; 1×1, 1024] ×23	[1×1, 256; 3×3, 256; 1×1, 1024] ×36
conv5_x	7×7	[3×3, 512; 3×3, 512] ×2	[3×3, 512; 3×3, 512] ×3	[1×1, 512; 3×3, 512; 1×1, 2048] ×3	[1×1, 512; 3×3, 512; 1×1, 2048] ×3	[1×1, 512; 3×3, 512; 1×1, 2048] ×3
	1×1	average pool, 1000-d fc, softmax				
FLOPs		1.8×10^9	3.6×10^9	3.8×10^9	7.6×10^9	11.3×10^9

Fig. 2 ResNet architectures

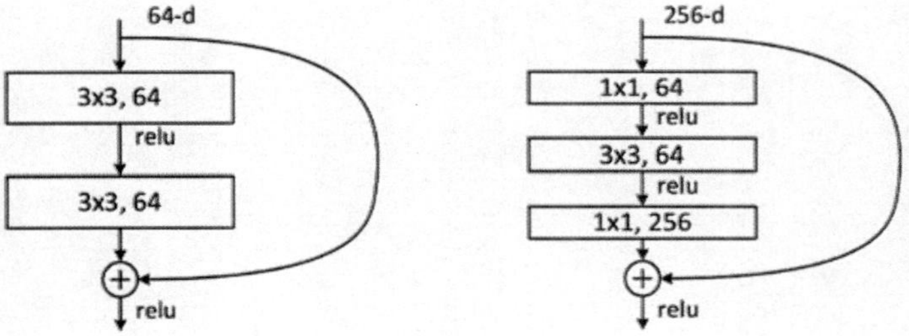

Fig. 3 Steps in ResNet

- Next convolution is 1 × 1, 64 kernel also follows a 3 × 3, 64 kernels at last 1 × 1, 256 kernels. These three layers are repetitive for three times, given as nine layers in this step.
- Next kernel is 1 × 1, 128 follows a 3 × 3, 128 last with 1 × 1, 512. These are repetitive for four times, given as 12 layers.
- Next kernel of 1 × 1, 256 with 2 more kernels is 3 × 3, 256, also 1 × 1, 1024 and this is repetitive for six times given as 18 layers.
- Last kernel is 1 × 1, 512 kernel with 3 × 3, 512 also 1 × 1, 2048 repetitive for three times given as nine layers.
- After all these layers of an average pool ended with a completely linked layer comprising 1000 nodes at the end of a softmax function given as one layer.
- The function of max or average pooling layers total gives a $1 + 9 + 12 + 18 + 9 + 1 = 50$ layers of Deep Convolutional Network.

4 Results

Figure 4 presents a few IDC (−) samples from the validation set and along with that also shows the model's predictions.

The top losses incurred by the model during the training process are as follows in Fig. 5.

Everyone can see that some samples are originally IDC (+), but the model predicts them as IDC (−). This is a staggering issue. One needs to be certainly cautious with the false negatives. One doesn't need to categorize somebody as "No cancer" when they are in fact "Cancer positive."

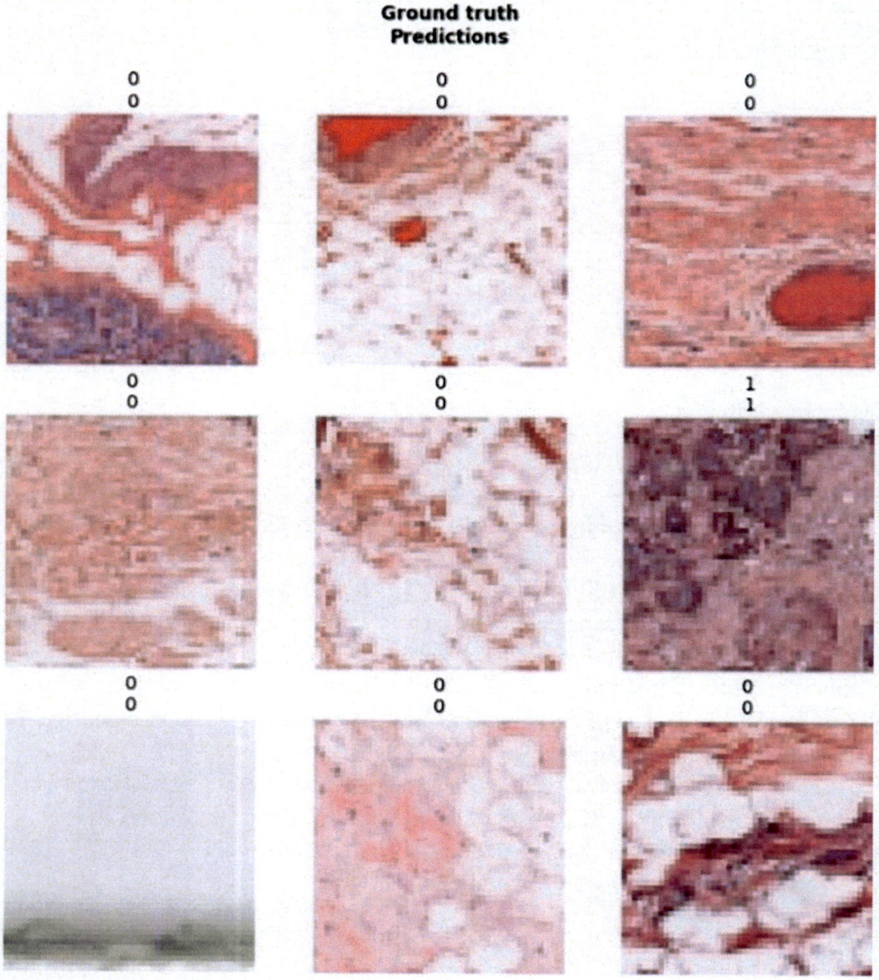

Fig. 4 Result of breast cancer detection

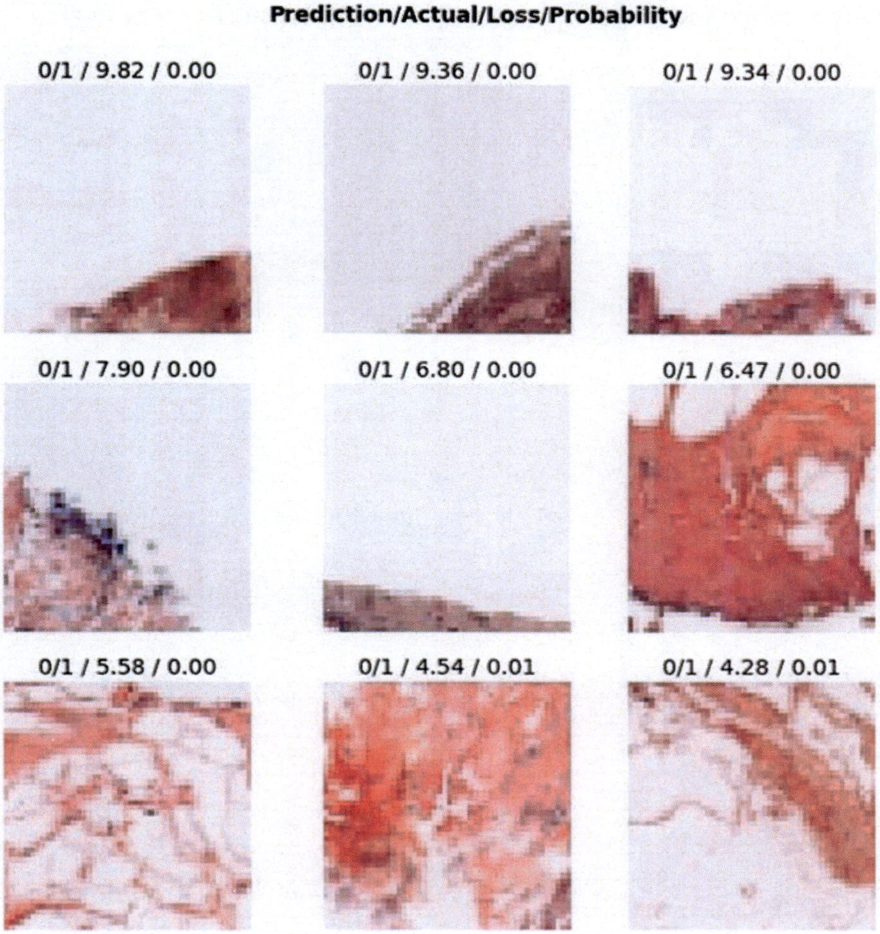

Fig. 5 Prediction/actual/loss/probability

The false-positive rate is equally significant—anyone doesn't need to erroneously categorize somebody as "Cancer positive" and then subject them to agonizing, exorbitant, and protruding treatments when they don't need them. To get a stronghold of how the model is doing on false positives and false negatives, one can plot the confusion matrix as shown in Fig. 6. It provides more emphasis on predicted and actual instances.

The accuracy of test data is 86.9% shown in Table 1. Other parameters are shown by the following classification report in Table 2.

Fig. 6 Confusion matrix

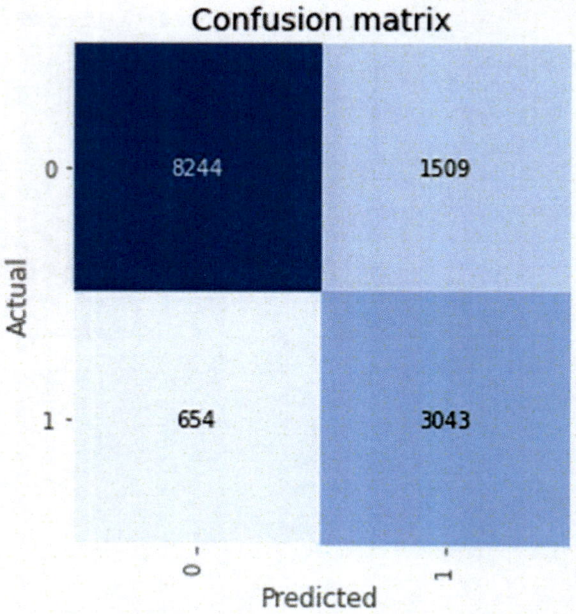

Table 1 Classification report

	Precision	Recall	$F1$-score	Support
0	0.94	0.86	0.90	9753
1	0.70	0.85	0.77	3697
Accuracy			0.86	13,450
Macro-average	0.82	0.86	0.83	13,450
Weighted average	0.87	0.86	0.86	13,450

5 Conclusion

So finally designed a model which is 86.9% accurate and has got an improved recall for both in case of the positive and the negative classes. Anyone still could have trained the network for more. Here, the model is trained with 7 epochs and it took approximately two hours. More fine-tuning could have been done. Sophisticated data augmentation and resolution techniques could have been applied as a Future Scope.

Table 2 Key comparisons

Property	Breast cancer detection using image processing	Breast cancer detection using deep learning
Steps	Pre-processing, image enhancement, algorithm for image analysis, segmentation, feature extraction, and classification	Deep neural network architecture with 48 layers of convolution, pooling, and flattering with full connection layer to the output layer for classification
Implementation support	Image processing libraries	Machine learning and deep learning libraries
Scope for improvements	The outputs and accuracies obtained by simple image processing techniques are fixed. The only way to improve the classification is to improve the algorithm	Deep learning models have a huge scope for improvements in accuracy as they train on more images and receive feedback
Accuracy ranges	65–80%	80–99%
Need for validation	The outputs may need to be validated by a physician. They majorly help in providing more insights for diagnosis	No manual intervention is required for highly accurate models
Initial wait period	There is no waiting period involved as the architecture does not need any training on the image. It can dynamically give the output based on the algorithm	The model needs to be trained before it can be used for detection. This time may vary from hours to a few days
Cost Involved	The computation cost is low	The computation cost is high as the model keeps improving as the dataset keeps increasing

References

1. Vaka AR, Soni B, Reddy S (2020) Breast cancer detection by leveraging machine learning. ICT Express 6(4):320–324. ISSN 2405-9595
2. Dabeer S, Khan MM, Islam S (2019) Cancer diagnosis in histopathological image: CNN based approach. Inf Med Unlocked 16:100231. https://doi.org/10.1016/j.imu.2019.100231. ISSN 2352-9148
3. Alanazi SA, Kamruzzaman MM, Nazirul Islam Sarker MD, Alruwaili M, Alhwaiti Y, Alshammari N, Siddiqi MH (2021) Boosting breast cancer detection using convolutional neural network. J Healthcare Eng 5528622:11
4. Saber A, Sakr M, Abo-Seida OM, Keshk A, Chen H (2021) A novel deep-learning model for automatic detection and classification of breast cancer using the transfer-learning technique. IEEE Access 9:71194–71209
5. Epimack M et al (2021) Breast cancer segmentation methods: current status and future potentials. BioMed Res Int 2021:9962109. https://doi.org/10.1155/2021/9962109
6. Han Z, Wei B, Zheng Y, Yin Y, Li K, Li S (2017) Breastcancer multi-classification from histopathologycal images with structured deep learning model. Sci Rep 7(1):4172

7. Salama WM, Aly MH (2021) Deep learning in mammography images segmentation and classification: automated CNN approach. Alexandria Eng J 60(5):4701–4709. https://doi.org/10.1016/j.aej.2021.03.048. ISSN 1110-168
8. Joshi SA, Bongale AM, Bongale AM (2021) Breast cancer detection from histopathology images using machine learning techniques: a bibliometric analysis. Library Philos Pract (e-J) 5376. https://digitalcommons.unl.edu/libphilprac/5376
9. Espanhol FA, Oliveira LS, Petitjean C, Heutte L (2016) Breast cancer histopathological image classification using convolutional neural networks. In: 2016 International joint conference on neural networks, IJCNN 2016, Vancouver, BC, Canada, July 24–29, 2016, pp 2560–2567
10. Solanki YS, Chakrabarti P, Jasinski M, Leonowicz Z, Bolshev V, Vinogradov A, Jasinska E, Gono R, Nami M (2021) A hybrid supervised machine learning classifier system for breast cancer prognosis using feature selection and data imbalance handling approaches. Electronics 10:699. https://doi.org/10.3390/electronics10060699
11. Agarap AF. On breast cancer detection: an application of machine learning algorithms on the Wis consin diagnostic dataset. CoRR abs/1711.07831
12. 2015. Breast cancer diagnosis using genetically optimized neural network model. Expert Syst Appl 42(10):4611–4620. https://doi.org/10.1016/j.eswa.2015.01.065
13. Ashraf OI, Siti MS (2018) Intelligent breast cancer diagnosis based on enhanced Pareto optimal and multilayer perceptron neural network. Int J Comput Aided Eng Technol Inderscience 10(5):543–556
14. Yu X, Zhou Q, Wang S, Zhang Y-D (2021) A systematic survey of deep learningin breast cancer. Int J Intell Syst 37(1):152–216. https://doi.org/10.1002/int.22622
15. Khamparia A, Bharati S, Podder P et al (2021) Diagnosis of breast cancer based on modern mammography using hybrid transfer learning. Multidim Syst Sign Process 32:747–765. https://doi.org/10.1007/s11045-020-00756-7

Voltage Stability Analysis for Distribution Network Using D-STATCOM

Babita Gupta, Rajeswari Viswanathan, and Guruswamy Revana

Abstract The evolution of transmission networks is driven by the complexity of today's electricity systems. This could be owing to the interconnection of multiple electrical utilities on a huge scale. D-STATCOM is being used to minimize harmonics in load may be on the horizon. Even if there is widespread demand for D-STATCOM, the expense of equipment may be a barrier to its widespread use. The use of D-STATCOM can be justified extremely well if various control functions are included on the same piece of equipment. This paper presents a D-STATCOM controller for any sort of system for Voltage stability analysis, as well as a Simulink examination of various power transmission challenges.

Keywords D-STATCOM · FACTS · Newton–Raphson · Gauss–Seidel

1 Introduction

Electricity demand is steadily increasing across all sectors, including residential, commercial, and industrial. Reactive power is also in great demand due to the majority of loads being reactive and imbalanced. In addition, because reactive power adjustment is in high demand inductive loads in power systems necessitate trailing reactive power. It is more cost effective to provide reactive power support near the load in the distribution system. To transport active power via wires while keeping a constant voltage, reactive power is required. As the reactive load on the distribution system grows, the voltage profile of the network decreases. Low voltage profiles will be found in a major portion of power networks, leading in voltage collapse. When the demand for reactive power rises, feeder loss rises, and the distribution network's power flow capacity falls. Flexible AC Transmission Systems—FACTS devices are increasingly being used by researchers to increase transmission network voltage stability. This may be accomplished by rewiring the network. D-STATCOM [1, 2] is a tool that improves the voltage profile of distributed wind energy. D-STATCOM

B. Gupta (✉) · R. Viswanathan · G. Revana
BVRIT HYDERABAD College of Engineering, Hyderabad, India
e-mail: babitagupta@bvrithyderabad.edu.in

K. A. Reddy et al. (eds.), *Proceedings of Fourth International Conference on Computer and Communication Technologies*, Lecture Notes in Networks and Systems 606,
https://doi.org/10.1007/978-981-19-8563-8_60

provides better voltage profiles at each bus of the distribution system, as well as increased reactive loading capabilities in all loading conditions, improved system stability, and reduced reactive power flow to reduce line losses. Using a D-STATCOM controller and Simulink, this work aims to [3] construct an indicator for analysing the voltage stability of distribution systems. The impact of load variation based on voltage is investigated in this research. With the presence of DG and D-STATCOM, the voltage critical stability limits will be calculated using the continuation load flow technique.

2 Methodology

Two methods can be used to perform a load flow analysis hypothetically.

2.1 Newton–Raphson Method

This is a method for judging successive superior estimations in order to obtain a real-value estimate.

$$x{:}f(x) = 0$$

The Newton–Raphson approach is explained in the following way: The digression line is used to approximate the purpose, and the x-intercept of this digression line is calculated. The procedure can be repeated because often, the x-intercept is a better approximation of the function's root than the initial assumption. Because collinear scaling and local quadratic approximation enhance quasi-Newton techniques when the function value is not fully employed in the Hessian matrix. This research creates a new collinear scaling factor since the collinear scaling factor in this study appears to be single. The method's global convergence is proven, and an improved collinear scaling technique based on local quadratic approximation is given to increase the method's stability. In addition, numerical results of neural network training using the unique collinear scaling strategy are presented which reveals that it is far more efficient than the conventional algorithm. Newton's technique is a numerical analytic method for discovering successively improved approximations to a real-valued function's roots (or zeroes).

$$x{:}f(x) = 0$$

The Newton–Raphson technique is implemented in one variable as follows: Given a function x defined over the reals and its derivative, we begin with a first guess of $\times$ 0 for a root of the function f. $\times$ 1 is a better approximation if the function fulfils all of the assumptions set forth in the formula's derivation. The phrase is

$$x_1 = x_0 - \frac{f(x_0)}{f'(x_0)}$$

2.2 G–S Method

Gauss–Seidel (G–S) is a more advanced variant of Gauss iterative. The change was made to keep the number of iterations to a bare minimum, which is appropriate for studying power flow in small-scale power systems. A solution vector is first anticipated based on the real-time data. To acquire the updated value of a given variable, one of the iterations involves substituting the current values of the other variables into one of the modelling equations. In respect to this variable, the solution vector is updated in real time, and the method is repeated until one iteration is complete. The solution vector will be iterated in this manner until it converges to the given precision.

2.3 Principle of D-STATCOM Controller

D-STATCOM, or Distribution Static Synchronous Compensator, is a FACTS instrument that prevents voltage instability, [4] flickers, and other problems from occurring when distribution lines are disturbed by injecting reactive power into the line to compensate for the reactive power [5]. There are two control modes in D-STATCOM: voltage control and current control. In the first mode of control, independent of the load or source or current, the dc bus voltage is made sinusoidal. In the second method of control, the source side currents must be balanced sinusoids, based on the generalized instantaneous power concept, which may be applied with sinusoidal or non-sinusoidal signals, three-phase power systems, balanced or unbalanced, with or without zero sequence components. The reference current is calculated using the theory of instantaneous power, which takes into account harmonics in current in deviated lines as well as reactive power. The controller is depicted as a drawing in Fig. 1. This category includes, for example, voltage source converters. A coupling transformer is a distribution network with DC energy storage device connected in shunt. The voltage source converter transforms the storage device's DC voltage to a three-phase AC output voltage. The reactance of transformer coupling integrates and phases these voltages with the AC system. In order for D-STATCOM to function properly, the following conditions must be satisfied.

(1) If $V_1 = V_{pcc}$, D-STATCOM will not draw or create any reactive power since the power exchange between the controller and the grid is zero.
(2) If $V_1 > V_{pcc}$, then D-STATCOM provides phantom power by acting as an inductive reactance linked at its terminals.

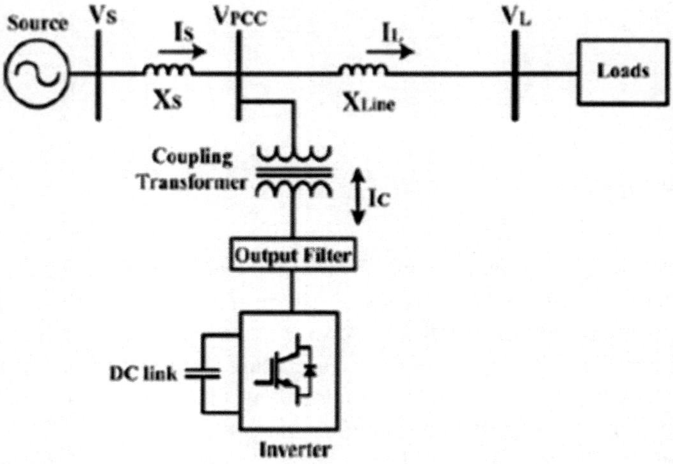

Fig. 1 Illustration of D-STATCOM controller

(3) $V_1 < V_{\text{pcc}}$, the gadget absorbs inductive reactive power after D-STATCOM has acted as capacitive reactive power.

Figure 2 shows the Simulink model of the Distribution Static Synchronous Compensator, D-STATCOM which manages on a 25-kV distribution network, the voltage is (21 and 2 km). The power factor on Bus B2 is rectified using a shunt capacitor. A 25kV/600V transformer connects the 600-volt load to Bus B3. The transformer depicts a plant that, like an arc furnace, absorbs constantly fluctuating currents, causing voltage fluctuation. The apparent power of the variable load current is modulated at a frequency of 5 Hz, with a range of 1 MVA to 5.2 MVA and a 0.9 lagging power factor and has a range of 1 MVA to 5.2 MVA. This load variation will be used to assess the D-STATCOM's efficacy in reducing voltage flicker. By absorbing or creating reactive power, the D-STATCOM regulates the voltage on Bus B3. Reactive power is delivered by establishing a secondary voltage in phase with the primary voltage through the coupling transformer's leaky reactance (network side). A voltage-sourced PWM inverter is used to generate this voltage. The D-STATCOM functions as an inductor, absorbing reactive power when the secondary voltage is less than the bus voltage. The D-STATCOM produces reactive power when the secondary voltage exceeds the bus voltage.

3 Simulation and Results

The study focuses on constructing an indicator for analysing the voltage stability of distribution networks using the Simulink platform and the D-STATCOM controller.

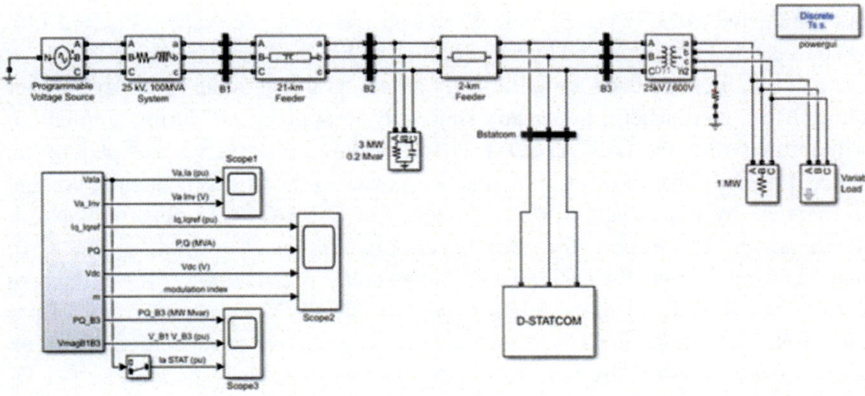

Fig. 2 Simulink model using D-STATCOM

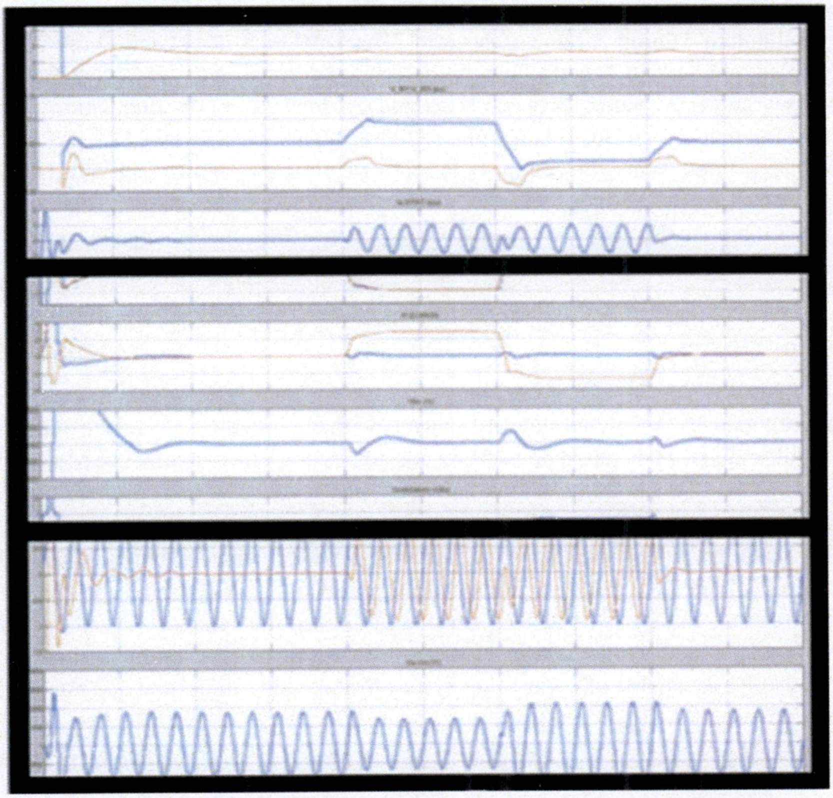

While load variations are kept constant, the controller's reaction to changes in the source voltage is detected with a change in time of [Ton Toff] = [0.15 1] *100 times longer than the simulation period. The internal voltage of the 25-kV equivalent line is changed using the source voltage. To keep the D-STATCOM floating at first, the input is adjusted to 1.077 pu ($B3$ voltage = 1 pu and reference voltage $V_{ref} = 1$ pu). The source voltage is improved by 6%, dropped by 6%, and increased by 6% in three phases of 0.2, 0.3, and 0.4 s and ultimately obtain its initial value (1.077 pu). After around 0.15 s of transient, the steady state will be approached during simulation. For the time being, the D-STATCOM will be declared inactive with respect to the source. It doesn't absorb or provide reactive power to the grid. The source voltage will increase by 6% after $t = 0.2$ s passes. The D-STATCOM absorbs network reactive power. The source voltage is lowered by 6%, or $Q = 0$, at $t = 0.3$ s. To maintain a 1 pu voltage, the D-STATCOM must create reactive power (Q shifts from +2.7 to −2.8 MVAR). The modulation index of the PWM inverter increases from 0.56 to 0.9 when the D-STATCOM switches from inductive to capacitive operation (output 4). As a result, the voltage of the inverter rises correspondingly. On D-STATCOM current, there is a rapid reversal of reactive power (output 1). During voltage flickering mitigation, the source voltage is unaffected, but the variable load is modulated, allowing the D-STATCOM [6–8] to detect voltage flickering mitigation. Bus B3 (output 1) voltage, as well as Buses $B1$ and $B3$ voltage (output 2), When the modulation Timing parameter is set to [Ton Toff] = [0.15 1] and Q regulation is enabled, changes in P and Q can be detected. In the absence of D-STATCOM, the bus voltage $B3$ varies between 0.96 and 1.04 pu (+/− 4% fluctuation). The voltage fluctuations on Bus B3 are reduced to less than 0.7% when the D-STATCOM controller is fitted. When the voltage falls below a certain threshold, the D-STATCOM compensates by injecting a 5 Hz-modulated reactive current (output 3) that varies between 0.6 pu capacitive and 0.6 pu inductive depending on the voltage.

4 Conclusions

The work focuses on voltage stability analysis with the D-STATCOM controller and Simulink analysis of different power transmission issues. The performance of a single D-STATCOM that can do both load balancing and reactive power compensation is being investigated. The controller's performance is assessed in a range of operational scenarios, with each case's resilience being documented. The D-STATCOM improves voltage regulation in the power system while also helping to lower fault current during fault circumstances since it stabilizes the reactive power need in power systems and works as a controlled reactive source. The findings show that the controller is capable of controlling system voltage in both normal and abnormal conditions.

References

1. Kumar P, Kumar N (2012) D-STATCOM for stability analysis. IOSR J Electr Electron Eng 1(2):2278–1676
2. Patel AA, Karan B (2017) Application of DSTATCOM for voltage regulation and power quality improvement. Int J Res Edu Sci Methods 5(3)
3. Hannan MA. Effect of DC capacitor size on D-STATCOM voltage regulation performance evaluation. Przegląd Elektrotechniczny. ISSN 0033-2097
4. Swaminathan HB (2017) Enhancing power quality issues in distribution system using D-STATCOM. Int J Recent Res Aspects 4(4):356–359. ISBN 2349-7688
5. Palod A, Huchche V (2015) Reactive power compensation using DSTATCOM. Int J Electr Electronic Data Communication 2320–2084:21–24 Special Issue
6. Singh B, Solanki J (2006) A comparative study of control algorithms for DSTATCOM for load compensation. In: IEEE International conference on industrial technology, Mumbai, pp 1492–1497
7. Singh B, Solanki J (2009) A comparison of control algorithms for DSTATCOM. IEEE Trans Ind Electron, 56(7):2738–2745. ISBN 1557-9948
8. Singh B, Jayaprakash P, Kothari DP, Chandra A, Al Haddad K (2014) Comprehensive study of DSTATCOM configurations. IEEE Trans Ind Inf 10(2). ISBN 1941-0050

E-Dictionosauraus

J. Naga Vishnu Vardhan, Ramesh Deshpande, Sreeya Bhupathiraju, K. Padma Mayukha, M. Sai Priyanka, and A. Keerthi Reddy

Abstract Reading is an essential part in everyone's life and is important to build a good reading habit. Be it a novel, a textbook, a research paper, or a newspaper and one rely on these modes know what's happening in the society or gain knowledge. Reading a favorite book or an article is one of the best ways to indulge ourselves to relax our mind and also to improve vocabulary, concentration, and many. Often, while reading a book or a newspaper, etc., usually find words that reader is not aware of. Also, in case of a person started learning, a language often faces this difficulty. This makes it challenging to understand the meanings of the words. There are many words that are never heard and generally feel to have a dictionary to find the meaning. Using/carry a physical dictionary every time when a reader comes across unknown words may not be possible. Moreover, they need to search from the first alphabet of a particular word to find the word and its meaning. More importantly, nowadays, more words are added day to day and all words may not be available in the dictionary compared to available in online search. This is exactly where E-Dictionosauraus comes to wonderful use. This smart web application using optical character recognition (OCR) identifies the location of the word pointed by the user

J. Naga Vishnu Vardhan (✉) · S. Bhupathiraju · K. Padma Mayukha · M. Sai Priyanka · A. Keerthi Reddy
Department of ECE, BVRIT HYDERABAD College of Engineering for Women, Bachupally, Hyderabad, India
e-mail: vishnu.j@bvrithyderabad.edu.in

S. Bhupathiraju
e-mail: 17wh1a0409@bvrithyderabad.edu.in

K. Padma Mayukha
e-mail: 17wh1a0424@bvrithyderabad.edu.in

M. Sai Priyanka
e-mail: 17wh1a0434@bvrithyderabad.edu.in

A. Keerthi Reddy
e-mail: 17wh1a0439@bvrithyderabad.edu.in

R. Deshpande
Associate Professor, Department of ECE, B V Raju Institute of Technology, Hyderabad, India
e-mail: ramesh.deshpande@bvrit.ac.in

K. A. Reddy et al. (eds.), *Proceedings of Fourth International Conference on Computer and Communication Technologies*, Lecture Notes in Networks and Systems 606,
https://doi.org/10.1007/978-981-19-8563-8_61

(usually, with the back of a pencil). It then extracts the word from the text, gathers its meaning. Additionally, it also gives the user its synonyms and antonyms. Further, it delivers the output in the form of audio that helps out how to pronounce. So, with the help of E-Dictionosauraus, one can effortlessly get their work done quickly.

Keywords Dictionary · Optical character recognition · Image recognition

1 Introduction

While reading a book or magazine or article english is widely used language across the globe. It has huge number of words and is quite impossible for a person to know the meaning of each. The dictionaries that are currently available are either a physical dictionary or a search engine such as Google, Firefox, etc., both of these options are wearisome. Searching for the meaning in a physical dictionary is a tedious job, and searching in the latter is time-consuming. Also, neither of them has readily available synonyms and antonyms. Further, effective usage of dictionary depends on user-friendliness of dictionaries and also on skills of the users.

Some electronics such as the kindle have an inbuilt dictionary. The software available in this technology gives the meaning of the word searched by the reader. Even though it gives the meanings of words on the system, it cannot help in the case of books or newspapers, but, the majority of the people use physical mode to read. So, we need an application to assist us in this situation. We need software that not only saves time but also makes the job simpler and effortless. That is exactly what "E-Dictionosauraus" do. With this software, reading can be made joyous and uninterrupted.

Using E-Dictionosauraus, when a reader comes across a foreign word, he/she can directly get the meaning of it, irrespective of where they are reading. The reader uses the back of the pencil to point to the word they are unaware. The system captures the image of the page and extracts the word that is being pointed. Inbuilt libraries such as Pydictionary and Wordnet help to draw out the meaning of the word along with its synonyms and antonyms. The output is displayed on the screen. In addition to this, an audio output of the meaning is also provided. This helps user to know how to pronounce a word.

2 Literature Survey

People across the globe developed different types of e-Dictionaries but are restricted to a narrow application like language of a particular country and few discussed about the usage of e-dictionary. In the paper [1], it is mainly focused and discussed on dictionary usage skills required to use modern digital dictionaries and how as we move from the usage of paper to electronic dictionaries.

An e-dictionary is developed to use with textbook Maharah al-Qiraah textbook at matriculation center is presented in [2]. The usage and efficacy of e-dictionaries of Japanese language by Japanese language learners was discussed in [3] along with the studies related to investigating the use by L2 learners of Japanese, comparative study examining the existing e-glossaries to explore the optimal level of support of reading Japanese e-texts.

Now days, with the development of technology and comfort in usage, pocket electronic dictionaries are gaining popularity. The colleges and universities of Chinese made an impact on the Chinese dictionary along with the study that compares patterns of use and perceptions of PEDs and paper dictionaries (PDs) were presented in [3]. The impact of usage of dictionary on vocabulary under the PED and PD conditions, different patterns of usage between PEDs and PDs, advantages and limitations are discussed [4].

Study related to explore the role of dictionary use in L2 vocabulary learning in reading context discussed in [5] involved the use of English-Chinese bi lingualized dictionaries with the help of paper BLD (PBLD) or an electronic BLD (EBLD), or without access. The latter showed advantage over the former for vocabulary retention. The intelligent English electronic dictionary system discussed in [6] is related to design and implementation of electronic dictionary in accord to the supremacy of Internet of Things. Software architecture, design, and implementation of client–server related technologies integrated in the development process of the electronic dictionary application are discussed.

The work presented in this paper E-Dictionosauraus address the usage of e-dictionary, for the international language "English" widely used across the globe. It also provides the technologies required along with the advantages. This smart web application using optical character recognition identifies the location of the word pointed by the user. Then, the pointed word will be extracted from the text and provides its meaning. Additionally, it also gives the user the synonyms and antonyms of the word detected. The application developed also speaks out the word and its meaning so that it helps how to pronounce the word. So, with the help of E-Dictionosauraus, one can effortlessly get their work done quickly.

3 Proposed System

Optical character recognition (OCR) is a technique, which is used to recognize the text from images and to converts into editable text form. Those images can be handwritten text, printed text such as documents, receipts, name cards, books, and newspapers. OCR is a two-step process. In the initial step of "text detection", the text in the image will be detected. The second step called "text recognition" is one where the detected text will be extracted from the image. Performing these two steps together is how text from the image is extracted (Fig. 1).

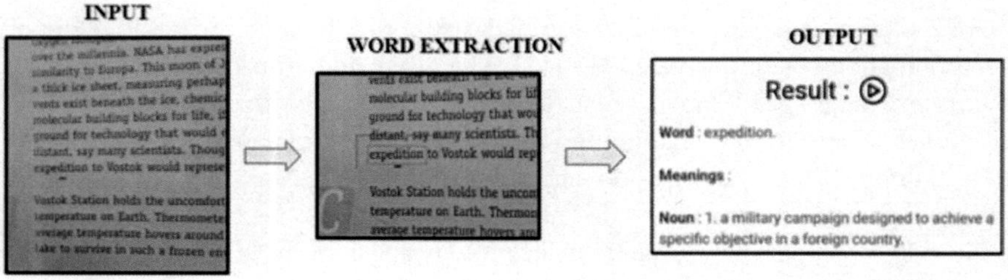

Fig. 1 Steps of OCR technique

3.1 Tesseract OCR

Tesseract is an open-source optical character recognition engine developed by Google. It cannot only detect more than 100 languages and can also process right-to-left literatures such as Arabic. The Version 4 adds LSTM and the ability to detect many additional languages, which made the total languages that can detect up to 116.

3.2 Color Spaces

The most commonly used color space—RGB is represented as their red, green, and blue components. RGB describes the color as a tuple of three components. Each component can take a value in between 0 and 255. The tuple (255, 255, and 255) represents white color, the tuple (0, 0, and 0) represents black color, and the tuple (255, 0, and 0) represents a red color. RGB is one of the five major color space models. There are so many color spaces because different color spaces have their unique purposes. HSV is a representation of hue, saturation, and brightness. These characteristics are specifically useful for identifying contrast in images.

3.3 Working

Firstly, the image of the page along with a pointer pointing to the word whose meaning is to be extracted is captured. This captured input image will be given as input to the application. The application extracts the color of the pointer, which is pointing to the word in the input image using the color detection method. After extracting the location of the pointer, a rectangular box with fixed length and breadth is drawn and the image will be cropped according to the rectangular box. A few image processing methods are applied on the cropped image like converting the image into grayscale and thresholding.

The found nearest contour will be taken as input to the by tesseract function, which extracts the word from the image. Then, the meaning, synonyms, and antonyms of

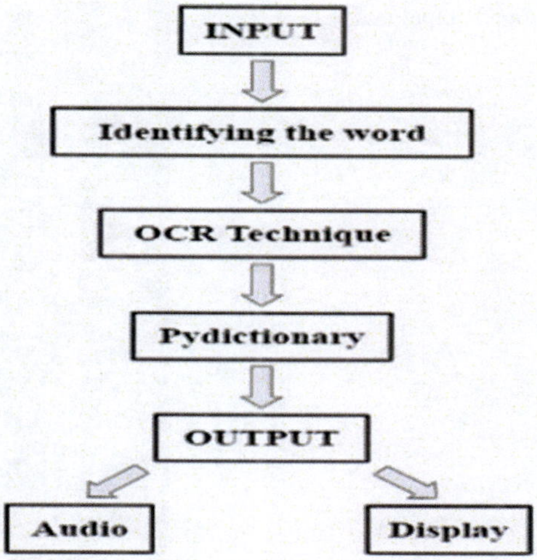

Fig. 2 Flow chart of E-Dictionosauraus

the extracted word are displayed and produced as audio on the output. The meaning, synonyms, and antonyms of the word are extracted using Pydictionary and Wordnet library (Fig. 2).

4 Results and Discussion

This section discusses the results for various test cases to check the outcome of the system designed in successfully pointing the targeted word and providing its meaning along with antonyms and synonyms with voice pop-up.

4.1 CASE (i): Example 1

Consider an image shown below where the image of the page is captured initially. Then, the word pointed is extracted.

Figure 3 shows the captured input picture that shows a word "successful" pointed with red color, whose meaning is to be produced. Figure 4 shows, the word "successful" which is to be extracted is detected and highlighted with a rectangular box.

Figure 5 is displaying the meaning, synonyms, and antonyms of the extracted word "successful". Audio output can also be produced by clicking on the play button, which is beside the result.

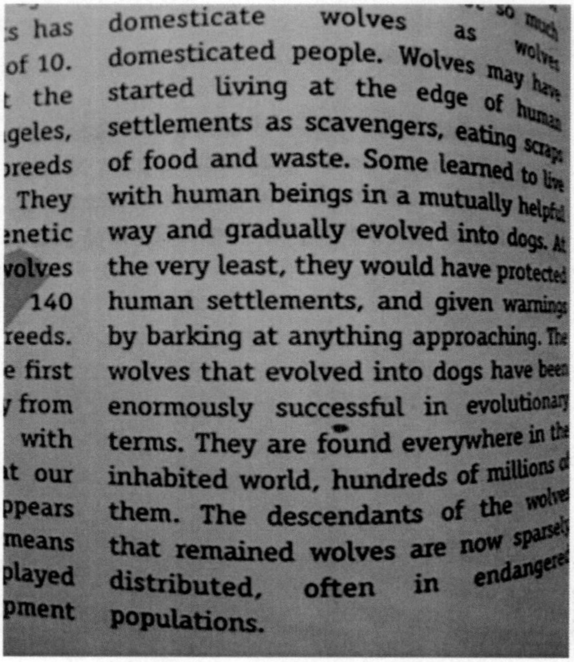

Fig. 3 Input image 1

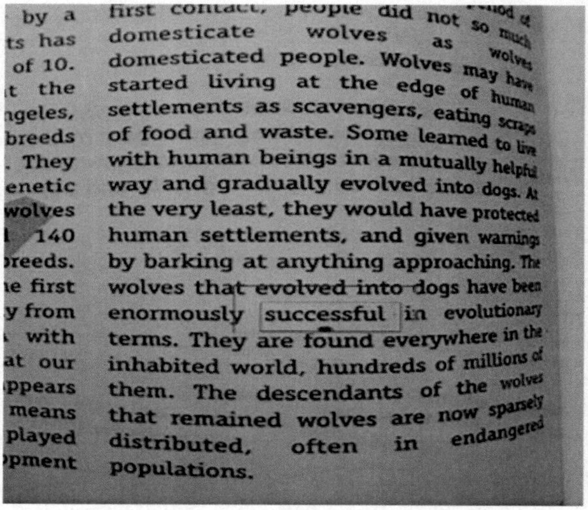

Fig. 4 Image in which the word is detected from input image 1

4.2 *Example 2*

The process is repeated to check the functionality of the application developed with another word.

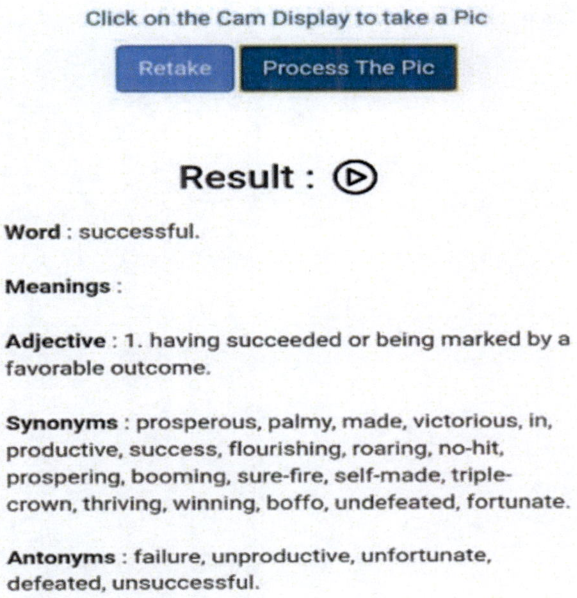

Fig. 5 Output of input image 1displaying the meaning along with synonyms and antonyms

Figure 6 shows the captured input picture that shows a word "expedition" pointed with red color, whose meaning is to be produced. Figure 7 shows, the word "expedition" which is to be extracted is detected and highlighted with a rectangular box.

Figure 8 is displaying the meaning, synonyms, and antonyms of the extracted word "expedition". Audio output can also be produced by clicking on the play button, which is beside the result.

4.3 *CASE (ii)*

From Fig. 9, one can observe that instead of giving a valid word or a text or even a red pointer is not found and some image was shown to the detector, then we get the output as "No Valid Word Detected. Please Try Again". This confirms the efficacy of the system designed. It can be concluded that only if a valid input is detected, the required output is generated, else no.

5 Conclusion

The purpose of the project is to help readers to find the meaning of the word along with their synonyms and antonyms instantaneously using the application developed. It makes their tasks easier and comfortable. The application has been tested successfully

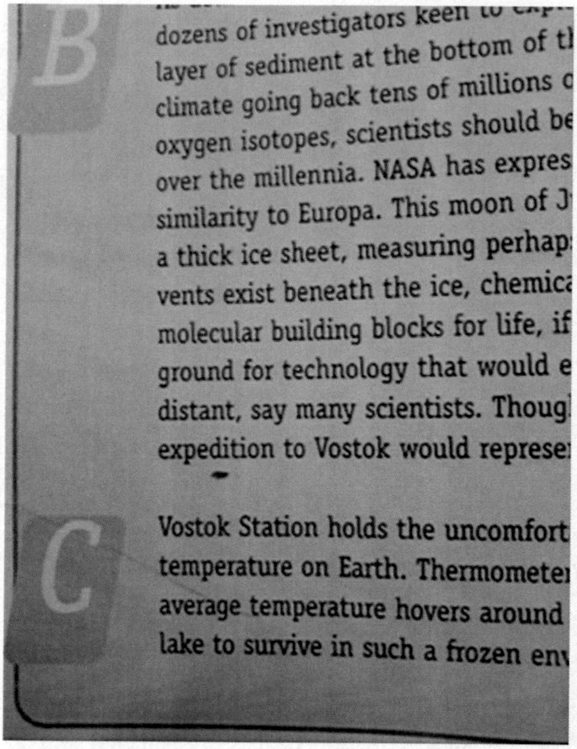

dozens of investigators keen
layer of sediment at the bottom of t
climate going back tens of millions
oxygen isotopes, scientists should b
over the millennia. NASA has expres
similarity to Europa. This moon of J
a thick ice sheet, measuring perhap
vents exist beneath the ice, chemic
molecular building blocks for life, if
ground for technology that would e
distant, say many scientists. Thoug
expedition to Vostok would represe

Vostok Station holds the uncomfort
temperature on Earth. Thermomete
average temperature hovers around
lake to survive in such a frozen en

Fig. 6 Input image 2

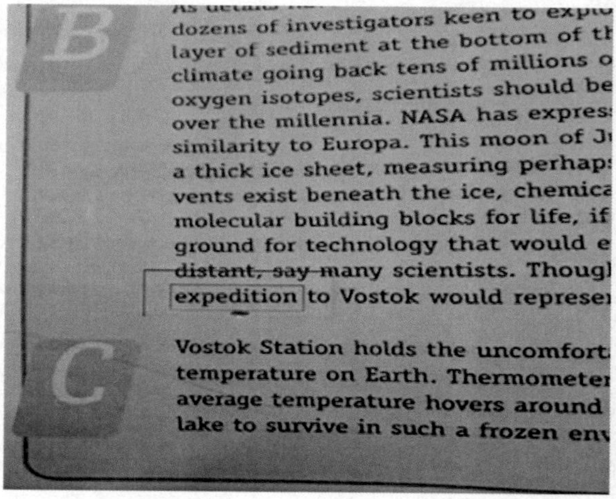

dozens of investigators keen to exp
layer of sediment at the bottom of t
climate going back tens of millions o
oxygen isotopes, scientists should be
over the millennia. NASA has expres
similarity to Europa. This moon of J
a thick ice sheet, measuring perhap
vents exist beneath the ice, chemica
molecular building blocks for life, if
ground for technology that would e
distant, say many scientists. Thoug
expedition to Vostok would represe

Vostok Station holds the uncomfort
temperature on Earth. Thermomete
average temperature hovers around
lake to survive in such a frozen en

Fig. 7 Image in which the word is detected (of input image 2)

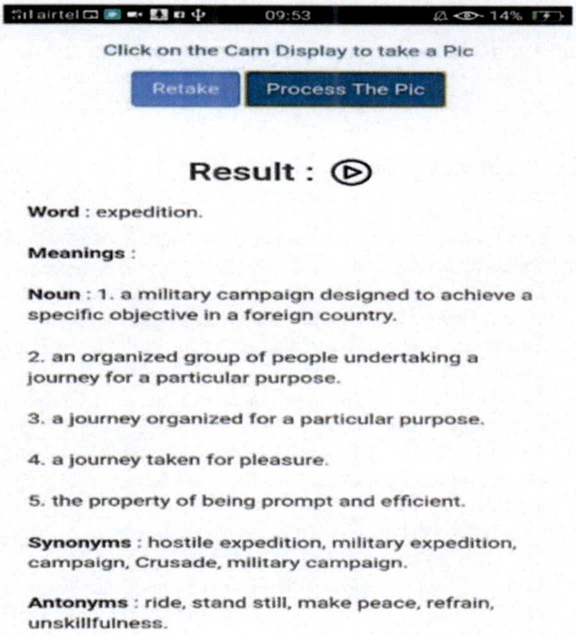

Fig. 8 Output of input image 2 displaying the meaning, synonyms, and antonyms

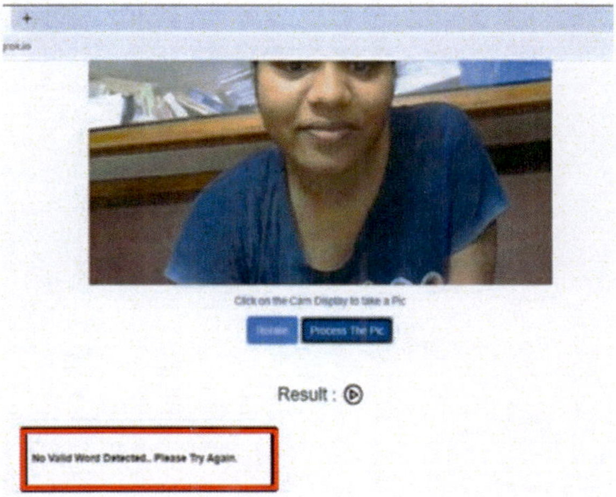

Fig. 9 Output image of input image 3

for various words where the outcome is effective. In case of non-text, like where image is given as input, it displays that no valid word is detected. This application developed helps the reader to get the meaning, synonyms, and antonyms of the word they are looking for easily and also saves time. The application presented also produces audio output by which the user can also know how to pronounce the word. The application developed is easy to use, saves time as we do not have to manually search for the word in a physical dictionary or on the Internet. However, pointer needs to be precise, user must be properly aware of the working of the device.

References

1. Lew R (2013) From paper to electronic dictionaries: evolving dictionary skills
2. Omar CAMBC, Dahan HBAM (2011) The development of E-Dictionary for the use with Maharah Al-Qiraah textbook at a matriculation Centre in a University in Malaysia. Omar Matriculation Center, International Islamic University, Petaling Jaya, Selangor, Malaysia. Faculty of Education, University of Malaya, Kuala Lumpur, Malaysia. TOJET: Turkish Online J Edu Technol 10(3)
3. Toyoda E (2016) Usage and efficacy of electronic dictionaries for a language without word boundaries. The University of Melbourne, Australia. EUROCALL Rev 24(2)
4. Chen Y (2010) Dictionary use and EFL learning. A contrastive study of pocket electronic dictionaries and paper dictionaries. Int J Lexicogr 23(3):275–306. https://doi.org/10.1093/ijl/ecq013
5. Chen Y (2012) Dictionary use and vocabulary learning in the context of reading. Int J Lexicogr 25(2):216–247. https://doi.org/10.1093/ijl/ecr031
6. Yong W (2021) Design and implementation of intelligent english electronic dictionary system based on internet of things. Wirel Commun Mob Comput 2021(5586662):11. https://doi.org/10.1155/2021/5586662

Effective Prediction Analysis for Cardiovascular Using Various Machine Learning Algorithms

M. Shanmuga Sundari, M. Dyva Sugnana Rao, and Ch Anil Kumar

Abstract Healthy life leads to healthy growth in a human's life. Health is more important than all other things in our life. The heart is the backbone of our life. Many types of research are going on in the medical field that finds new treatments in the healthcare industry. Machine learning is an important technology that helps predict the accuracy of disease. Machine learning takes the attributes from the health industry and analyzes the data using many algorithms. Based on the trained data, the experiment will explore the analysis prediction in the given dataset. It helps in many ways for humans in their life. In this research, we predicted cardiovascular disease before the early stage. We collected healthcare data and applied various machine learning algorithms. We analyzed the accuracy of the support vector machine (SVM), logistic regression (LR), and stochastic gradient descent (SGD) algorithms. Finally, research proves support vector machines as the best algorithm for predicting cardiovascular disease in the early stage of our lifespan.

Keywords Cardiovascular disease · Logistic regression · Stochastic gradient descent · Support vector machine

1 Introduction

Cardiovascular disease [1] is a collection of different resources that affect the blood circulation in the body and will affect the heart. This leads to coronary artery disease, heart arrhythmias, and heart failure. The heart patients will disturb their minds and life. Patients need support and encouragement from neighbor relatives and friends.

M. Shanmuga Sundari (✉) · M. Dyva Sugnana Rao · C. A. Kumar
BVRIT HYDERABAD College of Engineering for Women, Bachupally, Hyderabad, India
e-mail: sundari.m@bvrithyderabad.edu.in

M. Dyva Sugnana Rao
e-mail: sugnanarao.m@bvrithyderabad.edu.in

C. A. Kumar
e-mail: anilkumar.ch@bvrithyderabad.edu.in

K. A. Reddy et al. (eds.), *Proceedings of Fourth International Conference on Computer and Communication Technologies*, Lecture Notes in Networks and Systems 606,
https://doi.org/10.1007/978-981-19-8563-8_62

You will find heart-friendly recipes, healthy living tips, and strategies for maintaining a healthy heart. Our goal of informing the public is to improve their heart health and avoid heart disease and stroke. Learn stress management and eating tricks to strengthen your heart and become one less statistic.

Nowadays, predicting heart disease is an essential role in the healthcare industry. It has a more vital role in giving treatment to patients in the early stages. Heart disease is also predictable with various factors and attributes. A healthy lifestyle produces fewer factors to create heart disease. If the patient has a high value of chronic parameters, then we can use machine learning technology to predict the problem earliest in life. Machine learning [2] is the technology that will help the heart disease using human activity and lifestyle. It will give accurate predictions and solutions in the health industry. It saves human resources in terms of heart diseases.

Machine learning is a field of data mining that handles all real-time and dynamic problems efficiently. In the health industry, machine learning is acting as a doctor that helps in many ways for researchers to predict and diagnose diseases. The main goal of this paper is to provide the prediction about heart disease in the early stage of life. It helps to predict whether humans will get heart problems or not at an early stage of our life. Machine learning is a good sense of technology that will analyze and produce accurate results based on a few parameters.

This paper explains the support vector machine [3], logistic regression [4], and stochastic gradient descent algorithms [5]. We implemented the model and applied three algorithms. This research found the best algorithm, which gives the accurate result of heart problems.

2 Literature Survey

Nowadays, various tools and technologies are available in the healthcare domain that leads the industry to the next level [2–5]. It works on machine learning techniques using algorithms such as logistic regression (LR), support vector machine (SVM), and stochastic gradient descent algorithms (SGD). In health care, many studies are going on predicting heart disease earlier. Many laboratories are developing a real-time analysis for heart prediction using machine learning algorithms. SVM is one of the algorithms to find the dependency values between attributes and analyze the disease and acute cardio effect in [3] and blood pressure in [6]. This research discussed cardiac problems in using SVM in [7]. The classical machine learning algorithms analyze with predictions [8]. SGD algorithm [9] uses to predict acute heart problems. Random forest algorithm used to predict cardiovascular disease [10]. SVM has used many research processes for heart disease in [11]. SVM, SGD, and LR are applied to predict the efficiency of survival rates in [12]. The author compared multilayer perceptron SVM to predict cardiac problems [13]. Clinical decision support systems are the collection of incorporated hospital datasets [3–6, 9, 10, 14]. From 2017 onwards, laboratories used to predict the process mentioned in [15]. Feature detection [16]

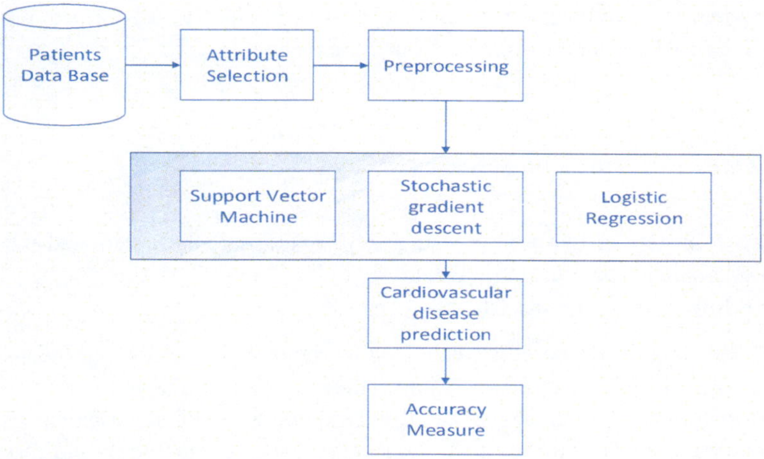

Fig. 1 Proposed model for cardiovascular heart disease

technique applies to detect the required features in the data preprocessing model. The brain tumor prediction [17] was implemented using CNN.

3 Working Procedure for Proposed System

Our proposed system shows to implement the prediction of heart disease. Figure 1 shows the working process flow of the recommended model. The research implemented the model to predict these heart diseases earlier. The dataset has a large volume of patients' details. Attribute selection helps to identify the relevant features to support cardiac disease research. After dataset collection, the preprocessing technique uses to clean our dataset. We implemented three algorithms in machine learning after extracting data from a dataset. The medical database consists of discrete information. Discrete data becomes a complex and tedious task in the prediction.

4 Implementation

4.1 Dataset

We collected a dataset from Kaggle [18] website to research heart disease. The dataset is collected from Residents of Framingham city in Massachusetts, USA. We implemented the prediction of the coronary heart dataset in the early stage. The dataset contains 4240 records and 16 attributes/columns. The collected dataset is in excel format that we changed to comma separated value (CSV) format. We used panda's library in Python to construct our code.

4.2 Data Preprocessing

There are four steps in data preprocessing [19]. We followed these steps and extracted the noiseless data from our original dataset.

Steps followed in preprocessing:

- Data cleaning: The dataset is cleaned up by removing the noisy and inconsistent data. We reduce 17 attributes to 14 attributes to fit in our analysis.
- Data integration: The data is collected and incorporated into the research analysis.
- Data reduction: The attributes were selected which are suitable for our research.
- Data transformation: The final dataset has the required attributes.

4.3 Correlation Matrix Between Attributes

We have 14 attributes after cleaning our dataset. The next step is finding the correlation [20] between two numerical values between the attributes to find an insight of relation. The dataset has many fields, and the matrix has correlation attributes. Figure 2 shows the correlation matrix for heart diseases.

4.4 Box Plot to Check the Outliers

The points that are away from the expected line are called outliers [21]. We need to delete all the outlier values in the matrix. The outliers produce noisy transactions in the process. Figure 3 shows the outlier of each attribute and gets the original dataset for our implementation.

5 Algorithm

5.1 Support Vector Machine (SVM)

SVM is the technique to investigate information and patterns for prediction. SVM includes two stages: the initial step is to prepare an informational index and build a model, and the second step is to utilize the SVM model to predict the testing dataset. Hyperplanes are segregated based on the related attributes. Our idea is to

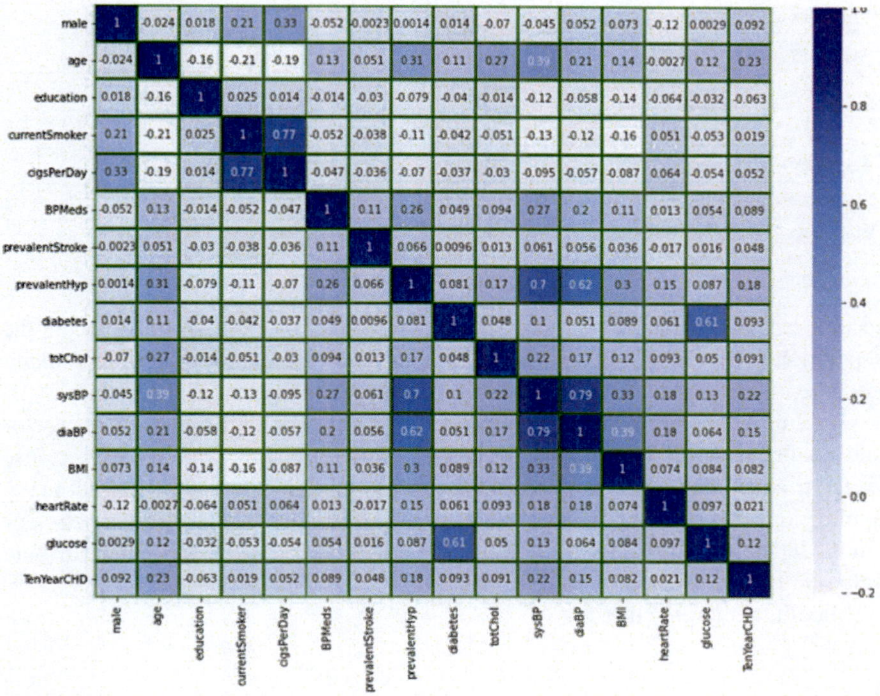

Fig. 2 Correlation matrix

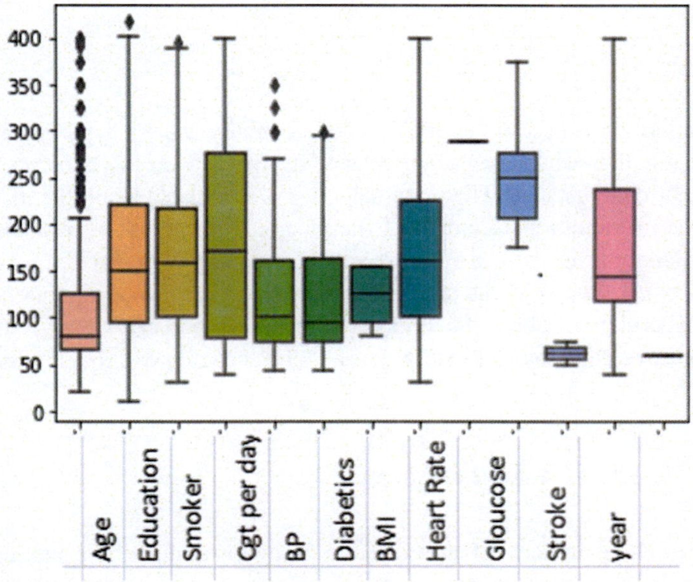

Fig. 3 Outlier's diagram

Table 1 Classification report using SVM

Accuracy attributes	Precision	Recall	F1-score	Support
0	0.86	0.99	0.92	725
1	0.54	0.06	0.10	123
Accuracy			0.86	848
Macro average	0.70	0.52	0.51	848
Weighted average	0.81	0.86	0.80	848

recognize a plane that is the most accurate hyperplane, i.e., we need to locate the separation between information purposes of the two classes. We need to increment, i.e., maximizing the edge separation of the information in particular certainty.

The support vector machine is the approach during non-separable support vector and nonlinear projection without depending on the cost function. We have to use kernel tricks to handle data. SVM is a powerful tool in health care to predict many applications and reduce health issues. It also plays a huge role in medicine composition. In the formula (1)–(3), β is a constant value, and X is the input attribute value.

Maximum margin classifier

$$\beta_0 + \beta_1 X_1 + \beta_2 X_2 > 0 \tag{1}$$

$$\beta_0 + \beta_1 X_1 + \beta_2 X_2 < 0 \tag{2}$$

$$\beta_0 + \beta_1 X_1 + \beta_2 X_2 = 0 \tag{3}$$

SVM algorithm is best to find the decision boundary using a hyperplane. It classifies the prediction value using a hyperplane. The model creates a hyperplane based on the vector data distance. The marginal value is calculated using the distance of vectors. The maximum margin of the hyperplane is the optimal hyperplane. Using this hyperplane concept, we can optimize whether the cardiovascular disease is within the margin or not. So, SVM can predict the accuracy of cardiovascular disease using marginal hyperplanes. Table 1 shows that classification report of SVM algorithm after applied cardiovascular database attributes, and the accuracy was found as 85.61%.

5.2 *Stochastic Gradient Descent*

Stochastic gradient descent (SGD) is used to fit the linear classifiers and regressors with convex loss functions. SGD is mainly used for large-scale and machine learning sparse data, which are encountered in text classification. In the training set, examples ($x1$, $y1$)... (xn, yn) where xi in R are considered as input attributes and yi in R ($yi \in$

Table 2 Classification report using SGD

Accuracy attributes	Precision	Recall	F1-score	Support
0	0.86	0.98	0.92	725
1	0.44	0.09	0.15	123
Accuracy			0.85	848
Macro average	0.65	0.54	0.53	848
Weighted average	0.80	0.85	0.81	848

−1,1 for classification), from that a linear scoring function is calculated as $f(x) = wTx + b$ using other parameters $w \in Rm$ and intercept $b \in R$. Binary classification prediction is calculated using sign function of x.

Using the below formula, we have to reduce the regularized training

$$(w, b) = \frac{1}{n}\sum_{i=1}^{n} L\big(y_{i,} f(x_i)\big) + \alpha R(w) \tag{4}$$

Table 2 shows that classification report of SGD algorithm after applied cardiovascular database attributes, and the accuracy was found to 85.14%.

5.3 Logistic regression

The logistic regression (LR) model works well in which two classes have a label with 0 and 1. The regression will get the minimized distance points in the hyperplane. The sigmoid function is used to predict values to probabilities. The limit of the probability is between 0 and 1. The curve looks like S, so it is called a sigmoid function. The threshold value is calculated using the sigmoid function and the curved line.

The logistic function is defined as

$$\text{logistic}(\eta) = \frac{1}{1 + \exp(-\eta)} \tag{5}$$

The equation gives the probabilities between 0 and 1 using the sigmoid equation. The output enforced from this gives the prediction of cardiovascular disease using is a constant value, and X is the input attribute value.

$$P\big(y^{(i)} = 1\big) = \frac{1}{1 + \exp\Big(-\big(\beta_0 + \beta_1 x_1^{(i)} + \cdots + \beta_p x_p^{(i)}\big)\Big)} \tag{6}$$

Table 3 shows the classification report of the LR algorithm after applied cardiovascular database attributes, and the accuracy found as 85.49%.

Table 3 Classification report using LR

Accuracy attributes	Precision	Recall	F1-score	Support
0	0.86	0.99	0.92	725
1	0.50	0.07	0.13	123
Accuracy			0.85	848
Macro average	0.68	0.53	0.52	848
Weighted average	0.81	0.85	0.81	848

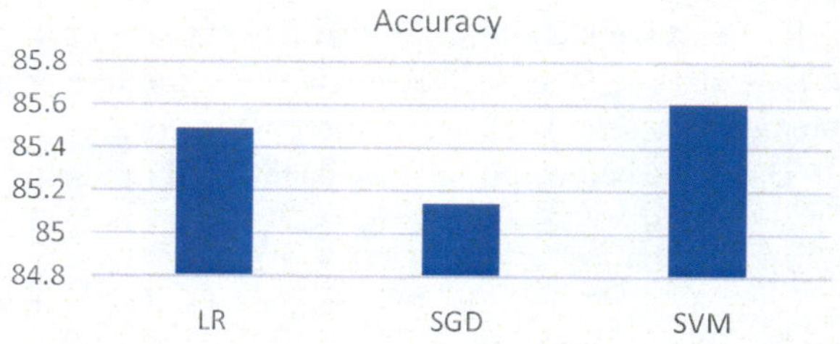

Fig. 4 Comparison bar chart for SVM, LR, and SGD

6 Comparative Analysis

We implemented three algorithms and found the accuracy value of cardiovascular attacks in the early stage. They are different accuracy levels with the given dataset parameters. Figure 4 shows the comparison bar chart for three algorithms. Despite using three algorithms, the support vector machine gave the best accuracy of the cardiovascular attack.

7 Conclusion and Future Scope

We find the optimal values of the parameters to find the minimum possible value of the given cost function. SGD produced 85.14% accuracy. Logistic regression is a supervised classification algorithm. We use a predictive analysis algorithm based on the concept of probability. It estimates the probabilities of the dependent variable and the one or more independent variables using a logistic function. LR algorithm produced 85.49% accuracy. Since we implemented three algorithms, we achieved the highest accuracy using the support vector machine, a supervised learning algorithm. Using the SVM algorithm, we reached an accuracy of 85.61%. In the future, we will add the health image attributes as a parameter in our dataset and find the accuracy of cardiovascular disease.

References

1. Rubini PE, Subasini CA, Vanitha Katharine A, Kumaresan V, Gowdham Kumar S, Nithya TM (2021) A cardiovascular disease prediction using machine learning algorithms. Ann Romanian Soc Cell Biol 904–912
2. El-Ganainy NO, Balasingham I, Halvorsen PS, Rosseland LA (2020) A new real time clinical decision support system using machine learning for critical care units. IEEE Access 8:185676–185687
3. Wu J, Guo P, Cheng Y, Zhu H, Wang XB, Shao X (2020) Ensemble generalized multiclass support-vector-machine-based health evaluation of complex degradation systems. IEEE/ASME Trans Mechatron 25(5):2230–2240
4. Ksantini R, Ziou D, Colin B, Dubeau F (2007) Weighted pseudometric discriminatory power improvement using a bayesian logistic regression model based on a variational method. IEEE Trans Pattern Anal Mach Intell 30(2):253–266
5. Costilla-Enriquez N, Weng Y, Zhang B (2020) Combining Newton-Raphson and stochastic gradient descent for power flow analysis. IEEE Trans Power Syst 36(1):514–517
6. Ghosh P, Azam S, Jonkman M, Karim A, Shamrat FJ, Ignatious E, Shultana S, Beeravolu AR, De Boer F (2021) Efficient prediction of cardiovascular disease using machine learning algorithms with relief and LASSO feature selection techniques. IEEE Access 9:19304–19326
7. Micek A, Godos J, Del Rio D, Galvano F, Grosso G (2021) Dietary flavonoids and cardiovascular disease: a comprehensive dose–response meta-analysis. Molec Nutrition Food Res 65(6):2001019
8. Zhou J, Qiu Y, Zhu S, Armaghani DJ, Li C, Nguyen H, Yagiz S (2021) Optimization of support vector machine through the use of metaheuristic algorithms in forecasting TBM advance rate. Eng Appl Artif Intell 97:104015
9. Isola G, Polizzi A, Alibrandi A, Williams RC, Lo Giudice A (2021) Analysis of galectin-3 levels as a source of coronary heart disease risk during periodontitis. J Periodontal Res 56(3):597–605
10. Sundari MS, Nayak RK (2020) Master card anomaly detection using random forest and support vector machine algorithms. Int J Crit Rev 7(09). ISSN 2394-5125
11. Reddy RR, Ramadevi Y, Sunitha KVN (2017) Enhanced anomaly detection using ensemble support vector machine. In: 2017 International conference on big data analytics and computational intelligence (ICBDAC), March 2017. IEEE, pp 107–111
12. Padmaja B, Prasad VR, Sunitha KVN (2016) TreeNet analysis of human stress behavior using socio-mobile data. J Big Data 3(1):1–15
13. Ji Y, Kang Z (2021) Three-stage forgetting factor stochastic gradient parameter estimation methods for a class of nonlinear systems. Int J Robust Nonlinear Control 31(3):971–987
14. Sun D, Xu J, Wen H, Wang D (2021) Assessment of landslide susceptibility mapping based on Bayesian hyperparameter optimization: a comparison between logistic regression and random forest. Eng Geol 281:105972
15. Dai R, Zhang W, Tang W, Wynendaele E, Zhu Q, Bin Y, De Spiegeleer B, Xia J (2021) BBPpred: sequence-based prediction of blood-brain barrier peptides with feature representation learning and logistic regression. J Chem Inf Model 61(1):525–534
16. Pasha SJ, Mohamed ES (2020) Novel feature reduction (NFR) model with machine learning and data mining algorithms for effective disease risk prediction. IEEE Access 8:184087–184108
17. Khatoon Mohammed T, Shanmuga Sundari M, Sivani UL (2022) Brain tumor image classification with CNN perception model. In: Soft computing and signal processing. Springer, Singapore, pp 351–361
18. https://www.kaggle.com/amanajmera1/framingham-heart-study-dataset/version/1
19. Sundari MS, Nayak RK (2020) Process mining in healthcare systems: a critical review and its future. Int J Emerg Trends Eng Res 8(9). ISSN 2347-3983
20. Nayak RK, Tripathy R, Mishra D, Burugari VK, Selvaraj P, Sethy A, Jena B (2021) Indian stock market prediction based on rough set and support vector machine approach. In: Intelligent and cloud computing. Springer, Singapore, pp 345–355

21. Tripathy R, Nayak RK, Das P, Mishra D (2020). Cellular cholesterol prediction of mammalian ATP-binding cassette (ABC) proteins based on fuzzy c-means with support vector machine algorithms. J Intell Fuzzy Syst (Preprint) 1–8

Multi-layered PCM Method for Detecting Occluded Object in Secluded Remote Sensing Image

B. Narendra Kumar Rao, Apparao Kamarsu, Kanaka Durga Returi, Vaka Murali Mohan, and K. Reddy Madhavi

Abstract To identify items that are relevant for military-grade or surveillance purposes, object detection is more useful in remote sensing photographs. When an object is obscured by clouds or light, it can be difficult to recognise it in a picture. Implementing, the partial configuration model (PCM), a new partial configuration layer called PCM is added; it is made up of partial configurations that are chosen based on likely occlusion patterns. This work involves preventing the transfer of occlusion impact when compared to the common single-layer Deformable Part-Based model (DPBM). A multistage object model for the best spatial arrangement in object detections is necessary for accurate object detection. For the purpose of object identification and recognition, R-CNN and rapid R-CNN neural network models are utilised. PCM achieves a better compromise between large DPM discriminative shape-capturing capability and tiny component deformation modelling flexibility. Experimental results using the datasets show the effectiveness of the approach on high-resolution remote sensing images.

Keywords Object detection · Occlusion · Neural networks · Partial configuration model · Remote sensing images

B. Narendra Kumar Rao (✉) · K. Reddy Madhavi
Mohan Babu University, Tirupati, Andhra Pradesh, India
e-mail: narendrakumarraob@gmail.com

A. Kamarsu
IT, MA&UD, Mangalagiri, Andhra Pradesh, India

K. D. Returi · V. M. Mohan
Malla Reddy College of Engineering for Women, Maisammaguda, Medchal, Hyderabad, Telangana, India

K. A. Reddy et al. (eds.), *Proceedings of Fourth International Conference on Computer and Communication Technologies*, Lecture Notes in Networks and Systems 606,
https://doi.org/10.1007/978-981-19-8563-8_63

1 Introduction

Detection of objects has become a simple yet challenging problem in the interpretation of remote sensing photographs in recent years. Sensors and post-processing technology have advanced dramatically in recent years. With a bare minimum 0.5 m spatial resolution, high-resolution secluded remote sensing images (HR-RS) were with quality and quantity. On the one hand, the abundance of spatial and spectral information allows more precise recognition of more complicated geographical objects, while on the other side, the crowded backdrop introduces more interference.

Various tasks involving detection of objects like detection of aeroplanes, aerodromes, vehicles, ships and buildings have been the focus of recent research. These techniques primarily deal with different object formats and a model built from parts, employing modern classifiers like support vector machine (SVM), features (SIFT and HOG) and random forest. The appearance of remote objects in HR-RS images can be significantly influenced by weather, light or cloud, which can be termed as environmental influence and intraclass changes (size, style or texture). It has been a significant obstacle to identify the appropriate.

In HR-RS object detection [1], the common problem is occlusion. Large-scale object detection, such as airport detection, is mainly concerned with occlusion in moderate/low-resolution pictures. This problem is getting more important for smaller things as the resolution of remote sensing photographs improves.

2 Partial Configuration Object Model

In traditional approaches, objects are represented using a single layer, and entities of object are represented by low-level attributes directly or representational components. When the basic representations of this system are distorted, it is readily influenced. Parts are utilised to represent an object in most DPBM models, and they are representations of small significant regions in feature form; if one component is adversely impacted, the impact is passed on to the object's parent node. As a result, when distortion occurs, such as typical occlusion in HR-RS images, the performance of these single-layer models will deteriorate.

An effective occlusion model should prevent this effect is transmitted to the object. To protect these occluded representational aspects, it is natural to use as many of the remaining pieces as possible. These single-layer models, on the other hand, are unlikely to do this for two reasons. The first is that most of their structures rely on linear scoring algorithms, which limit their capacity to locate occluded pieces that alone protect them even when the mechanism is simply set to maximum scoring mode. The presence of an object is not supported by a robust reaction from a single element. The second consideration is that the representation element's capacity to infer the complete object independently in many single-layer models is also limited.

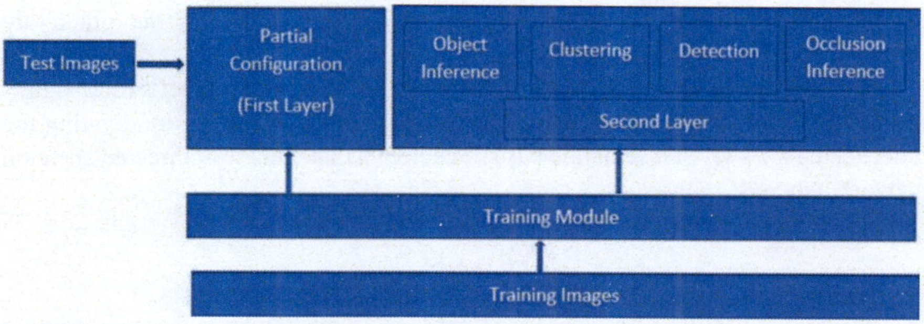

Fig. 1 Framework for object detection

When parts in object polling are absent, the performance of the remaining pieces is also affected.

To reduce this effect, a buffer layer is constructed. Only this layer is capable of passing this influence on to the buffer layer below. This layer essentially consists of a collection of frequently used single-layer models, each of which is capable of inferring the full-object entity and mimics the state of elements that are blocking one another. As a result, the impact would be reduced with minimal performance loss.

The suggested PCM model is made up of two layers, each of which represents distinct levels of object properties. The first layer, partial configuration, is a collection of semantic pieces that capture local object properties, while the second layer is a graph-structured arrangement from the first layer's incomplete configurations. The model's initial layer gives it flexibility, enabling it to take deformation within item categories into consideration. The second layer employs semi-global representation to capture the appearances of objects and shapes with a wider area of coverage. The model keeps both deformation modelling and shape-capturing properties. It is important to note that the model presented here is direction-dependent, and it only works with things that move in one way. To detect things in all directions, a group of models will be necessary. Through the PCM model in more using the aeroplane object as a sample, you can add more detail to this section. Figure 1 represents a framework for object detection.

2.1 First Layer

The DPBM definition, which is dynamically generated in the object's most significant location, differs from the semantic sections in the first layer. Beginning and defining is the set of n semantic parts for the object category. Semantic parts are arranged using the item's matching undirected skeleton graph, where $V = pi1$.

$$E\{(x, y)|x, y\,V, x = y\}$$

For object categories, this description of semantic pieces and the graph are interconnected.

The DPBM definition, which is automatically located in the object's most significant location, differs from the semantic sections in the first layer, by defining the object category's set of n semantic parts. The item's matching is undirected skeleton network, where $V = pi1$.

$$E\{(x, y)|x, y\ V, x = y\},$$

The above is useful to arrange semantic pieces. In this semantic piece description and item categories, the skeleton graph is linked.

2.2 Second Layer

In the second layer, partial configurations are organised using a graph structure. Each partial configuration casts a vote for the object's existence and is capable of deducing the bounding box of the object on its own. The graph's edges display the spatial connections between incomplete setups and items. This layout makes it relaxed to avoid the problem of high occlusion rates degrading detector performance. The presence of the object will be aided by the unaffected partial configurations in our model. The occlusion pattern is captured in a partial configuration. The full-object DPBM receives the greatest score, whereas the single-object DPBM performs badly. To account for the deformation of the item caused by intraclass geometric variance, the spatial relationship is used. In DPBM, to account the deformation of an item is similar to a "spring" in pictorial structural models.

When it comes to constellation models, through a similar approach, when easily approximated Gaussian distribution captures the spatial organisation, it's worth noting but only describes the partial configurations and the interaction of object, incomplete configurations interactions. Although the later interactions will benefit the model, the process of detection will be significantly simplified.

As a result, there is no need to re-annotate the bounding box. Furthermore, the partial configuration's spatial interdependence with the object can be directly approximated using the partial configuration's MBR and the full object's bounding box. Using the suggested weighted continuous NMS technique, before being clustered for final discoveries, the incomplete configuration hypotheses from the primitive layer are transformed into objects with a wide range of functionality. This section also includes an occlusion inference method. The detection rotation problem is solved by combining [2]; the results of a gathering of models are from different vectorised bins. Using DPBM's is a multiresolution detection capability to handle objects of various sizes.

3 Partial Configuration Object Model for Object Detection Framework Learning

The merging of partial configurations is a major challenge in the PCM model, because each partial configuration votes in favour of the object's existence. These partial-configuration-like items were handled similarly in prior part-based methods. Our objective is to explain why some partial configurations are more discriminative than others by definition, particularly when there is a lot going on in the background. Hence, some DPBMs during first layer are trained inequitably by poor training data. Resolving this by maintaining a healthy balance in partial configurations is by assigning weights. The appearance of partial configurations clearly reflects the differences in weights. These parameters are based on appearance replies or partial configuration scores. Positive samples must have a higher weighted partial configuration, while negative samples should have a lesser valued partial configuration, according to the concept and vice versa. This problem is considered as two groups of samples are involved in a categorisation challenge. Utilising a maximum margin SVM architecture, the weights are established by learning the scores to be used as the first layer on validation samples. The total of these mistake words should be reduced by at least two. The hyperplane and the reciprocal of the shortest distance between two points are sampled as closely as possible.

3.1 Detection of Objects

The first layer is a basic DPBM model with only one component. The first layer is a basic DPBM model with only one component. Many partial configuration hypotheses are created by using the first layer to the HR-RS pictures. Many partial configuration hypotheses are created by using the first layer to the HR-RS pictures. A generalised Hough transform framework is used to fuse these hypotheses because several of them represent the same object.

The NMS technique is employed prior to clustering to reduce recurrent or high interference detection for propositions from the same partial configuration, making the clustering easy and to function with only one candidate of one type of partial configuration in the local region. Based on the characteristics of the hypothesis, clustering is subdivided into clustering based on scores and bounded box clustering.

Clustering of scores

Learning of scores and clustering is through each premise and various proportional weights. Linear combination function is not applied to ensure that the scoring process is not influenced by the blockage. Instead, the partial configurations with the highest weighted score are measured.

Object bounding box clustering

Partially configured spaces based on the position and size relationships can be used to calculate the bounding box of the entire object. The hypotheses are divided into various categories. Each one depicts a possible object. Eventually, through using the hypotheses into categories leads to final conclusions.

Algorithm: Continuous Weighted NMS method.
Input: Images from Object Bounding Box with scores and weights.
Output: Clustered Bounding Boxes.

Step-1 *Calculate object's full bounding box per image from Object Bounding Box*
Step-2 *Sort Images in object's full bounding box based on weights*
Step-3 *Repeat the following until an image is identified from the image from object's full bounding box with the largest weight without overlap*
Step-4 *Perform Step-3 for each group and identify an image from the group*

3.2 Occlusion Inference

For a better comprehension of the image at the object level, occlusion inference is useful. Every incomplete configuration, according to our definition, indicates a potential occlusion pattern. As a result, based on partial configuration combination, the status of occlusion can be deduced, i.e. the discovered item's occluded region. The retrieved characteristics for occlusion inference are the partial configuration scores of an identified hypothesis. The occluded pattern is incorporated into the incomplete configuration response. Training happened on a one-vs-rest multiclass linear classifier [3, 4] SVM to tackle the classification problem in order to solve the occlusion inference problem. To accomplish so, the notion is that for the same occlusion pattern, the score distribution in N-dimensional space is similar.

The inferring scores are derived from the intra-group clustering intermediate outcomes The scores ($s1$, …, si,…, sN) of N partial configurations in a single bounding box output create a group. The undetected partial configuration score is set to 1. The trained classifier is able to estimate the output's occlusion status based on the partial configuration's direction. For N partial configurations, $N + 1$ classes are trained, along with a completely visible class, with every class indicating a partial configuration of an occlusion state. In this stage, the classifier was trained using the extra occlusion dataset. The occluded states and its true partial configuration ID are signified by each sample's label.

3.3 Datasets

Using publicly available picture datasets is to train the recommended PCM model [2]. To increase the amount of instances accessible, flip training dataset samples in various directions. Negative samples are randomly removed randomly by using training photographs and not including objects.

4 Datasets of Aeroplane Detection

A simple object recognition system is developed occlusion framework for generic HR-RS photographs. To increase the model's performance, there is still additional work to be done. The very first layer's partial configurations can also be integrated into a single DPBM model which distributes portions for speed. Our work is based on a partial configuration pick that is made automatically mechanism as part of the model. Furthermore, the issue of rotational and scaled variance is not sufficiently addressed by our technique, which is also our long-term strategy. For comparison, our model is tested on multiclass dataset's aeroplane class [5]. It consists of 90 photographs including a total of 750 aeroplanes (Kaggle dataset is referred) with a GSD of 0.50–2.0 m. In addition, difficult occlusion dataset is presented. This collection contains 45 photographs with a total of 180 aeroplanes, 96 of which are obscured shortened by the image border or truncated by cloud or hangar. For convenience, the size (10,000 × 10,000) of photographs is converted into 24 smaller ones. The ground truth is clearly marked. There are 370 pieces in this test set, all of which have distinct sizes, colours and styles. The image border and buildings obscure the test set has a total of 17 objects.

4.1 Experimental Setup

To account for rotational variability, the circle was split into different right-handed bins and the north is fixed to 0°. Aside from bounded boxes, there are a few other things to consider. As the projected direction, through the direction bin's centre degree, which is a max-scored model reflects. Zero-pad the occlusion pictures in aeroplane occlusion dataset is 200 pixels wide, while the ship and automobile datasets are 100 pixels wide to adjust for truncation of objects and enhance detection results visibility. In Algorithm 1, the least overlap t is set to 0.55. Five semantic components are applied for the category of ship, which are structured using a line skeleton graph, yielding three partial configurations each of which is made up of three semantic elements that are close together. Experiment establishes four semantic parts for the automobile category which are aligned using a skeleton graph and in the form of a line as two partial configurations selected from three related semantic elements.

4.2 Criteria for Evaluation

In the tests, a result is considered valid if the overlapping ratio is less than one, (the crossing ratio is used to calculate this) in between bounding box that was discovered and the ground truth is greater than 0.6; otherwise, it is called false positive (FP), false negatives (FNs) are the missing things, whereas genuine negatives are the rest (TNs).

Through precision–recall curve (PRC) to represent the choosing among precision and recall is a trade-off to measure detection performance. In trials, it also provides the average precision (AP), and this is used to determine the area covered by PRC. The elevated values of AP represent the better output. Furthermore, it is usually understood that recall and accuracy are trade-offs, with different thresholds resulting in a wide range of returns and accuracies to account for this the $F1$ score obtained is used to select the appropriate threshold leading to the maximum $F1$ in the subsequent detection. It is similar to a weighted sum and precise to recall values. In general, larger ideal $F1$ score indicates that the method is more effective.

4.3 Evaluation and Comparing Performance

Through testing dataset the model's detection performance can be evaluated. To make the item, a sliding window method suggestion during the test phase, different angles and sizes were used, and to apply the recommended settings for DPBMs. On the same test datasets, the model is tested against two cutting-edge object identification algorithms, exemplar SVMs and the original DPBM approach. To account for rotation variation, using the same method much other variety of models from various angles can be trained. Comparison is made by utilising identical training dataset and testing datasets for all algorithms, and using their finest parameters.

4.4 Scheme Influence Analysis

On the aeroplane datasets, detection technology can be applied to the test in a variety of scenarios. Through alternative models and aspects could lead to have a significant impact on the final outcome.

5 Weights in Partial Configuration

The work is based on initial examination of partial configuration characteristics which are based on their scores and then examining the partial configuration weights that have an impact on our model. The motivation for addressing partial configuration weights stems from the belief that their ability to reflect the object's corresponding local features that are essentially different due to structural variances [6].

To test the intuition, the planned ten north-direction partial configurations are referred. On the validation dataset, for both favourable and unfavourable samples, by analysing respective models determine their mean scores. After normalisation, the appropriate fine-tuned weights and the performance of a good partial configuration should be good. On the bright side, there are some positive samples while performing poorly on negative test data samples and vice versa. Poor partial configurations are connected with the fifth, sixth and ninth partial configurations and have their weights reduced, whereas greater weights are given to the first, fourth and eighth partial configurations. As a result, the weights learning approach has captured this property.

Partially configured semantic parts with four semantic parts have a higher average weight than partial configurations with three semantic parts, relies on partial configuration concept. It partially supports the assumption that "pieces" with a larger surface area are more discriminative than those with a smaller surface area. This phenomenon can be noticed, since partial configurations comprise in comparison with partial configurations, entire arms are more discriminative and just contains the head. The fourth partial arrangement demonstrates how the wings can be combined in a variety of ways and the tail is more discriminatory than the head, but spanning a similar area ratio. In future, these phenomena could be used to guide part selection for aeroplane identification tasks. The performance difference on dataset is negligible, but it increases on the occlusion dataset, showing that the partial configuration difference is amplified when just a small number of partial configured clusters are used. Finally, it may be said that weighted limited configurations more accurately depict the object's response, leading to outcomes in detection that are more consistent and dependable.

6 The Quantity of Partial Configurations Has an Impact

The partial configurations are meant to cover parts of things that aren't covered, the number of whom will have an impact on our PCM model's final performance. Through comparing the complete model (group A + B) as a baseline, versus the partial configuration A and B groups. This calculation is based on our occlusion datasets and the effective execution of PRC and AP.

It supports the idea that group A is more discriminatory than group B when the object is fully visible because larger partial configurations are occluded partly on the occlusion dataset. Small partial configurations collect more localised un-occluded information for the full object, whereas larger partial configurations capture more global unoccluded information. This helps to explain why group A performs so poorly, whereas group B and the complete model perform similarly. On the occlusion dataset, it turns out that more partial configuration coverage does not always mean higher performance because The entire model is made up of groups A and B.

There are performance gains compared to either group, but at the cost of increased computational time.

6.1 Occlusion Inference Results

Our model can estimate an object's occlusion state, or the occluded region of its bounding box, based on its detection and direction. By using intermediate results, this is the sum of all scores in occlusion inference. This model can estimate an object's occlusion state, or the occluded region of its bounding box, based on its detection and direction. The first layer intermediate results, which scores all partial configurations, and use the trained inference model and make inferences about them. The anticipated occlusions are extremely near to the ground truth, thanks partly due to well-predicted object trajectories. The percentage index represents our inference model's overall performance, which is about 60% entirely correct and 11% partially correct. Approximately a total of 29% of detected items are incorrectly inferred and many of them have poor scores from the first layer, making them unstable to predict. The results, on the other hand, are pleasing and very instructive for future inference. Precision and recall curves for the proposed PCM model for different object detection based on datasets is as follows (Fig. 2). Red curve (PCM) shows good performance over other SVM approach. FP-False Positives, TP-True Positives, FN-False Negatives. Table 1 is indicative of occlusion inference and prediction accuracy.

$$\text{Precision} = \frac{\text{TP}}{\text{TP} + \text{FP}}$$

$$\text{Recall} = \frac{\text{TP}}{\text{TP} + \text{FN}}$$

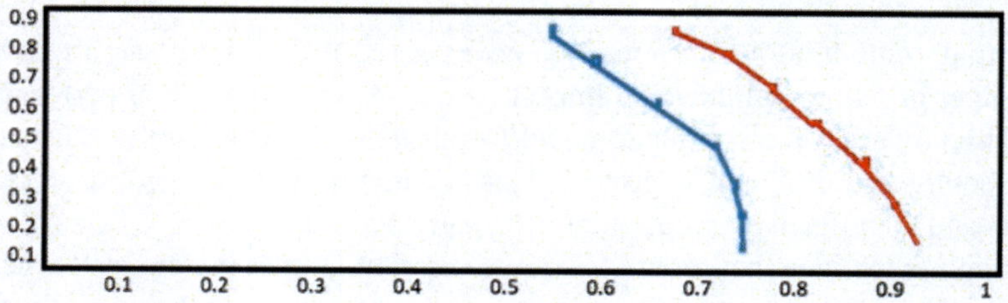

Fig. 2 Precision and recall curve

Table 1 Occlusion inference results

#Total TPs (%)	#Correct (%)	#Partially correct (%)	#Incorrect (%)
100	50	10	30

7 Conclusion

A Multilayer or two-layer PCM model is presented in a unique occluded object recognition approach in HR-RS pictures. The first layer built based on semantic components represents partial configurations which are in line with potential occlusion patterns, to address the problem of on occlusion datasets, DPBM performance drops. The second layer, which is built on a suggested weighted-continuous NMS and employs empirical spatial interdependence, clusters the results from the first layer. To ensure the detection accuracy for partially obscured objects high, the second layer employs a variety of partial configurations. The efficiency of experiments on three different types of item categories has proven that PCM is accurate. In addition, from detecting data, our approach may be used to predict occlusion situations, with a satisfactory inference result [7, 8]. Despite the fact that for general HR-RS pictures with occlusion, a simple object recognition framework is developed, to increase the model's performance, there is still additional work to be done. The first layer's partial configurations can be integrated into a single DPBM model that shares portions for speed. Our method is based on an automatic partial configuration selection mechanism that is built into the model. Furthermore, the issue of rotating and scale variation is not sufficiently addressed by our technique, which is also our plan for the future.

References

1. Cheng G, Han J, Zhou P, Guo L (2014) Multi-class geospatial object detection and geographic image classification based on collection of part detectors. ISPRS J Photogramm Remote Sens 98:119–132
2. Yu Y, Guan H, Zai D, Ji Z (2016) Rotation-and-scale-invariant airplane detection in high-resolution satellite images based on deep-Houghforests. ISPRS J Photogramm Remote Sens 112:50–64
3. Han J et al (2014) Efficient, simultaneous detection of multi-class geospatial targets based on visual saliency modeling and discriminative learning of sparse coding. ISPRS J Photogramm Remote Sens 89:37–48
4. Cheng G et al (2013) Object detection in remote sensing imagery using a discriminatively trained mixture model. ISPRS J Photogramm Remote Sens 85:32–43
5. Yao X, Han J, Guo L, Bu S, Liu Z (2015) A coarse-to-fine model for airport detection from remote sensing images using target-oriented visual saliency and CRF. Neurocomputing 164:162–172
6. Bai X, Zmultig H, Zhou J (2014) VHR object detection based on structural feature extraction and query expansion. IEEE Trans Geosci Remote Sens 52(10):6508–6520
7. Ranjana R, Narendra Kumar Rao B, Nagendra P, Sreenivasa Chakravarthy S (2022) Broad learning and hybrid transfer learning system for face mask detection. Telematique 21(1):182–196
8. Narendra Kumar Rao B, Naseeba B, Challa NP, Chakrvarthi S (2022) Web scraping (IMDB) using python. Telematique 21(1):235–247

Author Index

K. A. Reddy et al. (eds.), *Proceedings of Fourth International Conference on Computer and Communication Technologies*, Lecture Notes in Networks and Systems 606,
https://doi.org/10.1007/978-981-19-8563-8